CHEMISTRY AND WATER

CHEMISTRY AND WATER

The Science Behind Sustaining the World's Most Crucial Resource

Edited by

SATINDER AHUJA

Ahuja Consulting, Calabash, NC, United States

ELSEVIER

Amsterdam • Boston • Heidelberg • London • New York • Oxford
Paris • San Diego • San Francisco • Singapore • Sydney • Tokyo

Elsevier
Radarweg 29, PO Box 211, 1000 AE Amsterdam, Netherlands
The Boulevard, Langford Lane, Kidlington, Oxford OX5 1GB, United Kingdom
50 Hampshire Street, 5th Floor, Cambridge, MA 02139, United States

Library of Congress Cataloging-in-Publication Data
A catalog record for this book is available from the Library of Congress

British Library Cataloguing-in-Publication Data
A catalogue record for this book is available from the British Library

ISBN: 978-0-12-809330-6

Publisher: John Fedor
Acquisition Editor: Kathryn Morrissey
Editorial Project Manager: Amy M Clark
Production Project Manager: Maria Bernard
Cover Designer: Maria Cruz

Typeset by TNQ Books and Journals

This book is dedicated to all people who take extraordinary steps to sustain water, the world's most crucial resource.

CONTENTS

LIST OF CONTRIBUTORS

R.R.M. Abarca
Mindanao State University-Iligan Institute of Technology, Iligan City, Philippines

S. Ahuja
Ahuja Consulting, Calabash, NC, United States

J.L.A. Andit
Mindanao State University-Iligan Institute of Technology, Iligan City, Philippines

H. Aris
University of Maryland Baltimore County, Baltimore, MD, United States

J. Bartram
University of North Carolina at Chapel Hill, Chapel Hill, NC, United States

L. Blaney
University of Maryland Baltimore County, Baltimore, MD, United States

M.D.G. de Luna
University of the Philippines, Diliman, Quezon City, Philippines

A.M. Dhalla
Nanyang Technological University, Singapore

K.A. Gibson
University of California, Davis, CA, United States

R. Gupta
Green Chemistry Network Centre, University of Delhi, Delhi, India

A. Hassan
Red Cross Crescent Center, Hague, Netherlands

K.D. Hristovski
Arizona State University, Mesa, AZ, United States

M.A. Islam
Curtin University Sustainability Policy Institute, Perth, Australia

S.L. Islam
Washington State University, Pullman, WA, United States

R.B. Labad
Mindanao State University-Iligan Institute of Technology, Iligan City, Philippines

M.C. Larsen
Smithsonian Tropical Research Institute, Balboa, Republic of Panama

M.-C. Lu
Chia-Nan University of Pharmacy and Science, Tainan, Taiwan

A.S. Ludtke
U.S. Geological Survey, Office of Water Quality, Reston, VA, United States

J. Luh
University of North Carolina at Chapel Hill, Chapel Hill, NC, United States

J.W. Macharia
University of Nairobi, Nairobi, Kenya

J. Markovski
University of Belgrade, Belgrade, Serbia
Arizona State University, Mesa, AZ, United States

D.N. Myers
U.S. Geological Survey, Office of Water Quality, Reston, VA, United States

S. Ngigi Mbugua
University of Nairobi, Nairobi, Kenya

L.W. Olson
Arizona State University, Mesa, AZ, United States

M.A. Otieno
University of Nairobi, Nairobi, Kenya

J.R. Peller
Valparaiso University, Valparaiso, IN, United States

C. Pohl
Thermo Fisher Scientific, Sunnyvale, CA, United States

R.S. Pusta, Jr.
Mindanao State University-Iligan Institute of Technology, Iligan City, Philippines

J. Real
RST Cellars, Davis, CA, United States

C.M. Rejas
Mindanao State University-Iligan Institute of Technology, Iligan City, Philippines

J.K. Schoer
Valparaiso University, Valparaiso, IN, United States

J.N. Seiber
University of California, Davis, CA, United States

R.K. Sharma
Green Chemistry Network Centre, University of Delhi, Delhi, India

U. Shashvatt
University of Maryland Baltimore County, Baltimore, MD, United States

Y. Shevah
H.G.M. Consulting Engineers & Planners Ltd., Netanya, Israel

L. Smith
Filters for Families, Wheat Ridge, CO, United States

K. Srinivasan
Thermo Fisher Scientific, Sunnyvale, CA, United States

V. Tiangco
RST Cellars, Davis, CA, United States

S. Tongesayi
Walden University Public Health, Minneapolis, MN, United States

T. Tongesayi
Monmouth University, West Long Branch, NJ, United States

S.O. Wandiga
University of Nairobi, Nairobi, Kenya

R.L. Whitman
Great Lakes Science Center, Porter, IN, United States

M. Yadav
Green Chemistry Network Centre, University of Delhi, Delhi, India

PREFACE

According to the World Economic Forum, the water crisis is the number one global risk, based on the impact to society as a measure of devastation. Water is crucial to our survival, as we can survive only a few days without it. Unfortunately, clean safe drinking water is scarce; only 0.06% of freshwater is readily available to us. As a result, nearly 1 billion people in the developing world do not have access to safe drinking water. On the other hand, those of us who live in the developed world take it for granted. We waste it, and we are willing to pay too much to drink water from plastic bottles. Our aging water infrastructure loses trillions of gallons of water every year.

Water is almost a universal solvent, as most compounds can be solubilized in it at various levels. For example, over 700 different chemicals have been found in US drinking water as it flows from the tap. The US Environmental Protection Agency (EPA) classifies 129 of these different chemicals as being particularly dangerous and has set standards for approximately 90 contaminants in drinking water. This means that contaminants in water need to be monitored at ultratrace levels (parts per billion level) to assure purity and safety for drinking.

This book deals with issues relating to water scarcity, quality, reclamation, and sustainability. Chapter 1 discusses these issues, including the impact of sanitation and climate, and it offers solutions in terms of monitoring water quality and reclamation to help achieve water sustainability. Chapter 2 discusses lessons learned from water monitoring, and Chapter 3 covers the impact of climate change on Bangladesh, which is likely to be severely affected by it. Chapter 4 provides examples of water and ecosystem services from Panama, Puerto Rico, Venezuela, and the US. Chapters 5–8 cover water challenges and solutions in India, Africa, the Middle East, and the Balkan states. Ion chromatography instrumentation and analyses of ionic compounds found in drinking water are described in some detail in Chapters 9 and 10. Chapter 11 discusses issues relating to management of watersheds. A detailed account of water purification systems is given in Chapter 12. Chapters 13 and 14 cover resource recovery from wastewater. Interesting information is provided in Chapter 15 on the judicious use of water supplies for wine production. Singapore provides an excellent example of a country that has solved its water shortage problems (Chapter 16), and Chapter 17 reminds us of the effects of water quality on public health. Finally,

Chapter 18 further elaborates the science behind achieving sustainable development goals.

I would like to thank all the experts for their contributions to this book, and I feel confident that this book will be of great help to scientists and engineers in addressing water quality and sustainability issues in academic, industrial, and regulatory fields.

November 4, 2016 **Satinder Ahuja**

CHAPTER ONE

Overview: Sustaining Water, the World's Most Crucial Resource

S. Ahuja
Ahuja Consulting, Calabash, NC, United States

1. OVERVIEW

Water is a crucial resource for human survival [1–6]. Unfortunately, clean safe drinking water is scarce, even though the Earth is made up of 70% water. Only 3% of freshwater is available to us, and, of that amount, only 0.06% is readily accessible. Nearly 1 billion people in the developing world cannot access safe drinking water; they may spend a better part of their days searching for it. On the other hand, those of us who live in the developed world take it for granted. We waste water and are willing to pay too much to drink it from plastic bottles, which have a large water footprint and add significant problems in terms of disposal. Water leaks drain city water supplies. According to the Wall Street Journal of June 22, 2016, p. A3, brittle, aging systems lose trillions of gallons of water every year.

Water is a simple molecule made up of two atoms of hydrogen and one of oxygen, with a molecular weight of 18. It can occur as liquid, solid (ice), and gas (steam). A molecule with such a low weight should be gaseous; however, hydrogen bonding makes it into a unique liquid that quenches our thirst, feeds our crops, and helps us produce energy. It is almost a universal solvent, as most compounds can be solubilized in it at various levels, depending on their polarity. As can be anticipated, polar compounds dissolve more favorably; however, nonpolar compounds can be solubilized at ultratrace levels (below parts-per-million). This means that contaminants in water need to be monitored at ultratrace levels to assure purity and safety for drinking. Hundreds of unregulated contaminants may be flowing from our taps; they are mostly invisible, tasteless, and difficult to detect [7]. Over 700 different chemicals have been found in US drinking water when it flows from the tap. The US Environmental Protection Agency (EPA) classifies 129 of these different chemicals as being particularly dangerous and has set standards for approximately 90 contaminants in drinking water.

Chemistry and Water
ISBN 978-0-12-809330-6
http://dx.doi.org/10.1016/B978-0-12-809330-6.00001-5

The water crisis is the number one global risk, based on the impact to society as a measure of devastation, according to the World Economic Forum [9]. Water sustains life; without water, life would not be possible. This is one of the reasons that our space program is constantly looking for water on various planets to detect potential life there. Here on Earth, we know that water is the most crucial resource for human survival, after air. Benjamin Franklin made this point crystal clear: "*When the well is dry, we learn the worth of water.*" Many Native American tribes recognized it early as they consider specific lakes to be the source of life, for example, Lake Tahoe (do-wa-ga, "center of existence" to the Washoe). In India, the Ganges and Yamuna rivers are considered holy. There was a time when you could drink the water of the Ganges without any health concerns; this is definitely not possible anymore.

All human beings need a safe and sustainable supply of water for drinking, washing/cleaning, cooking, and growing food. Unfortunately, governments worldwide do not ensure that their citizens are provided with this essential material (one could argue that it is a basic human right). In many countries around the world, taps, wells, and pipes simply do not exist. Even where they do exist, they are often not affordable for the poorest people or are not designed to last. It has been suggested that conflicts in the 21st century will be fought over water rather than over oil. According to a 2003 UN report, 507 documented "conflictive events" occurred over the last 50 years, with 37 of them involving violence and 21 resulting in military action [8].

1.1 Water Scarcity

Water scarcity is defined as "the point at which the aggregate impact of all users impinges on the supply or quality of water under prevailing institutional arrangements to the extent that the demand by all sectors, including the environment, cannot be fully satisfied." Hydrologists typically assess scarcity by looking at the population—water equation. Water stress is experienced when annual water supplies drop below 700 m^3 per person. When annual water supplies drop below 1000 m^3 per person, the population faces water scarcity, and below 500 m^3, "absolute scarcity." Water scarcity is a relative concept and can occur at any level of supply or demand. Scarcity may be due to social reasons (a product of affluence, expectations, and customary behavior) or the consequence of altered supply patterns that may be affected by climate change [9].

Global water scarcity is depicted in Fig. 1.1, where the US, Canada, and a few other countries have an abundant supply of water, while a large area of

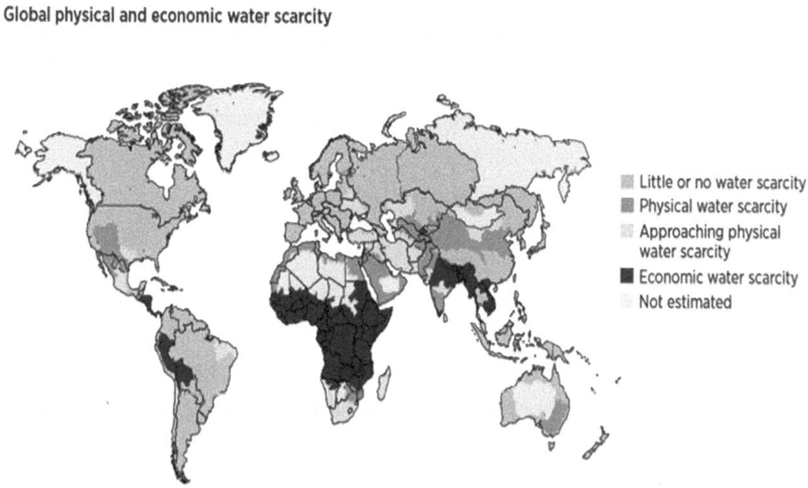

Figure 1.1 Global water scarcity. *World Water Development Report 4. World Water Assessment Programme (WWAP); March 2012.*

the world suffers from physical water scarcity or lacks economic means to secure water [10]. With the existing climate change scenario, almost half the world's population will be living in areas of high water stress by 2030 (including between 75 million and 250 million people in Africa alone). Sub-Saharan Africa has the largest number of water-stressed countries of any region in the world. In addition, water scarcity in some arid and semiarid areas could displace as many as 700 million people.

Water scarcity already affects every continent. Around 1.2 billion people, or almost one-fifth of the world's population, live in areas of physical water scarcity, and 500 million people are approaching this situation. Another 1.6 billion people, or almost one-quarter of the world's population, face economic water shortages (where countries lack the necessary infrastructure to take water from rivers and aquifers).

Data on water consumption in the world are available from the United Nations (UN/UNESCO). Worldwide water consumption is estimated to be around 914,546 billion liters per year. Agriculture accounts for 70% of all water consumption, industrial usage accounts for 20%, and domestic usage is 10%. In highly industrialized countries, however, manufacturing consumes more than half of the available water. In Belgium, for example, industries use up to 80% of the available water. Over the last 50 years, freshwater withdrawals have tripled. The demand for freshwater is increasing by

64 billion cubic meters per year (1 m^3 = 1000 L) because of the following reasons:

- The world's population is growing by roughly 80 million each year.
- Changes in lifestyles and eating habits in recent years require more water consumption per capita.
- Water demand is rapidly increasing because of accelerated energy demand.
- The manufacture of energy from alternate sources such as biofuels has a major impact on water demand because 1000–4000 L of water are needed to produce just 1 L of biofuel.
- By 2025, the UN estimates two-thirds of the global population will live under water-stressed conditions. This problem is further compounded by the fact that nearly one-third population of the world has no toilets; human waste can affect water supplies and cause several diseases from bacteria and parasites.

The water scarcity problem is further compounded by inadequate sanitation for almost 2.5 billion people. Globally, one-third of all schools lack access to safe water and adequate sanitation. Poor sanitation affects water quality; nearly 80% of diseases in developing countries are associated with water quality. Water shortages in this century are among the main problems faced by many societies. Water use has been growing at more than twice the rate of population increase in the last century, and although there is no global water scarcity as such, an increasing number of regions are chronically short of water. Water scarcity is both a natural and human-made phenomenon. There is enough freshwater on the planet for 7 billion people, but it is distributed unevenly and too much of it is wasted, polluted, and unsustainably managed.

A review of rural water system sustainability in eight countries in Africa, South Asia, and Central America found an average water project failure rate of 20–40%. Recently, Gina McCarthy, EPA Chief, delivered a dire warning: the US water supply infrastructure is aging, and states are not prepared to face current and future water challenges, which include scarcity and threats from emerging contaminants [11].

1.2 Water Quality

The world's seas are inundated by a variety of water pollution problems [12]. With a warming planet and acidifying oceans, species from corals to lobsters and fish are succumbing to pathogenic infections. Table 1.1 shows the most acute problem in major bodies of water.

Table 1.1 Pollution Problems in Major Water Bodies

Water Bodies	Most Acute Problem
Baltic Sea (North Europe), Yellow Sea (northeast coast of China), Bohol Sea (in South Pacific), Lake Victoria (Central Africa)	Eutrophication (nitrogen and phosphorus pollution)
Gulf of Mexico, Lake Victoria (Central Africa)	Microbiological
Caribbean Sea, Congo Basin, Victoria Lake, Benguela Current (west coast of Africa)	Solid wastes
Benguela Current, Pacific Islands	Radionuclides

The extensive use of plastics and their careless disposal has led to the pollution of various water bodies. Paul Ahuja, from La Paz, Mexico, reports that a group of more than 1000 volunteers have collected more than 10 tons of trash in coastal waters in just 1 year. In hopes of educating the public, the leader did in-class presentations to more than 3000 youngsters. Large parts of the Pacific Ocean are referred to as "plastic oceans," where enormous gyres, about the size of Texas, are covered with plastic debris. The Pacific is the largest ocean realm on our planet; approximately the size of Africa—over 10 million square miles—is the home of two very large gyres. The Atlantic Ocean contains two more gyres, and other plastic oceans exist in other bodies of water. Microplastics can stunt fish growth and alter their behavior.

People need to be prudent when they drink water in Africa, Asia, and Latin America. The rivers in these areas are frequently considered the most polluted in the world. There are three times as many bacteria from human waste as the global average and 20 times more lead than rivers in developed countries. Drinking water comes mainly from the following sources: rivers, lakes, wells, and natural springs. These sources are exposed to a variety of conditions that can cause water contamination. Bottled water is not necessarily safe either. In 2008, Environmental Working Group found arsenic, acetaminophen, caffeine, and nitrates in 10 brands of bottled water.

In 2004, water from half of the tested sections of China's seven major rivers was found to be undrinkable because of pollution. The Yangtze, China's longest river, is "cancerous" with pollution. The pollution from untreated agricultural and industrial waste could turn the Yangtze into a "dead river" within a short time. This would make it impossible to sustain marine life or provide drinking water to the booming cities along its banks. Almost half of China's water sources are polluted. Wells and aquifers are contaminated with fertilizers, pesticide residues, and heavy metals such as arsenic and manganese from mining, the petrochemical industry, and domestic and industrial waste. More than three-fourths (76.8%) of 800 wells monitored in nine provinces and autonomous regions and municipalities, including Beijing, Shanghai, and Guangzhou, failed to meet standards for groundwater in a 2011 national evaluation [13].

The quality of water in Europe's rivers and lakes worsened between 2004 and 2005. Almost one-third of Ireland's rivers are polluted with sewage or fertilizer. The Sarno River in Italy is the most polluted river in all of Europe, featuring a mix of sewage, untreated agricultural waste, industrial waste, and chemicals. The Rhine, which flows through many European countries, is regarded by many as the dirtiest large river; almost one-fifth of all the chemical production in the world takes place along its banks. The King River is Australia's most polluted river there, suffering from a severe acidic condition related to mining operations. Canadian rivers are also polluted.

In the US, nearly 40% of the rivers are too polluted for fishing, swimming, or aquatic life. The Mississippi River drains nearly 40% of the soil and water of continental US, including its central farmlands, and it carries an estimated 1.5 million metric tons of nitrogen pollution into the Gulf of Mexico every year. Nearly 1.2 trillion gallons of untreated sewage, storm water, and industrial waste are discharged into US waters annually. Lakes are even worse; 46% of them are extremely polluted. Two-thirds of US estuaries and bays are either moderately or severely degraded from eutrophication (nitrogen and phosphorus pollution). Even the most advanced country, like the US, is facing a water crisis. Most experts agree that the US water policy is in chaos. Decision-making about allocation, repair, infrastructure, and pollution is spread across hundreds of federal, state, and local agencies.

Various sources of contamination are listed alphabetically below.

- combustion products of oil (gasoline) and coal
- detergents
- disinfectants
- drugs (pharmaceuticals including endocrine disruptors and illicit drugs)
- fertilizers
- gasoline and its additives
- herbicides
- insecticides/pesticides
- phthalates
- radionuclides
- volatile and semivolatile compounds

Volatile organic contaminants (VOCs) and semivolatile contaminants may enter directly into our water resources from various spills, by improper disposal, or from the atmosphere in the form of rain, hail, and snow. In general, VOCs have high vapor pressures, low to medium water solubilities, and low molecular weights. These properties allow them to move freely between air and water [5]. Fog plays an important role in cycling neurotoxic mercury species among coastal systems [14]. Fog droplets contain high amounts of monomethylmercury, 3–4 ng/L as compared to 0.1 ng/L typically seen in raindrops. Both point and nonpoint source pollution occur for a variety of reasons.

About 12,500 tons of antimicrobials and antibiotics are administered to healthy animals on US farms each year. A 2002 US Geological Survey found pharmaceuticals (hormones and other drugs) in 80% of the streams sampled in 30 states. These contaminants are suspected in the increase of fish cancer, deformities, and feminization of male fish. By releasing into the environment antibiotic-containing wastewater, drug companies are fostering an emergence of deadly resistant bacteria [15].

There can be 50 different drugs in water at one place and time [16]. Zoloft has been detected in water samples and fish tissue in the US and Canada. Drugs can accumulate as they work their way up the food chain, thus exposing predators to higher levels. Fluoxetin (Prozac), an antidepressant, is excreted unchanged by humans and is environmentally stable. It can cause reduced libido and decreased appetite. Unmetabolized drugs like metformin and others that break down into various metabolites are polluting water; however, the EPA does not regulate a single human drug in drinking water.

Herbicides in drinking water can be deleterious to human health [17]. Three herbicides—amitrole, isoproturon, and trisulfuron—will be banned in the EU effective September 30, 2016, because of potential groundwater contamination and risks to aquatic life.

In 2015, Marc Edwards' team at Virginia Tech detected lead at high levels in the Flint, Michigan, water supply. In order to save money, instead of buying Lake Huron water from Detroit, Flint started drawing its water from the Flint River in April 2014. Residents started complaining about burning skin, hand tremors, hair loss, and even seizures. Lead is particularly harmful to kids, as it impacts their rapidly growing brains. For almost 19 months, the problem was ignored as the Flint River water corroded the city's decades-old pipes and leached lead into Flint's water supply. The crisis exposed as many as 8000 youngsters under age 6 to unsafe levels of lead. This may be the most serious contamination threat facing the country's water supplies, but it is hardly the only one. In Sebring, Ohio, routine laboratory tests in August 2015 found unsafe levels of lead in the town's drinking water after workers stopped adding a chemical to keep lead water pipes from corroding. Five months passed before the city told pregnant women and children not to drink the water and before it shut down taps and fountains in schools [18]. Unsafe levels of lead have turned up in tap water in city after city: in Washington, DC in 2001; in Columbia, South Carolina in 2005; in both Durham and Greenville, North Carolina in 2006; in Jackson, Mississippi in July 2016, as well as in scores of other places in recent years. Nearly 3.6 million people in the US were served by their local or regional drinking water systems' exceeding the federal lead standard at least once between January 01, 2013, and September 2015 [19].

Contamination of drinking water supplies by industrial chemicals has been examined by three states [Vermont, New Hampshire, and New York (*WSJ*, April 26, 2016)]. For decades, factories have used perfluorooctonoic acid as a plastic coating and to make consumer products such as Teflon nonstick pans, waterproof jackets, and pizza boxes. The problem was first discovered in 2014 near the border of New York and Vermont. The EPA has set 70 parts per trillion as the maximum allowable quantity, and the water utilities companies must inform the population when the values exceed this limit individually or combined [20]. Contamination of drinking water from perfluorochemicals has been reported from industrial sites where they were produced and near military bases and airports [20a]. Domoic acid (DA) was detected at high levels in Dungeness crabs and rock crabs by the California Department of Fish and Wildlife [21]. DA is produced by the marine alga *Pseudo-nitzschia*, which is eaten by shellfish and some small fish. It can cause nausea, diarrhea, and dizziness. At higher concentrations, it can cause seizures, coma, and death.

The impact of water pollution due to a variety of sources, including disinfectants, herbicides, coal ash, fracking, and radionuclides, has been discussed at length in earlier texts [2,5,6,22].

1.3 Sanitation and Water Quality

Water, sanitation, and hygiene (WASH) are essential to human health and development. Nearly 2.5 billion people live without basic sanitation. Two-thirds of the 94 countries included in this study [23] recognize that both drinking water and sanitation are human rights. Critical gaps in monitoring impede decision-making and progress for the poorest. National financing for WASH is insufficient. Only 17% of the countries apply financial measures to reduce disparities in access to sanitation for the poor, compared to 23% for drinking water. In 2010, a UN resolution declared the human right to "safe and clean drinking water and sanitation." We need to find ways to dispose of waste without water. Better yet, we have to find ways to use human waste more effectively.

1.4 Monitoring Water Contaminants

Water is a clear liquid, though a layperson might describe water's color as white or blue. The fact is that many colors that have been ascribed to water relate to the materials that may be present in it (as follows). For example, blue water generally refers to ocean water, which gets its color from the reflection of the color of the sky; and we have seas that are described by various other colors because of their appearance: Red Sea, Yellow Sea, Black Sea, and White Sea. Water generated from activities such as laundering, dishwashing, and bathing is described as gray water, and water has come into contact with fecal matter is called black water. Frequently, municipal water may have an odor, and that odor relates to the chlorination of water. A musty odor in drinking water may be the result of by-products of blue—green algae.

Our civilization has managed to pollute water sources to the point where we must disinfect water for drinking purposes. To assure water purity, we need to monitor contaminants from arsenic to zinc [4,6,24]. In the 1978 Metrochem meeting, the author emphasized the need to analyze very low levels of various contaminants in a paper entitled "In Search of Femtogram" (a femtogram is 10^{-15} g, or one part per quadrillion), to fully understand their impact on human beings. For example, dioxin (2,3,7,8-tetrachlorodibenzodioxin) can cause abortions in monkeys at the 200 parts-per-trillion (ppt) level [25]; PCBs (polychlorobiphenyls) at the 0.43 ppb (parts-per-billion) level can weaken the backbones of trout [26]. This suggests that ultratrace analysis is necessary to monitor materials like PCBs and dioxin [27—29].

A simple definition of potable water is "any water that is clean and safe to drink." We have known for some time now that water that we call potable may actually contain many trace and ultratrace contaminants, as exemplified by an analysis of Ottawa drinking water [30]. It contained insecticides like α-BHC (an isomer of lindane), lindane, and aldrin at ppt levels. In addition, it contained phthalates at significantly higher concentrations.

National Primary Drinking Water Regulations control water quality in the US in response to public concern about degraded water quality and the widespread view that pollution of our rivers and lakes was unacceptable. The Clean Water Act (CWA) became law in 1972. Control of point source contamination, traced to specific "end of pipe" points of discharge or outfalls, such as factories and combined sewers, was the primary focus of the CWA. Other nations adopted similar measures and have seen improvement in point source contamination as well. In the US, potable water must be cleaner than the maximum contaminant level mandated by local, state, and federal guidelines (US EPA national primary drinking water regulations). The tests commonly carried out on drinking water are turbidity, total organic carbon, chlorite, chlorine dioxide, fluoride, sulfate, and orthophosphate [31]. Some unregulated substances are also investigated. Surprisingly, testing of arsenic is not performed regularly.

1.5 Climate Disruptions

The impact of climate disruptions can no longer be ignored. Global surface temperature analysis reveals that global trends are higher than those reported by the Intergovernmental Panel on Climate Change [32]. The World Meteorological Organization says that warming effects from greenhouse gases increased by 36% from 1990 to 2014. So far, 160 countries have individually pledged to carry out specific actions to control emissions. The UN Environmental Programmes evaluation shows that this effort will fail to hold temperature rise of 2°C by 2100. In January 2016, the Department of Housing and Urban Development announced grants totaling $1 billion for 12 states to help communities adapt to climate change by building stronger levees, dams, and drainage systems [33]. One of those grants, $48 million for Isle de Jean Charles, Louisiana, is for moving an entire community struggling with the impacts of climate change. We need to recognize the massive problems the world could face in the coming decades as it confronts a new category of displaced people who are known as climate refugees [34].

Lake Oroville, in California, is at 39% of capacity, and Nevada's Lake Meade is at 1081.8 ft above sea level as compared to its high of 1200 ft. Mandatory water curbs have been introduced in California. That state and other drought-prone states will have to focus on expanding water infrastructure. It appears that the western US drought, which began in Texas over 15 years ago, is indicative of a long-term climate pattern. The potential effects of drought will reach far beyond the borders of these states.

Despite the snow in the Sierra Nevada, the water-filled Lake Shasta, the rapids in the Kern River, and the high water level in Lake Mendocino, California are still in a state of drought (see http://www.scwa.ca.gov/files/images/water-supply/reservoir-storage-graph.pdf). Water battles are heating up in Texas, where officials are suing New Mexico and Oklahoma over river water to quench the thirst of its booming population. A water shortage has forced Texas to recycle sewage to drinking water. The facility in Big Spring has the capacity to produce 2 million gallons of Texas sewage to drinking water [35]. Drought in Australia in 2003 forced big cities to develop new water supplies. A desalination plant powered by solar and wind farm energy now provides half of the drinking water in Perth.

1.6 Water Reclamation and Sustainability

Pollution knows no borders. Up to 90% of wastewater in developing countries flows untreated into rivers, lakes, and highly productive coastal zones, threatening health, food security, and access to safe drinking water and bathing water.

Water reclamation, the act or process of recovering water, is an absolute necessity because we have polluted our surface water, and even groundwater in some cases, to a point that it needs to be purified for drinking (see Chapter 1 in Ref. [6]). The limited water availability issues discussed in this chapter also dictate the need for water reclamation. The take-home message: we have to use water judiciously and reclaim contaminated water.

It is important to note that over 80% of used water worldwide is not collected or treated. The need for groundwater recharge may ultimately limit how much water farmers can have from surface irrigation systems, even in years of abundance. In states where irrigation rights have been zealously guarded for generations, such limitations may not go down easily. The treatment of wastewater requires significant amounts of energy, and demand for power to do so is expected to increase globally by 44% between 2006 and 2030 [14], especially in countries not covered by the Organization of Economic Cooperation and Development, where wastewater currently receives little or no treatment.

To achieve water sustainability, we must ensure that we meet our water needs and avoid compromising the ability of future generations to meet their needs [8]. We will have to address scientific, technical, economic, and social issues [4] to attain this objective. We don't have an unlimited supply of safe water for our needs, so water reclamation is absolutely necessary [22]. Some examples of wastewater reclamation are discussed in the following.

Water purification systems that improve drinking water at the point of use are a good fit in many areas. In Kenya, Bolivia, and Zambia, water purifiers have been shown to reduce diarrheal disease by 30–40%. Fewer than 5% of Chinese homes currently have these purifiers, despite a unit's low cost of about 1500–2000 renminbi. Treated gray water and black water are increasingly used in China for industrial and irrigation purposes, and for flushing toilets in new residences. But this type of recycling is impractical for most existing households because of the high cost and the disruption in the home while installing the necessary plumbing. Detailed discussion on water reclamation and sustainability is available in several books [22–24].

Globally, we face many water challenges in terms of availability, quality, and sustainability. This book provides information on various global water challenges and solutions, and it explores the important role water plays to assure our survival and maintain our lives.

2. PROGRESS AND LESSONS LEARNED FROM WATER QUALITY MONITORING NETWORKS

Stream quality monitoring networks in the US were initiated and expanded after passage of successive federal water pollution control laws from 1948 to 1972 (Chapter 2). Initial networks addressed information gaps on the extent and severity of stream pollution and served as early warning systems for spills. From 1965 to 1972, monitoring networks expanded to evaluate compliance with stream standards, track emerging issues, and assess water quality status and trends. After 1972, concerns arose regarding the ability of monitoring networks to determine if water quality was getting better or worse, and why. As a result, monitoring networks adopted a hydrologic systems approach targeted to key water quality issues accounted for human and natural factors affecting water quality, innovated new statistical methods, and introduced geographic information systems and models that predict water quality at unmeasured locations. Despite these improvements, national-scale monitoring networks have declined over time. Only about 1%, or 217, of more than 36,000 US Geological Survey monitoring sites sampled from 1975 to 2014 have been operated throughout the four decades since passage of the 1972 CWA. Efforts to sustain monitoring networks are important because these networks are providing information that will help us evaluate future trends.

3. IMPACT OF CLIMATE CHANGE ON THE GANGES–BRAHMAPUTRA–MEGHNA RIVER BASIN

The situation analysis of the Ganges–Brahmaputra–Meghna (GBM) River Basin landscape provides an understanding of the potential prospects and emerging challenges of the region in the face of changing climate and environmental pollution that affects water regime and its quality (Chapter 3). It reveals that strategic water management at local, national, and river basin scales can be the driving force to address the emerging challenges, common to the region, particularly through river conservation, effective governance, and agreed regional policy to be implemented by the governments of the basin countries. The importance of effective regional cooperation through data sharing, confidence building, and a water sharing treaty in water management for multiple benefit sharing are highlighted toward achieving the peoples' welfare of the greater basin. The river systems are interdependent; hence, integrated and collaborative policies with investment in adaptive technologies are required for a sustainable solution. This chapter sets objectives to identify and appraise the issues, problems, and prospects for common understanding, and to sensitize the policymakers of the GBM countries on managing water under uncertain climate risks, so that water-centered sustainable development through an integrated approach is encouraged in new ways of benefit sharing within the horizon of climate change.

4. FORESTED WATERSHEDS, WATER RESOURCES, AND ECOSYSTEM SERVICES IN THE US, PANAMA, AND PUERTO RICO

Forested watersheds provide critically important ecosystem services, as sources of high quality water for drinking, agriculture, and industry (Chapter 4). Montane watersheds provide additional services and benefits, including hydroelectric energy, wood products, recreation, esthetic values, and hazard mitigation. However, landowners in most countries are compensated at minimal rates, if at all, for these services to society, with a result that services are not provided at optimal levels. In some countries, including the US, modest public and private programs offer payments to landowners and serve to maintain and restore these ecosystem services.

The provision of ecosystem services is further challenged by climate change and global urbanization, particularly in regions of rapid economic growth. In some cases, a lack of effective stewardship of river floodplains and upstream forests compromises water, food, hydroelectric energy, wood products, carbon sequestration, maintenance of biodiversity, and a less well-studied but important ecosystem service: reduction of natural hazards and vulnerability. Examples of ecosystem services and their valuation are presented for areas of the US, Panama, and Puerto Rico.

5. WATER QUALITY AND SUSTAINABILITY IN INDIA

In India, rapid industrialization, urbanization, and population expansion over decades has given rise to a number of environmental problems, water pollution being the major one (Chapter 5). This has led to deterioration in both the quality and quantity of surface- and groundwater, thereby affecting the net availability of water for consumption. Despite numerous steps by government and local communities, India continues to be deprived of safe drinking water. Safe water provisions and environmental sanitation are critical for protecting the environment, improving health, alleviating poverty, and bringing safe hygiene practices to make India a safer place. Therefore, there is a tremendous need to create other avenues that need to be stimulated, intensified, and, above all, integrated, to fill in the gaps in the existing structure. This chapter discusses various challenges and opportunities related to water quality and sustainability in India, and emphasizes the need to prioritize them.

6. CHALLENGES AND SOLUTIONS TO WATER PROBLEMS IN THE MIDDLE EAST

Setbacks associated with the regional water scarcity and variability in respect to availability of water play a direct role in Middle East politics and economy (Chapter 6). The region is chronically suffering from depleted water resources coupled with bad management and degradation of water resources to a degree that parts of them are not safe for human consumption. The fast-growing population, urbanization, and food requirements exert further pressure on the dwindling resources. Global warming contributes to the severe persistent drought and also to stronger sector and interstate battles for water and to agricultural collapse. This situation leads to social unrest, mass environmental migration, and armed conflicts. Facing old and new challenges, the region has to restructure dominant water management practices and adopt a new strategy to confront the complicated linkages between water scarcity, global warming, and transboundary water conflicts. The state of water resources, the management practices, and the consequences of the projected impacts and strategies that can be implemented to alleviate the regional water crisis are reviewed to identify barriers, set new standards, and recommend management guidelines.

7. CHALLENGES AND SOLUTIONS TO WATER PROBLEMS IN AFRICA

Africa has one of the largest arid and semiarid lands, and one of the fastest growing population and urban centers. Water scarcity is acute in the arid areas and in semiurban communities (Chapter 7). Water supply to these areas of critical need will not be met through engineering waterworks. An alternative point-of-use water purification system is discussed in this chapter. The chapter reports work on clay ceramics made locally by women potters and frustum made in the laboratory, combining *Moringa oleifera* wood to make pores in both clay works. Next, two-nanoparticle composite catalysts of TiO_2-No_3 and TiO_3-WO_3 were synthesized and studied for efficiency in *Escherichia coli* disinfection, dye discoloration, and pesticide mineralization. The effectiveness of the clay ceramics and nanoparticle composite catalysts are evaluated in this chapter.

8. COMPARATIVE ANALYSIS OF EXISTING WATER RESOURCES DATA IN THE WESTERN BALKAN STATES

Vulnerability of water resources in the western Balkan states of Bosnia and Herzegovina, Macedonia, Montenegro, and Serbia is analyzed in this study (Chapter 8). Because of the absence of comprehensive data sets, the pool of available information was examined to define gaps and inconsistencies that hinder the initiation of in-depth water resource analyses. Comparative analysis of these four non-EU and former Yugoslav republics indicated that lack of reliable data is a major issue that hinders estimation of existing and future state of water resources. Nevertheless, the results suggest high vulnerability of water resources. Defined circumstances, however, hindered any coordinated efforts to adequately handle the existing water management system suitability to handle more frequent severe weather events predicted by climate changes.

9. ION CHROMATOGRAPHY INSTRUMENTATION FOR WATER ANALYSIS

Ion chromatography (IC) is the premier established method for analyzing ions in water samples derived from a variety of sources, such as surface water, groundwater, and well water (Chapter 9). The concentration of ions present in drinking water is highly dependent on the source of the water. Rainwater results in surface water that is usually free from major ionic contaminants. However, falling leaves and other debris result in the accumulation of organic matter in the water and can be transformed to ionic components when reacted with disinfectant products during the water processing step. As opposed to surface waters, groundwater has relatively higher amounts of ions dissolved due to equilibration with the ground matrix and typically has high levels of divalent ions, such as calcium and sulfate. Usually the presence of organic contaminants is low in the groundwater. Drinking water, depending on the availability of suitable sources of water, can be a blend of ground-, surface, and well waters. From an analysis perspective, water poses challenges because of the diversity of the concentration of the common inorganic ions and the presence of trace components, such as disinfectant by-products. Analyzing trace components in the presence of high levels of matrix ions is challenging. The focus of this chapter is on the required instrumentation, sample pretreatment, and setup required for pursuing water analysis.

10. USE OF ION CHROMATOGRAPHY FOR MONITORING IONIC CONTAMINANTS IN WATER

From its earliest days, IC has been the analysis of ionic components in a wide variety of sample types (Chapter 10). IC has been used in analytical applications, ranging from the analysis of biological fluids to air monitoring. Even the first paper published by the inventor of IC in 1975 described its use with a number of different applications. This paper pioneered the use of an ion exchange separation in conjunction with a "suppressor column" and conductivity detection as a general approach to the analysis of ionic species (technical properties of the suppressor column will be discussed in more detail in later sections of this chapter). In spite of the broad analytical application range of IC, use of the technique for the analysis of ionic contaminants in drinking water and wastewater represents the largest single application. IC provides the combined benefits of high sensitivity, broad dynamic range, and general suitability to the analysis of ionic compounds, making it ideally suited for this area of analytical chemistry.

11. IMPROVING SCIENCE AND COMMUNICATION IN WATERSHED MANAGEMENT

Interested citizens, environmental advocates, volunteers, planners, managers, political officials, scientists, and other professionals all have a role to play in the planning, remediation, engineering, management, monitoring, and research of a problem watershed (Chapter 11). Each entity brings an important perspective on the problem and the collective ability for achieving desired outcomes. In most cases, when the experiences and training are integrated with one another, the outcome is a more balanced strategy toward the ultimate solutions. For example, scientists rarely have a good feel for the political, financial, logistics, and construction costs of remediation, but they do have experience in estimating the cost of monitoring and study completion in addition to the required scientific approach. An interdisciplinary team is necessary to create a well-balanced, effective program for public benefit with efficient use of resources. The authors argue that environmental groups or volunteers may engage in scientific work, but not without adequate training or supervision by the scientific community. The lack of professional direction often leads to poor planning, site selection, analyses, and most of all, interpretation and suggestion of management alternatives. Likewise, researchers are poorly equipped to make management, policy, financial, or politically sensitive decisions. It is suggested that projects should be longer term so that temporal variation can be accounted for and long-term trends established. Often, officials want solutions resolved in the duration of their tenure or when the pressure of the issue subsides. The motivation, funding, and approach to local environmental projects have been reviewed and found inadequate in many cases. Finally, it has been suggested that local environmental managers should establish a new paradigm that incorporates continuous monitoring and correction when problems arise. These are areas where scientists are best suited to plan and advise. Adaptive management, a process based on continual learning and adjustments, may be the most reasonable way to approach long-term monitoring. It incorporates necessary corrective actions before they become issues that are difficult or impossible to correct.

12. WATER PURIFICATION SYSTEMS

Providing water for 7.4 billion people is a staggering challenge, let alone meeting the overwhelming demand for safe water (Chapter 12). Today more than 1.1 billion people living in emerging countries need safe drinking water. Additionally, those with compromised drinking water systems in other countries and the 200 million people affected by natural disasters each year suggest the problem is catastrophic. Ironically, bad water is as much a product of overpopulation and climate change as it is from technology. Anthropogenic chemicals, engineered antibiotics, excessive hormone use, vast areas of natural arsenic pollution coupled with microbial contamination require special media for purification. The enormity of household filters available can be difficult to sort out. This chapter is divided into three parts: (1) historical accounts of water technology; (2) the science behind elemental purification materials (bone char, charcoal, clay, copper, and silver); and (3) an evaluation of the common household purification technologies.

13. EVALUATION OF ANIMAL MANURE COMPOSITION FOR PROTECTION OF SENSITIVE WATER SUPPLIES

Chapter 13 identifies ecological, public health, and economic drivers for implementation of nutrient recovery operations in agriculture. The need for creating sustainable supplies of phosphorus is paramount in the coming decades because of population growth and diminishing supplies of phosphate rock. The potential for phosphorus recovery from poultry, swine, and dairy cow manure was described in detail using log C-pH diagrams to show the release of critical nutrients at acidic pH and the potential recovery at weakly basic pH. Phosphorus can be recovered at greater than 97% from manure slurries without the addition of external chemicals (i.e., ammonium, magnesium, calcium, etc.). The most common recovery products include struvite and hydroxyapatite, although a number of other minerals (e.g., monetite, newberyite, covellite, manganese hydrogen phosphate, and sphalerite) also precipitate under certain conditions. The recovered products have value as agrochemicals for both food crop and animal production operations. It is reasonable to think that increased emphasis of nutrient recovery from animal waste will lead to positive externalities, including the following: decreased eutrophication, hypoxia, and harmful algal blooms; improved fish/shellfish habitats; water security; food security; and political security.

14. EFFECT OF UPFLOW VELOCITY ON NUTRIENT RECOVERY FROM SWINE WASTEWATER

The swine breeding industry has developed rapidly in recent years; it generates sizable quantities of wastewater containing high concentrations of nitrogen and phosphorus (Chapter 14). A laboratory-scale setup was carried out to recover these nutrients by fluidized-bed magnesium ammonium phosphate, or struvite, crystallization. In the experiment, the fluidized-bed crystallizer used synthetic swine wastewater and the same crystal as substrate for growth. The effect of upflow velocity on the phosphate removal was investigated. Fluidized-bed crystallization was conducted at a phosphate concentration of 150 mg/L, 10.0 g/L seed dose, pH 9.53, Mg/P molar ratio of 1.3, N/P molar ratio of 4, and upflow velocity ranging from 20 to 50 cm/min. Results show that as the upflow velocity is increased from 20 to 35 cm/min, the phosphate recovery also increased from 93% to 95%; however, an additional increase will lower the recovery to 94%. Ammonium recovery increased with increasing upflow velocity. Maximum ammonium recovery of 90% was achieved at 50 cm/min. From the sieve analysis, small crystals are of higher quantity than coarse crystals for all levels; however, the relative amount of small crystals is high for elevated upflow velocities.

15. DROUGHT-INSPIRED ECONOMIC USE OF WATER IN WINE PRODUCTION

The increase in wine consumption has fostered an increase in the number of vineyards and wineries in California over the past two to three decades (Chapter 15). This increased grape and wine production, coupled with the current severe drought in the US Southwest, has pushed the wine and agricultural industry to practice efficiency, technology substitution, and water reuse. A number of water conservation measures are now in common practice, including use of drought-tolerant grape rootstock, drip and microsprinkler irrigation, soil moisture monitoring, and pest control (including fungicides and herbicides). Water is an essential component in the winery as well, mainly used for cleaning and purification. The effects of these measures and prospects for future water conservation methods are discussed. Efficient use of water is clearly a key part of sustainability in the wine industry, along with energy efficiency, emission and waste reduction, automation, and other factors.

16. SINGAPORE'S WATER INNOVATIONS

Singapore's path to economically viable water self-sufficiency by 2061 consists of various innovative strategies, including (Chapter 16):

- an ambitious goal to increase rain catchment to ~90% of its land surface;
- exploring the use of underground caverns for water storage;
- increasing NEWater capacity and output; and
- increasing desalination output.

In addition to these strategies, which are related to increasing the efficacies of the current sources of water, an increasingly important area of focus is industrial wastewater reuse. Unlike domestic wastewater, which is fairly uniform, and hence can be consolidated and treated, industrial wastewater coming from different industries varies significantly, depending on its point of origin. For example, process wastewater from refining would be very different from pharmaceutical process wastewater. Hence, consolidation and treatment would be limited to industrial clusters with similar products, and even this would lead to a fairly complex mixture of ingredient species.

Economics plays a big part in any industry's decision on whether to recycle and reuse its process water or to treat it to the extent mandated by law and subsequently discharge it. While there is a certain amount of industrial wastewater that is currently treated and reused, Singapore is actively exploring opportunities to do more, correspondingly saving on the supply of freshwater to industry. With admirable foresight and planning, and equally important, its pristine execution, Singapore has not only overcome the significant challenges associated with its water security, but has emerged as a world leader in the field. It has been at the forefront of adopting cutting-edge technologies, and is constantly developing innovations to take its global hydrohub to the next level.

17. WATER QUALITY AND PUBLIC HEALTH: ROLE OF WASTEWATER

The quest to sustain modern civilization and food security amid a world population explosion has resulted in unprecedented levels of water contamination from anthropogenic sources (Chapter 17). Also, natural phenomena such as climate change are set to worsen the already grave situation. Public health is under threat from emerging and reemerging water-related illnesses. In developed countries, thriving economies and advances in technology had rendered waterborne illnesses a thing of the past. The problems were left for developing countries to deal with, but not anymore. Water contamination is once again a global issue because of an increase in pollution in the midst of an increase in demand for water, poor economies, and faltering infrastructures across the globe. Donald Hopkins of the Carter Center once described the situation almost perfectly: "Throughout parts of the developing world, some people work hard 'just to get dirty water' for drinking, cooking, and for other personal needs.... Although the overall quality of water is far better, developed countries also face challenges in maintaining drinking water quality standards." Streams of raw wastewater in streets and residential areas are now a common sight in some countries, both developed and developing, and water is inevitably getting contaminated by a slew of environmental pollutants. A former EPA administrator once said of the US drinking water: "For years, people said that America has the cleanest drinking water in the world...." That was true 20 years ago. But people don't realize how many new chemicals have emerged and how much more pollution has occurred. This chapter discusses the source and health impacts of water contaminants. The biogeochemistry, speciation, toxicity of selected heavy metal(loid)s, as influenced by wastewater and microplastics, are also discussed.

18. ACHIEVING THE SUSTAINABLE DEVELOPMENT GOALS: WATER TREATMENT

The Sustainable Development Goal (SDG) Target 6.1 is to "achieve universal and equitable access to safe and affordable drinking water for all" by 2030, where the indicator of safe drinking water requires that the water source be compliant with fecal and priority chemical standards (Chapter 18). With increasing global population, decreasing availability of freshwater resources, and increasing pollution of existing water sources from industrial and agricultural activities, the global scale of water in need of treatment is immense. This chapter describes the challenges faced in providing safe drinking water, and the scale of treatment needed to achieve the SDG drinking water target is characterized.

CONCLUSIONS

After air, water is the most crucial resource for human survival. To achieve water sustainability, we will have to deal with its scarcity and quality, and find ways to reclaim it from various sources. Worldwide water problems and solutions are discussed, and analytical methods are provided to monitor water quality. Water reclamation techniques of interest are included to help us meet our needs, now and in the future.

REFERENCES

[1] Ahuja S. Water is indeed a crucial resource. In: American Chemical Society Meeting, Denver; March 22–26, 2015.
[2] Ahuja S. Handbook of water purity and quality. Amsterdam: Elsevier; 2009.
[3] Ahuja S. Arsenic contamination of groundwater: mechanism, analysis, and remediation. New York: Wiley; 2008.
[4] Ahuja S, Hristovski K. Novel solutions to water pollution. Washington (DC): American Chemical Society; 2013.
[5] Ahuja S. Monitoring water quality: pollution assessment, analysis, and remediation. Waltham (MA): Elsevier; 2013.
[6] Ahuja S. Water reclamation and sustainability. Amsterdam: Elsevier; 2014.
[7] Gorman R. Good housekeeping. March 2012. p. 131.
[8] UNESCO; 2003.
[9] World Economic Forum; January 2015.
[10] World Water Development Report 4. World Water Assessment Programme (WWAP); March 2012.
[11] C&EN; March 28, 2016. p. 17.
[12] UNEP SEO report, 2004–2005: the scientist. July 2016. p. 23–9.
[13] China City Statistics. Urban water supply statistics yearbook. 2012.
[14] C&EN; January 25, 2016.
[15] C&EN; October 05, 2015. p. 18.
[16] C&EN. The scientist, August 2015. November 02, 2015. p. 34.
[17] C&EN; April 25, 2016.
[18] Michael Wines, John Schwartz. NYT; February 08, 2016.
[19] C&EN; April 18, 2016. p. 16.
[20] C&EN; March 23, 2016.[a] . C&EN; August 22, 2016. p. 18.
[21] C&EN; November 16, 2015.
[22] Ahuja S. Food, energy, and water: the chemistry connection. Amsterdam: Elsevier; 2015.
[23] UN-water global analysis and assessment of sanitation and drinking water report, GLAAS 2014 Report.
[24] Ahuja S. Assuring water purity by monitoring water contaminants from arsenic to zinc. In: American Chemical Society Meeting, Atlanta; March 26–30, 2006.
[25] Chem Eng News August 07, 1978.
[26] Chem Eng News September 25, 1978.
[27] Ahuja S. CHEMTECH 1980;11:702.
[28] Ahuja S. Ultratrace analysis of pharmaceuticals and other compounds of interest. New York: Wiley; 1986.
[29] Ahuja S. Trace and ultratrace analysis by HPLC. New York: Wiley; 1992.
[30] McNeil EE, Otson R, Miles WF, Rahabalee FJM. J Chromatogr 1977;132:277–85.
[31] Ahuja S. Water sustainability and reclamation. In: American chemical society meeting, San Diego; March 25–29, 2012.
[32] Science; June 26, 2015.
[33] C&EN; November 16, 2015.
[34] New York Times; May 03, 2016.
[35] Star News; August 11, 2011.

CHAPTER TWO

Progress and Lessons Learned from Water-Quality Monitoring Networks

D.N. Myers, A.S. Ludtke
U.S. Geological Survey, Office of Water Quality, Reston, VA, United States

You could not step twice into the same river; for other waters are ever flowing on to you.

Heraclitus

1. INTRODUCTION

Water-quality monitoring networks in rivers and streams have achieved successes and faced challenges since their inception in the United States in the early 1950s [1]. Successes include providing regional- and national-scale information on status and trends in water quality to guide policy development and management actions. Technical challenges include the selection of the most appropriate elements including temporal and spatial sampling strategies and network designs, data collection methods, and data analysis strategies that will facilitate meaningful characterizations of water quality. Water-quality monitoring is a costly activity and difficult tradeoffs are necessary to balance the objectives of geographic description using many monitoring sites and temporal description using sufficiently frequent sampling so that year-to-year variations and seasonal patterns are not confused with long-term trends.

The purpose of this chapter is to present lessons learned from long-term water-quality monitoring to better support future activities. Efforts to sustain monitoring networks are important because these networks have collected information crucial to the description of water-quality trends over time. These data are irreplaceable. Once time and opportunities to collect samples have passed, it is not possible go back and collect these data to establish the baseline against which to evaluate current and future trends.

Chemistry and Water
ISBN 978-0-12-809330-6
http://dx.doi.org/10.1016/B978-0-12-809330-6.00002-7

The topic of monitoring networks is extensive and well represented in scientific literature. This chapter focuses primarily on national-scale long-term water-quality monitoring in rivers and streams over nearly seven decades. There are many examples of high-quality regional and statewide monitoring networks but including more than anecdotal information about them for context is beyond the scope of the chapter. The objectives of the chapter are to examine characteristics of long-term stream-quality monitoring networks such as monitoring objectives and designs. Design characteristics include criteria for site selection, geographic distribution of monitoring sites, frequency of observations, and types of physical, chemical, and aquatic biological analyses. Lastly, the chapter describes how these characteristics have changed over time in the number of sites providing these observations. Fixed-site networks of the US Geological Survey (USGS) are examined in greater detail than other national-scale networks because of their longevity.

Key questions addressed in this chapter include:

- What is the history and current status of long-term water-quality monitoring in the United States?
- How successful are long-term monitoring networks in achieving their objectives to detect policy-relevant changes and trends in water quality?
- What network designs have been successful in addressing key policy questions?
- How have long-term water-quality monitoring networks changed over time and what are the decadal and multidecadal patterns in water-quality monitoring?
- What have been the causes and consequences of these changes?

2. BACKGROUND AND DEFINITION OF TERMS

Definitions of terms such as monitoring, assessment, network, surveys, status, trends, change, long-term, and other terms are important for a clear understanding of the topic. Monitoring is defined as long-term, standardized measurement, observation, evaluation, and reporting of the aquatic environment in order to define status and trends [2]. Long-term monitoring as defined in this document is decadal to multidecadal in temporal scale. The definition of a "trend" is the general direction in which something is developing or changing. Conversely, the definition of status is the state or position of something at a particular time.

Water-quality monitoring networks are defined as sets of related sites that are sampled over time in an organized and perennial manner [2].

This type of monitoring involves repeated collection of samples at fixed or varying time intervals at fixed locations. Sites can be located at various spatial scales from small streams to large rivers. For rivers and streams, these water-quality monitoring activities are collocated with measurements of streamflow to provide a hydrologic context. These types of monitoring networks, often with no projected ending point, are known as ambient, fixed-site, long-term water-quality monitoring networks [3].

The purpose of status and trend networks is to describe current conditions and temporal changes and trends across water resources areas. Ancillary data on human and natural factors such as climate, geology, land use, chemical use, water use, and regulation help to explain the quality of water delivered from watersheds affected by those factors. Reference networks are located in relatively undisturbed watersheds and are mostly affected by natural factors [4]. Reference networks provide a means to compare hydrologic condition and trends in watersheds affected by natural factors to watersheds predominantly influenced by human activities. Often reference conditions are called background conditions [2]. Conversely, a baseline is an initial set of critical observations or data used for comparison or as a control. Baseline networks provide a yardstick to compare sites affected by a current or future human disturbance [2].

Surveillance is defined as continuing and repeated, specific measurements, observations, and reporting for the purpose of water-quality management and operational activities [2]. Surveillance networks are for monitoring water quality for management purposes such as compliance with water-quality criteria, other regulations, or goals such as limitations on the transport of mass (loads) delivered to sensitive receiving waters [2]. In addition, there are several other types of networks designed to meet specific objectives such as spill detection and warning.

To be successful, long-term networks and other monitoring activities need a clear statement of monitoring objectives, which are essential for assuring that monitoring activities meet their goals. Networks can have single or multiple objectives. Some networks have a single objective to address one purpose such as measuring pesticide occurrence, while other networks have multiple objectives such as a surveillance objective for monitoring the ongoing suitability of water for public, industrial, and agricultural supply, recreational quality, fisheries, and aquatic life, and to detect trends. Another single objective is to provide a baseline for measuring the status and trends in the quality of the water environment. Monitoring objectives should influence the network design and criteria for sampling locations, sampling frequencies, water-quality measurements, and samples for collection and constituent analysis. Assessments can be geographic addressing occurrence and distribution of a constituent at many sites and or temporal, addressing occurrence over short to long periods at single sites. Assessments are appraisals with interpretations of water-quality conditions and or trends. Assessments are important for addressing water-resource issues as they emerge and to track effectiveness of water policies and resource investments over time.

3. METHODS

A literature review is combined herein with an analysis of available data to put the history of water-quality monitoring networks into perspective and to meet the stated goals, objectives, and scope of this chapter.

3.1 Literature Review

A literature review was undertaken to determine the history, successes, and challenges of long-term water-quality networks and monitoring across a wide range of governmental programs at the national level. Journal articles, agency reports, and other published sources were reviewed to describe national-scale, long-term monitoring networks. Other common types of water-quality monitoring designs are also described to provide context and milestones for important steps in the evolution of monitoring. The review includes descriptions of monitoring agencies, goals and objectives of monitoring networks, site-selection criteria, constituents monitored, period of record, and frequency of sample collection. Also included is a review of monitoring successes, challenges, and policy relevance of monitoring results to water-quality management and protection. Descriptions of water-quality coordination activities are included. The time period of the review extends from the 1930s through 2015.

3.2 Methods of Analysis of Long-Term Water-Quality Data

To quantitatively determine the monitoring characteristics of USGS long-term fixed sites, a retrieval of all water-quality results for one physical property and five chemical constituents representing a broad spectrum of water-quality sampling activities that are well represented in USGS and other long-term monitoring networks was undertaken. Screening criteria were developed to determine what river and stream networks and sites are typical of long-term water-quality monitoring. Retrievals of numbers of sites by decade and by constituent were made from the USGS National Water Information System (NWIS), which is the computerized water-information system of the USGS.

Long-term water-quality sites in the NWIS were identified based on screening criteria for water-quality constituents including sampling frequency, methods of sample collection and analysis, quality control, and periods of record. The definition of a record is a database entry that may contain one or more values such as a series of results at a particular site for total phosphorus, chloride, sulfate, and atrazine. Groups of records are stored in an electronic database table. Records are often called rows since each new record creates a new row in the table [5]. A period of record is defined as the time period of sampling activities at a site from initiation of sampling activities to completion of activities after which the site is no longer sampled. Period of record at a site can range from a single point in time on a single day to decades. Water-quality sampling site types designated in the NWIS as stream, canal, ditch, tidal stream, or springs were included in the database retrievals. All efforts were made to remove samples from the analysis if the purpose code or network designation was a focused study for a special purpose, such as "Mount St. Helens," "irrigation effects," or "tracer test" because these type of investigations were not considered to be representative of national-scale long-term river- and stream-quality monitoring networks.

Concentrations of selected constituents were required to have an associated streamflow value, either a direct measurement at the time of sample collection or be located at or near a USGS streamflow gauging station with available streamflow measurements. Water-quality records not having associated streamflow measurements were dropped from further consideration. All sites in the NWIS are required to have a station identification number, station name, location described by latitude and longitude, and date and time of sampling. These data were critically important to the following analysis of the NWIS results. In addition, the NWIS contains records of water-quality data and metadata including parameter codes. A parameter code is 5-digit numerical code used to identify the constituent measured and the units of measure [5]. Parameter codes can be used to aggregate water-quality results that may be comparable.

Results for the six water-quality constituents were retrieved from the NWIS by station number, date, and time of sampling, over the period of record from 1975 to 2014, and by parameter code. Use of the parameter code provided a method to accommodate evolving analytical methods and sampling procedures over time. Therefore, results obtained from USGS methods for each constituent were combined and retrieved by parameter code. Filtered sample results include chloride and sulfate in the nonmetal inorganic group and atrazine in the pesticide group. Unless otherwise noted, filtered constituents represent samples analyzed after filtration through a 0.45 μm pore-size filter. Concentrations retrieved from the database for whole water samples (unfiltered) include total phosphorus in the nutrient group and suspended sediment in the physical properties group. Filtered and unfiltered results were retrieved and combined for nitrite plus nitrate as nitrogen in the nutrient group.

Many changes took place in analytical methods, sampling, and sample-processing methods from 1975 to 2014 in the USGS fixed-site networks. For example, two parameter codes are used by the networks for atrazine; parameter code 39632 is reported in micrograms per liter and parameter code 65065 is reported in nanograms per liter. These two parameter codes were combined and counted toward a site that sampled for atrazine. The parameter codes and other metadata in the NWIS in this case indicate that the method of analysis is the same; only the reporting units are different between the two parameter codes. Only one parameter code each was used to retrieve data from the NWIS for concentrations of suspended sediment (80154) and total phosphorus (00625). In the case of nitrite plus nitrate as indicator of total nitrogen, the parameter codes used by the networks for filtered (00630) and unfiltered (00631) samples were combined. In 1992, a procedural change was made in USGS long-term network operations from analysis of unfiltered samples to analysis of filtered samples for determinations of nitrite plus nitrate as nitrogen [6,7]. Data for both parameter codes were combined without investigation into the possible effects of filtration changes on sample results. However, changes in methods and procedures can be very important for the analysis of trends over time and for comparisons among sites.

A sufficient period of record and sampling frequency are needed to qualify as a long-term stream-quality monitoring site. For this study, data were retrieved for a 40-year period of record from 1975 to 2014. A water year begins October 1 and ends September 30 of the following year. For example, water year 2015 began on October 1, 2014 and ended on September 30, 2015. All further references to year in this chapter mean water year.

Long-term stream-quality monitoring sites were differentiated from short-term sites in the NWIS by using a set of screening criteria and associated algorithms. Each combination of site, constituent, and year is categorized into a temporal sampling frequency category of daily, weekly, monthly, seasonal, quarterly, semiannually to annually, or other frequency for the period of record of the site from 1975 to 2014. The weekly, monthly, seasonal, and quarterly categories of sampling approximate those commonly used by the USGS and other fixed-site monitoring networks. The category designated as other frequency contains sites that are sampled daily or sampled at intervals greater than annually.

Specifically, the sampling frequency algorithm was implemented by ordering the samples chronologically by sampling date and then calculating the days elapsed between each consecutive sample measured for each year of record. This measurement resulted in a range of elapsed days or lag days between samples that reflects sample-collection frequency. We describe a range of lag days for each temporal category (Table 2.1). Oftentimes, sampling networks combine seasonal and (or) hydrologically based sampling with fixed frequency sampling, which results in more than 12 samples per year but is not sufficiently frequent to qualify as weekly sampling. To capture those types of variations in sample numbers due to seasonal and hydrologically based sampling combined with fixed-frequency sampling, categories were defined by a combination of minimum number of samples in each category and a range in lag days (Table 2.1).

Within any given year, a site needed six or more monthly samples with a range in lag days from 15 to 45 to be categorized as monthly. Quarterly required at least three samples per year with a range of lag days of 46–150. Seasonal sites were differentiated from weekly, monthly, and quarterly by a minimum of six samples per year and a maximum of 270 lag days in the interval between samples (Table 2.1). Replicate samples collected for quality-control purposes are counted as only one sample for a given site on a given day. These situations were identified by a lag day of zero between samples and also are binned in the category described as other frequency.

Table 2.1 Sampling Frequency Category Is Determined by the Minimum Number of Samples and Range in Lag Days Between Samples for a Water Year

Sampling Frequency Category	Minimum Number of Samples per Water Year	Range in Lag Days
Daily	182	1
Weekly	26	2–14
Monthly and bimonthly	6	15–45
Seasonally[a]	6	2–270
Quarterly	3	46–150
Semiannual and annual	2	151–400
Infrequent	1	>400

[a]Seasonal category has six irregularly timed samples per water year. Lag days based on actual data set.

Sites that met the minimum completeness criteria in the weekly, monthly, bimonthly, seasonal, or quarterly sampling frequency category in any given year for any of the six constituents were further tested to determine if they also qualified as long-term sites at the decadal scale. Long-term monitoring sites in the five sampling frequency categories were identified for each constituent in one or more of four decades: 1975–84, 1985–94, 1995–2014, and 2005–14. A site qualified as decadal and thereby long-term if there was a minimum period of record in 6 out of 10 years within one or more of the four decades. Sites that met the completeness criteria for all four decades are considered 40-year long-term water-quality sites. Sites that met the completeness criteria for three, two, and the most recent decade are considered 30-year, 20-year, and 10-year long-term sites, respectively.

The algorithm used herein produces results that include more sites than would be included had an algorithm been adopted that required strict adherence to stated network sampling frequencies. Therefore a long-term site could be defined using other algorithms and other methods different from those used for this analysis. For this analysis, a compromise was made so that important long-terms sites would not be excluded due to slight differences between the stated sampling frequency in network publications and what was accomplished in field campaigns. Fig. 2.1 shows an example of a site with a sampling frequency that varies over decades. Sample collection for sulfate at the site began in 1978 and continued through 1986 at monthly to bimonthly frequencies. Sampling was discontinued from 1987 to 1995. Sampling resumed in 1996 and continued for the next 18 years through 2014. During the two most recent decades, sampling was monthly with additional high flow and low flow (hydrologic) or seasonally based samples; however, one year in this time period was sampled semiannually and another year was sampled quarterly. A strict algorithm would exclude this site for sulfate concentrations due to low sampling frequencies in 2 years and gaps in data in the 1990s, when in fact, these data have value for water-quality assessment and are part of a former long-term water-quality network.

4. HISTORY OF NATIONAL-SCALE, FIXED-SITE WATER-QUALITY MONITORING NETWORKS IN RIVERS AND STREAMS

The history of long-term water-quality monitoring is described in four major decadal periods from 1975 to 2014. These decadal periods are

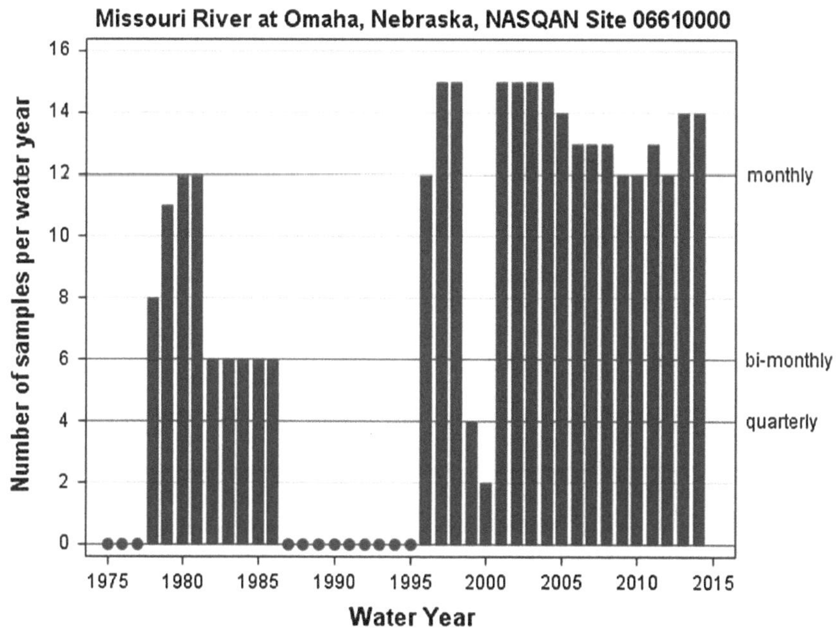

Figure 2.1 Frequency of sampling for concentrations of sulfate at a long-term US Geological Survey monitoring site, water years 1975–2014. *NASQAN*, National Stream Quality Accounting Network.

consistent with analysis of monitoring sites retrieved from the NWIS and described in Section 3. Results of the data analysis are incorporated into the decadal descriptions of water-quality monitoring activities for six constituents. These results were used to create Table 2.2 and maps in Figs. 2.5–2.9. Chloride and sulfate are commonly sampled together and their maps are nearly identical; therefore chloride is not shown in this sequence of maps.

4.1 The First National Water-Quality Monitoring Networks, 1935–64

Only 23 federally operated water-quality monitoring sites in 10 states were sampled at regular intervals over time in 1936. Commonly determined water-quality constituents—major and minor ions, alkalinity, hardness, dissolved solids, and suspended materials—supported decision making on water-resource development or provided an improved understanding of the geochemistry of natural water [8,9]. Federal agencies carrying out monitoring at the time included the USGS, US Bureau of Fisheries, US Public Health Service (PHS), US Bureau of Reclamation, US Army Corps of

Table 2.2 Patterns of Change by Sampling Frequency and by Decade in the Total Number of Sites, Number of Long-Term Sites, and Percentage Change in Number of Long-Term Sites for Selected Constituents, 1975–2014

Constituent Group	Constituent Name (Units Are in Milligrams per Liter Unless Otherwise Noted)	Sampling Decade	Number of Sites at Which a Selected Water-Quality Constituent Was Sampled at the Indicated Sample-Collection Frequency Within Each Decade. Numbers in Parentheses ( ) Indicate the Number of Long-Term Sites Within Weekly, Monthly to Bimonthly, Seasonal, and Quarterly Categories.							Percent Change in No. of Long-Term Sites Relative to No. of Long-Term Sites in 1975–84 (%)
			Weekly	Monthly and Bimonthly	Seasonally	Quarterly	Annually and Semiannually	Other Frequencies[a]	Total	
Nutrient	Nitrite plus Nitrate as Nitrogen	1975–84	27 (2)	2,233 (896)	145 (4)	1,452 (328)	8,390	118	12,365 (1,230)	0
		1985–94	40 (6)	1,088 (452)	129 (3)	1,645 (707)	4,385	315	7,602 (1,168)	−5
		1995–2004	47 (26)	1,467 (306)	229 (23)	1,261 (386)	4,560	689	8,253 (741)	−40
		2005–14	36 (11)	728 (298)	245 (14)	872 (395)	2,592	584	5,057 (718)	−42
	Total Phosphorus	1975–84	27 (4)	2,073 (830)	134 (4)	1,187 (269)	5,061	115	8,597 (1,107)	0
		1985–94	53 (9)	938 (407)	123 (5)	1,477 (689)	2,805	185	5,581 (1,110)	−1
		1995–2004	68 (38)	1,537 (298)	234 (20)	1,267 (349)	3,976	566	7,648 (705)	−36
		2005–14	53 (13)	836 (342)	255 (12)	949 (454)	2,331	539	4,963 (821)	−26
Inorganics, Major, Nonmetals	Chloride	1975–84	63 (31)	2,075 (848)	120 (6)	1,566 (393)	10,360	138	14,322 (1,278)	0
		1985–94	29 (6)	794 (276)	82 (0)	1,595 (684)	4521	427	7,448 (966)	−24
		1995–2004	44 (20)	802 (198)	101 (6)	1,067 (264)	4,260	676	6,950 (488)	−49
		2005–14	25 (12)	571 (220)	174 (15)	852 (328)	2,239	545	4,406 (575)	−55
	Sulfate	1975–84	55 (30)	2,066 (854)	103 (8)	1,657 (405)	11,092	126	15,099 (1,297)	0
		1985–94	27 (6)	792 (283)	66 (0)	1,502 (689)	4,667	444	7,498 (978)	−23
		1995–2004	33 (19)	758 (184)	87 (6)	1,188 (284)	4,302	761	7,129 (493)	−62
		2005–14	6 (1)	434 (198)	144 (8)	796 (342)	2,123	531	4,034 (549)	−58

Organics, Pesticide	Atrazine (micrograms per liter)	1985–94	14 (0)	43 (0)	34 (0)	76 (0)	1,017	3	1,187 (0)	0
		1995–2004	8 (0)	243 (90)	70 (10)	164 (26)	2,146	134	2,765 (126)	0%
		2005–14	0 (0)	125 (62)	118 (6)	70 (15)	528	117	958 (83)	−34
Sediment	Suspended sediment	1975–84	168 (24)	1,397 (501)	311 (30)	877 (155)	4,901	99	7,753 (710)	0
		1985–94	40 (10)	616 (133)	181 (12)	976 (428)	1,778	217	3,808 (583)	−18
		1995–2004	55 (24)	782 (238)	180 (25)	585 (112)	2,668	497	4,767 (399)	−44
		2005–14	78 (35)	559 (194)	309 (30)	472 (131)	1,423	472	3,313 (390)	−45

[a]Other contains replicate sites, sites sampled daily, and sites sampled at less than annual frequency.

Engineers, US Forest Service, and US Soil Conservation Service. The PHS was exclusively concerned with research into water-pollution control and drinking water treatment [1]. The USGS was almost exclusively concerned with research and monitoring of the quality of water for water supply and the natural geochemistry of water. The USGS began to sample rivers and streams in water-quality surveys in the early 1900s and began to work cooperatively with state and local agencies around 1928 [1].

A report by Congress in 1936 recommended for the first time that a federal stream-quality monitoring network be established at streamflow gauging stations [9]. A network of 200 sites was proposed for operation by federal agencies and a network of 400 sites was proposed as a federal/state cooperative effort. However, the proposed network was not realized. Almost all federal and state water-quality monitoring activities at the time were short term, lasting fewer than 3 years [9]. Data were regularly published by the USGS in cooperation with state and local agencies in a series of 134 USGS water supply papers from 1941 to 1976. These reports contain available water-quality data under the general title, Quality of Surface Waters in the United States [10]. Water-quality data were also published in state and other agency reports.

In 1948, Congress passed the first temporary Federal Water Pollution Control Act (FWPCA, P.L. 80−845). Due to a lack of monitoring programs at the federal level there was a large gap in quantitative information on the extent and severity of water pollution and its impacts on important water uses [11]. The FWPCA of 1948 was a driver for the creation of the first long-term water-quality monitoring networks. The earliest known long-term water-quality monitoring network was initiated by the Ohio River Valley Water Sanitation Commission (ORSANCO) [12]. A network of fixed-location fixed-frequency sites at streamflow gauging stations was created to address water-pollution problems. The ORSANCO surveillance and spill notification network began in 1951 with 11 sites that expanded to 35 sites on the Ohio River and its 19 major tributaries [12] (Fig. 2.2).

By 1960, the untreated sewage from 25 million people in 3500 cities and towns in the United States was discharged to surface waters [13]. Only two-thirds of 7500 cities and towns were served by secondary treatment systems; with the remaining third served by primary treatment systems [13]. Primary treatment can reduce the biochemical oxygen demand (BOD) of influent wastewater in a treatment plant by 20−30% and total suspended solids by 50−60%. Secondary treatment can reduce BOD and total suspended solids by 80−90%. The remaining fractions were released to

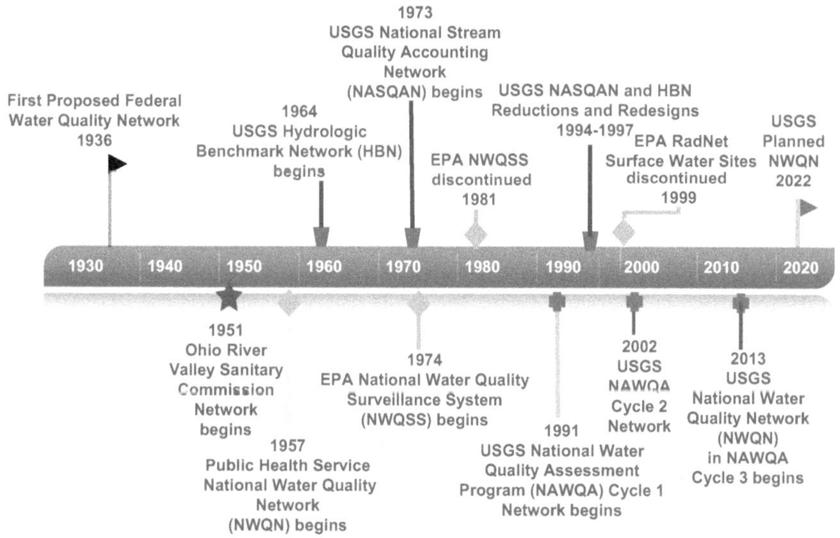

Figure 2.2 Timeline and milestones of selected fixed-site water-quality networks, 1936–2022.

receiving waters to be decomposed and diluted with time and distance downstream. Without disinfection of the effluent, pathogens were also released in wastewater discharges. Inadequately treated wastewater discharged to receiving waters in areas of concentrated urban populations resulted in severe water-quality impairments such as total or near total loss of concentrations of dissolved oxygen, elevated levels of bacteria and pathogens, and ammonia levels lethal to fish and other desirable aquatic organisms. These conditions made receiving waters neither fishable nor swimmable nor suitable as a source of drinking water supplies.

To address the problem of water pollution from untreated and partially treated wastewater from cities and towns, the first permanent FWPCA (P.L—660) was enacted in 1956. The 1956 FWPCA established grants to states for construction of wastewater-treatment plants to begin to address water-pollution problems [14]. This program became known as the Construction Grants Program. The FWPCA of 1956 authorized the creation of the first federal water-quality monitoring network.

In 1956, nationwide there were about 220 water-quality sites sampled on a regular schedule over a period of time. Most of these sites were operated by states [15]. For example, the state of Georgia in cooperation with the USGS sampled water quality in a state-wide survey of 125 streams from 1957 to 1958 and published the results in Georgia Geological Survey

Bulletin No. 69 [16]. A similar reconnaissance study was completed in New Jersey at 150 sites sampled quarterly [17]. The PHS established the first National Water Quality Network (PHS NWQN) in 1957 to collect basic data on water pollution at 50 fixed-frequency fixed-location sampling sites in 15 major river basins in the conterminous United States and Alaska [18] (Fig. 2.2). In 1963, the PHS NWQN was renamed the National Water Pollution Surveillance Network (NWPSN) and the number of sites increased to 125 (Fig. 2.3; Table 2.3; [18]).

The Atomic Energy Commission's Radiation Alert Network (RAN) and PHS Tritium Surveillance System (TSS) were established in 1956 and 1964, respectively, as part of a monitoring and alert system for the detection of radioactive contamination in streams and drinking water from above-ground nuclear weapons testing and leaks, spills, and other emissions from nuclear power plants and federal installations [19] (Tables 2.3 and 2.4).

The first national-scale reference or background network, the Hydrologic Benchmark Network (HBN), was established by the USGS and began operations in 1964 (Fig. 2.2). The HBN sites are located in watersheds minimally affected by water management and other human activities where these conditions are expected to remain relatively undisturbed into the foreseeable future [4,20]. As stated by Luna Leopold of the USGS in establishing the

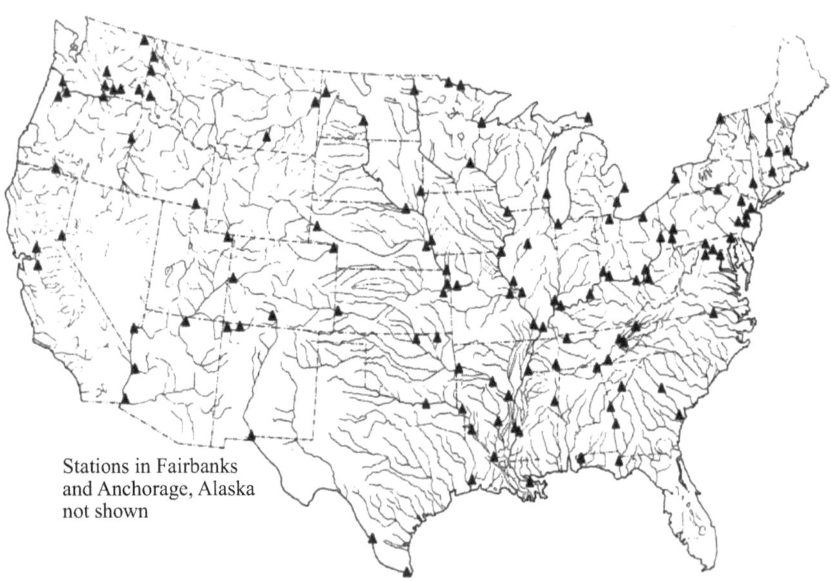

Figure 2.3 Map of the conterminous United States showing water-quality monitoring stations in the US Public Health Service National Water Quality Network (PHS NWQN). The PHS NWQN operated from 1958 to 1963 [18].

Table 2.3 Major Characteristics of National Fixed-Site Water-Quality Monitoring Networks; Network Name and Number, Years of Operation, Number of Sites, Objectives, and Criteria for Site Selection, 1956–2015

No.	Organization, Network Name, Years of Operation, and Reference(s)	Network Objectives and Criteria for Site Selection
Multiple objective status, trends, spill detection, and surveillance networks (1957–2016)		
1a	US Public Health Service (USPHS) National Water Quality Network 1957–65 [18]	*Objectives*: Collect, interpret, and disseminate basic information on changes in water quality at key points in river systems affected by changes in water use and development. Provide continuous information on the nature and extent of pollutants affecting water quality. Provide data which will be useful in the development of comprehensive water-resources programs, and provide data which will assist in water-pollution control programs and in the selection of sites for water uses. *Criteria*: Sites located on major waterways used for public water supply, propagation of fish and wildlife, recreational purposes, and agricultural, industrial, and other legitimate uses; on interstate, coastal, and international boundary waters, on waters affected by activities of the federal government.
1b	Federal Water Pollution Control Admin. National Water Pollution Surveillance Network (NWPSN) 1966–70 [153]	Operation of the network was transferred from the USPHS to the Federal Water Pollution Control Administration (FWPCA) in 1966 and later to the Federal Water Quality Administration (FWQA) in 1970. Operation of the NWPSN was variable by region with data collection intermittent at some sites. *Objectives*: Assist the states in the collection, standardization, and dissemination of water quality data, and to meet its own needs and those of other federal water resources agencies. *Criteria*: Sampling sites on major streams, at water supply intakes, interstate and coastal waters, and the Great Lakes.

(*Continued*)

Table 2.3 Major Characteristics of National Fixed-Site Water-Quality Monitoring Networks; Network Name and Number, Years of Operation, Number of Sites, Objectives, and Criteria for Site Selection, 1956–2015—cont'd

No.	Organization, Network Name, Years of Operation, and Reference(s)	Network Objectives and Criteria for Site Selection
2	US Environmental Protection Agency (EPA) Ambient Water Quality Network 1971–74 [61]	In 1971, all water quality monitoring activities in the FWQA were transferred to EPA. *Objective*: Sites transferred from the FWQA were operated for various purposes including monitoring, surveys, and surveillance. *Criteria*: Variety of criteria depending on objective of data collection.
3	US Environmental Protection Agency National Water Quality Surveillance System 1974–81 [51,63]	*Objective*: Obtain a long-term base for determination of trends, establishment of relationships between land use and water quality, and evaluation of the effectiveness of pollution-control efforts. *Criteria*: The system consisted of 96 sets of paired sites throughout the United States. Paired network sites were located upstream and downstream of major nonpoint pollution sources such as urban areas and major agricultural areas.
4	US Environmental Protection Agency—State Basic Water Monitoring Program 1980–85 [51]	*Objectives*: Accurately inventory and determine the status of all waters of the Nation. Provide a coordinated nationwide assessment of water quality to be used in decision-making processes, to educate the public, and inform Congress. The ambient sites were operated by the states primarily to determine national trends in water use areas, problem areas, land use areas, and where future development may impact water quality. *Criteria*: Paired sites bracketing land-use areas; single sites in small and homogenous subbasins with specific water uses; major rivers and their largest tributaries; at outlets from watersheds and lakes, at river mouths near estuaries or coastal areas, and on major interstate and intrastate rivers.

5a	US Geological Survey (USGS) Hydrologic Benchmark Network (HBN) 1964–97 [20,154]	*Objectives*: Provide consistent baseline data on hydrology, water quality, and related factors in representative undeveloped watersheds nationwide and to provide on a continuing basis to compare and contrast conditions to watersheds more influenced by human activities. *Criteria*: Watersheds ranging in size from 5 to 5200 km^2 in which there is minimal human disturbance such as no artificial storage, regulation, or diversion of water in the past or foreseeable future. Groundwater within the watershed is not affected by pumping from wells and disturbance from natural factors is small. In the early 1990s, eight stations were discontinued and two sites were replaced. Because of budgetary restrictions, water-quality sampling was discontinued in October 1997 except at five sites in the northeastern United States.
5b	Hydrologic Benchmark Network Radionuclide Study 1968–71 [32]	*Objective*: Monitor radiochemical concentrations at stream sites. *Criteria*: Operated as a special study of background levels of radionuclides in streams. Streamflow based; once during high and low streamflow periods from October 1968–September 1970. One sample in 1971 at low streamflow.
5c	Hydrologic Benchmark Network 2003–14 [97]	*Objectives*: All sites deactivated in 1997 were reevaluated to determine if upstream development had taken place or if the watersheds had remained undisturbed. The focus of the network became the detection of water-quality trends in streams as influenced by atmospheric deposition. All of these data are used to study long-term trends in surface water flow, water chemistry, aquatic biology, and soil chemistry and as a bench mark against which to compare changes in flow and chemistry in developed watersheds. *Criteria*: Watershed sizes range from 5 to 658 Km^2 with a single site at 5180 Km^2. In 2011, measurements of aquatic biology and soil chemistry were added to many of the sites in the network.

(*Continued*)

Table 2.3 Major Characteristics of National Fixed-Site Water-Quality Monitoring Networks; Network Name and Number, Years of Operation, Number of Sites, Objectives, and Criteria for Site Selection, 1956–2015—cont'd

No.	Organization, Network Name, Years of Operation, and Reference(s)	Network Objectives and Criteria for Site Selection
6a	US Geological Survey National Stream Quality Accounting Network 1973–80 [64]	*Objectives*: Satisfy national and regional needs for streamflow and water quality data. Account for the quantity and quality of 90% of the water moving within and from the United States, depict geographic variability of water quality, detect changes in stream quality, and lay the groundwork for future assessments of changes in stream quality. *Criteria*: Sites located at the terminus of major river basins in watershed with average size of 10,400 Km^2 that "account" for 90% or more of the water and represent the water quality originating in the accounting unit. Special studies of pesticides at one-third of the sites (1975–80) and radionuclides at 50 sites in the 1970s.
6b	National Stream Quality Accounting Network 1981–85 [154]	*Objectives*: Obtain information on the quality and quantity of water moving within and from the United States through a systematic and uniform process of data collection, summarization, analysis, and reporting to describe the areal variability of water quality, detect changes or trends with time in the pattern of occurrence of water quality characteristics, provide a nationally consistent database useful for water-quality assessment and hydrologic research. *Criteria*: A major funding cut of 20% in 1981 resulted in a reduction in sites and constituents monitored. Fewer sites but no change in site criteria from previous design.

6c	National Stream Quality Accounting Network1986–95 [154]	*Objectives*: Objectives changed to assess water quality trends nationally, and to the extent possible, related those trends to upstream land and water use; account for the mass transport of selected constituents from the continent; determine the transport of selected constituents to estuaries and (or) other important water bodies of national significance; and determine the quality of water crossing international boundaries. *Criteria*: A major funding cut of 20% cut in 1994 resulted in a reduction in sites.
6d	National Stream Quality Accounting Network 1996–2007 [77,154]	*Objectives*: Redesigned as NASQAN-II in 1993–95 to provide an ongoing characterization of the concentrations, trends, and flux of sediment and chemicals in the nation's largest river basins, determine regional source areas for these materials, and assess the effects of human influences on observed concentrations and flux. NASQAN-II network no longer accounted for water quality of streamflow within the United States. *Criteria*: Replaced previous national network with four regional networks on large rivers ranging from 20,000 to 70,000 mi^2; examining nutrients in the Mississippi River, salinity in the Rio Grande River, trace elements in the Colorado River, and pesticides in the Columbia River. The Yukon River was sampled for 5 years. This design resulted in gaps in water-quality monitoring of tributaries discharging to several important -coastal areas and estuaries.

(*Continued*)

Table 2.3 Major Characteristics of National Fixed-Site Water-Quality Monitoring Networks; Network Name and Number, Years of Operation, Number of Sites, Objectives, and Criteria for Site Selection, 1956–2015—cont'd

No.	Organization, Network Name, Years of Operation, and Reference(s)	Network Objectives and Criteria for Site Selection
6e	National Stream Quality Accounting Network 2008–12 [131,132]	*Objectives*: NASQAN was redesigned and once more regained its national design. New objectives were to provide data for flux (load) estimation from and within large river networks within the United States and off the continent to large estuaries and bays, provide data for time trend analysis in concentrations and linkages to human activities such as land use and chemical use, provide data to support modeling efforts, and provide data to address marine issues like hypoxia in the Gulf of Mexico and harmful algal blooms in receiving waters, supports development of indicators of water quality conditions and trends. *Criteria*: Established a network of large rivers that deliver 80% of the total discharge of streamflow, nitrogen, phosphorus, and suspended sediment to coastal waters of the United States.
7	US Geological Survey National Monitoring Network (NMN) 2008–12 [131,132]	*Objectives*: Similar to contemporary NASQAN design; provide data for mass transport (load) estimation from large river networks within the United States and off the continent to large estuaries and bays. *Criteria*: Five sites added to fill geographic gaps in NASQAN design for coastal and Mississippi River subnetworks to estimate concentrations and loads of nutrients, pesticides, sediment, and selected inorganic constituents.

8a	US Geological Survey National Water Quality Assessment Program (NAWQA) Cycle 1 1991–2000 [87]	*Objectives*: Data from fixed sites in rivers and streams are the primary source of information for meeting water-column assessment objectives of status and trends. Decadal-scale trends and changes in water-quality conditions are assessed by using a combination of existing historical data, periodic intensive assessments, and selected long-term monitoring strategies. See NAWQA Cycle 2 below for more information on trends. *Criteria*: Three types of water-column sampling activities—continuous monitoring, fixed-interval sampling, and extreme-flow sampling—were conducted for 2–3 years during the first cycle of investigations. Reach assessments were completed for aquatic ecological conditions and bed-sediment and tissue surveys of trace elements and bioaccumulative organic contaminants were completed at these sites. Study Units usually had 8-12 fixed sites; with 4–8 indicator sites and 3–5 integrator sites. Indicator sites were fixed stream sampling sites located at outlets of drainage basins with homogeneous land use and physiographic conditions, typically about 50–500 Km^2. Integrator sites are stream sampling sites located downstream of drainage basins that are large and complex and often contain multiple environmental and land-use settings. Most integrator sites are on major streams with drainage basins that include a substantial portion of the Study Unit area (typically 10–100%).

(*Continued*)

Table 2.3 Major Characteristics of National Fixed-Site Water-Quality Monitoring Networks; Network Name and Number, Years of Operation, Number of Sites, Objectives, and Criteria for Site Selection, 1956–2015—cont'd

No.	Organization, Network Name, Years of Operation, and Reference(s)	Network Objectives and Criteria for Site Selection
8b	National Water Quality Assessment Program Cycle 2 2001–12 [130]	*Objectives*: Cycle 2 investigations in 42 Study Units emphasized an increased focus on understanding the factors that affect water quality and establish links among sources, transport, and effects of contaminants in hydrologic systems. A total of 13 Study Units were discontinued. Trend detection became an objective. *Criteria*: For trend assessment, long-term monitoring was established in eight major river basins encompassing the 42 Study Units. From 84 to 113 stream sites were monitored at which there were more than 10–15 years of consistent monitoring data. The fixed monitoring sites were a subgroup of original sites from the first Cycle. Similar to Cycle 1, sites fell into several categories as integrator and indicator sites and were monitored from 12 to 24 times per year. A total of 83 sites were sampled one in four years, 44 sites samples every year, and 15 sites sampled every other year.

8c	National Water Quality Assessment Program Cycle 3 National Water Quality Network 2013—Current Year [131,132]	*Objectives*: Assessments of the distributions and trends in contaminants, nutrients, and sediment. Reference sites included for evaluating effects of changing climate and human activities on water quality and aquatic ecosystems. Long-term monitoring for trends in water quality to determine whether conditions are getting better or worse. Consistent and systematic information collected over many years helps to distinguish long-term trends from short-term fluctuations. Analysis of long-term trends is essential for assessing how environmental controls and best-management practices are working and for choosing cost-effective strategies for the future. *Criteria*: This network suffered substantial declines in the number of sites and the frequency of sampling between Cycles 1 and 2. To achieve a stable and consistent network all USGS NAWQA, NASQAN, NMN, and HBN sites are harmonized into one National Water Quality Network. A substantial increase was made in the number of samples collected per year—from 6 to 25— and continuous water-quality monitoring for nutrients included at selected sites. Critical improvements include 1) updated contaminant coverage; 2) expanded reference-site monitoring for tracking climate change and evaluating ecological background conditions; and (3) expanded assessment of contaminant, nutrient, and sediment loading to inland and coastal waters.

(Continued)

Table 2.3 Major Characteristics of National Fixed-Site Water-Quality Monitoring Networks; Network Name and Number, Years of Operation, Number of Sites, Objectives, and Criteria for Site Selection, 1956–2015—cont'd

No.	Organization, Network Name, Years of Operation, and Reference(s)	Network Objectives and Criteria for Site Selection
Single objective networks for monitoring the status and trends of persistent contaminants or radionuclides; 1956–2005		
9a	Federal Water Pollution Control Admin. National Pesticide Monitoring Network Water and Sediment 1967–71 [48,49]	*Objectives*: Provide information on the overall extent of pesticide contamination of the nation's water resources. Develop the minimum program that will enable an adequate assessment of conditions in the major drainage rivers of the United States through a nationwide network of sampling locations. Over a period of years, it is expected that data obtained from this network will reflect important changes and long-term trends in pesticide levels. *Criteria*: Monitoring sites were located near the mouths of rivers that represent major river drainages throughout the country. Samples collected at sites where there was known organochlorine and chlorophenoxy acid herbicide contamination or where water quality was being affected by use of these pesticides; at or near stream-gauging sites; at sites where other kinds of water-quality data were collected; and where historical data in the form of carbon filter extracts were available.
9b	US Environmental Protection Agency National Pesticide Water Monitoring Network 1974–81 [51]	*Objective*: Established in response to a legislative mandate under Public Law 92–516 to provide information on the presence of specific pesticides in the environment. The objectives of the National Pesticide Water Monitoring Program are: provide a continuing nationwide assessment of the general levels of pesticide residues in the environment, and provide the location of possible problem areas within specific segments of the environment. *Criteria*: A sub-network of NWQSS for water and sediment samples

9c	US Fish and Wildlife Service National Pesticide Monitoring Network 1967–85 [55]	*Objective*: Established in response to a legislative mandate under Public Law 92–516 to provide information on the presence of specific pesticides in the environment. The objectives of the National Pesticide Monitoring Program are to provide a continuing nationwide assessment of the general levels of pesticide residues in fish and wildlife, and provide the location of possible problem areas within specific segments of the environment. *Criteria*: Samples of water and fish collected at 50 surface water sites also sampled by FWPCA, FWQA, and USGS.
9d	US Fish and Wildlife Service National Pesticide Monitoring Network 1986–95 [47]	*Objectives*: Monitor the concentrations of pesticides, other long-lived toxic contaminants, and trace elements and their temporal and geographic trends in all segments of the environment. *Criteria*: Periodically measured contaminant concentrations in freshwater fish and birds at large river sites.
9e	National Biological Service/US Geological Survey Biomonitoring Status and Trends Network 1995–2009 [59,155]	*Objectives*: Examined fish health in multiple river basins by using a suite of organismal and suborganismal endpoints to monitor and assess the effects of organochlorine pesticides, their degradates, polychlorinated biphenyls, and selected trace elements contaminants in fish. The network is unique in that it utilizes biomarkers to evaluate less persistent chemicals in the environment and to detect molecular-level changes before population effects may be evident. *Criteria*: Sites on major rivers and their major tributaries including the Colorado River Basin (14), Columbia River Basin (15), Mississippi River Basin (48); Rio Grande River Basin (10), Southeastern River Basins (13), Yukon River basin (10).

(*Continued*)

Table 2.3 Major Characteristics of National Fixed-Site Water-Quality Monitoring Networks; Network Name and Number, Years of Operation, Number of Sites, Objectives, and Criteria for Site Selection, 1956–2015—cont'd

No.	Organization, Network Name, Years of Operation, and Reference(s)	Network Objectives and Criteria for Site Selection
10a	Atomic Energy Commission/US Public Health Service Radiation Alert Network and Surveillance System 1956–64 [19]	The Atomic Energy Commission (AEC) followed by the PHS operated several networks to monitor the environment for radiation, including: the Radiation Alert Network (RAN) and the Tritium Surveillance System (TSS). Radiation was measured in surface water, at locations in the Interstate Carrier Drinking Water Network, and in the Pasteurized Milk Network (PMN). *Criteria*: Sites were located on major rivers downstream of nuclear facilities. Other sites were sited to measure background levels and levels in treated drinking water from 68 water treatment plants.
10b	US Environmental Protection Agency Environmental Radiation Ambient Monitoring Systems and RadNet 1973–99 [19]	*Objective*: To directly assess the population's intake of radioactive pollutants and fallout. Detect specific radioactive materials in the water environment. Radiation Alert Network (RAN) *Criteria*: Grab samples are collected at sites located downstream from nuclear facilities or at background sites and at drinking water treatment plants. Sampling ended in 1999 to avoid duplicating state sampling programs around nuclear facilities.

No., number.

Table 2.4 Network Name and Number, Number of Sites Sampled Over Time, Frequency and Type of Data Collection, and Changes Due to Network Redesign, 1957–2015

Network and Number From Table 2.3	Number of Sites Sampled Over Time; Frequency and Type of Data Collection and Changes Due to Network Redesign
1a Public Health Service National Water Quality Network	*Range in No. Sites over time*; 50–125 *Data and Collection Frequency*: *Weekly*: BOD, DO, COD, chlorine demand, NH_4–N, pH, color, turbidity,°C, alkalinity, hardness, Cl, SO_4, PO_4, TDS, CB. *Semimonthly*: Plankton populations. *Monthly*: Organic chemicals. *Quarterly*: Na, K, F, and trace elements. Suspended and dissolved gross alpha and gross beta activities. *Composite samples*: Trace metal concentrations and Sr–90 activity were analyzed quarterly from composites of weekly samples.
1b National Water Pollution Surveillance System	*No. Sites*: 154 *Data and Collection Frequency*: Monthly for °C, pH, DO, turbidity, BOD, COD, TDS, Cl, alkalinity, hardness, NH_3, PO_4, SO_4, CB, phytoplankton, zooplankton.
2 Surface Water Monitoring Network	*No. Sites*: 120–800 *Data and Collection Frequency*: Daily: Q, °C, pH, SC, DO turbidity, Monthly: BOD, COD, Cl, NH_4–N, SO_4, hardness, $NO_3 + NO_2$—N, PO_4, TC
3 National Water Quality Surveillance System	*Range in No. Sites Over Time*: 98 paired sites, 188 total sites *Data and Collection Frequency*: Daily: Q; semimonthly to Monthly: °C, SC, pH, transparency or turbidity, DO, HCO_3, CO_3, TDS, TSS, Ca, Mg, Na, K, Cl, SO_4, TP, total $NO_3 + NO_2$—N, total NH_4–N, total Fe, TOC, COD
4 EPA/State Basic Water Monitoring Network	*No. Sites*: about 1000 *Data Collection Frequency*: Daily: Q; Monthly: °C, SC, pH, DO, FCB, TKN, $NO_3 + NO_2$—N, TP, COD, and TSS; *annually*: Chlorinated hydrocarbon pesticides, PCBs, As, Cd, Cr, Cu, Pb, and Hg in fish and shellfish tissues.
5a Hydrologic Benchmark Network	*Range in No. Sites Over Time*: 22–57 *Data Collection Frequency*: Continuous: Q; Monthly from 1963 to 82, quarterly 1986–97: °C, hardness, pH, Si, Fe, Mg, Ca, Na, K, HCO_3, CO_3, SO_4, Cl, F, $NO_3 + NO_2$—N, TDS, SSC *Special studies or biannually*: Chlorinated hydrocarbon pesticides in water, sediment and fish tissue; minor elements As, Ba, Cr^{6+}, Co, Cu, Pb, Hg, Se, Ag, Zn, and radiochemicals (see next).

(*Continued*)

Table 2.4 Network Name and Number, Number of Sites Sampled Over Time, Frequency and Type of Data Collection, and Changes Due to Network Redesign, 1957–2015—cont'd

Network and Number From Table 2.3	Number of Sites Sampled Over Time; Frequency and Type of Data Collection and Changes Due to Network Redesign
5b Hydrologic Benchmark Network	*No. Sites*: 55 *Data and Collection Frequency*: Semiannual, Q and DO in the field with pH, and SC measured in the field and laboratory. Water samples partitioned into filtered water and solid fractions. Filtered water was evaporated and dissolved solids analyzed for uranium, radium, equivalent natural uranium for gross alpha and as strontium-90/yttrium-90 or cesium-137 equivalent for gross beta. Values are also given for suspended gross alpha radioactivity in terms of natural uranium.
5c Hydrologic Benchmark Network	*No. of Sites*: 15–22 *Data and Collection Frequency*: Sites were equipped in 2003 with refrigerated automatic samplers to collect about 24 samples per year to cover all flow conditions. Samples collected for pH, Ca. Mg, Na, K, Cl, SO_4, DOC, DON, ANC, Si, Al. In 2012 added: TP, PO_4, continuous water temperature, fish, aquatic invertebrates, and aquatic habitat.
6a National Stream Quality Accounting Network	*Range in No. Sites Over Time*: 50–518 *Data Collection Frequency*: Continuous or daily: Q, °C, SC, pH, DO; *Monthly*: FCB, SB, HCO_3, CO_3, total hardness, Mg, F, Na, K, TDS, Si, turbidity, Cl, SO_4, TP, total $NO_3 + NO_2$—N, TKN, SSC, total phytoplankton, periphyton; *quarterly*: Unfiltered and filtered water samples for As, Cd, Cr, Co, Cu, Fe, Pb, Mn, Hg, Se, Zn, periphyton biomass and ash-free dry weigh, chlorophyll a and b, TOC. Periphyton discontinued in 1980. *Biannually*: Radiochemicals and pesticides
6b National Stream Quality Accounting Network	*No. Sites*: 434 *Data Collection Frequency*: Changed from monthly and bimonthly to bimonthly to quarterly in 1982. *Discontinued*: Continuous measurements of Q (unless collected by others), °C, SC, pH, and DO. Discontinued samples of unfiltered $NO_3 + NO_2$—N, NH_4–N, filtered samples of TKN, TOC, DOC (1981), and phytoplankton. Pesticides dropped in 1983 and unfiltered samples for trace elements dropped in 1982. *Added*: unfiltered samples of TKN, filtered samples for PO_4, semiannual analysis of total trace inorganic constituents in fine fraction of bottom materials.

6c National Stream Quality Accounting Network	*Range in No. Sites over time*: 434–140. A total of 91 sites were discontinued, 26 sites became inactive, and 33 sites were added to replace discontinued sites or better represent land use categories. Additional sites dropped after 1990. Shift toward smaller sites that remained. *Data Collection Frequency*: Bimonthly to quarterly. Daily streamflow available for most sites but not supported by the network. Ca, Mg, Na, K, Cl, SO_4, TKN, NH_4–N, NO_3 + NO_2–N, TP, PO_4, SSC, FCB SB; *Biannually*: Radiochemicals. *Dropped*: Filtered samples for trace elements in 1991.
6d National Stream Quality Accounting Network	*Range in No. Sites over time*: 39–66 A total of 66 sites were retained from the original network and prioritized; 33 highest priority sites were continued in the network; 18 sites in the Mississippi River Basin; 9 in the Rio Grande River, 8 in the Colorado River, and 6 in the Columbia River. The Yukon River was sampled from 2001 to 2005. The lower priority sites were sampled quarterly in 1996 and then discontinued in 1997. *Data Collection Frequency*: Daily: Q; Bimonthly plus 1–3 high-flow samples per year: °C, SC, pH, and DO, TOC, DOC, DIC; Ca, Cl, SO_4, NH_4–N, NO_3 + NO_2–N, TP, PO_4, 47 water-soluble commonly used pesticides; SSC, coarse and fine particles, suspended and filtered fractions for Pb, Cd, Cu, Zn.
6e National Stream Quality Accounting Network	*No. Sites*: 32 sites; 13 sites on the largest rivers discharging to coastal areas, St. Lawrence River, and 19 sites in the Mississippi River Basin. *Data Collection and Frequency*: Seasonally weighted sampling with a base frequency of bimonthly with 6 additional fixed interval samples during periods of high variability in instantaneous loads. °C, SC, pH, and DO, alkalinity, As, B, Ca, Cl, F, Fe, Li, Mg, K, TDS, Se, Si, Na, SC, Sr, SO_4, turbidity, V, DIC, DOC, TOC, TPC, NH_4–N, NH_4–N + organic ammonia, NO_3 + NO_2–N, TP, PO_4, UV absorbing substances, SSC, 85 current use pesticides and pesticide degradates.
7 National Monitoring Network	*Range in No. Sites over time*: 4–5 *Data Collection and Frequency*: Same as NASQAN 2008–12

(*Continued*)

Table 2.4 Network Name and Number, Number of Sites Sampled Over Time, Frequency and Type of Data Collection, and Changes Due to Network Redesign, 1957–2015—cont'd

Network and Number From Table 2.3	Number of Sites Sampled Over Time; Frequency and Type of Data Collection and Changes Due to Network Redesign
8a National Water Quality Assessment Program	*No. Sites*: 496 Within each Study Unit; 6–7 sites on small streams in agricultural, urban, and reference watersheds upstream of 2–3 sites on larger rivers in mixed land use settings. *Data Collection Frequency*: Small streams in predominately urban and agricultural areas were sampled up to 24 times per year targeting periods of runoff after land applications of fertilizers and pesticides. Larger streams in mixed land use settings were sampled 15–18 times per year.
8b 2002–06 National Water Quality Assessment Program	*No. Sites*: 84–145 *Data Collection and Frequency*: Seasonally weighted sampling with a base frequency of bimonthly with 6 additional fixed interval samples during periods of high variability in instantaneous loads; 12–26 samples per year.
2007–12 National Water Quality Network	*Range in No. Sites over time*: Total of 142; 113 (NAWQA), 4 (NMN), 25 (NASQAN) *Data Collection and Frequency*: Seasonally weighted sampling with a base frequency of bimonthly with 6 additional fixed interval samples during periods of high variability in instantaneous loads; 83 sites were sampled one year out of every four years; 15 sites were sampled every other year, and 44 sites were sampled each year. Ecological sampling at 58 sites and 19 reference sites. Seasonally weighted sampling with a base frequency of bimonthly with 6 additional fixed interval samples during periods of high variability in instantaneous loads; 12–26 samples per year.
8c 2015 National Water Quality Network	*Data Collection and Frequency*: Seasonally weighted sampling with a base frequency of bimonthly with 6 additional fixed interval samples during periods of high variability in instantaneous loads; 12–26 samples per year.

Single objective networks for monitoring the status and trends of persistent contaminants or radionuclides, 1956–2005

9a	*No. Sites*: 156
National Pesticide Monitoring Network	*Data and Collection Frequency*: Quarterly to semi-annually: Water (quarterly) and sediment (semi-annually) analyzed for DDT, DDD, DDE, aldrin, dieldrin, endrin, heptachlor and its epoxide, lindane, chlordane, toxaphene, endosulfan, phosphorothioates, polychlorinated biphenyls, 2, 4-dichlorophenoxy) acetic acid, 2, 4, 5-trichlorophenoxy acetic acid, silvex
9b	*No. Sites*: 156
National Pesticide Monitoring Network	*Data and Collection Frequency*: Quarterly to semi-annually DDT, DDD, DDE, aldrin, dieldrin, endrin, heptachlor and its epoxide, lindane, chlordane, toxaphene, endosulfan, phosphorothioates, polychlorinated biphenyls, 2, 4-dichlorophenoxy) acetic acid, 2, 4, 5-trichlorophenoxy acetic acid, silvex
9c	*No. Sites*: 50
National Pesticide Monitoring Network	*Data and Collection Frequency*: Three composites, preferably the same three species, were to be collected each spring (April, May) and fall (October, November) at each of the sites for DDE, DDE, DDT, dieldrin, aldrin, endrin, lindane, heptachlor, heptachlor epoxide, chlordane, and toxaphene
9d	*No. Sites*: 112
National Pesticide Monitoring Network	*Data and Collection Frequency*: Biennially for bottom feeding and sportfish for DDE, DDE, DDT, dieldrin, aldrin, α-BHC, chlordane, dacthal, endrin, lindane, heptachlor, heptachlor epoxide, hexachlorobenzene, nonachlor, oxychlordane, and toxaphene. Trace elements include As, Cd, Cu, Pb, Hg, Se, and Zn.
9e	*No. sites*: 100: Colorado River basin (14), Columbia River basin (15), Mississippi River Basin (48); Rio Grande River Basin (10), Southeastern River Basins (13), Yukon River basin (10).
National Pesticide Monitoring Network	*Data and Collection Frequency*: Contaminant concentrations: elemental and organochlorine pesticides, polychlorinated biphenyls and fish health indicators—condition factor, somatic indices, health assessment index, macrophage aggregate parameters; and reproductive biomarkers such as steroid hormones, vitellogenin, and gonadal histology.

(Continued)

Table 2.4 Network Name and Number, Number of Sites Sampled Over Time, Frequency and Type of Data Collection, and Changes Due to Network Redesign, 1957–2015—cont'd

Network and Number From Table 2.3	Number of Sites Sampled Over Time; Frequency and Type of Data Collection and Changes Due to Network Redesign
10a Radiation Alert Network and Tritium Surveillance System	*No. Sites*: 39 *Data and Collection Frequency*: Unknown
10b Environmental Radiation Ambient Monitoring Systems and RadNet	*No. Sites*: 55 downstream from nuclear facilities or at background sites and 76 sites for monitoring drinking water *Data and Collection Frequency*: Quarterly: Tritium; Annually: Gamma emitting radionuclides

ANC, acid neutralizing capacity; *As*, arsenic; *B*, boron; *BOD*, biochemical oxygen demand; °C, water temperature; *CB*, coliform bacteria; *Cd*, cadmium; *Cl*, chloride; *Co*, cobalt; CO_3, carbonate; *COD*, chemical oxygen demand; *Cr*, chromium; *Cu*, copper; *DIC*, dissolved inorganic carbon; *DO*, dissolved oxygen; *DON*, dissolved organic nitrogen; *F*, fluoride; *FCB*, fecal coliform bacteria; *Fe*, iron; HCO_3, bicarbonate; *Hg*, mercury; *K*, potassium; *Mn*, manganese; *Na*, sodium; NH_4-N, ammonia nitrogen; *No.*, number; $NO_3 + NO_2-N$, nitrate plus nitrite as nitrogen; *Pb*, lead; *pH*, hydrogen ion concentration; PO_4, phosphate; *Q*, daily streamflow; *SB*, streptococci bacteria; *SC*, specific conductance; *Se*, selenium; *Si*, silica; SO_4, sulfate; *Sr*–90, strontium–90; *SSC*, suspended sediment concentration; *TDS*, total dissolved solids; *TKN*, total Kjeldahl nitrogen; *TOC*, total organic carbon; *TP*, total phosphorus; *TSS*, total suspended solids; *V*, vanadium; *Zn*, zinc.

HBN, "A recurring question of our times, and one that we anticipate will be increasingly vexing to posterity, is to know how much of the change in our environment is caused by man and how much is natural." Routine monthly monitoring at HBN sites originally included field measurements, nutrients, major and minor elements, and suspended sediment at 56 sites. The HBN and NWPSN also monitored for other constituents of national interest such as pesticides and radioactive substances during short-term studies (Tables 2.3 and 2.4).

The growth of water monitoring prompted several advances at the national level. In 1964, the Bureau of the Budget, now the Office of Management and Budget (OMB), issued Circular A-67, calling on federal agencies to establish common guidelines for collecting and analyzing water and water-quality data [21]. The OMB Circular A-67 gave the responsibility to the US Department of Interior, delegated to the USGS, "for exercising leadership in achieving coordination of national networks and standardized water-data acquisition activities" [21].

Another advance in monitoring was the introduction of quality assurance and quality-control activities for sample collection, processing, and laboratory analysis to improve data reliability. Quality assurance is defined as a system of activities whose purpose is to provide to the user of a service with assurance that it meets defined standards of quality with a stated level of confidence [22]. In contrast, quality control is defined as the overall system of activities whose purpose is to control the quality of a service so that it meets the needs of users. The aim is to provide quality that is satisfactory, adequate, dependable, and economic [22].

Activities of the USGS Branch of Quality Systems

The Branch of Quality Systems (BQS) and its predecessor organizations were established and sustained within the USGS Office of Water Quality to quality assure laboratory and field measurements and supply reference materials to USGS water-quality programs and projects. The consistent application of quality-assurance assessment to USGS water-quality monitoring activities provides a measure of comparability that is essential for long-term trend analysis. The consistent quality assurance of laboratories and field activities contributing water-quality data to the USGS minimizes, to the degree possible, variation in results due solely to changes over time in methods of sample collection and laboratory analysis. The independent laboratory and field proficiency testing activities of the BQS are not a substitute for quality-control activities in the field offices and laboratories contributing water-quality results to the NWIS.

(Continued)

Activities of the USGS Branch of Quality Systems (*cont'd*)

To independently quality assure laboratory results for the USGS, the Standard Reference Samples Project (SRS) was started in 1962. Participation is required for all laboratories contributing data to USGS databases. The SRS formulates reference samples using both natural-matrix and deionized water and determines known concentrations from multiple laboratories' results [23]. For example, SRS results from 1965 to 1982 described what laboratory methods were used and the performance of those laboratories over time [23]. The importance of the SRS is that it serves as a consistent external measure of performance of laboratories contributing results to the NWIS.

The majority of inorganic and organic analyses at the USGS principal water-quality laboratory—the National Water Quality Laboratory (NWQL)—are quality assured through three activities of the BQS: the Inorganic Blind Sample Project, which began in 1973, the Organic Blind Sample Project, which began in 1994, and the Blind Blank Project for which records are maintained back to 1998. These projects frequently and regularly submit proficiency testing samples of known concentration to the analytical systems of the NWQL for purposes of performance testing [24]. Quality-assurance results from the blind samples are used to quantify and qualify laboratory results on a continuing basis but are not substituted for quality-assurance and quality-control activities at the NWQL.

The National Field Quality Assurance (NFQA) project was created in 1979 to provide quality-assurance reference samples to field personnel who make field measurements of alkalinity, pH, and specific conductance measurements [25]. Sample-collection methods are another source of variability. Water-quality samples from rivers and streams have been collected by the USGS using relatively similar methods starting in the 1970s. Publications on laboratory and field methods [26,27] and technical memoranda required the collection of samples using a depth and width integrated collection method for water-quality analyses. Near-surface grab samples are known to provide a poor estimate of the amount of suspended constituents, such as total phosphorus and suspended sediment, passing through a river cross-section under most conditions. Since the 1970s, this policy has been continuously in place and documented in the USGS National Field Manual for the Collection of Water Quality Data [28].

Over the entire period of record one analytical method has been used by all USGS sediment laboratories for the determination of suspended sediment concentration [29]. The Sediment Laboratory Quality Assurance Project (SLQA) began in 1996 to provide proficiency samples to all laboratories providing suspended sediment data to the USGS [30]. The laboratory evaluation project began in 1987 and provides technical assistance to the USGS projects by supplying BQS reference materials, and where applicable, laboratory performance data [31].

Several early successes were achieved with data from the RAN and TSS as well as data from the HBN and NWPSN. These networks were able to document concentrations of strontium-90 and other radioactive isotopes that were deposited from the atmosphere to watersheds and rivers as a result of above-ground nuclear weapons testing in the mid-20th century [18,19,32]. In 1963, mounting evidence from these networks and from studies on the dangers of radioactive substances to human health and the environment helped to pass the first Limited Test Ban Treaty. The Treaty prohibited nuclear weapon testing underwater, above ground, and in outer space by the United States, Great Britain, and the Soviet Union.

4.2 Expansion of Networks, Monitoring Challenges, and Pesticide Issues, 1965–74

In 1970, the population of the United States was just over 205 million. About 44 million of those people were underserved by municipal wastewater treatment facilities, which provided only primary levels of treatment prior to discharge to surface waters [33]. The wastes of an additional 10 million people, 5% of the total population, received no treatment at all [33].

The Water Quality Act of 1965 (P.L. 89–234) expanded the Construction Grants Program from $50 million to $100 million per year [14]. Untreated municipal sewage was not the only concern; industries discharged 80% of the pollutant loading to surface water of which 29% was untreated [13]. The 1965 Water Quality Act required the states to establish water-quality standards for surface waters by 1968 [34]. Water-quality monitoring was expanded in the amendments to the FWPCA of 1966 known as the Clean Water Restoration Act (P.L. 89–753).

Fixed-site water-quality monitoring became the most common approach to determining compliance with water-quality standards [35]. For example, the Georgia Water Quality Control Act established the Water Quality Control Board in 1964. In 1968, the Georgia Water Quality Control Board began a monthly water-quality monitoring program at a fixed network across the state in cooperation with the USGS. The USGS began to cooperate with the state of New Jersey in 1964 to monitor rivers and streams in a fixed-site network [17,36,37].

In 1966, the responsibility for the NWPSN was transferred from the PHS in the Department of Health Education and Welfare to the Department of Interior under the newly formed Federal Water Pollution Control Administration (FWPCA). The FWPCA was renamed the Federal Water Quality Administration in 1970. Under a reorganization plan of federal agencies, water-quality monitoring activities including the NWPSN were transferred in December 1970 to the newly created US Environmental Protection Agency (EPA).

Between 1964 and 1968, two national computerized databases, the EPA's STOrage and RETrieval system (STORET) and the USGS WATer data STOrage and REtrieval (WATSTORE) system, were developed [38,39]. Computerized databases were a major technological advance that made water-quality data accessible within and outside the government.

The need to demonstrate the effectiveness of water pollution-control investments and the associated improvements in water quality became increasingly important [40]. Short-term water-quality surveys in rivers and streams across the nation had long been the approach to water-quality monitoring and assessment before 1950 [1]. As a result, few sites with long-term data were available for trend analysis. In an analysis of available data researchers [40] found it difficult to demonstrate conclusively that water quality had improved after water-pollution-control practices had been put in place. Wolman questioned the ability of water-quality monitoring to answer trend questions of national interest such as, "Is water quality getting better or worse and are there measurable trends in the observed data?"

The first national study to examine the utility of available water-quality data for trend analysis was undertaken at the request of the Council on Environmental Quality (CEQ), a federal agency established in 1969 to advise the President on environmental issues [41]. Statistical trends were evaluated for concentrations of dissolved oxygen, biochemical and chemical oxygen demand, and total organic carbon; total, organic, and inorganic forms of nitrogen and phosphorus; total dissolved and suspended solids; and levels of specific conductivity and turbidity. Results were reported for which there was a minimum of 8 years of data and concurrent measurements of streamflow. A criterion that limited the utility of many sampling sites was the lack of corresponding streamflow measurements to provide a hydrologic context for changes in concentrations and mass transport (loads) of water pollutants [41]. The study reported that of the approximately 70,000 water-quality sites in STORET in 1970, only 142 met a set of criteria deemed necessary to calculate trends and progress in water-pollution control [41]. In another report, the USGS found that only 88 sites in WATSTORE had concurrent data of sufficient length of record for analysis of trends in streamflow, water temperature, and specific conductance [42].

Data gaps for national-scale trend analysis necessitated aggregating data from multiple agencies to obtain sites and sampling periods over time for the analysis of trends. One method was used to aggregate results from many agencies to increase the numbers of sites and improve geographic coverage [41]. Where temporal gaps were found, results for the same constituents collected at the same site, at different times, and for different purposes were aggregated to obtain a longer period of record [42]. Data gaps could be filled when one project stopped and another project started collecting data at the same site. This scenario is not uncommon because most stream-quality samples are collected at streamflow gauging stations, if possible. Answers to key questions about whether the nation's water quality was getting better or worse remained anecdotal and unsatisfying to policy-makers despite the growing amount of data and number of monitoring programs [11].

Monitoring information from the PHS NWQN, HBN, and other fixed-site networks were successfully applied to the problem of pesticide contamination beginning in the 1960s. Annual usage of pesticides in the United States tripled from the 1930s to the 1970s and reached nearly 1.5 billion pounds in the late 1970s [43]. In response to concerns over pesticide use, pesticide monitoring began in 1957 at 100 surface-water sites in the PHS NWQN. The widescale use of organochlorine pesticides and their observed effects on fish and wildlife led to the publication in 1962 of the book *Silent Spring* by Rachel Carson [44]. Using new methods and instruments such as the gas chromatograph, the PHS NWQN documented the contamination of surface water by DDT, DDD, DDE, dieldrin, and other organochlorine pesticides [45]. Additional pesticide monitoring began in 1965 in western rivers [46]. Despite challenges, identification of the adverse effects of DDT and other pesticides on the environment was achieved, leading to regulations to eliminate the use of these pesticides.

Silent Spring and results from the NWQSN prompted the creation of the National Pesticide Monitoring Program (NPMP) in 1964 to determine the extent of contamination of surface waters, fish, wildlife, drinking water, and foods with organochlorine pesticide residues [47]. Creation of the NPMP meant that pesticide assessment became an important and perennial monitoring objective in several single and multiple objective networks (Tables 2.3 and 2.4). Initially, water and sediment samples were monitored by the FWPCA and USGS [48–50]. After its creation, pesticide monitoring began at the EPA in cooperation with the USGS [51].

A survey of organochlorine insecticides and chlorophenoxy acid herbicides in water and sediments at HBN sites in the early 1970s demonstrated that these compounds were detected in relatively undisturbed watersheds, indicating the breadth and scope of the contamination of surface waters [52]. Mounting evidence from the NPMP led to the elimination by EPA of all uses of DDT in the United States in 1972 followed by the phase-out of most other organochlorine pesticides for farm and home use by the late 1980s [43,53]. The Federal Insecticide, Fungicide and Rodenticide Act (FIFRA, Public Law 92−516), originally enacted in 1947 was amended in 1972 to mandate the NPMP and to create a registration program for pesticides [51] (Table 2.3).

Evaluation of pesticide data from the NPMP also informed future monitoring efforts [54]. Because organochlorine pesticides are hydrophobic compounds, they tend to accumulate in sediment and bioaccumulate in fish tissue. As a result, monitoring networks preferentially shifted from sampling water to sampling sediment and tissues in order to improve the ability to detect organochlorine compounds in the environment. Over time, pesticide residues in fish and wildlife were monitored by the US Fish and Wildlife Service (FWS) [55−57]. These responsibilities were later carried out by the National Biological Service [58] and the USGS [59].

The USGS began sampling water soluble pesticides in the 1960s [54]. Organophosphate insecticides and chlorophenoxy herbicides are relatively water-soluble, making water the preferred sampling medium. Targeting the location of monitoring sites to areas of greatest pesticide use was driven by observations of the positive correlation between pesticide use, detection, and concentrations. Other factors affecting detection are pesticide persistence in the environment and the ability of analytical methods to detect compounds at low levels that are environmentally relevant [54]. Pesticide monitoring for change and trend detection plays an important role in human and ecosystem exposure assessments and in decision making on the registration, reregistration, and restrictions of pesticide use [53].

4.3 Monitoring and Assessment for the Clean Water Act, 1975−84

Successive laws such as the Federal Water Pollution Control Act of 1972 (Clean Water Act), the Clean Water Act of 1977 (P.L. 95−217), the Municipal Wastewater Treatment Construction Grants Amendments of 1981 (P.L. 97−117), and the Water Quality Act of 1987 (P.L. 100−4) provided assistance to cities and towns in the form of grants and loans to improve municipal wastewater-treatment infrastructure. The goals of the Clean Water Act are to make the waters of the United States fishable and swimmable. To do so, federal and state investments were greatly increased for construction of municipal wastewater-treatment systems to reduce oxygen-demanding wastes, eliminate pathogens, and reduce and control excessive levels of ammonia nitrogen [60].

Expectations of water-quality monitoring networks after passage of the Clean Water Act in 1972 went well beyond the previous monitoring objectives of the 1965 Water Quality Act (P.L. 89–234). The 1965 Act required assessment of surface waters for the purposes of evaluating attainment of water-quality standards. The new expectation included effluent monitoring, intensive surveys for measurement of the effectiveness of wastewater treatment, and trend detection to determine if water quality was getting better or worse. States continued to rely on fixed-site networks, in part, to address these new monitoring requirements.

Water-quality monitoring and assessments were expanded across the nation to address the issue of point-source discharges with a focus on the measurement of concentrations of dissolved oxygen, fecal bacteria, suspended solids, and nutrients for purposes of evaluating the effectiveness of wastewater treatment on the quality of receiving waters. Oxygen-demanding wastes containing carbon, nitrogen, and phosphorus were targeted for reduction.

After 10 years of operation, there were over 102,000 monitoring sites in STORET in 1973, although most sites were monitored over the short-term and had become inactive [61]. The EPA estimated that in 1973 it operated over 800 fixed-location monitoring sites and the states operated about 6000 sites of a total of 19,000 active sites in STORET [61]. Of the 800 EPA sites, 400 were operated in cooperation with the USGS, 280 were monitored by the EPA in cooperation with states, and 120 were operated solely by the EPA. For example, in 1972, the Environmental Protection Division (EPD) of the Georgia Department of Natural Resources was established and a cooperative water-quality program between the USGS and the EPD was initiated at a statewide fixed-site network. In New Jersey, the Ambient Surface Water Quality Monitoring Network began as a cooperative effort with the USGS at 112 fixed sites in 1976 [36,37]. In Connecticut, a cooperative statewide water-quality monitoring network was initiated in 1974 with the USGS at 33 sites on 15 rivers [62].

The EPA developed the National Water Quality Surveillance System (NWQSS) in 1974, discontinuing the previous network of more than 800 fixed sites [51] (Table 2.3; Fig. 2.2). The NWQSS objectives were to determine trends, establish relationships between land use and water quality, and

Figure 2.4 Map of the conterminous United States showing the US Environmental Protection Agency's National Water Quality Surveillance System (NWQSS). The NWQSS was operated at 96 paired sites in the conterminous United States from 1974 to 1981 [51,63].

evaluate the effectiveness of pollution-control efforts [51]. The NWQSS used a paired design at 98 fixed-site locations with one site located upstream and the other downstream from a pollutant source or land-use area (Fig. 2.4;[35]).

The USGS established the National Stream Quality Accounting Network (NASQAN) in 1973 as a fixed-site, fixed-frequency network to address the deficiencies in previous monitoring networks identified by Wolman [40] and by the CEQ [41,64] (Table 2.3; Fig. 2.2). The NASQAN was designed for long-term monitoring of water-quality trends, to describe temporal and geographic changes over multidecadal periods of time, and to lay the groundwork for future assessments of changes in stream quality [64]. An important characteristic of the NASQAN network was the consistency with which it was operated. The same constituents were monitored at the same sampling frequency and location, using the same sample-collection and analysis methods and laboratory. The NASQAN sites were located at the terminus of the nation's major inland and coastal rivers [64]. The design of NASQAN accounted for the concentrations and discharge of about 90% of the streamflow and inorganic compounds, nutrients, trace elements, and suspended sediment from the major rivers of the United States and their tributaries to coastal waters and the Great Lakes (Tables 2.3 and 2.4). At its peak in 1978, the NASQAN operated at 518 sites in all 50 states and Puerto Rico.

Table 2.2 summarizes information on the USGS water-quality monitoring sites for six constituents that are representative of long-term monitoring networks such as the NASQAN and HBN in each of four decades from 1975 to 2014. In Table 2.2, the total number of water-quality monitoring sites in the NWIS are further differentiated by sampling frequency. Long-term sites, those that were sampled for a minimum of 6 of the 10 years of each decade, are shown in parentheses along with the total number of sites. The remaining sites with periods of record lasting less than 6 years per decade were likely sampled for short-term surveys and studies.

The USGS long-term monitoring sites for the decade 1975 to 1984 are shown in Fig. 2.5. Less than half of the long-term sites shown in Fig. 2.5 were part of the NASQAN and HBN networks. The remaining long-

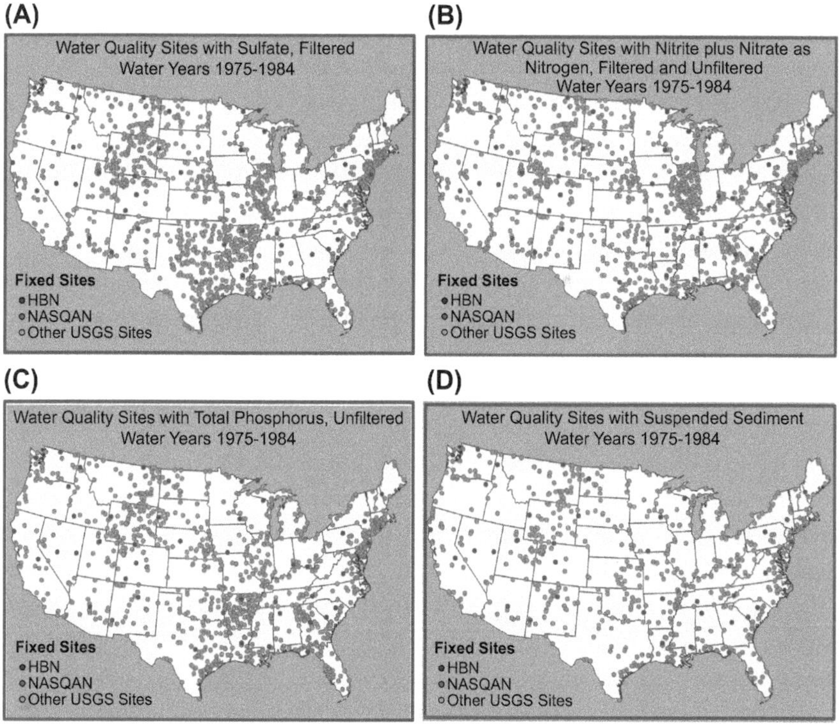

Figure 2.5 US Geological Survey water-quality sites in the conterminous United States that were sampled at weekly, monthly, quarterly, or seasonal frequencies for water years 1975–84 for concentrations of (A) sulfate, (B) nitrite plus nitrate as nitrogen, (C) total phosphorus, and (D) suspended sediment. Sites in Alaska, Hawaii, and Puerto Rico are not shown. *HBN*, Hydrologic Benchmark Network; *NASQAN*, National Stream Quality Accounting Network.

term monitoring sites were part of other federal agency, state, and local monitoring projects with the USGS. Fig. 2.5A–D and Table 2.2 show that from 1975 to 1984 (A) 1297 long-term sites were monitored for concentrations of sulfate; (B) 1230 long-term sites were monitored for concentrations of nitrite plus nitrate as nitrogen; (C); 1107 long-term sites were monitored for concentrations of total phosphorus; and (D) 710 long-term sites were monitored for concentrations of suspended sediment. Sites were well distributed across the conterminous United States with relatively higher densities of sites in a few states (Fig. 2.5A–D). No long-term sites were monitored for concentrations of atrazine from 1975 to 1984.

The EPA–State Basic Water Monitoring Program, initiated in 1980, included the operation of about 1000 fixed water-quality sampling sites nationwide and also included 53 NWQSS and 110 NASQAN sites [65]. Data from these networks formed the basis for biennial reporting in the *National Water Quality Inventory*, which is required by section 305b of the Clean Water Act to report on the status of the nation's water quality and progress toward attaining the fishable and swimmable goals of the Clean Water Act [35,65].

To meet expected monitoring objectives of the Clean Water Act, a greater variety of monitoring approaches were needed in addition to fixed-site networks. The Clean Water Act created requirements for effluent monitoring of discharges from industrial and municipal wastewater treatment plants. The Clean Water Act amendments of 1977 placed further emphasis on monitoring and control of industrial discharges [60]. In a 1986 study of the characteristics of water-quality monitoring programs in Colorado and Ohio, it was estimated that the most monitoring resources were expended on regulatory monitoring compared to monitoring rivers, streams, and groundwater. About 46% of total monitoring costs in Colorado and 84% in Ohio were spent on regulatory monitoring [66].

Intensive surveys were developed to replace fixed-site monitoring networks, that were deemed too expensive due to the necessity for sampling at high frequencies for several years [67]. Intensive stream surveys of short duration became a commonly used approach to assess the impacts to surface waters from discharges of oxygen-demanding wastes, pathogens, and nutrients such as ammonia nitrogen. Intensive surveys are designed to diagnose water-quality problems, to inform river basin development plans, to determine levels of wastewater treatment required to improve water quality, and to provide confidence in decisions on investments in wastewater treatment [35,65,68]. Trace elements and trace organic substances such as polychlorinated biphenyls were added to intensive surveys and to fixed-site networks in the 1970s and 1980s as a way to improve knowledge about the occurrence of toxic industrial compounds in the environment [35]. Intensive surveys eventually were found to have higher than anticipated costs compared to other types of monitoring and were not easily aggregated for regional and national assessments [65].

The US General Accounting Office (GAO) issued a series of reports critical of the effectiveness of water-quality monitoring networks in measuring progress under the Clean Water Act. A thorough discussion of the criticisms is documented in Ref. [69]. In 1981, a two-part GAO report titled *Better Monitoring Techniques Are Needed to Assess the Quality of Rivers and Streams* evaluated the NASQAN, the EPA-State Basic Water Monitoring Program, and the NWQSS for their ability to detect status, change, and trends in water quality [70]. The report recommended that the USGS and EPA discontinue their fixed-site networks in favor of using only intensive surveys for monitoring to evaluate progress under the Clean Water Act. This conclusion was disputed by the EPA, the USGS, and others [65,71,72]. In particular, the USGS and the Department of Interior disagreed with a major premise and a conclusion of the report pertaining to the NASQAN. The GAO report stated that the NASQAN was designed to evaluate the effectiveness of pollution-control measures and then concluded that the NASQAN was not able to achieve this objective [70]. The USGS response pointed out that the NASQAN had not been designed to evaluate pollution-control measures [70]. The GAO findings and the USGS response point out the importance of monitoring objectives and how they relate to the overall success of water-quality monitoring activities. This was perhaps the beginning of a realization that one monitoring approach cannot answer all questions.

A second GAO report publication critical of monitoring was titled *Key Unanswered Questions about the Quality of Rivers and Streams* [73]. The GAO evaluated five major agency reports published from 1982 to 1984 that examined the status of the nation's water quality. The GAO examined the reports to determine if they answered questions about the status and trends of the nation's water quality, sources of pollution affecting water quality, and effectiveness of government investments in the Construction Grants Program. The reports examined by the GAO included the 1982 *National Water Quality Inventory*, which relied in part on monitoring results from the NWQSS and the EPA–State Basic Water Monitoring Program [33]; the 1982 *National Fisheries Survey*, which relied on results of a field campaign by FWS to assess the condition of recreational fisheries [74]; and *The National Water Summary*, which relied on results from the NASQAN [75]. Two other reports were also evaluated by the GAO but these reports did not rely on data from national networks or other monitoring activities [76].

The GAO findings noted that the three assessment reports that relied on water-quality monitoring data could be used to answer questions about status, trends, and changes in water quality at the national scale [73]. However, the GAO also found that there was insufficient information in those reports to evaluate the specific benefits of the Construction Grants Program [73].

After release of the 1981 and 1986 GAO reports, the NWQSS and the fixed-site monitoring component of the EPA–State Basic Water Monitoring Program were reduced or in some cases eliminated and replaced by intensive surveys. By 1980, flat funding over time resulted in reductions to the HBN from 54 to 15 long-term sites followed by discontinuation of all monitoring in 1997 for a period of 6 years. The growing costs of network operation throughout the 1980s led to a reduction in the NASQAN from 518 to 400 long-term sites and a reduction in sampling frequency from monthly to bimonthly or quarterly (Table 2.4). Downsizing in the NASQAN occurred gradually, and a complete redesign was undertaken in the following decade [77]. At the national scale, only the USGS maintained its national-scale multipurpose, long-term, fixed-site networks such as the NASQAN after 1986. The single-purpose EPA RadNet (formerly ERAMS and RAN) surface-water network for monitoring radionuclides was discontinued in 1999 because monitoring of this type was being done by the states [19] (Fig. 2.2).

Water-quality trends in the nation's rivers were described for a 10-year period following the 1972 enactment of the Clean Water Act using streamflow and 24 water-quality constituents collected in the NASQAN and NWQSS networks [63]. Widespread declines in fecal-indicator bacteria and lead and widespread increases in nitrate and chloride were reported [63]. Decreases in fecal-indicator bacteria and lead were attributed to improvements in wastewater treatment and to declines in the use of leaded gasoline, respectively. Nonpoint sources of fertilizers and road salt were inferred as the cause of uptrends in nitrate and chloride levels in rivers, respectively [63]. Although improvements in water quality were detected for selected sites, a demonstration of the effectiveness of specific polluation-control measures was not [11]. Monthly to semimonthly sampling of relatively large rivers was deemed insufficient by itself to identify the causative factors of water-quality trends [11,65].

A conclusion of all these evaluations of monitoring was that many networks suffered from a syndrome of being "data rich and information poor" [78]. More specifically, the problem was defined by the lack of statistical analysis and interpretation of water-quality monitoring data followed by regular publication of assessments to address the needs of policymakers and water-quality managers [78]. Researchers concluded that purposeful dedication of monitoring resources toward statistical data analysis, interpretation of results in a policy context, and publication of water-quality assessments were needed to improve routine, fixed-site monitoring [78]. A major focus was placed on improving the design of monitoring networks and their objectives so that data interpretation and reporting could be carried out efficiently and effectively to meet the information needs of policymakers and water-quality managers [78]. The solution to the syndrome of "data rich and information poor" resulted in improvements to designs of routine monitoring systems to include "not only the where, what, and when of sampling, but also why" [78]. Indeed, this was a sufficiently important topic at the time to generate the publication of several textbooks on and reviews of water-quality monitoring designs. A few examples are provided in Refs. [3,79–81].

Given the need for definitive answers to water-quality status and trend questions, it became a goal of fixed-site monitoring networks to explain the causative and associative factors and governing processes related to conditions and trends. Many new and revised monitoring program designs were aided by new technologies such as geographic information systems (GIS) and advances in statistical approaches [67,82,83].

Twenty years after the initial passage of the Clean Water Act of 1972, policymakers remained dissatisfied with the state of knowledge about the effectiveness of $61 billion in expenditures by the Construction Grants Program from 1970 to 1995 and an additional $16.1 billion from 1988 to 1999 by the State Revolving Loan Fund [11,84]. The EPA performed a comprehensive analysis of the benefits of water-quality investments was made for major rivers where there had been severe water-quality impairments due to point-source loadings of oxygen-demanding wastes prior to the passage of the Clean Water Act in 1972. The report documented improvements of at least 2 milligrams per liter (mg/L) of dissolved oxygen in more than 36% of 311 river reaches investigated nationwide. Sources of dissolved-oxygen monitoring data were fixed sites on streams near pollutant sources sampled before versus after improvements in wastewater treatment. Evidence was also gathered from nine case studies where intensive surveys had been completed. This report demonstrated the benefits of water-pollution control investments under the Construction Grants Program and State Revolving Loan Program of the Clean Water Act [84,85].

4.4 Improved Monitoring Designs Through a Hydrologic Systems Approach, 1985–94

After passage of the Water Quality Act of 1987 (P.L. 100–4), policymakers in Congress began to address nonpoint sources of water pollution with voluntary measures under Section 319 [11,60]. Nonpoint sources are defined as those that contribute pollutants from numerous and widespread locations and are not released from a point source such as a discharge pipe [86]. Many nonpoint sources are related to land-use activities such as applications of fertilizers and pesticides on agricultural and urban lands. Section 319 made grants available to states for control of nonpoint sources from agricultural lands, feed lots, urban areas, and suburban developments. Congress relied on a voluntary approach to control nonpoint-source pollution because at first the magnitude of the nonpoint-source pollution problem was not known [11].

In a study of 107 watersheds nationwide, nonpoint sources of nitrogen such as from fertilizer, manure, atmospheric deposition, sewage, and industry were found to vary geographically and with climate, land use, hydrology, population, and natural features of the landscape [86]. Point sources were commonly a major source of nitrogen to streams near large cities, whereas nonpoint sources were commonly a major source of nitrogen in agricultural areas because of the application of nitrogen fertilizers to crops [86]. The atmosphere was a major source of nitrogen from air pollution in areas where urban and agricultural sources were not as important [86].

As a result of continuing questions concerning the nation's water quality and if it was getting better or worse, particularly with regard to nonpoint sources of water-quality impairment, the USGS initiated the National Water Quality Assessment (NAWQA) program with the intent of making improvements over previous monitoring and assessment approaches [87,88] (Fig. 2.2). The goals of the program were to:

- Provide a nationally consistent description of current water-quality conditions for a large part of the nation's water resources;
- Define long-term trends (or lack of trends) in water quality; and
- Identify, describe, and explain, as possible, the major factors that affect observed water-quality conditions and trends.

The NAWQA program began in 1986 with a pilot program in seven hydrologic systems to test the design concepts based on a targeted hydrologic systems approach; four pilot studies were initiated in surface water study areas and three were in groundwater study areas. The NAWQA program was reviewed after its 4-year pilot phase was concluded by the National Research Council (NRC) [89] of the National Academies of Science and Engineering. Following that review, the Program was revised and fully implemented in 1991 with the activation of 20 of the planned 60 study unit investigations. A NAWQA study unit was defined as a hydrologic system in which water-quality monitoring and assessment studies were focused. Study units were defined geographically by a combination of groundwater and surface-water features and usually encompassed more than 10,000 square kilometers of land area. Study units collectively covered more than 50% of the nation's population and water use and were diverse hydrologic systems that differed widely in natural features and human activities that affect water quality [90]. The NRC has evaluated the NAWQA program periodically to assess progress in meeting its objectives [91—93].

The NAWQA study unit design takes a targeted approach to water-quality monitoring and assessment [87,88]. A targeted approach is one in which the focus of study is on specific areas and constituents selected for investigation of topical water-quality issues [88]. Specifically, targeted sampling of known or suspected water-quality problem areas was a major design feature used to examine the geographic distribution of selected water-quality constituents [88]. Results from study units were aggregated into periodic regional and national synthesis assessments of nutrients, pesticides, trace elements, volatile organic contaminants, and other contaminants to evaluate water-quality conditions, trends, and factors affecting water quality (Tables 2.3 and 2.4).

The NAWQA was designed as a perennial program planned to be completed in sequential 10-year cycles [88]. Cycle 1 of NAWQA investigations, from 1991 to 2000, assessed water-quality status and began trend-monitoring networks (Fig. 2.2; Table 2.3). Cycle 1 established a regional and national design eventually replicated in 51 of the 60 originally planned study units that served as building blocks of regional and national assessments. In each study unit, a 3-year intensive sampling period was followed by 6 to 7 years of lower-intensity data collection, data analysis, and report publication prior to reactivation in the next decade of investigations [87]. About one-third of the study units were actively monitoring water quality from 1991 to 1993, another one-third from 1994 to 1996, and the remaining one-third from 1997 to 1999.

Another assessment component in Cycle 1 was retrospective data analyses within each of the 51 study units to examine what information and insights into the status of water-quality conditions could be gained by analysis of historical data collected by the USGS and other agencies. Much like the effort by the CEQ [41], the NAWQA analysis gathered data from various sources to improve geographic and temporal coverage with available water-quality sites. Data were inventoried for statistical analyses from the STORET and NWIS. Ancillary data sets were assembled on land use, population, climate, and agricultural crops within the boundaries of each study unit to help explain observations of monitored water-quality conditions.

One of the national retrospective publications was a synthesis of nutrients using data through 1992 at more than 300 river and stream sites with analysis from more than 22,000 samples [94]. Much like the CEQ report in 1972, data were aggregated to improve spatial and temporal coverage of monitoring sites. The analysis was limited by poor distribution of sites among the NAWQA study units and among land-use types. Other limitations included a lack of monitored data for all the desired nutrient forms, geographic gaps in sample-collection sites and gaps in temporal records, and dissimilar sampling frequencies [94]. The limitations and challenges of using retrospective data were noted in a report by the GAO titled *National Water Quality Assessment: Geological Survey Faces Formidable Data Management Challenges* [95]. Experiences from the retrospective data analysis underscore the importance of monitoring objectives, consistency in sampling and analysis methods, and careful operation of water-quality networks to yield useful results for water-quality assessments.

A fixed-site network was one component of the much broader national assessment plan for surface-water quality in the NAWQA study units that also included special short-term studies of stream chemistry and algae, aquatic invertebrates, fish, and in-stream and near-stream habitats [87]. Within each study unit, fixed-monitoring sites were targeted to locations to assess the effects of natural features and human activities in the monitored watershed. Indicator sites were located at the outlets of predominantly urban or agricultural watersheds with relatively homogeneous natural features. Integrator sites were located at the outlets of large and complex watersheds that commonly contain multiple types of natural features and land-use settings. A small number of relatively undeveloped watersheds, about 20, were included to establish baseline conditions. Where conditions permitted, a nested network was developed. A nested design is one in which smaller indicator and reference watersheds are monitored at sites located upstream of larger integrator sites. In this manner, complex water-quality patterns at the larger downstream site are better explained by sampling at nested upstream sites in smaller, relatively homogenous watersheds [87].

The NAWQA fixed-site design departed from the approach of using fixed monthly sampling frequencies. Instead, sampling frequencies were stratified to cover a range in hydrologic conditions with intensive seasonal sampling to coincide with applications of fertilizers and pesticides in urban and agricultural watersheds. Another important addition was the inclusion of ancillary data on explanatory factors such as land use, chemical use, and natural features such as geology and climate within monitored watersheds obtained by the use of digital maps in a GIS.

The USGS long-term monitoring sites for the decade 1985 to 1994 are shown in Fig. 2.6. Slightly less than half of the long-term sites shown in Fig. 2.6, or 46%, were part of the NASQAN and HBN networks. The NAWQA fixed-site water-quality network was active at a total of 496 sites in Cycle 1 from 1992 to 2001 [87] (Tables 2.3 and 2.4). None of the

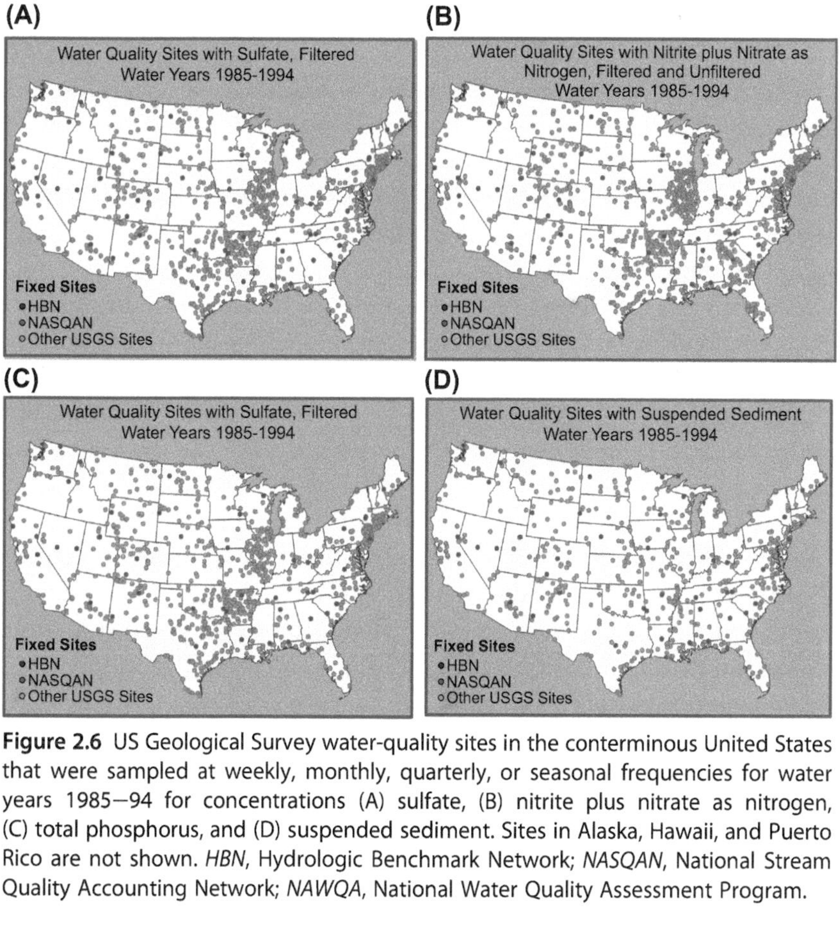

Figure 2.6 US Geological Survey water-quality sites in the conterminous United States that were sampled at weekly, monthly, quarterly, or seasonal frequencies for water years 1985–94 for concentrations (A) sulfate, (B) nitrite plus nitrate as nitrogen, (C) total phosphorus, and (D) suspended sediment. Sites in Alaska, Hawaii, and Puerto Rico are not shown. *HBN*, Hydrologic Benchmark Network; *NASQAN*, National Stream Quality Accounting Network; *NAWQA*, National Water Quality Assessment Program.

NAWQA sites can be considered as long-term sites in this decade because the network was started in 1992. The remaining long-term monitoring sites were part of other federal agency, state, and local cooperative monitoring projects with the USGS. Fig. 2.6A–D and Table 2.2 show that from 1985 to 1994 (A) 978 long-term sites were monitored for concentrations of sulfate; (B) 1168 long-term sites were monitored for concentrations of nitrite plus nitrate as nitrogen; (C); 1100 long-term sites were monitored for concentrations of total phosphorus; and (D) 583 long-term sites were monitored for concentrations of suspended sediment. Sites were well distributed across the conterminous United States with relatively higher densities of sites in a few states. No long-term sites were identified in the NWIS for concentrations of atrazine from 1985 to 1994.

4.5 Monitoring Coordination and Network Redesigns for the 21st Century, 1995–2004

During this period, the GAO issued several more reports on water-quality monitoring that shaped the future efforts of federal agencies and fostered improved coordination among monitoring agencies. The OMB issued Memorandum No. 92–01 to form the Intergovernmental Task Force on Monitoring Water Quality (ITFM) [96]. The ITFM was a federal/state partnership of 10 federal agencies, nine state and interstate agencies, and one American Indian tribe. The EPA chaired the ITFM and the USGS served as vice chair and Executive Secretariat within the USGS Water Information Coordination Program under the Advisory Committee on Water Information (ACWI). The ITFM carried out a review of water-quality monitoring activities in 1992 and recommended improvements in how water-quality studies were designed, how samples were collected and analyzed, and how data were reported [96].

The ITFM was reconstituted in 1997 with both public and private sector representatives as the National Water Quality Monitoring Council (NWQMC) [96]. The NWQMC gathers and summarizes monitoring methods and guidelines through the Methods Board and promotes support for institutional collaboration through meetings and biennial national monitoring conferences. The NWQMC also provides information on methods of quality assurance and quality control, strategies for data management and sharing, data interpretation techniques, and training. The EPA and USGS in cooperation with the NWQMC support products that can be used by monitoring programs nationwide.

From 1995 to 1997, the USGS fixed-site networks such as the NASQAN and HBN were undergoing major changes. A realization had developed over time for the HBN that even the most remotely located reference sites may be affected by human activities from outside the watershed including long-range transport of atmospheric deposition. In the mid-1990s, the USGS conducted a complete review of the HBN, discontinued most sites, and selected five eastern stations to conduct a pilot study to assess the optimum sampling strategy for assessing long-and short-term trends. Data from these few remaining HBN sites were used to demonstrate improving trends in stream sulfate concentrations resulting from reductions in nonpoint sources of atmospheric emissions of sulfur dioxide and other acidic gases from coal-fired powerplants [97].

All HBN watersheds were reevaluated in 2002 to determine whether upstream development or other factors had made them unsuitable as reference watersheds. As a result, a redesigned HBN was restarted in 2002. The 36 sites that best met the network criteria were selected for continued streamflow monitoring and water-quality sampling was reinitiated at 15 of those 36 sites. In 2003, the new design provided for tracking trends in water quality at a range of river-flow conditions. The sampling frequency for the 15 HBN sites was increased relative to previous approaches by use of automated samplers to capture the rapid response of small streams to atmospheric deposition. More recently, data from HBN sites were used to document sulfate declines from 1970 to 2010 at stream stations in the northeastern United States. These declines in stream sulfate concentrations were related to larger declines in atmospheric deposition of sulfate and declines in sulfur-dioxide emissions to the atmosphere mandated by the Clean Air Act amendments of 2007 [98]. By 2005, over 40 scientific studies using HBN data had been published over the 50-year period of operation [97].

In 1995, the NASQAN was redesigned as regional networks of research, monitoring, and assessment at 39 sites on the nation's largest rivers. The four regional networks monitored streamflow, concentrations, and mass transport of pesticides, nutrients, organic carbon, and sediment in the Mississippi River, salinity in the Rio Grande River, trace elements in the Colorado River, and pesticides in the Columbia River [99] (Tables 2.3 and 2.4). A fixed-site approach was retained in the redesign of NASQAN, however; key design features were improved. Monthly and quarterly sampling frequencies were increased to include seasonal and high-flow sampling. A nested design was developed with sampling sites on the main stem and on major tributary streams. To support assessments of status, change, and trends, the network began to collect ancillary data on land-use conditions within major watersheds [99].

The total number of USGS monitoring sites for concentrations of five constituents remained stable or slightly increased compared to the previous decade despite the major reductions and redesigns of the NASQAN and HBN. Declines in the NASQAN and HBN resulted in a decline in the number of long-term sites for the selected constituents. The only major increase was the number of sites monitored for concentrations of atrazine from 1995 to 2004. Most of the atrazine samples were collected by the NASQAN and NAWQA fixed-site networks. About one-third of all the USGS long-term sites were part of the NAWQA network. Fig. 2.7A–D shows that from 1995 to 2004 (A) 493 long-term sites were monitored for concentrations of sulfate; (B) 741 long-term sites were monitored for

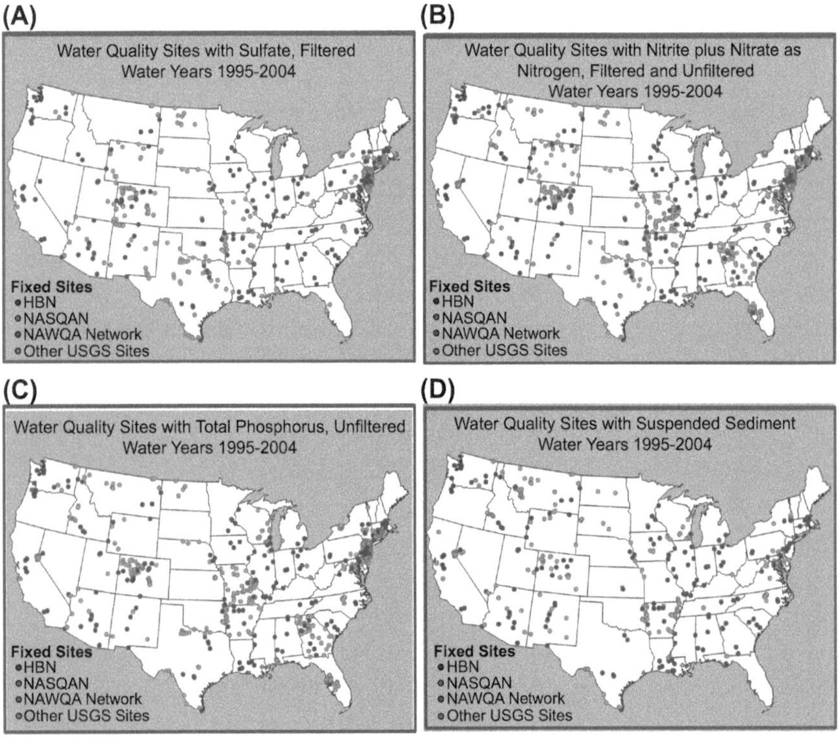

Figure 2.7 US Geological Survey water-quality sites in the conterminous United States that were sampled at weekly, monthly, quarterly, or seasonal frequencies for water years 1995–2004 for concentrations of (A) sulfate, (B) nitrite plus nitrate as nitrogen, (C) total phosphorus, and (D) suspended sediment. Sites in Alaska and Puerto Rico are not shown. *HBN*, Hydrologic Benchmark Network; *NASQAN*, National Stream Quality Accounting Network; *NAWQA*, National Water Quality Assessment Program.

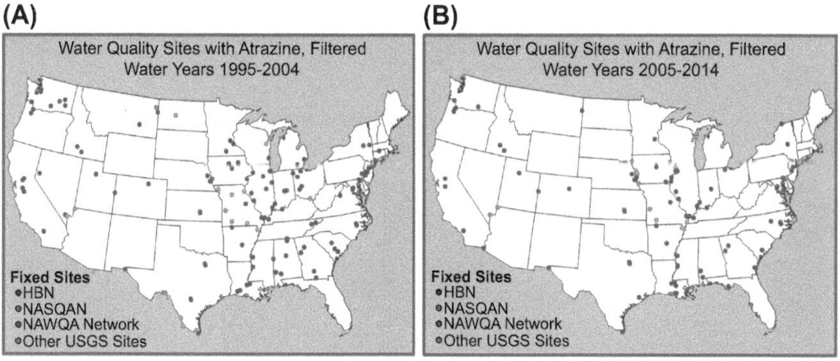

Figure 2.8 US Geological Survey water-quality sites in the conterminous United States that were sampled for concentrations of atrazine at weekly, monthly, quarterly, or seasonal frequencies for water years (A) 1995–2004 and (B) 2005–14. *HBN*, Hydrologic Benchmark Network; *NASQAN*, National Stream Quality Accounting Network; *NAWQA*, National Water Quality Assessment Program.

concentrations of nitrite plus nitrate as nitrogen; (C) 705 long-term sites were monitored for concentrations of total phosphorus; and (D) 399 long-term sites were monitored for concentrations of suspended sediment. Fig. 2.8A shows 126 long-term sites were monitored for concentrations of atrazine from 1995 to 2004. The geographic coverage of all long-term monitoring sites changed from the relatively well-distributed set of sites across the country for the period from 1985 to 1994 for which slightly less than half were the NASQAN and HBN (Fig. 2.6) to the uneven and patchy distribution if sites due to the reduction of the NASQAN sites from 1995 to 2004 and the increase in a smaller number of the long-term NAWQA sites (Figs. 2.7–2.9). The remaining short-term sites with sample collection lasting less than 6 years were likely monitored for intensive surveys and other short-term studies.

The redesign of the NASQAN as a regional network was consistent with the goals of the Environmental Monitoring and Research Task Force (EMRT) of the Committee on Environmental and Natural Resources in the White House Office of Science and Technology Policy [77]. The EMRT recommended the integration of environmental monitoring and research networks. A monitoring and research framework was recommended to optimize the large expenditures of funds for environmental monitoring programs [100]. A comprehensive environmental monitoring and research program, to be effective, should be able to determine the status, change, and trends in resources, provide an improved understanding of

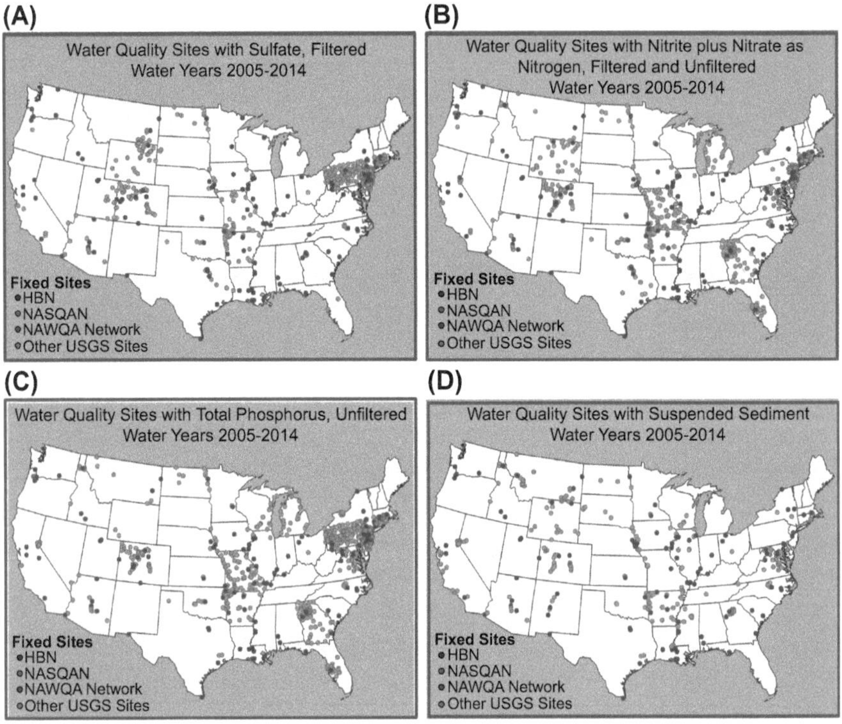

Figure 2.9 US Geological Survey water-quality sites in the conterminous United States that were sampled at weekly, monthly, quarterly, or seasonal frequencies for water years 2005–14 for concentrations of (A) sulfate, (B) nitrite plus nitrate as nitrogen, (C) total phosphorus, and (D) suspended sediment. Sites in Alaska, Hawaii, and Puerto Rico are not shown. *HBN*, Hydrologic Benchmark Network; *NASQAN*, National Stream Quality Accounting Network; *NAWQA*, National Water Quality Assessment Program.

ecological processes, provide early warning of emerging problems, and measure the effectiveness of environmental policies [101].

In following recommendations of the EMRT, interpretation of NASQAN data was supported by assessments conducted by the NAWQA in smaller rivers and streams [100]. A nested network of 23 NAWQA sites in the Mississippi River watershed was analyzed in conjunction with the NASQAN regional network of 18 large river sites on the main stem and major tributaries of the Mississippi River [77,102]. Selected NASQAN sites were also sampled during this time by the FWS in the NCBP and by the USGS within the Biomonitoring of Status and Trends program. Samples were collected for analysis of organochlorine pesticides and other bioaccumulative substances such as total polychlorinated biphenyl (PCB) residues in fish at the NASQAN sites in the Colorado, Columbia, Rio Grande, and Mississippi River basins [58,59,103].

The results of sampling NASQAN and NAWQA sites in the Mississippi River watershed led to the finding that excessive nutrients were being discharged to the northern Gulf of Mexico near the outlets of the Mississippi and Atchafalaya rivers and to the Mississippi River from the Upper Mississippi and Ohio rivers [104]. These discharges of nitrogen were causing seasonal hypoxia in the northern Gulf of Mexico [105,106]. These data and assessments led to an improved understanding of the role of nitrogen in the formation of the hypoxic zone in the northern Gulf of Mexico [107]. The nitrogen discharged from the Mississippi River watershed is primarily the result of applications of fertilizers to agricultural fields and is greater in years with higher than normal precipitation and streamflows than in years with lower precipitation and streamflows [107]. Urban runoff, wastewater, and animal agriculture were also found to be important sources of nitrogen and phosphorus in the Mississippi River Basin [108].

The Mississippi River/Gulf of Mexico Watershed Nutrient Task Force, created in 1997, developed an Action Plan based on scientific assessments describing the role of excessive nitrogen in the formation of the hypoxic zone [109]. A goal of reducing nitrogen and phosphorus delivery by 45% by 2020 was established in a second Action Plan in 2008 [110]. The annual mass flux of nitrate plus nitrite nitrogen or total nitrogen are reported each year for each of 21 small subbasins and five large subbasins within the Mississippi and Atchafalaya rivers and their tributaries [111,112]. The springtime flux in nitrate nearest the outlets of the Mississippi and Atchafalaya rivers to the Gulf of Mexico are used to annually estimate the size of the mid-summer hypoxic zone in the northern Gulf of Mexico [111,112]. A recent multidecadal trend analysis found a 10% increase from 1980 to 2008 in nitrate flux to the Gulf of Mexico [113]. The size of the hypoxic zone is measured by the Louisiana Universities Marine Consortium each summer in July when the zone is anticipated to be at its greatest extent and compared to predictions from nitrogen loads [114–117].

Although progress was being made on measuring the impacts of nonpoint sources of pollution in places like the Mississippi River Basin, new approaches to monitoring the status of the nation's waters was being developed. In the report titled *Key EPA and State Decisions Limited by Inconsistent and Incomplete Data* the GAO stated that the assessments presented in the *National Water Quality Inventories* did not accurately represent national water-quality conditions. Furthermore, the GAO stated that the assessments were unreliable for the Total Maximum Daily Load (TMDL) program under Section 303(d) of the Clean Water Act [118]. For the 2002 reporting cycle, only 19% of approximately 3.7 million miles of streams had been assessed and reported by states to evaluate water-quality status according for the *National Water Quality Inventory*. The remaining 81% of river miles were not assessed during the reporting cycle [119].

As a result of concerns about statistically representing the status of the nation's water quality, a complete review of the EPA's environmental monitoring strategy was undertaken within and outside of the EPA [120,121]. The Environmental Monitoring and Assessment Program (EMAP), a research program in the EPA's Office of Research and Development, was initiated from 1990 to 2006 in response to recommendations by the EPA's Science Advisory Board regarding the EPA's system for evaluating the status of the nation's environment [121]. Under the EMAP, a new probabilistic design to monitor and assess the status of aquatic resources was developed [122–124].

The probabilistic design was developed to fulfill the specific objectives of Section 305b of the Clean Water Act to assess the status of the nation's aquatic resources. A probabilistic sampling method utilizes a random selection process for sampling water chemistry, aquatic biota, and in-stream habitat. In one of the first pilot studies, approximately 1000 reaches on small western streams were randomly selected for field sampling of water chemistry and aquatic invertebrates [125]. During the 1990s and early 2000s similar surveys were designed for large rivers, lakes, coastal areas, and wetlands. The GAO identified probabilistic water-quality monitoring as statistically valid and useful for meeting the needs of the *National Water Quality Inventory* [118].

In 2004, almost 1400 stream reaches were sampled for the first Wadeable Streams Assessment (WSA) using a probabilistic design [119]. With inferences to the nation's wadeable streams as a whole, the 2004 WSA reported that 42% of the nation's small streams were in poor ecological condition, 25% were in fair condition, and 28% were in good condition [119]. The 2004 WSA was estimated to represent, by statistical inference, about 644,000 kilometers (400,000 miles) of small perennial streams [119]. At the same time and using their own methods such as fixed-site monitoring, intensive surveys, rotating basin assessments, and effluent monitoring, states identified 45% of the assessed miles as being impaired or not supporting one or more of their designated uses for the same time period. A much smaller and less costly level of effort is necessary to complete a probabilistic survey compared to a combination of other approaches.

The nation's aquatic resources are now monitored every 5 years using probabilistic surveys [119]. The 2008 and 2009 National Rivers and Streams Assessment represented 1.2 million miles of rivers and streams in the United States [126]. The survey found that 46% of rivers and streams are in poor biological condition, 25% are in fair condition, and 28% are in good condition [126].

Beginning with the 2004 reporting cycle for the *National Water Quality Inventory*, a list of all waters not meeting water-quality standards in each state was included to address Section 303d of the Clean Water Act [119]. The TMDL assessments, as called for in Section 303d, are evaluated using multiple targeted monitoring approaches within watersheds including fixed-site water-quality monitoring, intensive water-quality surveys, rotating basin assessments, and water-quality modeling. The availability of timely, reliable, and complete data about the nation's waters is critical for determining which waters do not meet states' standards and must therefore be targeted for potentially expensive cleanup [118].

The emphasis on TMDLs resulted in a focus on the watershed as a unit for water-quality planning and management. To improve water-quality nationwide, point and nonpoint sources of water pollution would need to be addressed comprehensively at the watershed scale. For example, an important regional monitoring effort, The Chesapeake Nontidal Monitoring Network, started in 1985 as a fixed-site monitoring network and expanded in 2004. In 2011, emphasis was placed on monitoring at 117 sites to assess progress under a TMDL for the Chesapeake Bay. Through this effort, nitrogen, phosphorus, and suspended-sediment loads and trends are determined from extensive water-quality sampling by multiple agencies. The USGS computes the loads and trends using advanced statistical analysis to determine if the nutrient reduction goals of the TMDL are being met.

The emphasis on TMDLs resulted in a new focus government-wide on improving water-data coordination because states did not have adequate data for development of TMDLs. The GAO produced a 2004 report titled *Watershed Management: Better Coordination of Data Collection Efforts Needed to Support Key Decisions* [127]. The GAO [127] was asked to determine "(1) the key entities that collect water quality and water quantity data including the types of data they collect, how they store their data, and how entities can access the data; and (2) the extent to which these entities coordinate their water quality and water quantity data-collection efforts." Specific to water-quality data, the GAO found that 15 of the 16 federal agencies interviewed for the report were collecting water-quality data. The GAO also found that coordination of these efforts was lacking, which was a reason that states did not have all the data need or available for purposes of developing TMDLs.

Seven barriers to better data sharing were identified of which three were common among federal, state, and local agencies. The first of the three most common barriers was that agencies were collecting water-quality data for their own purposes and were reluctant to modify their activities to make the information more comparable or broadly available. Second, many agencies stated that differences among agencies in methods and practices of collecting water-quality data resulted in data that could not be aggregated for analysis. Third, many agencies stated that they were not aware of each other's water-quality monitoring programs.

The GAO recommended that Congress clearly define authority for coordinating the collection of water-quality data nationwide by formally designating a lead organization for this purpose. Among its responsibilities, the organization would "(1) support the development and continued operation of regional and state monitoring councils, (2) coordinate the development of an Internet-based clearinghouse to convey what entities are collecting what types of data, and (3) coordinate development of clear guidance on metadata standards so that data users can integrate data from various sources" [127].

As the identified lead federal organization for water-quality monitoring and coordination, the NWQMC through the joint leadership of the USGS and the EPA, provided a forum to address all three recommendations from the GAO report. The USGS and EPA in coordination with the NWQMC began to develop the Internet clearinghouse to address the problem of data availability. After many years of effort, the Water Quality Portal (WQP) was made available in 2012 to serve the EPA's modern STORET, the USGS NWIS, and the Natural Resources Conservation System's STEWARDS database [128]. The WQP provides a single, user-friendly web interface to locate water-quality data collected by federal, state, local, and tribal partners. In 2015, the WQP contained over 150 million publicly available water-quality data records that can be accessed and downloaded in a variety of formats.

The NWQMC also established a multiagency Methods and Data Comparability Board (MDCB), which addressed the third major recommendation from the 2004 report by the GAO. The result of the MDCB was the development in 2002 of the National Environmental Methods Index (NEMI), a major design component of which is to make comparisons of methods, and therefore the data generated from those methods, more straightforward [129]. The NEMI is a searchable database that allows scientists and managers to find and compare analytical and field methods for all phases of environmental monitoring. The WQP is making monitoring methods information from NEMI available with data retrievals. Now water managers and data analysts can quickly link a sample result to the full method that was used to determine that result.

4.6 A Time of Increasing Complexity: Trends, Models, and Indicators, 2005–14

Cycle 2 of the NAWQA (Cycle 2) was carried out from 2002 to 2012 [130] (Fig. 2.2). In Cycle 2, several study units were consolidated or discontinued resulting in a decrease from 51 to 42. Eventually, the 42 study units were aggregated further into large regional areas within eight major river basins in the conterminous United States [131,132]. The NAWQA discontinued many of the original 496 sites in the fixed-site network for budgetary reasons and because of a shift in emphasis to shorter-term topical studies on issues of national importance. Three national-scale surface-water topical studies were completed in Cycle 2: nutrient effects on aquatic ecosystems, mercury effects on aquatic ecosystems, and the effects of urbanization on stream ecosystems. These national topical studies added explanatory power to descriptions of patterns in status and trends in water quality and aquatic life at the fixed sites.

During this time period, the NAWQA began to redesign its fixed-site network to emphasize trend analysis because a greater emphasis was placed on water-quality trends in Cycle 2 than in Cycle 1. The Cycle 2 fixed-site network for long-term trend sampling initially started with 147 fixed sites in rivers and streams that were sampled in Cycle 1. Through successive adjustments after 2004, the network was reduced to a range of 84 to 113 sites through 2012 [132] (Tables 2.3 and 2.4). Features of the fixed-site network design such as sampling to cover ranges in streamflow, seasonal sampling for constituents that are seasonal in occurrence, and continued sampling at indicator and integrator watersheds yielded valuable information to address earlier criticisms of fixed-site monitoring approaches.

The National Monitoring Network for Coastal Waters and Their Tributaries (NMN) was designed by the NWQMC and more than 80 stakeholders in response to a recommendation by the US Commission on Ocean

Policy in 2004 to fill monitoring gaps in inland, estuarine, and coastal monitoring networks maintained by the EPA, the National Oceanic and Atmospheric Administration, and the USGS [133]. The NMN integrates biological, chemical, and physical features and links uplands to the coastal ocean. The riverine component of the NMN was proposed as a network of 386 sites; 258 sites to monitor 90% of freshwater outflow and constituent loads discharged to the ocean from hydrologic unit code (HUC)-6 watersheds; 72 sites located to monitor 97% of freshwater inflow to specific high-priority estuaries; and 56 sites located to monitor the outflow of basins draining 250 square miles or more to the Great Lakes in the United States.

The USGS began a process in 2007 to combine operations and to harmonize data-collection and analysis methods in NAWQA, NASQAN, and the NMN. The NASQAN regional design was modified in 2007 to create a set of coastal sites near the terminus of rivers that are direct tributaries to coastal areas and a nested subnetwork within the Mississippi River watershed with sites near the outlet of the Mississippi River and its major tributaries. The objectives of the redesign are to monitor concentrations and for some constituents, loads, of suspended sediment, nutrients, pesticides, and inorganic compounds to marine coastal waters from inland rivers and from the Great Lakes.

The redesign reduced main stem and tributary monitoring sites within the sub-networks of the Colorado, Columbia, and Rio Grande rivers to a single site located near the outflow of each large watershed. The nested sub-network within the Mississippi River Basin was retained with a reduced number of main stem sites on the Ohio and Missouri rivers. Several large monitoring sites near the outflows of coastal rivers were added by selecting sites previously sampled in the original NASQAN network that met the design requirements recommended by the NMN [134]. Five previously sampled NAWQA and NASQAN sites on major rivers were added to complete a national coastal sub-network for large rivers discharging directly to the Atlantic Ocean, to the Gulf of Mexico, and to the Pacific Ocean in California. The current network represents about 80% of the discharge of streamflow and provides information on concentrations and loads of nutrients, pesticides, organic carbon, and suspended sediment to marine coastal waters.

The maps for the time period 2005 to 2014 (Figs. 2.8B and 2.9A–D) and Table 2.2 show major reductions in the total number of water-quality monitoring sites for all six constituents compared to the previous three decades. However, the number of long-term monitoring sites remained about the same or increased slightly compared to the previous decade indicating that some stabilization of the network had taken place during the two most recent decades. However, the geographic distributions of constituents are uneven and there are large areas of the conterminous United States that have poor or no coverage with long-term sites (Fig. 2.9A–D). Fig. 2.9A–D shows that from 2005 to 2014 (A) 549 long-term sites were monitored for concentrations of sulfate; (B) 718 long-term sites were monitored for concentrations of nitrite plus nitrate as nitrogen; (C) 821 long-term sites were monitored for concentrations of total phosphorus; and (D) 390 long-term sites were monitored for concentrations of suspended sediment. Fig. 2.8B shows 83 long-term sites were monitored for concentrations of atrazine.

Regional and national synthesis assessments of pesticides, nutrients, and other constituents have been published periodically with data from the USGS fixed-site networks and topical studies. These synthesis products have provided critical information on occurrence, distribution, seasonality, changes, and long-term trends in pesticides and nutrients in relation to land use, chemical use, and the influence of natural factors such as streamflow, climate, geology, and soil characteristics [53,135,136].

Results from the NAWQA and NASQAN fixed-site networks were used to develop chemical indicators of aquatic ecosystem conditions and trends in the 2002 and 2008 reports on the *State of the Nation's Ecosystems* by the H. John Heinz III Center for Science, Economics and the Environment [137,138]. Design principals of the Heinz indicators include a focus on condition and trends, relevance to contemporary policy issues, use of a balanced and unbiased array of indicators, and use of high-quality data that are national in scale and have a reasonable likelihood of being collected in the future. Indicators addressed four ecosystem types: forests, fresh waters, grasslands and scrublands, and urban and suburban landscapes. The USGS water-quality data from the NAWQA and NASQAN contributed to 2 of 10 core indicators; one on Movement of Nitrogen and the other on Contaminants. These core indicators were calculated from more than 400 NASQAN sites and a similar number of NAWQA sites [137]. Chapter 8 on Freshwaters in The Heinz Center report, indicated that USGS water-quality data contributed to an additional 3 of 17 indicators on chemical contaminants [137].

In 2005, in a report titled *Status of Federal Data Programs that Support Ecological Indicators*, the GAO reviewed 20 federal data programs to determine whether federal agencies responsible for the programs anticipate that changes during fiscal years 2005 and 2006 related to funding, shifting priorities, or other factors would affect the ability of the programs to (1) continue to generate data comparable with data from past years, and (2) continue providing data used in an updated nationwide ecological indicator study by the Heinz Center planned for release in 2008 [139].

For the 2008 Heinz Center *State of the Nation's Ecosystems* report, monitoring results from the NASQAN had been reduced from 400 to 35 sites, a 92% decrease [139]. Reasons given for the loss of sites with which to compute indicators included a combination of reductions in NASQAN operating funds from 1990 to 2005 and increases in operational costs [139]. In 2005, the NAWQA reduced the number of long-term monitoring sites from 113 to 84 [139]. Likewise, some of the NAWQA data on contaminants in stream-bed sediments and fish tissues in the 2002 Heinz Center report were not available for the 2008 report due to a change in emphasis of the NAWQA program. The change in emphasis was due in part to the continuing decline in bioaccumulative and persistent organic chemicals in bed sediments and fish tissues following the phase-outs of usage of these compounds in the United States and as a result of international phase-out policies under the Stockholm Convention on Persistent Organic Pollutants in 2004 [140]. Other reasons for change included program reductions and increases in operational costs [139]. Although the NASQAN and NAWQA reduced the number of fixed monitoring sites, the high quality of the data continued to be maintained.

Results from the USGS fixed-site networks were critical to the Heinz Center reports and other reports at the national scale. These results have consistently shown that certain pesticides are frequently detected in streams at concentrations related to the magnitude of use in agricultural and mixed land-use watersheds [53]. Concentrations of water-soluble pesticides in urban and agricultural watersheds also are detected at downstream large river sites [53,136,141]. Insecticides continue to be detected over time in urban rivers and streams at levels that can cause adverse effects on aquatic organisms [53]. Pesticide concentrations in streams change over annual to decadal periods in response to climate, crops, economic factors, and the arms race between pests, pesticide resistance, and improvements in chemical controls. Information from NAWQA publications have been used to support pesticide registrations, reregistrations, and cancellations of certain uses by the EPA [53,141].

The NAWQA national assessments use results from the USGS fixed-site networks to describe nutrient occurrence, key sources, relevance to drinking water and aquatic life, and trends and changes [135]. Results show that excessive nutrient enrichment is a widespread cause of eutrophication and associated consequences in streams. Despite major federal, state, and local efforts to control the sources and transport of nutrients, concentrations have remained the same or increased in many streams across the nation [135].

The NAWQA Program and the USGS have a long history of publishing national, regional, statewide, and watershed focused trend reports based on data from continuing and discontinued USGS fixed-site networks and results from the networks of state agencies where appropriate. About 75 publications from the NAWQA Program are available in journal articles or USGS reports describing water-quality trends in rivers, streams, and lakes [112]. Most of the publications concerning trends for rivers and streams make use of results from the USGS fixed-site networks. Trends have been reported for concentrations of alkalinity and concentrations and loads of chloride, nitrogen, pesticides, phosphorus, suspended sediment, other trace organic contaminants, and trace elements. Some of these publications use data from fixed sites dating as far back as the early part of the 20th century [142]. Most publications on trends cover time periods after the passage of the Clean Water Act in 1972.

Recently, investigations of multidecadal to century-scale water-quality trends have taken on increasing importance [143–146]. For example, a recent multidecadal analysis of nitrate trends from long-term water-quality monitoring sites including current and discontinued USGS fixed-sites and data available in the NWIS from the first half of the 20th century show that increasing trends in nitrate concentrations and mass transport from the Mississippi River and other rivers in the Midwest increased rapidly from 1945 to 1980, then leveled off or decreased slightly from 1982 to 2008 [146]. Trend analysis of water-quality and streamflow-monitoring data at the half-century or longer scale provides context for shorter-term temporal trends at the decadal scale [146]. These reports demonstrate that long-term trends can be discerned if the trend signal is greater than the variability in results due to changes over time in sample-collection and analysis methods, changes in frequencies of sample collection, and temporal gaps in data collection. However, the number of known sites meeting criteria for century or half-century scale trends is relatively small, on the order of a few dozen sites [145].

Status, change, and trend analyses are the most common use of fixed-site monitoring. However, fixed-site monitoring is serving as a basis for a new monitoring and modeling approach [131,132]. Knowledge of how watersheds function acquired after many decades of monitoring and assessment serve as the basis for geospatial models. Water quality, source information on nutrients, and geospatial data on land use, point sources, the river network, and other natural factors come together in statistical and other types of models to determine the relative importance of various sources of nutrients or pesticides, respectively, and to predict constituent concentrations or loads at unsampled locations. The models also provide for scenario testing of best management practices [131,132].

The SPAtially Referenced Regressions On Watershed attributes (SPARROW) use geospatial data about the stream network and the geology, topography, soils, land use, and nutrient inputs, along with mean load estimates at monitoring sites to build a statistical model to predict loads at any stream location in the studied region. The model relies on water-quality sampling results and streamflow information from monitoring sites in the USGS and from as many as 72 other agencies. These data are available for calculating long-term mean annual nutrient loads required for SPARROW model calibration [147]. Regional and national SPARROW models require water-quality and streamflow data from monitoring sites. Suitable monitoring sites for modeling with SPARROW required streamflow measurements at or nearby the sampling site and a minimum number of 20 or more samples collected over a range of streamflows representative of the site. SPARROW models published in 2011 were used to predict the mass flux of total nitrogen and total phosphorus for the year 2002 in six regions of the country. After screening over 124,500 sites, only 2739 or 2% of the total number of sites from 73 agencies were identified as having suitable data for SPARROW modeling in the six regions [148].

Pesticides were modeled in the geospatial Watershed Regression for Pesticides model (WARP) [149], which relies on monitoring results from the USGS fixed-site networks. The USGS networks are almost the only consistent source of these data. For example, chlorpyrifos, an insecticide used commonly on cotton, corn, citrus, and almond crops was modeled using WARP for 2012 conditions in streams in the Midwest, central Texas, southwest Florida, and the Central Valley in California. Many streams were predicted to have concentrations of chlorpyrifos that are likely to exceed the acute aquatic life benchmark for fish [149].

Major improvements and innovations over time have been made in fixed-site monitoring networks. The improvements provide for scientific understanding of watershed processes and factors that influence water quality.

The targeted approach for watershed understanding includes sampling during different hydrologic conditions, over different seasons, among different land uses, in different geologic or climatic settings, and include biological characteristics, and over long time periods [150]. Monitoring programs with these characteristics provide an understanding of watershed functions as a basis for status and trend assessments. A watershed approach, when coupled with information on human and natural factors affecting water quality, produces trend results that have the explanatory power on causal and associative factors and drivers of change and trends. The targeted monitoring approach when coupled with geospatial information on causal factors and watershed processes also provides a basis for predictive models that contribute information on sources, fate, transport, and effects of water contaminants on water quality. These predictive models provide tools for water-quality protection and management decisions today and in the future [150].

The second decade of the 21st century is a time of increasing complexity in water-quality monitoring and assessment. More than six decades of progress have been made since the ORSANCO network was initiated in 1951. Experience has demonstrated that different questions require different kinds of monitoring and that one design cannot answer all questions. However, the fixed-site approach provides results for evaluating long-term trends and changes in rivers and streams to answer the question "Is water quality getting better or worse and why?" The temporal and spatial designs of fixed-site networks and the ancillary data on human and natural factors provide data-rich measures of temporal variability in major factors affecting water-quality conditions and trends over long periods of time.

Cycle 3 of NAWQA began in 2013 and required a fresh evaluation of all the USGS national fixed-site networks in rivers and streams including sites from the NAWQA, NMN, NASQAN, and HBN [131,132] (Fig. 2.2). The NAWQA Cycle 3 design was carefully developed and reviewed by a host of agencies and the National Research Council [93]. The product is the USGS National Water Quality Network (NWQN). The USGS NWQN is the only national-scale network of fixed sites because other national-scale monitoring networks have been merged or discontinued over time. All monitoring sites in the NWQN share a core set of objectives, methods, and water-quality constituents for collection and analysis [131,132]. Some differences are preserved from former USGS fixed-site networks in the NWQN such as sampling frequency, which is dependent on the scale of the watershed contributing to the monitoring site and necessary for each individual subnetwork to achieve its individual goals and objectives. The

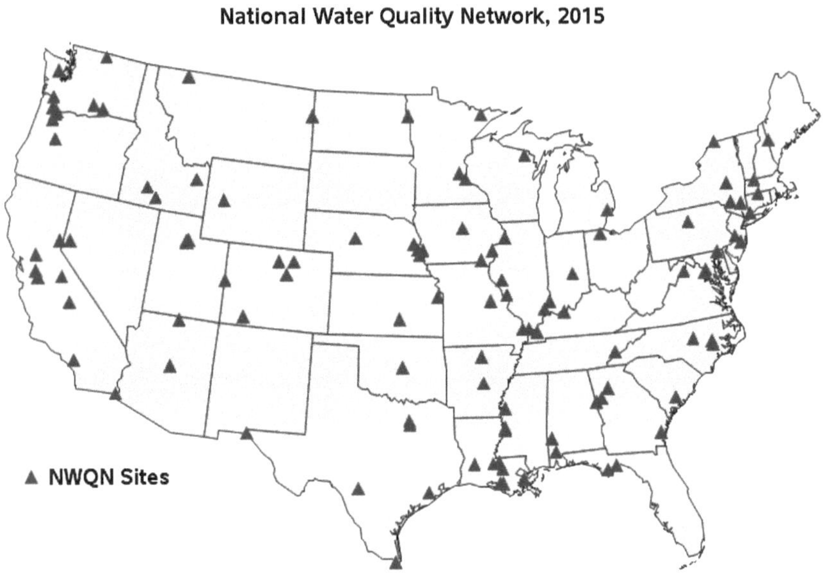

Figure 2.10 US Geological Survey fixed water-quality sites in the conterminous United States that are part of the National Water Quality Network (NWQN), water year 2015. Sites in Alaska and Hawaii are not shown.

resulting NWQN currently is composed of 113 sites operated entirely by the USGS and 4 additional sites operated cooperatively with a state agency as of 2015 (Fig. 2.10).

4.7 Whither Long-Term Water-Quality Monitoring Networks?

The NWIS contains nearly 3 million individual records for 36,031 unique water-quality sites sampled from 1975 to 2014 for the 6 constituents described herein. Of the 36,031 unique sites, 2873 or 8% were found to have a long-term sampling period in at least 1 of the 4 decades and for at least 1 of the 6 constituents. Less than 1% of long-term sites, 217 in total, were monitored for 1 or more constituents for the entire 40-year period since passage of the Clean Water Act in 1972. Fig. 2.11 displays six groups of bar graphs showing the total number of sites sampled for each constituent for each decade contrasted with the number of sites meeting the long-term criteria within individual decades and over the 40-year period. Table 2.5 contains a list of the water-quality monitoring sites with 40 years of record by NWIS site number, name, location, and types of constituents sampled over time.

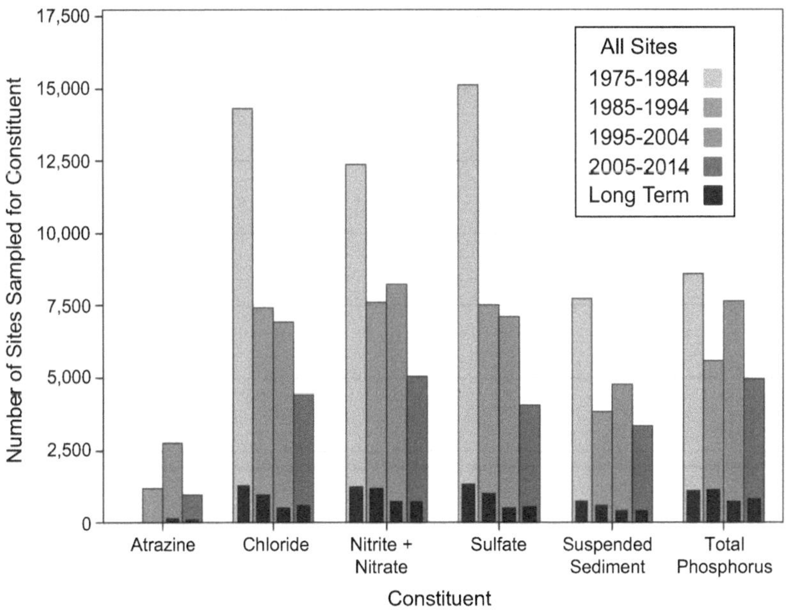

Figure 2.11 Total number of water-quality sites and number of long-term sites sampled by the US Geological Survey for concentrations of atrazine, chloride, nitrite plus nitrate as nitrogen, sulfate, suspended sediment, and total phosphorus by decade, 1975–2014.

Several national-scale long-term fixed-site networks were discontinued by 1986, and after 1999 only the USGS was monitoring water quality in national networks of fixed sites such as the NASQAN, NAWQA, NMN, and HBN. The percentage of the total number of long-term USGS sites monitored over the 40-year time period for 5 constituents has declined over a range from 26% to 58% (Table 2.2). The specific declines in the relative numbers of long-term monitoring sites over the four decades were for concentrations of sulfate (58%), chloride (55%), suspended sediment (45%), nitrite plus nitrate nitrogen (42%), and total phosphorus (26%). For the most recent 20-year time period, there was a 34% decline in the total number of long-term sites monitored for concentrations of atrazine. Although it is clear in Table 2.2 that all water-quality monitoring has declined since the decade 1975 to 1984, declines in monitoring at weekly, monthly, bimonthly, quarterly, and seasonal sampling frequencies best represent the declines in fixed-site monitoring.

The category with the largest number of sites assessed for this study are those sampled annually or semiannually, indicating sampling frequencies of

Table 2.5 US Geological Survey Site Number, Site Name, Hydrologic Unit Code, Site Latitude, Site Longitude, Fixed-Site Network Name, and Sites With Weekly, Monthly, Quarterly, or Seasonal Sampling for Sulfate, Chloride, Suspended Sediment, Total Organic Carbon, Nitrite Plus Nitrate as Nitrogen, and Total Phosphorus: 1975–2014

						Site With Weekly, Monthly, Quarterly, or Seasonal Sampling for a Constituent From 1975–2014				
USGS Site Number	Site Name	Hydrologic Unit Code	Site Latitude	Site Longitude	Fixed-Site Network Name	Sulfate	Chloride	Suspended Sediment	Nitrite Plus Nitrate as Nitrogen	Total Phosphorus
01112900	BLACKSTONE RIVER AT MANVILLE, RI	01090003	41.97121	−71.47006	NAWQA				X	X
01118500	PAWCATUCK RIVER AT WESTERLY, RI	010909005	41.38371	−71.83312	NASQAN		X		X	X
01119375	WILLIMANTIC R AT MERROW, CT	01100002	41.83538	−72.31008	Unknown	X	X		X	X
01122610	SHETUCKET R AT SOUTH WINDHAM, CT	01100002	41.68232	−72.16591	NASQAN	X	X		X	X
01127000	QUINEBAUG RIVER AT JEWETT CITY, CT	01100001	41.59749	−71.98409	NASQAN	X	X		X	X
01184000	CONNECTICUT RIVER AT THOMPSONVILLE, CT	01080205	41.98732	−72.60537	NAWQA	X	X	X	X	X
01188000	BUNNELL BROOK NEAR BURLINGTON, CT	01080207	41.78621	−72.916483	Unknown	X	X		X	X
01189030	PEQUABUCK R AT FARMINGTON, CT	01080207	41.71677	−72.83982	Unknown	X	X		X	X
01189995	FARMINGTON RIVER AT TARIFFVILLE, CT	01080207	41.90828	−72.75935	Unknown	X	X		X	X
01193500	SALMON RIVER NEAR EAST HAMPTON, CT	01080205	41.55232	−72.44925	Unknown	X	X		X	X

01196222	QUINNIPIAC R NR MERIDEN, CT	01100004	41.52926	−72.86343	Unknown	X	X		X	X
01196500	QUINNIPIAC RIVER AT WALLINGFORD, CT	01100004	41.45026	−72.84128	Unknown	X	X		X	X
01200600	HOUSATONIC RIVER NEAR NEW MILFORD, CT	01100005	41.59315	−73.44957	NAWQA	X	X		X	X
01203000	SHEPAUG R NR ROXBURY, CT	01100005	41.54982	−73.32984	Unknown	X	X		X	X
01205500	HOUSATONIC RIVER AT STEVENSON, CT	01100005	41.3838	−73.16664	NASQAN	X	X		X	X
01208500	NAUGATUCK RIVER AT BEACON FALLS, CT	01100005	41.4423	−73.06228	Unknown	X	X		X	X
01208990	SAUGATUCK RIVER NEAR REDDING, CT	01100006	41.29454	−73.39512	Unknown	X	X		X	X
01367770	WALLKILL RIVER NEAR SUSSEX, NJ	02020007	41.19389	−74.57528	Unknown	X	X		X	X
01377000	HACKENSACK RIVER AT RIVERVALE, NJ	02030103	40.99917	−73.98917	Unknown	X	X		X	X
01381800	WHIPPANY RIVER NEAR PINE BROOK, NJ	02030103	40.845	−74.34722	Unknown	X	X		X	
01382000	PASSAIC RIVER AT TWO BRIDGES, NJ	02030103	40.89722	−74.27278	NAWQA	X	X		X	
01387500	RAMAPO RIVER NEAR MAHWAH, NJ	02030103	41.09806	−74.16278	Unknown	X	X		X	
01389500	PASSAIC RIVER AT LITTLE FALLS, NJ	02030103	40.88472	−74.22611	NASQAN	X	X		X	
01391500	SADDLE RIVER AT LODI, NJ	02030103	40.89028	−74.08056	Unknown	X	X		X	X
01394500	RAHWAY RIVER NEAR SPRINGFIELD, NJ	02030104	40.6875	−74.31167	Unknown	X	X		X	

(Continued)

Table 2.5 US Geological Survey Site Number, Site Name, Hydrologic Unit Code, Site Latitude, Site Longitude, Fixed-Site Network Name, and Sites With Weekly, Monthly, Quarterly, or Seasonal Sampling for Sulfate, Chloride, Suspended Sediment, Total Organic Carbon, Nitrite Plus Nitrate as Nitrogen, and Total Phosphorus: 1975–2014—cont'd

						Site With Weekly, Monthly, Quarterly, or Seasonal Sampling for a Constituent From 1975–2014				
USGS Site Number	Site Name	Hydrologic Unit Code	Site Latitude	Site Longitude	Fixed-Site Network Name	Sulfate	Chloride	Suspended Sediment	Nitrite Plus Nitrate as Nitrogen	Total Phosphorus
01396660	MULHOCKAWAY CREEK AT VAN SYCKEL, NJ	02030105	40.6475	−74.96889	Unknown	X	X		X	
01399780	LAMINGTON RIVER AT BURNT MILLS, NJ	02030105	40.63472	−74.68667	Unknown	X	X		X	
01405340	MANALAPAN BROOK AT FEDERAL ROAD NEAR MANALAPAN, NJ	02030105	40.29611	−74.39778	Unknown		X			
01408500	TOMS RIVER NEAR TOMS RIVER, NJ	02040301	39.98639	−74.22333	NASQAN				X	
01409387	MULLICA RIVER AT OUTLET OF ATSION LK AT ATSION, NJ	02040301	39.74028	−74.72667	Unknown	X	X			
01409416	HAMMONTON CREEK AT WESCOATVILLE, NJ	02040301	39.63389	−74.71778	Unknown	X	X		X	
01409815	WEST BRANCH WADING RIVER AT MAXWELL, NJ	02040301	39.675	−74.54083	NASQAN	X	X		X	X
01410150	EAST BRANCH BASS RIVER NEAR NEW GRETNA, NJ	02040301	39.62306	−74.44139	Unknown	X	X			

01411110	GREAT EGG HARBOR RIVER AT WEYMOUTH, NJ	02040302	39.51389	−74.77972	Unknown	X	X		X	
01411500	MAURICE RIVER AT NORMA, NJ	02040206	39.49556	−75.07694	NASQAN	X	X		X	X
01412800	COHANSEY RIVER AT SEELEY, NJ	02040206	39.4725	−75.25556	Unknown	X	X		X	
01443500	PAULINS KILL AT BLAIRSTOWN, NJ	02040105	40.98083	−74.95333	Unknown	X	X			X
01457400	MUSCONETCONG RIVER AT RIEGELSVILLE, NJ	02040105	40.5925	−75.18611	Unknown	X	X			
01463500	DELAWARE RIVER AT TRENTON, NJ	02040105	40.22167	−74.77806	NAWQA	X	X	X	X	X
01464515	DOCTORS CREEK AT ALLENTOWN, NJ	02040201	40.17694	−74.59889	Unknown	X	X			
01466500	MCDONALDS BRANCH IN BYRNE STATE FOREST, NJ	02040202	39.885	−74.50528	HBN	X	X		X	X
01477120	RACCOON CREEK NEAR SWEDESBORO, NJ	02040202	39.74056	−75.25917	NAWQA	X	X		X	X
01482500	SALEM RIVER AT WOODSTOWN, NJ	02040206	39.64389	−75.33028	Unknown	X	X		X	
01491000	CHOPTANK RIVER NEAR GREENSBORO, MD	02060005	38.99719	−75.78581	NASQAN			X	X	X
01545600	YOUNG WOMANS CREEK NEAR RENOVO, PA	02050203	41.38951	−77.69082	HBN	X	X			

(*Continued*)

Table 2.5 US Geological Survey Site Number, Site Name, Hydrologic Unit Code, Site Latitude, Site Longitude, Fixed-Site Network Name, and Sites With Weekly, Monthly, Quarterly, or Seasonal Sampling for Sulfate, Chloride, Suspended Sediment, Total Organic Carbon, Nitrite Plus Nitrate as Nitrogen, and Total Phosphorus: 1975–2014—cont'd

						Site With Weekly, Monthly, Quarterly, or Seasonal Sampling for a Constituent From 1975–2014				
USGS Site Number	Site Name	Hydrologic Unit Code	Site Latitude	Site Longitude	Fixed-Site Network Name	Sulfate	Chloride	Suspended Sediment	Nitrite Plus Nitrate as Nitrogen	Total Phosphorus
01578310	SUSQUEHANNA RIVER AT CONOWINGO, MD	02050306	39.65791	−76.17418	NAWQA	X	X	X	X	X
01594440	PATUXENT RIVER NEAR BOWIE, MD	02060006	38.95592	−76.69369	NASQAN			X	X	X
01646580	POTOMAC RIVER AT CHAIN BRIDGE, AT WASHINGTON, DC	02070010	38.92956	−77.11692	NAWQA	X	X	X	X	X
01668000	RAPPAHANNOCK RIVER NEAR FREDERICKSBURG, VA	02080104	38.30846	−77.52915	NASQAN				X	
01673000	PAMUNKEY RIVER NEAR HANOVER, VA	02080106	37.76764	−77.3322	NASQAN				X	
02035000	JAMES RIVER AT CARTERSVILLE, VA	02080205	37.67098	−78.08583	NASQAN			X	X	
02041650	APPOMATTOX RIVER AT MATOACA, VA	02080207	37.22515	−77.47526	NASQAN				X	
02089500	NEUSE RIVER AT KINSTON, NC	03020202	35.25778	−77.58556	NAWQA	X	X	X	X	X
02091500	CONTENTNEA CREEK AT HOOKERTON, NC	03020203	35.42889	−77.5825	NAWQA	X	X	X	X	X

02175000	EDISTO RIVER NR GIVHANS, SC	03050206	33.02795	−80.39149	NAWQA	X	X	X	X	X
02177000	CHATTOOGA RIVER NEAR CLAYTON, GA	03060102	34.81398	−83.30599	Unknown				X	X
02198500	SAVANNAH RIVER NEAR CLYO, GA	03060109	32.52851	−81.26844	NASQAN				X	X
02202190	OGEECHEE RIVER AT GA 24, NEAR OLIVER, GA	03060202	32.49462	−81.55566	Unknown				X	X
02208005	YELLOW RIVER AT GA 212, NEAR STEWART, GA	03070103	33.45428	−83.88074	Unknown				X	X
02215500	OCMULGEE RIVER AT LUMBER CITY, GA	03070104	31.92018	−82.67403	Unknown				X	X
02218000	OCONEE R AT BARNETT SHOALS RD, NR WATKINSVILLE, GA	03070101	33.85595	−83.32655	Unknown				X	X
02223600	OCONEE RIVER AT I-16, NEAR DUBLIN, GA	03070102	32.48017	−82.8582	Unknown				X	X
02226010	ALTAMAHA RIVER NEAR GARDI, GA	03070106	31.62355	−81.76511	Unknown				X	X
02226582	SATILLA RIVER AT GA 15&121, NEAR HOBOKEN, GA	03070201	31.21689	−82.16234	Unknown				X	X
02301000	NORTH PRONG ALAFIA RIVER AT KEYSVILLE, FL	03100204	27.88391	−82.10009	Unknown				X	X
02301300	SOUTH PRONG ALAFIA RIVER NEAR LITHIA, FL	03100204	27.7967	−82.11759	Unknown				X	X

(Continued)

Table 2.5 US Geological Survey Site Number, Site Name, Hydrologic Unit Code, Site Latitude, Site Longitude, Fixed-Site Network Name, and Sites With Weekly, Monthly, Quarterly, or Seasonal Sampling for Sulfate, Chloride, Suspended Sediment, Total Organic Carbon, Nitrite Plus Nitrate as Nitrogen, and Total Phosphorus: 1975–2014—cont'd

						Site With Weekly, Monthly, Quarterly, or Seasonal Sampling for a Constituent From 1975–2014				
USGS Site Number	Site Name	Hydrologic Unit Code	Site Latitude	Site Longitude	Fixed-Site Network Name	Sulfate	Chloride	Suspended Sediment	Nitrite Plus Nitrate as Nitrogen	Total Phosphorus
02302500	BLACKWATER CREEK NEAR KNIGHTS, FL	03100205	28.13972	−82.15	Unknown				X	X
02303000	HILLSBOROUGH RIVER NEAR ZEPHYRHILLS, FL	03100205	28.15029	−82.23175	NASQAN					X
02314500	SUWANNEE RIVER AT US 441, AT FARGO, GA	03110201	30.68056	−82.56056	Unknown				X	X
02328200	OCHLOCKONEE RIVER NEAR CALVARY, GA	03120003	30.73158	−84.23657	Unknown				X	X
02338000	CHATTAHOOCHEE RIVER NEAR WHITESBURG, GA	03130002	33.47706	−84.90077	NAWQA				X	X
02344400	FLINT RIVER AT GA 92, ABOVE GRIFFIN, GA	03130005	33.30928	−84.39326	Unknown				X	X
02353000	FLINT RIVER AT NEWTON, GA	03130008	31.30694	−84.33889	NASQAN				X	X
02387000	CONASAUGA RIVER AT TILTON, GA	03150101	34.66675	−84.92828	Unknown				X	X
02388520	OOSTANAULA RIVER AT ROME, GA	03150103	34.26926	−85.17273	Unknown				X	X
02392000	ETOWAH RIVER AT CANTON, GA	03150104	34.23993	−84.49467	Unknown				X	X

02394980	ETOWAH RIVER AT HARDIN BRIDGE RD, NR EUHARLEE, GA	03150104	34.18898	−84.92522	Unknown				X	X
02398037	CHATTOOGA RIVER AT CHATTOOGAVILLE, GA	03150105	34.33565	−85.44524	Unknown				X	X
02469762	TOMBIGBEE R BL COFFEEVILLE L&D NEAR COFFEEVILLE	03160203	31.7571	−88.12501	NAWQA	X	X	X	X	X
03085000	MONONGAHELA RIVER AT BRADDOCK, PA	05020005	40.39118	−79.8581	NAWQA	X	X			X
03567340	WEST CHICKAMAUGA CR AT GA 146, NEAR LAKEVIEW, GA	06020001	34.9573	−85.20551	Unknown				X	X
03612500	OHIO RIVER AT DAM 53 NEAR GRAND CHAIN, IL	05140206	37.20311	−89.04174	NAWQA	X	X	X	X	X
04193500	MAUMEE RIVER AT WATERVILLE OH	04100009	41.50005	−83.71271	NAWQA	X		X		
05114000	SOURIS RIVER NR SHERWOOD, ND	09010008	48.99002	−101.9582	Unknown	X	X		X	X
05120000	SOURIS RIVER NR VERENDRYE, ND	09010003	48.15973	−100.7296	Unknown	X	X		X	X
05325000	MINNESOTA RIVER AT MANKATO, MN	07020007	44.16886	−94.00329	Unknown			X		
05420500	MISSISSIPPI RIVER AT CLINTON, IA	07080101	41.78059	−90.25207	NAWQA	X	X		X	X
05427948	PHEASANT BRANCH AT MIDDLETON, WI	07090001	43.10333	−89.51167	Unknown			X		

(Continued)

Table 2.5 US Geological Survey Site Number, Site Name, Hydrologic Unit Code, Site Latitude, Site Longitude, Fixed-Site Network Name, and Sites With Weekly, Monthly, Quarterly, or Seasonal Sampling for Sulfate, Chloride, Suspended Sediment, Total Organic Carbon, Nitrite Plus Nitrate as Nitrogen, and Total Phosphorus: 1975–2014—cont'd

						Site With Weekly, Monthly, Quarterly, or Seasonal Sampling for a Constituent From 1975–2014				
USGS Site Number	Site Name	Hydrologic Unit Code	Site Latitude	Site Longitude	Fixed-Site Network Name	Sulfate	Chloride	Suspended Sediment	Nitrite Plus Nitrate as Nitrogen	Total Phosphorus
05465500	IOWA RIVER AT WAPELLO, IA	07080209	41.17809	−91.18209	NAWQA	X	X	X	X	X
05474000	SKUNK RIVER AT AUGUSTA, IA	07080107	40.75365	−91.27709	NAWQA			X		
05586100	ILLINOIS RIVER AT VALLEY CITY, IL	07130011	39.70338	−90.6454	NAWQA	X	X	X	X	X
06089000	SUN RIVER NEAR VAUGHN MT	10030104	47.52578	−111.511	NASQAN				X	
06279500	BIGHORN RIVER AT KANE, WY	10080010	44.75857	−108.1815	NAWQA			X		
06305500	GOOSE CREEK BELOW SHERIDAN, WY	10090101	44.8215	−106.9616	Unknown				X	
06313400	SALT CREEK NEAR SUSSEX, WY	10090204	43.62192	−106.3684	Unknown	X	X			
06313500	POWDER RIVER AT SUSSEX, WY	10090202	43.69709	−106.3062	Unknown	X	X			
06317000	POWDER RIVER AT ARVADA, WY	10090202	44.64998	−106.1275	Unknown	X	X			
06324970	LITTLE POWDER RIVER AB DRY CREEK, NEAR WESTON, WY	10090208	44.92693	−105.3533	NAWQA	X	X		X	

06326500	POWDER RIVER NEAR LOCATE, MT	10090209	46.42944	−105.3103	NAWQA	X	X	X	X	X
06329500	YELLOWSTONE RIVER NEAR SIDNEY, MT	10100004	47.67741	−104.1554	NAWQA	X	X	X	X	X
06426500	BELLE FOURCHE RIVER BELOW MOORCROFT, WY	10120201	44.32193	−104.9405	NASQAN				X	
06452000	WHITE R NEAR OACOMA, SD	10140204	43.74833	−99.55649	NASQAN			X		
06486000	MISSOURI RIVER AT SIOUX CITY, IA	10230001	42.48583	−96.41392	NASQAN			X		
06734900	OLYMPUS TUNNEL AT LAKE ESTES, CO	10190006	40.37376	−105.4737	Unknown	X	X		X	X
06752260	CACHE LA POUDRE RIVER AT FORT COLLINS, CO	10190007	40.58808	−105.0692	Unknown				X	
06803525	SALT CR BL STEVENS CR NR WAVERLY NE	10200203	40.905	−96.58612	Unknown	X	X		X	X
06805500	PLATTE RIVER AT LOUISVILLE, NE	10200202	41.01528	−96.15778	NAWQA	X	X	X	X	X
06818000	MISSOURI RIVER AT ST. JOSEPH, MO	10240011	39.75325	−94.85683	NASQAN	X	X	X	X	X
06902000	GRAND RIVER NEAR SUMNER, MO	10280103	39.64003	−93.27369	NASQAN	X	X		X	X
06918070	OSAGE RIVER ABOVE SCHELL CITY, MO	10290105	38.05586	−94.14542	NASQAN				X	X
06926510	OSAGE RIVER BELOW ST. THOMAS, MO	10290111	38.42144	−92.20825	NASQAN				X	X

(*Continued*)

Table 2.5 US Geological Survey Site Number, Site Name, Hydrologic Unit Code, Site Latitude, Site Longitude, Fixed-Site Network Name, and Sites With Weekly, Monthly, Quarterly, or Seasonal Sampling for Sulfate, Chloride, Suspended Sediment, Total Organic Carbon, Nitrite Plus Nitrate as Nitrogen, and Total Phosphorus: 1975–2014—cont'd

						Site With Weekly, Monthly, Quarterly, or Seasonal Sampling for a Constituent From 1975–2014				
USGS Site Number	Site Name	Hydrologic Unit Code	Site Latitude	Site Longitude	Fixed-Site Network Name	Sulfate	Chloride	Suspended Sediment	Nitrite Plus Nitrate as Nitrogen	Total Phosphorus
06930450	BIG PINEY RIVER AT DEVILS ELBOW, MO	10290202	37.8481	−92.06239	Unknown				X	X
06930800	GASCONADE RIVER ABOVE JEROME, MO	10290203	37.92004	−91.976	NASQAN	X	X		X	X
06934500	MISSOURI RIVER AT HERMANN, MO	10300200	38.70981	−91.4385	NAWQA	X	X	X	X	X
07014500	MERAMEC RIVER NEAR SULLIVAN, MO	07140102	38.15853	−91.10847	Unknown	X	X		X	X
07022000	MISSISSIPPI RIVER AT THEBES, IL	07140105	37.2216	−89.46298	NAWQA	X	X		X	X
07040000	ST. FRANCIS RIVER AT FISK, MO	08020203	36.78061	−90.20288	Unknown			X		
07040070	WILHELMINA CUT-OFF NEAR CAMPBELL, MO	08020203	36.51472	−90.15722	Unknown			X		
07040100	ST. FRANCIS RIVER AT ST. FRANCIS, AR	08020203	36.45556	−90.13778	Unknown			X		
07040110	ST. FRANCIS RIVER NEAR PIGGOTT, AR	08020203	36.39694	−90.07833	Unknown			X		
07040130	ST.FRANCIS RIVER @ HOLLY ISLAND, AR	08020203	36.23639	−90.13028	Unknown			X		

07040450	ST. FRANCIS RIVER AT LAKE CITY, AR	08020203	35.82083	−90.4325	Unknown			X		
07046600	RIGHT HAND CHUTE OF LITTLE RIVER @ RIVERVALE, AR	08020204	35.67194	−90.33806	Unknown			X		
07060500	WHITE RIVER AT CALICO ROCK, AR	11010004	36.11667	−92.14306	Unknown				X	
07060710	NORTH SYLAMORE CREEK NEAR 56, AR	11010004	35.99167	−92.21389	NAWQA	X	X	X	X	X
07103700	FOUNTAIN CREEK NEAR COLORADO Springs, CO.	11020003	38.85471	−104.878	Unknown				X	
07105500	FOUNTAIN CREEK AT COLORADO Springs, CO	11020003	38.81638	−104.8228	Unknown				X	
07105530	FOUNTAIN CR BLW JANITELL RD BLW COLO. Springs, CO	11020003	38.80305	−104.7958	Unknown				X	
07186480	CENTER CREEK NEAR SMITHFIELD, MO	11070207	37.15561	−94.60301	NAWQA				X	X
07194800	ILLINOIS RIVER AT SAVOY, AR	11110103	36.10306	−94.34444	Unknown	X	X			X
07196900	BARON FORK AT DUTCH MILLS, AR	11110103	35.88	−94.48639	Unknown	X	X			X
07227500	CANADIAN RV NR AMARILLO, TX	11090105	35.47033	−101.8796	Unknown	X	X		X	
07241550	NORTH CANADIAN RIVER NEAR HARRAH, OK	11100302	35.50034	−97.19392	NAWQA	X	X		X	

(Continued)

Table 2.5 US Geological Survey Site Number, Site Name, Hydrologic Unit Code, Site Latitude, Site Longitude, Fixed-Site Network Name, and Sites With Weekly, Monthly, Quarterly, or Seasonal Sampling for Sulfate, Chloride, Suspended Sediment, Total Organic Carbon, Nitrite Plus Nitrate as Nitrogen, and Total Phosphorus: 1975–2014—cont'd

						Site With Weekly, Monthly, Quarterly, or Seasonal Sampling for a Constituent From 1975–2014				
USGS Site Number	Site Name	Hydrologic Unit Code	Site Latitude	Site Longitude	Fixed-Site Network Name	Sulfate	Chloride	Suspended Sediment	Nitrite Plus Nitrate as Nitrogen	Total Phosphorus
07249400	JAMES FORK NEAR HACKETT, AR	11110105	35.1625	−94.40694	Unknown	X	X			X
07250550	AR RIVER AT JAMES W. TRIMBLE L&D NR VAN BUREN, AR	11110104	35.34889	−94.29833	NASQAN	X	X	X	X	X
07263620	AR RIVER @ DAVID D TERRY L&D BELOW LITTLE ROCK, AR	11110207	34.68111	−92.15139	NAWQA	X	X	X	X	X
07295100	MISSISSIPPI RIVER AT TARBERT LANDING, MS	08060100	31.00851	−91.62373	Unknown			X		
07362000	OUACHITA RIVER AT CAMDEN, AR	08040102	33.59639	−92.81806	NASQAN	X	X	X	X	X
07373420	MISSISSIPPI R NR ST. FRANCISVILLE, LA	08070100	30.75852	−91.39595	NAWQA	X	X	X	X	X
07381490	ATCHAFALAYA RIVER AT SIMMESPORT, LA	08080101	30.9825	−91.79833	NASQAN			X		
07381495	(COE) ATCHAFALAYA RIVER AT MELVILLE, LA	08080101	30.69074	−91.73623	NAWQA	X	X		X	X
07381590	WAX LAKE OUTLET AT CALUMET, LA	08080101	29.69799	−91.37289	NAWQA			X		

07381600	LOWER ATCHAFALAYA RIVER AT MORGAN CITY, LA	08080101	29.69282	−91.21194	NAWQA			X		
08028500	SABINE RV NR BON WIER, TX	12010005	30.74715	−93.60851	Unknown	X	X			
08033500	NECHES RV NR ROCKLAND, TX	12020003	31.025	−94.39944	Unknown	X	X		X	
08049500	W FK TRINITY RV AT GRAND PRAIRIE, TX	12030102	32.7625	−96.99444	Unknown	X	X		X	
08057410	TRINITY RV BL DALLAS, TX	12030105	32.70763	−96.73583	NAWQA	X	X		X	X
08062500	TRINITY RV NR ROSSER, TX	12030105	32.42653	−96.46304	Unknown	X	X		X	
08065350	TRINITY RV NR CROCKETT, TX	12030201	31.33851	−95.65634	NASQAN	X	X		X	
08251500	RIO GRANDE NEAR LOBATOS, CO	13010002	37.07861	−105.7569	NAWQA	X	X		X	X
08313000	RIO GRANDE AT OTOWI BRIDGE, NM	13020101	35.8745	−106.1424	NAWQA			X		
08319000	RIO GRANDE AT SAN FELIPE, NM	13020201	35.44458	−106.4398	Unknown			X		
08330000	RIO GRANDE AT ALBUQUERQUE, NM	13020203	35.08917	−106.6807	Unknown			X		
08332010	RIO GRANDE FLOOD, WAY NEAR BERNARDO, NM	13020203	34.41701	−106.8006	Unknown			X		
08353000	RIO PUERCO NEAR BERNARDO, NM	13020204	34.41028	−106.8544	NAWQA			X		

(Continued)

Table 2.5 US Geological Survey Site Number, Site Name, Hydrologic Unit Code, Site Latitude, Site Longitude, Fixed-Site Network Name, and Sites With Weekly, Monthly, Quarterly, or Seasonal Sampling for Sulfate, Chloride, Suspended Sediment, Total Organic Carbon, Nitrite Plus Nitrate as Nitrogen, and Total Phosphorus: 1975–2014—cont'd

						Site With Weekly, Monthly, Quarterly, or Seasonal Sampling for a Constituent From 1975–2014				
USGS Site Number	Site Name	Hydrologic Unit Code	Site Latitude	Site Longitude	Fixed-Site Network Name	Sulfate	Chloride	Suspended Sediment	Nitrite Plus Nitrate as Nitrogen	Total Phosphorus
08354900	RIO GRANDEFLOODWAY AT SAN ACACIA, NM	13020203	34.25639	−106.8908	Unknown			X		
08358400	RIO GRANDEFLOODWAY AT SAN MARCIAL, NM	13020203	33.67908	−106.997	NAWQA			X		
08364000	RIO GRANDE AT EL PASO, TX	13030102	31.80288	−106.5408	NAWQA	X	X	X	X	X
08475000	RIO GRANDE NR BROWNSVILLE, TX	13090002	25.87675	−97.45443	NAWQA	X	X	X	X	X
09013000	ALVA B ADAMS TUN AT E PORTAL, NR ESTES PARK, CO.	10190006	40.32776	−105.5781	Unknown	X	X		X	X
09163500	COLORADO RIVER NEAR COLORADO-UTAH STATE LINE	14010005	39.13276	−109.0271	NAWQA			X		
09180500	COLORADO RIVER NEAR CISCO, UT	14030005	38.81054	−109.2934	NASQAN	X	X			
09211200	GREEN RIVER BELOW FONTENELLE RESERVOIR, WY	14040103	42.02097	−110.0498	Unknown				X	X

09217000	GREEN RIVER NEAR GREEN RIVER, WY	14040106	41.51635	−109.449	NASQAN	X	X			
09217010	GREEN RIVER BELOW GREEN RIVER, WY	14040106	41.49608	−109.4387	Unknown				X	X
09224050	HAMS FORK NEAR DIAMONDVILLE, WY	14040107	41.75161	−110.5332	Unknown				X	
09251000	YAMPA RIVER NEAR MAYBELL, CO	14050002	40.50275	−108.0334	NASQAN	X	X			
09306200	PICEANCE CREEK BL RYAN GULCH, NR RIO BLANCO, CO.	14050006	39.92108	−108.2976	Unknown	X	X		X	
09315000	GREEN RIVER AT GREEN RIVER, UT	14060008	38.98608	−110.1512	NASQAN	X	X			
09380000	COLORADO RIVER AT LEES FERRY, AZ	14070006	36.86471	−111.5882	NAWQA	X	X	X	X	X
09415000	VIRGIN RV AT LITTLEFIELD, AZ	15010010	36.89164	−113.9244	NASQAN	X	X		X	
09421500	COLORADO RV BLW HOOVER DAM, AZ −NV	15030101	36.01526	−114.7386	NASQAN	X	X		X	X
09427520	COLORADO RIVER BELOW PARKER DAM, AZ−CA	15030104	34.29557	−114.1402	Unknown	X	X		X	X
09429490	COLORADO RIVER ABOVE IMPERIAL DAM, AZ−CA	15030104	32.88365	−114.4674	NASQAN	X	X		X	X
09448500	GILA RIVER AT HEAD OF SAFFORD VALLEY, NR SOLOMON,	15040005	32.8684	−109.5112	Unknown	X	X			X

(Continued)

Table 2.5 US Geological Survey Site Number, Site Name, Hydrologic Unit Code, Site Latitude, Site Longitude, Fixed-Site Network Name, and Sites With Weekly, Monthly, Quarterly, or Seasonal Sampling for Sulfate, Chloride, Suspended Sediment, Total Organic Carbon, Nitrite Plus Nitrate as Nitrogen, and Total Phosphorus: 1975–2014—cont'd

						Site With Weekly, Monthly, Quarterly, or Seasonal Sampling for a Constituent From 1975–2014				
USGS Site Number	Site Name	Hydrologic Unit Code	Site Latitude	Site Longitude	Fixed-Site Network Name	Sulfate	Chloride	Suspended Sediment	Nitrite Plus Nitrate as Nitrogen	Total Phosphorus
09498500	SALT RIVER NEAR ROOSEVELT, AZ	15060103	33.61949	−110.9215	NAWQA	X	X		X	X
09504000	VERDE RIVER NEAR CLARKDALE, AZ	15060202	34.85224	−112.066	Unknown	X	X		X	X
09522000	COLORADO RIVER AT NIB, ABOVE MORELOS DAM, AZ	15030108	32.71866	−114.7188	NAWQA	X	X		X	X
10172200	RED BUTTE CREEK AT FORT DOUGLAS, NEAR SLC, UT	16020204	40.77995	−111.806	NAWQA	X	X	X	X	X
11074000	SANTA ANA R BL PRADO DAM CA	18070203	33.88335	−117.6453	NAWQA	X	X	X	X	X
11152300	SALINAS R NR CHUALAR CA	18060005	36.55357	−121.5494	NASQAN	X	X		X	X
11264500	MERCED R A HAPPY ISLES BRIDGE NR YOSEMITE CA	18040008	37.73131	−119.5591	NAWQA	X	X			
11303500	SAN JOAQUIN R NR VERNALIS CA	18040002	37.67604	−121.2663	NAWQA	X	X		X	X
11447650	SACRAMENTO R A FREEPORT CA	18020109	38.45602	−121.5013	NAWQA	X	X	X	X	X

13154500	SNAKE RIVER AT KING HILL ID	17050101	43.00222	−115.2025	NAWQA	X	X	X	X	X
13213000	BOISE RIVER NR PARMA ID	17050114	43.78167	−116.9728	NASQAN				X	X
14211720	WILLAMETTE RIVER AT PORTLAND, OR	17090012	45.5175	−122.6692	NAWQA	X	X	X	X	X
310355091411500	OLD RIVER OUTFLOW CHANNEL NR KNOX LANDING	08040301	31.06518	−91.68567	Unknown			X		
310552091361200	MISSISSIPPI RIVER AT COOCHIE, LA	08060100	31.09795	−91.60345	Unknown			X		
50010500	RIO GUAJATACA AT LARES, PR	21010002	18.29721	−66.87348	Unknown				X	X
50011000	CANAL DE DERIVACION AT LAGO DE GUAJATACA, PR	21010002	18.39752	−66.92713	Unknown				X	X
50011400	RIO GUAJATA, CA ABV MOUTH NR QUEBRADILLAS, PR	21010002	18.47328	−66.9624	Unknown				X	X
50027250	RIO GDE. DE ARECIBO BLW LAGO DOS BOCAS NR FLORIDA	21010002	18.34523	−66.66684	Unknown				X	X
50028000	RIO TANAMA NR UTUADO, PR	21010002	18.29873	−66.78263	Unknown				X	X
50029000	RIO GRANDE DE ARECIBO AT CENTRAL CAMBALACHE, PR	21010002	18.45353	−66.70207	Unknown				X	X

(Continued)

Table 2.5 US Geological Survey Site Number, Site Name, Hydrologic Unit Code, Site Latitude, Site Longitude, Fixed-Site Network Name, and Sites With Weekly, Monthly, Quarterly, or Seasonal Sampling for Sulfate, Chloride, Suspended Sediment, Total Organic Carbon, Nitrite Plus Nitrate as Nitrogen, and Total Phosphorus: 1975–2014—cont'd

						Site With Weekly, Monthly, Quarterly, or Seasonal Sampling for a Constituent From 1975–2014				
USGS Site Number	Site Name	Hydrologic Unit Code	Site Latitude	Site Longitude	Fixed-Site Network Name	Sulfate	Chloride	Suspended Sediment	Nitrite Plus Nitrate as Nitrogen	Total Phosphorus
50035950	RIO CIALITOS AT HWY 649 AT CIALES, PR	21010002	18.33634	−66.47406	Unknown				X	X
50038100	RIO GRANDE DE MANATI AT HWY 2 NR MANATI, PR	21010002	18.42897	−66.52647	NASQAN				X	X
50039500	RIO CIBUCO AT VEGA BAJA, PR	21010002	18.44533	−66.37399	Unknown				X	X
50043000	RIO DE LA PLATA AT PROYECTO LA PLATA, PR	21010005	18.15801	−66.2285	Unknown				X	X
50046000	RIO DE LA PLATA AT HWY 2 NR TOA ALTA, PR	21010005	18.40959	−66.26057	NASQAN				X	X
50047600	RIO DE BAYAMON NR AGUAS BUENAS, PR	21010005	18.24218	−66.14378	Unknown				X	X
50048510	RIO DE BAYAMON AT, FLOOD CHANNEL AT BAYAMON, PR	21010005	18.40884	−66.15211	Unknown				X	X
50048800	RIO PIEDRAS NR RIO PIEDRAS, PR	21010005	18.36884	−66.06072	Unknown				X	X
50049100	RIO PIEDRAS AT HATO REY, PR	21010005	18.40755	−66.06894	Unknown				X	X

50050300	QUEBRADA BLASINA NR CAROLINA, PR	21010005	18.38884	−65.97405	Unknown	X	X
50055000	RIO GRANDE DE LOIZA AT CAGUAS, PR	21010005	18.2407	−66.0092	Unknown	X	X
50055250	RIO CAGUITAS AT HWY 30 AT CAGUAS, PR	21010005	18.25107	−66.0235	Unknown	X	X
50071000	RIO FAJARDO NR FAJARDO, PR	21010005	18.29696	−65.69343	NASQAN	X	X
50082000	RIO HUMACAO AT HWY 3 AT HUMACAO, PR	21010005	18.14496	−65.82655	Unknown	X	X
50091000	RIO MAUNABO AT MAUNABO, PR	21010005	18.00469	−65.90488	Unknown	X	X
50092000	RIO GRANDE DE PATILLAS NR PATILLAS, PR	21010004	18.03219	−66.03211	NASQAN	X	X
50114000	RIO CERRILLOS BLW LAGO CERRILLOS NR PONCE, PR	21010004	18.07087	−66.58101	NASQAN	X	X
50115000	RIO PORTUGUES NR PONCE, PR	21010004	18.07718	−66.63323	Unknown	X	X
50124700	RIO GUAYANILLA AT CENTRAL RUFINA, PR	21010004	18.01191	−66.78406	Unknown	X	X
50138000	RIO GUANAJIBO NR HORMIGUEROS, PR	21010003	18.14069	−67.14837	Unknown	X	X
50143000	RIO GRANDE DE ANASCO NR LARES, PR	21010003	18.25579	−66.91768	Unknown	X	X
50144000	RIO GRANDE DE ANASCO NR SAN SEBASTIAN, PR	21010003	18.28217	−67.05058	NASQAN	X	X

HBN, Hydrologic Benchmark Network; *NASQAN*, National Stream Quality Accounting Network; *NAWQA*, National Water Quality Assessment Program; *USGS*, US Geological Survey.

once or twice per year for the selected constituents (Table 2.2). The specific objectives of projects collecting one or two samples per year were not assessed in this study. The value of a single sample for evaluating a water-quality or drinking water standard is limited to those situations in which a single-sample is meaningful. For example, a single result for nitrate at a concentration greater than 10 mg/L is relevant in a source of drinking water. A single-sample exceedance of the maximum concentration level (MCL) for nitrate in drinking water can result in the issuance of an advisory by the water provider to consumers to avoid drinking the water until concentrations are at or below the MCL. An exceedance from a single sample could also result in the need for additional steps by the water provider to reduce nitrate concentrations in finished water. Single samples for constituent analysis may also be collected in synoptic studies at the time of collection of aquatic biota. Probabilistic designs such as those used by the EPA and states may rely on a single sample collected with the sampling of aquatic biota during a specific sampling period called an index period. There may be low-flow networks where samples are collected infrequently but targeted to certain hydrologic conditions.

Lessons learned from the past demonstrate the large variability in sample concentrations in rivers, the importance of a streamflow value with every water-quality sample, as well as seasonal and hydrologically based sampling to characterize water-quality conditions and trends. As stated many years ago, "The water in a stream varies so much that an opinion based on the results of an examination of a single sample of water from a given river would nearly always be very different from an opinion based on the results of a series of analyses, where the samples were taken regularly for some time" [151]. The reliance on annual or semiannual monitoring approaches deserves further investigation given the preponderance of sites monitored over time with annual or semiannual sampling frequencies.

The growing cost of monitoring over time coupled with flat or declining budgets, competition for resources with other types of monitoring, and increasing operational costs have contributed to the decline in fixed-site monitoring. However, long-term trends require long periods of consistent monitoring because there does not seem to be a substitute in river systems for time and length of record. A well-designed and perennial network can answer questions regarding whether the nation's water quality is getting better or worse and why. A recent report examined the type of monitoring needed to detect improvements in agricultural best management practices to reduce phosphorus delivery to Lake Erie [152]. The study found that decadal monitoring, at a minimum, as well as ancillary data on causative and correlative factors are needed to detect major changes in these river systems and inform policymakers with insights that provide improved water-quality information for decision making [152].

CONCLUSIONS

The history of water-quality monitoring and assessment is one of continual innovation and modification of networks and designs to accommodate changing policy questions on status and trends. Long-term monitoring networks were successful from the start in evaluating water-quality standards, in providing warning signals of emerging water-quality problems, and for measurement of water-quality status and trends. Environmental laws and regulations drive many decisions on why, where, what, how, and when to monitor. A large emphasis has been placed on how to improve monitoring for the Clean Water Act as indicated by several GAO studies that addressed this topic. A combination of approaches including fixed-site monitoring, synoptic or special short-term surveys, and models can answer many questions on Clean Water Act programs and policies and their benefits. It can take a decade or more of monitoring to be able to measure these improvements due to the great variability of water quality in rivers and streams.

No one monitoring design can answer all questions. In fact, a combination of designs to include fixed-site monitoring networks, short-term synoptic and topical studies, and probabilistic approaches all have their place and have proven successful for different objectives. However, for long-term trends, fixed-site monitoring is the most relied upon approach to date and is sensitive to change given the richness of the data collected geographically, temporally, and with the support of information on factors and influences at the landscape and watershed scales. Efforts to sustain monitoring networks are important because these networks have collected information crucial to the description of water-quality trends over time and are providing information against which to evaluate future trends.

ACKNOWLEDGMENTS

The authors wish to acknowledge the assistance of Mike Canova, Dave Dupre', Melanie Clark, and Dave Wolock of the US Geological Survey for their assistance with the National Water Information System aggregated database. The authors wish to thank USGS reviewers, Denise Argue, Jeff Deacon, Bob Hirsch, Gary Rowe, and Lori Sprague, who contributed helpful suggestions for improvement of the draft manuscript.

REFERENCES

[1] Myers DN. Foundations of water quality monitoring and assessment in the United States. In: Ahuja S, editor. Food, energy and water–the chemistry connection. Elsevier; 2015. p. 21–79.

[2] Helmer R. Water quality monitoring: national and international approaches. In: Proceedings of the Rostov-on-Don symposium, vol. 219. International Association of Hydrological Sciences; 1994. p. 3–17.

[3] Ward RC, Loftis JC, McBride GB. Design of water quality monitoring systems. New York (NY): Van Nostrand Reinhold; 1990. 10 pp.

[4] Leopold LB. A National Network of Hydrologic bench Marks. U.S. Geological Survey; 1962. Circular 460-B, 4 pp.

[5] Dupre' DH, Scott JD, Clark ML, Canova MG, Stoker YE. User's Manual for the National Water Information System of the U.S. Geological Survey: Water-Quality System. Version 5.0. U.S. Geological Survey; 2013. Open-File Report 2013-1054, 730 pp.

[6] U.S. Geological Survey Office of Water Quality. Technical memorandum 93.04: discontinuation of the National Water Quality Laboratory determinations for "total" nitrite, "total" nitrite plus nitrate, "total" ammonia, and "total" orthophosphate. U.S. Geological Survey; 1992. Accessed at: http://water.usgs.gov/admin/memo/QW/qw93.04.html.

[7] Patton CJ, Truitt EP. U.S. Geological Survey Nutrient Preservation experiment: Nutrient Concentration Data for Surface-, Ground-, and Municipal-Supply Water Samples and Quality-Assurance Samples. U.S. Geological Survey; 1995. Open-File Report 95-141, 140 pp.

[8] Durum WH, Haffty J. Implications of the minor element content of some major streams of the world. Geochem Comochimica Acta 1963;27:1–11.

[9] National Resources Committee. Report of the special advisory committee on standards and specifications for hydrologic data, deficiencies in basic hydrologic data. Washington (DC): Water Resources Committee of the National Resources Committee; 1936. 72 pp.

[10] Durum WH. Historical profiles of quality of water laboratories and activities, 1879–1973. U.S. Geological Survey; 1978. Open-File Report 78-432, 235 pp.

[11] Knopman DS, Smith RA. 20 Years of the Clean Water Act. Environment 1993;35:16–41.

[12] Cleary EJ. Perspective on river-quality diagnosis. J Water Pollut Control Fed 1978;50:825–32.

[13] Andreen WL. The evolution of water pollution control in the United States—State, local, and federal efforts, 1789–1972: Part I. Stanf Environ Law J 2003;22:145–200.

[14] Andreen WL. The evolution of water pollution control in the United States—State, local, and federal efforts, 1789–1972: Part II. Stanf Environ Law J 2003;22:215–94.

[15] Everts Jr CM, Dahl AH. The Federal Water Pollution Control Act of 1956. Am J Public Health Nations Health 1957;47(3):305–10.

[16] Cherry RN. Chemical quality of water of Georgia streams, 1957–1958: a reconnaissance study. Atlanta (GA): Georgia Geological Survey; 1961. Bulletin 60, 100 pp.

[17] Anderson PW, George JR. Water-quality characteristics of New Jersey streams. U.S. Geological Survey; 1966. Professional Paper 1819-G, 1 Pl, 48 pp.

[18] U.S. Public Health Service. In: Water pollution surveillance system annual compilation of data, vol. 9. Washington (DC): U.S. Department of Health Education and Welfare; 1963. C1–17.

[19] U.S. Environmental Protection Agency. Technical report on historical uses of RadNet data. Washington (DC): U.S. Environmental Protection Agency; 2008. EPA-402-R-08–007; 31 pp.

[20] Cobb ED, Beisecker JE. The National Hydrologic Benchmark Network. U.S. Geological Survey; 1971. Circular 460-D, 38 pp.
[21] U.S. Geological Survey. Goals of the U.S. Geological Survey. U.S. Geological Survey; 1986. Circular 1010; 17 pp.
[22] Taylor JK. Quality assurance of chemical measurements. Chelsea (MI): Lewis Publishers; 1987. p. 2.
[23] Friedman LC, Fishman MJ. Evaluation of methods used from 1965 through 1982 to determine inorganic constituents in water samples. U.S. Geological Survey; 1989. Water Supply Paper 2293, 126 pp.
[24] Ludtke AS, Woodworth MT, Marsh PS. Quality-assurance results for routine water analyses in U.S. Geological Survey Laboratories, water year 1998. U.S. Geological Survey; 2000. Water Resources Investigations Report 2000-4176, 198 pp.
[25] Stanley DL, Boozer TM, Schroder LJ. Summary of the U.S. Geological Survey National Field Quality Assurance Program from 1979 through 1997. U.S. Geological Survey; 1998. Open-File Report 98-392, 11 pp.
[26] Rainwater FH, Thatcher LL. Methods for collection and analysis of water samples. U.S. Geological Survey; 1960. Water Supply Paper 1454, 301 pp.
[27] Skougstad MW, Fishman MJ, Friedman LC, Erdmann DE, Duncan SS. Methods for determination of inorganic substances in water and fluvial sediments. Techniques of Water Resources Investigations of the U.S. Geological Survey; 1979. Book 5, Chap A1, Laboratory Analysis, 626 pp.
[28] U.S. Geological Survey. Collection of water samples (ver. 2.0). U.S. Geological Survey; 2006. Techniques of Water-Resources Investigations, Book 9, Chap. A4, variously paginated.
[29] Guy HP. Laboratory theory and methods for sediment analysis. U.S. Geological Survey; 1969. Techniques of Water Resources Investigations, Book 5, Chap. C1, 58 pp.
[30] Gordon JD, Newland C. U.S. Geological Survey Quality-Assurance Project for Sediment Analysis. U.S. Geological Survey; 2000. Fact Sheet 031-00, 4 pp.
[31] Erdmann DE. USGS Laboratory Review Program ensures analytical quality. U.S. Geological Survey; 1995. Fact Sheet 187-95, 1 pp.
[32] Janzer VJ, Saindon LG. Radiochemical analyses of surface water from US Geological Survey Hydrologic bench-mark stations. U.S. Geological Survey; 1972. Open-File Report 72-191, 41 pp.
[33] U.S. Environmental Protection Agency. National water quality inventory: 1982 report to Congress. Washington (DC): U.S. Environmental Protection Agency; 1984. 440/2-84-006, 191 pp.
[34] Federal Water Pollution Control Administration. Water quality criteria; report of the National Technical Advisory Committee to the Secretary of the interior. Washington (DC): Federal Water Pollution Control Administration; 1968. 234 pp.
[35] U.S. Environmental Protection Agency. Basic water monitoring program. Washington (DC): U.S. Environmental Protection Agency; 1976. EPA 440/9-76-025, 51 pp.
[36] State of New Jersey. New Jersey Integrated Water Quality Assessment Report. Trenton (NJ): Department of Environmental Protection; 2014. 7 p.
[37] State of New Jersey. Ambient Surface Water Quality Monitoring Network (ASWQMN). 2016. Accessed at: http://www.state.nj.us/dep/wms//bfbm/surfacewater.html. 81616.
[38] Green RS. The storage and retrieval of data for water quality control. U.S. Public Health Service; 1964. 36 pp.
[39] Hutchison NE. WATSTORE: National Water Data Storage and Retrieval System of the U. S. Geological Survey; user's guide. U.S. Geological Survey; 1975. Open File Report 1975-426, 956 pp.

[40] Wolman MG. The nation's rivers: problems are encountered when appraising trends in water and river quality. Science 1971;174:905–18.
[41] Enviro Control, Inc. National assessment of trends in water quality. U.S. Department of Commerce, National Technical Information Service. 1972. PB-210, 69 pp.
[42] Steele TD, Gilroy EJ, Hawkinson RO. An assessment of areal and temporal variations in streamflow quality using selected data from the National Stream Quality Accounting Network. U.S. Geological Survey; 1974. Open-File Report 74-217, 210 pp.
[43] Aspelin AL. Pesticide usage in the United States: trends during the 20th century. Center for Integrated Pest Management. Raleigh (NC): North Caroline State University; 2003. 10 pp.
[44] Carson R. Silent spring. New York (NY): Houghton Mifflin Harcourt; 1962. 378 pp.
[45] Breidenbach AW, Gunnerson CG, Kawahara FK, Lichtenberg JJ, Green RS. Chlorinated hydrocarbon pesticides in major river basins, 1957–65. Public Health Rep 1967;82(2):139–56.
[46] Brown E, Nishioka YA. Pesticides in selected western streams—a contribution to the national program. Pestic Monit J 1967;1(2):38–46.
[47] Jacknow J, Ludke JL, Coon NC. Monitoring fish and wildlife for contaminants: National Contaminant Biomonitoring Program. Washington (DC): U.S. Department of Interior, Fish and Wildlife Service; 1986. Leaflet 4, 18 pp.
[48] Feltz HR, Sayers WT, Nicholson HP. National monitoring program for the assessment of pesticide residues in water. Pestic Monit J 1971;5(1):57–62.
[49] Greene RS, Love SK. Network to monitor hydrologic environment covers major drainage rivers. Pestic Monit J 1967;1(1):13–6.
[50] Lichtenberg JJ, Eichelberger JW, Dressman RC, Longbottom JE. Pesticides in surface water of the United States—a five year summary, 1964–1968. Cincinnati (OH): U.S. Department of Interior, Federal Water Pollution Control Administration; 1969. 37 pp.
[51] Crim RL, Murray TL. Water monitoring: a dynamic process. Environ Int 1978;1: 183–94.
[52] Biesecker JE, Leifeste DK. Water quality of hydrologic benchmarks; an indicator of water quality in the natural environment. U.S. Geological Survey; 1975. Circular: 460-E, 21 pp.
[53] Gilliom RJ, Barbash JE, Crawford CG, Hamilton PA, Martin JD, Nakagaki N, Nowell LH, Scott JC, Stackelberg PE, Thelin GP, Wolock DM. The quality of our nation's waters—pesticides in the nation's streams and ground water; 1992–2001. U.S. Geological Survey; 2006. Circular 1291, 172 pp.
[54] Gilliom RJ, Alexander RB, Smith RA. Pesticides in the nation's rivers, 1975–1980, and implications for future monitoring. U.S. Geological Survey; 1985. p. 1–11. Water Supply Paper, 2271.
[55] Henderson C, Johnson WL, Inglis A. Organochlorine insecticide residues in fish—national pesticide monitoring program. Pestic Monit J 1969;3(3):145–71.
[56] Henderson C, Inglis A, Johnson WL. Organochlorine insecticide residues in fish—fall 1969 National Pesticide Monitoring Program. Pestic Monit J 1971;5(1):1–11.
[57] Schmitt CJ, Zajicek JL, Ribick MA. National pesticide monitoring program: residues of organochlorine chemicals in freshwater fish, 1980-81. Arch Environ Contam Toxicol 1985;14:225–60.
[58] Schmitt CJ, Bunck CM. Persistent environmental contaminants in fish and wildlife in our living resources: a report to the nation on the distribution, abundance, and health of U.S. plants, animals, and ecosystems. In: LaRoe ET, editor. Washington (DC): U.S. Department of the Interior, National Biological Service; 1995. http://dx.doi.org/10.5962/bhl.title.4172.

[59] Bauch NJ, Schmitt CJ, Crawford CG. Development of an approach for integrating components of the U.S. Geological Survey Biomonitoring of Environmental Status and Trends (BEST) and National Water Quality Accounting Network (NASQAN) Programs for Large Rivers. U.S. Geological Survey Scientific Investigation Report 2005-5083. 2005. 53 pp.

[60] Copeland C. Clean Water Act: A Summary of the Law. Congressional research service report. RL30030. 2010. p. 7−5700. Washington (DC), 11 pp.

[61] Blacker SM, Burton JS. Environmental Protection Agency's Monitoring Programs. Washington (DC): U.S. Environmental Protection Agency; 1973. 34 pp.

[62] State of Connecticut. Ambient monitoring strategy for rivers and streams and rotating basin approach. Hartford: Conn.; 1999. 23 pp.

[63] Smith RA, Alexander RB, Wolman MG. Water-quality trends in the nation's rivers. Science 1987;235:1607−15.

[64] Ficke JF, Hawkinson RO. The National Stream Quality Accounting Network (NASQAN) − some questions and answers. U.S. Geological Survey; 1975. Circular 719, 23 pp.

[65] van Belle G, Hughes JP. Monitoring for water quality: fixed stations versus intensive surveys. J Water Pollut Control Fed 1983;55(4):400−4.

[66] Hren J, Chaney TH, Norris JM. Childress, CJO. Water-quality data-collection activities in Colorado and Ohio: Phase I Inventory and Evaluation of 1984 Programs and Costs. U.S. Geological Survey; 1987. Water Supply Paper 2295, 67 pp.

[67] Ward RC. Regulatory water-quality monitoring—a systems perspective. Las Vegas (NV): U.S. Environmental Protection Agency Office of Research and Development; 1978. EPA-600/7-78-228; 23 pp.

[68] Rickert DA, Hines WG. A practical framework for river-quality assessment. U.S. Geological Survey; 1975. Circular 715-A, 17 pp.

[69] Ward RC. Water quality monitoring: where's the beef? Water Resour Bull 1996; 32(4):673−80.

[70] U.S. General Accounting Office. Better monitoring techniques are needed to assess the quality of rivers and streams−volumes I and II. 1981. Report CED4HI40, Gaithersburg, Md., [variously paginated].

[71] Loftis JC, Ward RC. Routine, fixed-station sampling−an integral part of regulatory water quality monitoring? Environ Manage 1982;6(4):279−80.

[72] Schaeffer DJ. Effective data use in regulatory water quality monitoring. Environ Manage 1982;6(4):280−2.

[73] U.S. General Accounting Office. The nation's water—key unanswered questions about the quality of rivers and streams. 1986. Report PEMD 86−6; Washington (DC), 163 pp.

[74] Judy RD, Seeley PN, Murray TM, Svirsky SC, Whitworth MR, Ischinger LS. National Fisheries Survey. In: Technical report: initial findings, vol. 1. U.S. Fish and Wildlife Service; 1984. FWS/OBS-84/06, 140 pp.

[75] U.S. Geological Survey National water summary. Hydrologic events, selected water-quality trends, and ground-water resources. U.S. Geological Survey; 1984. Water Supply Paper 2275; 467 pp.

[76] Association of State and Interstate Water Pollution Control Administrators. America's clean water: the States' evaluation of progress 1972−1982, vol. 2. Washington (DC): Association of State and Interstate Pollution Control Administrators; 1984.

[77] Hooper RP, Goolsby DA, Rickert DA, McKenzie SW. NASQAN−A program to monitor the water quality of the nation's largest rivers. U.S. Geological Survey; 1997. Fact Sheet 055-97, 6 pp.

[78] Ward WC, Loftis JC, McBride GB. The "data-rich but information-poor" syndrome in water quality monitoring. Environ Manag 1986;10(3):291−7.

[79] Harmancioglu NB, Fistikoglu O, Ozkul SD, Singh VP, Alpaslan MN. Water quality monitoring network design. Norwell (MA): Kluwer Academic Publishers; 1999. 290 pp.
[80] Sanders TG, Ward RC, Loftis JC, Steele TD, Adrian DD, Yevjevich V. Design of networks for monitoring water quality. Highlands Ranch (CO): Water Resources Publications; 1983. 328 pp.
[81] Strobl RO, Robillard PD. Network design for water quality monitoring of surface freshwaters: a review. J Environ Manage 2008;87:639–48.
[82] Hirsch RM, Slack JR, Smith RA. Techniques of trend analysis for monthly water quality data. Water Resour Res 1982;18:107–21.
[83] Lettenmaier DP. Detection of trends in water quality data records with dependent observations. Water Resour Res 1976;12:1037–46.
[84] U.S. Environmental Protection Agency. Progress in water quality: an evaluation of the national investment in municipal wastewater treatment. 2002. EPA-832-R-00–008, variously paginated.
[85] U.S. Environmental Protection Agency. Research strategy environmental monitoring and assessment program. (NC): Research Triangle Park; 2002. EPA 620/R-02/002, variously paginated with appendices.
[86] Puckett LJ. Nonpoint and point sources of nitrogen in major watersheds of the United States. U.S. Geological Survey; 1994. Water-Resources Investigations Report 94-4001, 9 pp.
[87] Gilliom RJ, Alley WM, Gurtz ME. Design of the National Water-Quality Assessment Program: occurrence and distribution of water-quality conditions. U.S. Geological Survey; 1995. Circular 1112, 33 pp.
[88] Hirsch RM, Alley WM, Wilber WG. Concepts for a National Water-Quality Assessment Program. U.S. Geological Survey; 1988. Circular 1021, 42 pp.
[89] National Research Council. A review of the USGS National Water Quality Assessment Pilot Program. Washington (DC): National Academy Press; 1990. 153 pp.
[90] Leahy PP, Rosenshein JS, Knopman DS. Implementation plan for the National Water-Quality Assessment Program. U.S. Geological Survey; 1990. Open-File Report 90-174, 10 pp.
[91] National Research Council. National Water Quality Assessment Program: the challenge of national synthesis. Washington (DC): National Academy Press; 1994. 51 pp.
[92] National Research Council. Opportunities to improve the U.S. Geological Survey National Water Quality Assessment Program. Washington (DC): National Academy Press; 2002. 238 pp.
[93] National Research Council. Preparing for the third decade of the National Water-Quality Assessment Program. Washington (DC): National Academy Press; 2012. 185 pp.
[94] Mueller DK, Hamilton PA, Helsel DR, Hitt KJ, Ruddy BC. Nutrients in ground water and surface water of the United States an analysis of data through 1992. U.S. Geological Survey; 1995. Water Resources Investigation Report 95-4031, 74 pp.
[95] U.S. General Accounting Office. National water quality assessment: Geological survey faces formidable data management challenges. 1993. Report GAO/IMTEC 93–30; Washington (DC), 24 pp.
[96] Advisory Committee on Water Information. Coordination of water resources information. Washington (DC). M-92-01. 2016. 6916, http://acwi.gov/m9201.html.
[97] Murdoch PS, McHale MR, Mast MA, Clow DW. The U.S. Geological Survey Hydrologic Benchmark Network. U.S. Geological Survey; 2005. Fact Sheet 2005–3135, 6 pp.

[98] Mast AM. Evaluation of stream chemistry trends in U.S. Geological Survey reference watersheds, 1970–2010. Environ Monit Assess 2013;185(11):9343–59.

[99] Hooper RP, Aulenbach BT, Kelly VJ. The National Stream Quality Accounting Network: a flux-based approach to monitoring the water quality of large rivers. Hydrol Process 2001;15:1089–106.

[100] National Science and Technology Council. Integrating the Nation's Environmental Monitoring and Research Networks and Programs: a proposed framework. Washington (DC): National Science and Technology Council; 1997. variously paginated.

[101] Bricker OP, Ruggiero MA. Toward a national program for monitoring environmental resources. Ecol Appl 1998;8(2):326–9.

[102] Coupe Jr RH, Goolsby DA. Monitoring the water quality of the Nation's largest rivers—Mississippi river basin NASQAN program. U.S. Geological Survey; 1999. Fact Sheet FS-055–99, variously paginated.

[103] Biological Science Report. In: Schmitt CJ, editor. Biomonitoring of Environmental Status and Trends (BEST) Program: environmental contaminants and their effects on fish in the Mississippi River Basin; 2002. USGS/BRD/BSR—2002—0004, 217 pp. appendices.

[104] Turner RE, Rabalais NN. Changes in Mississippi river water quality this century and implications for coastal food webs. Bioscience 1991;41(3):140–7.

[105] Rabalais NN, Turner RE, Justic D, Dortsch Q, Wiseman Jr WJ, Sen Gupta BK. Nutrient changes in the Mississippi river and system responses on the adjacent continental shelf. Estuaries 1996;19:386–407.

[106] Turner RE, Rabalais NN. Coastal eutrophication near the Mississippi river delta. Nature 1994;368:619–21.

[107] Goolsby DA, Battaglin WA. Long-term changes in concentrations and flux of nitrogen in the Mississippi River Basin, U.S.A. Hydrol Process 2001;15:1209–26.

[108] Alexander RB, Smith RA, Schwarz GE, Boyer EW, Nolan JV, Brakebill JW. Differences in phosphorus and nitrogen delivery to the Gulf of Mexico from the Mississippi River Basin. Environ Sci Technol 2008;42(3):822–30.

[109] Mississippi River/Gulf of Mexico Watershed Nutrient Task Force. A science strategy to support management decisions related to hypoxia in the Northern Gulf of Mexico and excess nutrients in the Mississippi River Basin. U.S. Geological Survey; 2004. Circular 1270, 58 pp.

[110] Mississippi River/Gulf of Mexico Watershed Nutrient Task Force. Gulf Hypoxia Action Plan 2008 for reducing, mitigating, and controlling hypoxia in the Northern Gulf of Mexico and improving water quality in the Mississippi River Basin. Washington, DC. 2008. 61 pp.

[111] Aulenbach BT, Buxton HT, Battaglin WA, Coupe RH. Streamflow and nutrient fluxes of the Mississippi-Atchafalaya river basin and subbasins for the period of record through 2005. U.S. Geological Survey; 2007. Open-File Report 2007-1080.

[112] U.S. Geological Survey. NAWQA publications: trends. 2016. Accessed at:http://water.usgs.gov/nawqa/bib/.

[113] Sprague LA, Hirsch RM, Aulenbach BT. Nitrate in the Mississippi river and its tributaries, 1980 to 2008: are we making progress? Environ Sci Technol 2011;45:7209–16.

[114] Scavia D, Rabalais NN, Turner RE, Justić D, Wiseman Jr WJ. Predicting the response of Gulf of Mexico hypoxia to variations in Mississippi river nitrogen load. Limnol Oceanogr 2003;48(3):951–6.

[115] Scavia D, Justić D, Bierman Jr VJ. Reducing hypoxia in the Gulf of Mexico—advice from three models. Estuaries 2004;27(3):419–25.

[116] Turner RE, Rabalais NN, Justić D. Predicting summer hypoxia in the northern Gulf of Mexico—Riverine N, P, and Si loading. Mar Pollut Bull 2006;52(2):139–48.

[117] Turner RE, Rabalais NN, Justić D. Gulf of Mexico hypoxia—alternate states and a legacy. Environ Sci Technol 2008;42(7):2323–7.
[118] U.S. General Accounting Office. Key EPA and state decisions limited by inconsistent and incomplete data. GAO/RCED-00–54. 2000. Washington (DC), 73 pp.
[119] U.S. Environmental Protection Agency. Wadeable streams assessment—a collaborative survey of the nation's streams. EPA 841-B-06–002. 2006. Washington (DC), 98 pp.
[120] National Research Council. Review of EPA's environmental monitoring and assessment program: overall evaluation. 1995. Washington (DC), 178 pp.
[121] U.S. Environmental Protection Agency. Future risk: research strategies for the 1990s—The report of the research strategies committee. Washington (DC): U.S. Environmental Protection Agency, Science Advisory Board; 1988. Report SAB-EC-88–040, 19 pp.
[122] Hughes RM, Paulsen SG, Stoddard JL. EMAP-surface waters: a multiassemblage, probability survey of ecological integrity in the U.S.A. Hydrobiologia 2000;422/423:429–43.
[123] Paulsen SG, Larsen DP, Kaufmann PR, Whittier TR, Baker JR, Peck DV, McGue JR, Hughes RM, McMullen D, Stevens D, Stoddard JL, Lazorchak J, Kinney W, Selle AR, Hjort R. EMAP-surface waters monitoring and research strategy: fiscal year. Corvallis (OR): U.S. Environmental Protection Agency; 1991. EPA/600/3–91/022, 184 pp.
[124] Paulsen SG, Hughes RM, Larsen DP. Critical elements in describing and understanding our nation's aquatic resources. J Am Water Resour Assoc 1998;34:995–1005.
[125] U.S. Environmental Protection Agency. Environmental monitoring and assessment program: western research strategy. 2001. 42 pp.
[126] U.S. Environmental Protection Agency. National rivers and streams assessment 2008–2009: a collaborative survey. 2016. EPA/841/R-16/007, Washington (DC), 116 pp.
[127] U.S. General Accounting Office. Watershed management: better coordination of data collection efforts needed to support key decisions. GAO-04–382. 2004. Washington (DC), 145 pp.
[128] National Water Quality Monitoring Council. Working together for clean water: 2015–2016 council highlights. Accessed at: http://acwi.gov/monitoring/pubs/index.html#brief on 8916.
[129] Keith LH, Brass HJ, Sullivan DJ, Boiani JA, Alben KT. An introduction to the national environmental methods index. Environ Sci Technol 2005;39(8):173A–6A.
[130] Gilliom RJ, Hamilton PA, Miller TA. The National Water-Quality Assessment Program—entering a new decade of investigations. U.S. Geological Survey; 2001. Fact Sheet 071-01, 6 pp.
[131] Rowe Jr GL, Gilliom RJ, Woodside MD. Tracking and forecasting the nation's water quality—priorities and strategies for 2013–2023. U.S. Geological Survey; 2013. Fact Sheet 2013–3008, 6 pp.
[132] Rowe Jr GL, Belitz K, Demas CR, Essaid HI, Gilliom RJ, Hamilton PA, Hoos AB, Lee CJ, Munn MD, Wolock DW. Design of Cycle 3 of the National Water-Quality Assessment Program, 2013–23: Part 2: Science Plan for Improved Water-Quality Information and Management. U.S. Geological Survey; 2013. Open-File Report 2013–1160, 110 pp.
[133] U.S. Commission on Ocean Policy. Creating a national monitoring network. In an ocean blueprint for the 21st century, final report. 2004. p. 226–35. Washington (DC), [chapter 15].
[134] Advisory Committee on Water Information and National Water Quality Monitoring Council. Executive Summary—A National Water Quality Monitoring Network for U.S. Coastal Waters and their Tributaries. Council on Environmental Quality, National Science and Technology Council, Joint Subcommittee on Ocean Science and Technology; 2006. 99 pp. http://acwi.gov/monitoring/network/design/.

[135] Dubrovsky NM, Burow KR, Clark GM, Gronberg JM, Hamilton PA, Hitt KJ, Mueller DK, Munn MD, Nolan BT, Puckett LJ, Rupert MG, Short TM, Spahr NE, Sprague LA, Wilber WG. The quality of Our nation's water—nutrients in the nation's streams and groundwater, 1992–2004. U.S. Geological Survey; 2010. Circular 1350, 174 pp.
[136] U.S. Geological Survey. The quality of Our nation's waters—nutrients and pesticides. U.S. Geological Survey; 1999. Circular 1225; 82 pp.
[137] The Heinz Center. The State of the nation's ecosystems. H. John Heinz III Center for Science, Economics, and the Environment. 2002. Washington (DC), 270 pp.
[138] The Heinz Center. The State of the nation's ecosystems: Highlights. H. John Heinz III Center for Science, Economics, and the Environment. 2008. Washington (DC), 34 pp.
[139] U.S. Government Accountability Office. Status of federal data programs that support ecological indicators. 2005. GAO-05–376, Washington (DC), 184 pp.
[140] Secretariat of the Stockholm Convention. Stockholm convention on persistent organic pollutants (POPs); text and annexes. The Stockholm convention. 2009. Stockholm, Sweden, 63 pp.
[141] Stone WW, Gilliom RJ, Ryberg KR. Pesticides in U.S. streams and rivers: occurrence and trends during 1992–2011. Environ Sci Technol 2014;48:11025–30.
[142] Robinson KW, Campbell JP, Jaworski NA. Water-quality trends in New England rivers during the 20th century. U.S. Geological Survey; 2003. Water-Resources Investigations Report 03-4012, 20 pp.
[143] Broussard W, Turner RE. A century of changing land-use and water-quality relationships in the continental U.S. Front Ecol Environ 2009;7(6):302–7.
[144] Kelly VG, Stets EG, Crawford CG. Long-term changes in nitrate conditions over the 20th century in two Midwestern Corn Belt streams. J Hydrol 2014;525:559–71.
[145] Stets EG, Kelly VG, Broussard W, Smith ET, Crawford CG. Century-scale perspective on water quality in selected river basins of the conterminous United States. U.S. Geological Survey; 2012. Scientific Investigations Report 2012-5225, 120 pp.
[146] Stets EG, Kelly VG, Crawford CG. Long-term trends in alkalinity in large rivers of the conterminous U.S. in relation to acidification, agriculture, and hydrologic modification. Sci Total Environ 2015;488:280–9.
[147] Preston SD, Alexander RB, Schwarz GE, Crawford CG. Factors affecting stream nutrient loads: a synthesis of regional SPARROW model results for the continental United States. J Am Water Resour Assoc 2011;47(5):891–915.
[148] Saad DA, Schwarz GE, Robertson DM, Booth NL. A multiagency nutrient dataset used to estimate loads, improve monitoring design, and calibrate regional nutrient SPARROW models. J Am Water Resour Assoc 2011;47(5):933–49.
[149] Stone WW, Crawford CG, Gilliom RJ. Watershed regressions for pesticides (WARP) models for predicting stream concentrations of multiple pesticides. J Environ Qual 2013;42:1838–51.
[150] Hirsch RM, Hamilton PA, Miller TL. U.S. Geological Survey perspective on water-quality monitoring and assessment. J Environ Monit 2006;8:512–8.
[151] Collins WD. Quality of the surface waters of Illinois. U.S. Geological Survey; 1910. Water Supply Paper 239, 102 pp.
[152] Betanzo EA, Choquette AF, Reckhow KH, Hayes L, Hagen ER, Argue DM, Cangelosi AA. Water data to answer urgent water policy questions: monitoring design, available data and filling data gaps for determining the effectiveness of agricultural management practices for reducing tributary nutrient loads to Lake Erie, Northeast-Midwest Institute Report. 2015. Washington (DC), 169 pp.
[153] Weber CI. Biological Monitoring of the Aquatic Environment by the Environmental Protection Agency. ASTM STP 528 1973. American Society for Testing and Materials. pp. 46–50.

[154] Alexander RB, Slack JR, Ludtke AS, Fitzgerald KK, Schertz TL. Data from selected U.S. geological survey national stream water quality monitoring networks. Water Resources Research 1998;34(9):2401–5.
[155] Hinck JE, Schmitt CJ, Ellersieck MR, Tillett DL. Relations between and among biomarkers in Black Bass (*Micropterus spp.*) and Common Carp (*Cyprinus carpio*) from large U.S. Rivers; 1995–2004. J Env Mon 2008:1499–518.

CHAPTER THREE

Impact of Climate Change on Water with Reference to the Ganges–Brahmaputra–Meghna River Basin

M.A. Islam[1], S.L. Islam[2], A. Hassan[3]
[1]Curtin University Sustainability Policy Institute, Perth, Australia
[2]Washington State University, Pullman, WA, United States
[3]Red Cross Crescent Center, Hague, Netherlands

1. INTRODUCTION

The water environment in the face of uncertain climate change is one of the global drivers predicted to have the greatest effect on human health, including demography, agriculture, energy, and infrastructure as reported in the United Nations World Water Development Report 4 [1]. The Ganges–Brahmaputra–Meghna (GBM) river basin is prominently related to these drivers and particularly vulnerable to climate change. This basin is well endowed with the highest reserve of precious freshwater sources that supports the habitats for rich biodiversity and the high density of population. One of the driving factors for sustainable development in this region emanates from timely availability, distribution, and quality of the water as the basic life-sustaining resource. Security of food, fiber, and fuel as well as growth and development associated with environmental health and poverty reduction centered on the conservation, quality, and continuous natural flow of the GBM rivers and normal seasonal distribution of rainfall without fluctuations. Future scenario responsive to integrated water resource planning and management in the basin could open up immense opportunities to harness potential socioeconomic and ecological benefits in the face of changing climate and rising water pollution challenges. This is a critical issue of the time that availability of quality water can be the key force for addressing challenges common to the region. In order to address this challenge, immediate attention on short-, medium-, and long-term mutually agreed-upon actions should be taken into consideration, particularly through ecosystem-based river conservation, effective governance, and regional

Chemistry and Water
ISBN 978-0-12-809330-6
http://dx.doi.org/10.1016/B978-0-12-809330-6.00003-9

harmonized policy implemented by the governments of the basin countries. Sustainability is a critical factor for river water management projects that store monsoon water, mitigate the effects of floods and droughts, augment river flows during the dry season, and expand irrigation and navigation facilities. These projects also include generating hydropower and strengthening food, energy and environmental security under uncertain climate risk. This chapter emphasizes the upmost importance of effective regional cooperation in water management to achieve these advantages. The river systems are interconnected, and hence, integrated and collaborative policies are required for a sustainable solution. In this context, one of the major challenges is that countries attach different valuations to water and do not approach the resource with a common understanding as reported in the United Nations World Water Development Report [2]. The Intergovernmental Panel on Climate Change (IPCC), in its Fifth Assessment Report (AR5)[1] for policymakers, highlighted that climate change will amplify existing risks and create new risks for natural and human systems. Risks are unevenly distributed and are generally greater for disadvantaged people and communities. The research question in this chapter is that climate change impact on the GBM watershed region cannot be addressed in a sustainable manner unless it follows the "basin-wide approach" in policy, planning, and programming. With this background in view, this chapter sets objectives to identify and appraise the issues, problems, and prospects for common understanding and sensitize the policymakers of the GBM countries on managing water under uncertain climate risks together so that water-centered sustainable development through an integrated approach is being encouraged in new ways within the horizon of climate change.

1.1 The Ganges–Brahmaputra–Meghna Water Basin

The Ganges–Brahmaputra Delta, also known as Ganges Delta or Bengal Delta, is the world's largest delta, which has a surface area of some 100,000 km^2. The GBM river system is the third largest freshwater source to the world's oceans [3]. The GBM is a transboundary river basin with a total area of approximately 1.7 million km^2, which is distributed (Table 3.1) between India (64%), China (18%), Nepal (9%), Bangladesh (7%), and Bhutan (3%). Nepal is completely within the Ganges river basin, and Bhutan is situated entirely inside the Brahmaputra river basin. The GBM comprises three

[1] IPCC. Synthesis report, Summary for Policy Makers; 2014.

Table 3.1 Country Areas in the Ganges–Brahmaputra–Meghna River Basin

	Area					
Basin	**km²**	**% of Southeast Asia**	**Countries Included**	**Area of Country in Basin (km²)**	**As % of Total Area of the Basin**	**As % of Total Area of the Country**
Ganges	1,087,300	5.2	India	860,000	79	26
			China	33,500	3	0.3
			Nepal	147,500	14	100
			Bangladesh	46,300	4	32
			Bhutan	—	—	—
Brahmaputra	543,400	2.7	India	195,000	36	6
			China	270,900	50	3
			Nepal	—	—	—
			Bangladesh	39,100	7	27
			Bhutan	38,400	7	100
Meghna	82,000	0.4	India	47,000	57	1
			China	—	—	—
			Nepal	—	—	—
			Bangladesh	35,000	43	24
			Bhutan	—	—	—
Total	1,712,700	8.3	Bhutan	1,102,000	64	33
			China	304,400	18	3
			Nepal	147,500	8	100
			Bangladesh	120,400	7	83
			Bhutan	38,400	3	100

From JRCB (Joint Rivers Commission Bangladesh). Basin map of the Ganges, the Brahmaputra, and the Meghna rivers; 2011.

big rivers the Ganges, the Meghna, and the Brahmaputra. Each river has unique features, and they flow through very different regions for most of their duration. They all connect just a few hundred kilometers upstream of the mouth in the Bay of Bengal. Individually, their tributaries play a key role in determining social and economic benefits as well as for water availability and use [4]. This delta is among the most economically productive regions in the world.

1.2 Development Challenges and Aspirations

Water is necessary for economic productivity and welfare of the economy. It is very important to protect the reliable source of water supply since it plays a key role in achieving higher economic growth. Water is vital to many sectors; therefore, investment in water resources often forms the basis for broad regional and national development [6]. According to Iyer [7], "Water is in fact the magic key to future prosperity in this GBM region."

AR5 of IPCC [8,20] identifies a set of key climate-related risks of South Asian countries. These are flood damage to infrastructure, livelihoods, and settlements. It flags the issue of riverine, coastal, and rising sea level and cyclones, which are potential threats for widespread damage to infrastructure, livelihoods, and settlements. The risk of crop damage, loss of life, and property associated with floods is the highest in India and Bangladesh. A large proportion of Asia's population lives in low-elevation coastal zones that are particularly at risk from climate change hazards, including sea level rise, storm surges, and cyclones/typhoons.

Global climate change is expected to aggravate the current scenario and impose further pressure on water resources in the near future. This may result in an increase in the intensity of natural disasters, such as drought and floods. It is anticipated that climate change will affect the availability of water resources through fluctuations in rainfall, soil moisture, glacier and ice/snow melt, and river and groundwater flows. In disaster forums, it is widely discussed that 90% of the natural hazards are caused by water-related issues, which will have serious consequences by acting as a barrier to advancing economic development in South Asia. It is predicted that this region will be threatened most by climate change—related food security issues by the end of 2030. Population increase along the river basin means there is a greater demand for food, fuel, and building materials. This region will face huge deforestation, which again reduces interception and increases runoff, leading to soil erosion. River channels get filled with soil; as a result, the capacity of the River Ganges and Brahmaputra is reduced, which, in turn, causes a higher level of inundation. Flooding is a common problem in the GBM river basin. Out of all the South Asian countries, Bangladesh, being lower riparian, is mostly affected by it. This occurrence of inundation is mostly due to anthropogenic activities.

The countries of GBM have hydrological linkages, which are connected to one another. Floods and erosion originating in one country may have a negative impact on others, i.e., water storage facilities exist in one country and water demand in another; hydropower potential exists in one country and the energy market in another. The data and information generated in one country can significantly reduce disasters from flooding and drought in another. These interdependencies require immediate cooperative efforts by each and every member of GBM in order to progress. As demonstrated by the World Bank's Ganges Strategic Basin Assessment reported by Wu et al. [9], the potential benefits from this cooperation are large, and the cost of noncooperation is huge. Even Nepal, the most water-rich country, suffers from a severe water crisis during the dry season. Economic growth and industrialization will be constrained as long as there exists a poor management of water, and eventually the society will have to bear the huge costs.

Poor access to safe water has serious implications for society and particularly for women and children, who are responsible for fetching water, especially in this South Asian region. Despite huge benefits, regional cooperation has remained lowest priority due to institutional rigidity, lack of trust, and a mindset of seeing water as a national property rather than as a shared resource to be managed with shared perspective, as reflected in the work of Gareth Price et al. [10]. Synergistic policies of transboundary water resources with multiple benefit sharing options for member countries is inevitable, not only to meet the growing challenges of water, food, and energy security, but also for peace and security of the South Asian region. These policies will be an aid to promote regional energy trade-off between the regions and pave the way for regional cooperation on water, hydropower, and risk mitigation. Policymakers in the region seem to be taking a much more positive and balanced view of the necessity of regional cooperation.

Regional cooperation can be in the form of joint water resource development programming, data sharing, flood management, hydropower development, and energy trade. These could lead to a starting point for a new revolution of joint regional development. Development of water resources should be based on scientific principles, as it has direct and indirect chain relation related to social, economic, ecological, and environmental implications, along with indigenous wisdom as well as views of diverse stakeholders and their engagement. Strategic environmental assessments, including detailed studies of technical, ecological, and economic feasibility, are required at the cross-boundary level with an independent expert mission, either through engagement of United Nations or South Asia Association for Regional Cooperation (SAARC), to identify areas for potential water conservation infrastructure and hydropower areas, and to demarcate fragile zones where heavy construction must be avoided, for example, at high altitude and in vulnerable watersheds [11]. Lessons learned revealed the fact that while developing water resources, it is not enough to develop physical infrastructure alone; development of institutional capacity and provision of operation and maintenance budgets are also critical for sustainability context. Past experience shows that weak institutional capacity not only poses a major obstacle for planning and implementation of complex transboundary project, but also causes serious damages, as happened in the breach of Kosi Dam in Eastern Nepal in 2008, which flows into India to join the Ganges.

Sustainable water development in river basins depends on large-scale measures to protect upstream water sources, forests, and soils in mountain areas as a part of watershed management in the basin areas. Protection and conservation of the mountain environment are thus critical for long-term sustainable economic development of downstream areas, which is explained in a study by Dore [12]. If these issues are addressed, enabling an environment for win–win opportunities on water cooperation could be the way out to bring positive change to the economic and social landscape of the region as well as herald for a means to improve trust and peacebuilding in the region.

1.3 State of Water Quality in Ganges–Brahmaputra–Meghna River Basin

The South Asian countries are faced with serious public health and ecosystem consequences due to the deterioration of quality of both surface and groundwater. The GBM river basin countries are yet to meet safe water supply and hygienic sanitation in both rural and urban areas. More than half of the morbidity in the GBM basin stems from the use of impure drinking water. A holistic approach is needed to monitor and evaluate the water quality in each of the GBM member countries, and together with regional initiatives, they can reduce further deterioration and bring amelioration in the quality of water. Monitoring water quality in the GBM rivers needs to be extensive in the case of the Ganges River in India and the Buriganga River in Bangladesh. Coordination of their actions in order to deal with transboundary transmission of pollution and evolving a mechanism for realtime water quality data exchange could lead to clean and efficient water quality management[2] quoted in United Nations Food and Agriculture Organization (FAO) Aquastat [13].

Projected future socioeconomic scenarios suggest a significant impact on water quality at the downstream end of the Ganga. A less sustainable future resulted in a deterioration of water quality due to the pressures from higher population growth, land use change, increased sewage treatment discharges, enhanced atmospheric nitrogen deposition, and water abstraction. However, water quality was found to improve under a more sustainable strategy, as envisaged in the Ganga cleanup plan.[3]

In the lower riparian of GBM, Bangladesh irrigation water quality has deteriorated due to pollution from agrochemicals, industrial waste, municipality waste, sewage, and other sources. Arsenic contamination of groundwater has been reported in many government and donor agency documents (Government of Bangladesh, United Nations Children's Fund, World Bank, and FAO). Arsenic concentration has been found to be at a maximum within the upper 50 m depth of aquifers in most regions of the country [14]. In many places, concentration of iron and arsenic in irrigation water has gone beyond the limit of safe water quality standards of Bangladesh Government and the World Health Organization. Some diseases and health hazards, such as arsenicosis, blindness, and physical disability, are reported as a result of arsenic toxicity to humans [15]. Throughout the country, about 1.44 million tube wells have been affected by arsenic contamination, and about 30 million people are exposed to arsenic toxicity [16]. The mitigation of the additional problems of salinity and arsenic in Bangladesh involves special action plans. The Bangladesh Arsenic Mitigation Water Supply Project is presently engaged in assessing the extent, dimensions, and causes of the arsenic problem with a view of developing a long-term strategy for supplying arsenic-free water [4].

Even in a lower riparian deltaic country like Bangladesh, during the dry season, groundwater has become an increasingly important source of water for irrigation, municipal, and industrial purposes. Some environmental hazards have been encountered in various areas, and a number of adverse effects

[2] FAO, Aquastat 2011, quoted from: Biswas AK. After 2006. Management of Ganges-Brahmaputra-Meghna system: way forward (Available at: http://www.thirdworldcentre.org/akbgangesbrah.pdf).

[3] Whitehead PG, Sarkar S, Jin L, Futter MN, Caesar J, Barbour E, Butterfield D, Sinha R, Nicholls R, Hutton C, Leckie HD. Dynamic modeling of the Ganga river system: impacts of future climate and socio-economic change on flows and nitrogen fluxes in India and Bangladesh. Environ Sci Process Impacts 2015;17(6):1082–97.

have emerged as a result of the overextraction of groundwater, such as lowering of water tables, reduction in dry season flows of rivers and streams, groundwater pollution, intrusion of saline water in coastal areas, ecological imbalance, and possible land subsidence. Study findings revealed the fact that the dangerous trend of lowering of groundwater levels happened particularly in the Dhaka metropolitan area, where the average annual decline in the groundwater level is about 3 m from the base level, as reported by Bangladesh Agricultural Development Corporation in 2006 [17], and this trend of overextraction was noticed in the northwest dry region of the country. This trend will further deteriorate in the face of climate change as water demand will increase over time.

The water quality of rivers in India is relatively good in the upper reach. However, the middle and lower reaches of most of the cross-boundary rivers are affected by major degradation of surface water quality. This is because of the importance of water use for cities, agriculture, and industries, and the lack of wastewater treatment plants. Groundwater is also affected by municipal, industrial, and agricultural pollutants. The pollution control action plan of the Ganges basin was formulated in 1984 and has been enforced by the Ganges Project Directorate, under the Central Ganges Authority, to oversee pollution control and the consequent cleaning of the Ganges River. Water quality in the middle stretch of the Ganges River, which had deteriorated to class C and D (the worst class is E, the best A), was restored to class B in 1990 after the implementation of the action plan [17].

2. CLIMATE CHANGE SCENARIOS OF THE GANGES–BRAHMAPUTRA–MEGHNA BASIN

The study by Hassan et al.[4] [18] on climate change and water availability in the GBM basin discussed the potential impact of the changed climate scenario on flooding in the basin. Monsoon flow was projected to increase in all three basins. One coauthor of this chapter, Hassan, in the same study, further reported that more than 60% of the yearly flow across the basin occurs during the monsoon, and this percentage is also projected to increase by the 2050s. The increase in flow could result in more frequent flood events of

[4] ICIMOD. In: Vidya RA, Sharma E, editors. Research insights on climate and water in the Hindu Kush Himalayas. One of the chapter contributors in this research is Hassan et al. on Climate Change and Water Availability in the Ganges-Brahmaputra-Meghna Basin: Impact on Local Crop Production and Policy Directives; 2014.

higher magnitudes. A frequency analysis of flood events under the climate change scenarios showed that in the Ganges basin, a 50-year flood will become a 20-year flood by the 2050s under the A2 and B1 scenarios, and a 10-year event under the A1B scenario. A similar tendency was also estimated in the Brahmaputra and Meghna basins, with a 50-year event becoming a 35-year event in the Brahmaputra basin and a 20-year event in the Meghna basin [19].

Effects of climate changes on the sources of GBM rivers in the Himalayan mountains is an established fact with scientific evidence because of melting of the glaciars and icecaps and their negative impact on water resources. The increase in melting, as well as the increased length of the melt season of glaciers, leads at first to increased river runoff and discharge peaks, while in the longer timeframe (decadal to century scale), glacier runoff is expected to decrease. Evidence for increased runoff in the last 20 years due to enhanced glacier melt has already been detected.

Some parts of South Asia face higher climate vulnerability compared to the rest of the region. Extreme events are expected to be more catastrophic for the people living in such districts. Although rapidly urbanizing, South Asia is still predominantly an agrarian society, where a majority of the population is dependent on agriculture for their livelihoods. Rural poverty is higher than urban poverty, reflecting the heavy dependence on natural resources that are directly influenced by changes in weather and climate [8].

2.1 Short-Term and Long-Term Climate Scenarios

Globally, sea levels have risen faster than at any time during the previous two millennia, and the South Asian region is mostly affected. Changing patterns of rainfall or melting snow and ice are altering freshwater systems, affecting the quantity and quality of water available in many regions, including South Asia. Climate change will have widespread impacts on South Asian society and South Asians' interaction with the natural environment [20]. Water availability, in terms of temporal as well as spatial distribution, is expected to be highly vulnerable to anticipated climate change. Growing populations and concentration of population in urban areas will exert increasing pressures on water availability and water quality, as reported by IPCC [21].

2.2 Potential Climate Impact on Lower Riparian Areas

The potential impact of climate change resulting from global warming are of great importance to lower riparian countries, such as Bangladesh. In broad terms, imbalance in surface water demand and supply, as well as evaporation-to-precipitation ratios, is expected to rise immensely, prompting an increase in irrigation water requirements unless offset by diversification toward dry-foot crops. While groundwater resources may be little affected, dry season flows toward the delta may decline. Dry regions will experience a further impact of slow-onset aridity hazards. Main river flooding may increase in duration, and flash flooding will tend to be more frequent, as will cyclones. In the coastal area of the deltaic part of the GBM basin areas susceptible to a comparatively greater impact on sea level rise (SLR), in terms of more coastal erosion, sedimentation in the estuary will further exacerbate drainage congestion, flooding, and disruption in navigability, and it will lead to cyclones with higher frequencies and intensities. The morphological changes would have severe implications on agriculture, food security, socioeconomic development, and general livelihood of the coastal communities. Already today, all major infrastructure projects, such as flood protection works, coastal structural and green defenses, and irrigation and drainage projects, should take active consideration of climate change factors to make resilient initiatives. To launch that cost-effective initiative, it will be necessary to quantify the climate change impact on runoff and saline intrusion, and quantify the morphological processes as well as population displacement and how these processes will be affected in the future by climate change and SLR as flagged in the scenario building.

2.3 Climate Change and Sea Level Rise

According to the Report of Delta Plan 2100 of the Planning Commission, the impacts of climate change for Bangladesh are most critical because of its geographical location, high variability of rainfall and river flow pattern, high population density, high levels of poverty, and the reliance of many livelihoods on climate-sensitive sectors, such as agriculture, fisheries, and ecosystems. Trend analyses of measured temperature (1948–2010) at 34 locations show that annual temperature is rising at a rate of about 1.2°C per century in Bangladesh. This trend has become stronger in recent years. The trend in recent mean annual temperature (1980–2010) is almost double the longer term, as observed in a study report of Comprehensive Disaster Management Programme (CDMP) [22]. Recent trend analyses of observed rainfall indicate that the distribution of rainfall over time is changing, as it is found comparatively higher in a shorter duration during the monsoon period. Future climate change (A1B, i.e., a medium emission scenario) has been projected as a difference from the baseline period by CDMP [22]. The differences of mean annual temperature and precipitation during 1911–41 and 1971–2100 from the baseline period of 1961–90 were computed. The mean annual and monsoon precipitation is likely to increase by 0.64 and 1.40 mm/day, respectively, during 2011–41 from the baseline period. However, mean winter precipitation may decrease by 0.05 mm/day during 2011–41 compared to the baseline period. According to Delta Plan 2100, the mean annual, monsoon, and winter precipitation is likely to increase by about 0.90, 1.43, and 0.03 mm/day, respectively, during 2071–2100.[5]

2.4 Relative Sea Level Change

Relative sea level change is the local change in sea level relative to the elevation of the land at a specific point on the coast. Relative sea level change is a combination of both global and local sea level change due to changes in estuarine and shelf hydrodynamics, regional oceanographic circulation patterns, hydrologic cycles (river flow), and local and/or regional vertical land motion (subsidence or uplift). On the eastern coast of India, along the Bay of Bengal, recent study reveals the fact that SLR is comparatively high and trends derived from altimeter data shows 5 mm+/year, which is higher than those estimated from tide gauge records over longer periods during 20th century as reported in the Current Science, India [44]. On the other hand, in the northern coast of the Indian Ocean, this rate of SLR during the 1993–2012 period reveals 3.2 mm/year, as estimated.[6] Thus, relative sea level change is variable along the coast of Bangladesh as well, depending upon the local and regional factors. Trend analysis of water level along the coast of Bangladesh shows an increase in annual mean water level in the southwest coast (Hiron Point), Meghna estuary (Khepupara), and Chittagong coast (Rangadia) by 6.8, 3.7, and 4 mm/year, respectively [23]. There is no detailed scientific research to establish relative mean SLR along the coast of Bangladesh for the climate scenarios presented in AR5.

It is also important to establish the projection on temperature and precipitation on representative concentration pathways (RCP) of green house gases emission scenarios. The projected SLR, temperature, and precipitation

[5] *Based on Bangladesh Bureau of Statistics and BMD data sources used in the preparation of Bangladesh Delta Plan 2010 Formulation Project: Coast and Polder Issues*, prepared for the General Economics Division, Planning Commission, Government of Bangladesh with Technical Assistance of the Royal Netherlands Government.

[6] http://www.currentscience.ac.in/Volumes/108/05/0966.pdf.

are needed to assess the likely impacts on flood, storm surge, water availability, salinity intrusion, agriculture, fisheries, infrastructure, etc. For the coastal area, one of the main questions will be how changes in climate will affect the frequency and severity of cyclonic storms. Furthermore, changed patterns of rainfall and temperature over the seasons will have an impact on the agricultural productivity.

The consequences of climate change have an impact on coastal polders and riverine islands (chars), as far as water management is concerned, may have implications such as overtopping of embankments, damage of drainage systems, waterlogging, crop damage, and decline of livelihood opportunities for farmers and fishers. Current water management practices may not be robust enough to cope with the impacts of climate change. It demands improved incorporation of information about current climate variability and climate change into planning and design of water infrastructure. Increase of precipitation and SLR in the changing climate may cause prolonged drainage congestions in the proposed project area if these issues are not considered in designing the drainage systems. The consequence of an SLR of 22 cm in 2050 and a 5% increase of cyclonic wind speed would be an increase of the crest level for sea-facing embankments of 31 cm [24].

2.5 Water Threats: Tropical Cyclone and Storm Surges

Because of its geographical location, Bangladesh is one of the global hotspots for tropical cyclones. These tropical cyclones can cause significant phenomena, which can adversely impact communities and the environment. The most common features are destructive winds, heavy rainfall, and cyclone-induced storm surges that can lead to flooding. The Bay of Bengal is one of the hotspots for the generation of tropical cyclones. About one-tenth of the global numbers of cyclones that form in different regions of the tropics occur in the Bay of Bengal [25,26]. In every 3 years on average, a severe tropical cyclone associated with tidal surge hits Bangladesh. The Fourth IPCC Assessment Report [27] indicates that the future tropical cyclones, storm surges, and their related saline water flood will be more severe and subsequently add vulnerability to the coastal towns and settlements in particular. Another study using the regional climate model (RCM) shows a significant increase in the frequency of highest storm surges (considering the current and future climates) for the Bay of Bengal, despite no significant change in frequency of cyclones [28].

2.6 Impacts of Sea Level Rise on Coastal Settlements

Projected SLR could flood the homes of millions of people living in the low-lying areas of South Asia. The IPCC estimates that almost 60% of the projected increase in the annual number of people flooded in coastal populations will occur in South Asia, along the coasts from Pakistan, India, Sri Lanka, and Bangladesh through to Myanmar. The potential impacts of a 1-m SLR include inundation of 5763 square kilometers in India; a 45-cm SLR in Bangladesh may dislocate about 35 million people from 20 coastal districts by 2050. The repercussions from such population displacements could spread well beyond the coastal regions. Around the Indian Ocean rim, millions of people are already living in urban slums without adequate employment or shelter. Climate migrants seeking refuge from more exposed regions could put enormous pressure on urban economies and infrastructure as well as on basic services such as water supply, power, health, and sanitation. Bangladesh, for example, is already experiencing higher levels of tidal inundation in its coastal districts, with troubling implications. The Sundarbans region, the largest continuous mangrove area in the world, provides direct employment to some 500,000 to 600,000 people for at least half the year, many of whom are engaged in harvesting the region's natural resources through fishing, woodcutting, and collecting thatching materials, honey, beeswax, and shells. By some estimates, just 10 cm of SLR would submerge 15% of the region, and the Sundarbans would disappear altogether if the sea level rose 60 cm [29]. However, dynamics of active delta formation process need further research on gaining knowledge to manage the sediment released by the GBM and may have potential factors to offset the impact of the SLR in this region.

2.7 Effects of Climate Change on the River Morphology of Bengal Delta

The drainage basin of the Bengal Delta covers parts of China, India, and the entire Bhutan, Nepal, and Bangladesh territories. The total drainage basin area is approximately 1,746,000 km^2, of which Ganges covers 1,078,000 km^2, Brahmaputra 577,000 km^2, and Meghna occupies 91,000 km^2. Bangladesh is subject to tectonic-induced subsidence, which generally is of the order of a few mm per year and reaches its maximum of about 15 mm/year in the Sylhet basin in northeastern Bangladesh. Today, the huge sediment load from the mighty rivers, which has been estimated to around 1 billion tons per year, can keep pace with this subsidence, so Bangladesh is slowly accreting. With SLR, a greater part of the upstream sediment load to the coastal zone will deposit near the coast, but it is not known to what extent this increased deposition can balance the sea level rise. Converting mangrove into forests afforestation along the coastline will bring further positive change in silt deposition at an accelerated rate, which is apprehended to offset the sea level rise. However, the subject of concern, how the dynamics of delta formation process, is managed in favor of such change, which depends on (1) the flow of water and sediment load from upstream, (2) the tidal flow and surges, and (3) the wave behavior in the Bay of Bengal. All these processes will be affected by climate changes, either directly or indirectly, and for the first process, also to a wide extent by human interference in the basin and in the delta itself. A key to predicting future developments is the understanding and evaluation of the present conditions and the past. Hence, an understanding of the past littoral sediment transports, governing near shore processes, and the morphological development of the dynamic delta needs to be established through a study.

The change in sediment transport capacity of the delta caused by sea level rise—induced changes of the flow (change in tidal constituents, ebb, flood dominance, etc.) and the increased penetration of saltwater and changes in tidal conditions, which will move the turbidity maximum and siltation of fine sediments upstream, will all have impact on the sediment budget. A reliable prediction of the long-term morphological development of the Bengal Delta is far beyond present understanding of the processes and the capability of predictive morphological—dynamic models. Today, comprehensive projects are being formulated, aimed at bridging this knowledge gap. Key components of these projects are comprehensive data collection programs in both the coastal zone and in the rivers and research programs aimed at improving and integrating mathematical models that describes flow and water levels, saline intrusion, waves, and transport of sand and mud. Salinity intrusion due to the reduction of freshwater flow from the Ganges, Meghna, and Brahmaputra basins, combined with the SLR, is a major concern in Bangladesh. The Flood Forecasting and Warning Center of Bangladesh, supported by the Institute of Water Modeling (IWM) in Dhaka and DHI (international corporate organization on water environment), has developed an operational salinity intrusion forecast model, enabling both short-term and long-term forecasting of salinity conditions in the delta.

2.8 Water Context: Socioeconomic Scenarios

One of the most socioeconomically vulnerable regions to climate change is South Asia's GBM region. Low-lying Bangladesh is at the front line of at-risk countries from climate change and could suffer annual losses of up to 9.4% of its economy by the end of this century [30]. The Asian Development Bank commissioned studies, which forecast that six countries—Bangladesh, Bhutan, India, the Maldives, Nepal, and Sri Lanka—will see an average economic loss of around 1.8% of their collective annual GDP by 2050, rising sharply to 8.8% by 2100 if the world continues on its current fossil fuel—intensive path [30].

The IPCC Report on Shared Socioeconomic Pathways (SSPs) strategy considers a set of socioeconomic pathways as a means on integrating social aspects of future change with climate change. There are five broad classifications for future conditions, namely SSP1 for Sustainability, SSP2 for business as usual (BaU), SSP3 for Fragmented World, SSP4 for Inequality, and SSP5 for Conventional Development in terms of energy sources. Within the Ecosystem Services for Poverty Alleviation (ESPA) Deltas project, three plausible socioeconomic scenarios have been considered that have been adapted from the five IPCC SSPs, namely BaU, plus two others, a more sustainable scenario (MS) and a less sustainable scenario (LS).[7] In order to assess these in terms of their impacts on the GBM flow and water quality, we need to quantify these scenarios. There are many socioeconomic factors that have the potential to affect flows and water quality in the GBM basin. These include: (1) population change, (2) sewage treatment works capacity and design for water quality control, (3) water demand for irrigation and public supply, (4) atmospheric nitrogen deposition, (5) land use change, and (6) water transfer plans [31].

In general, the socioeconomic changes considered had minimal impact on flows. However, the magnitude of these changes is also uncertain. Of most concern is the potential development of large-scale dams and water transfer schemes upstream of Bangladesh. These could seriously threaten the flows in the river system, significantly reducing both peak flows and low flows, with some major consequences for water availability, public policy, and poverty alleviation in the delta region. However, through negotiation and cooperation between countries, the impact of these developments can be minimized and the benefits equitably shared [31].

Upstream developments (e.g., Farakka Barrage and other dams in the catchments of the Ganges, Brahmaputra, and Meghna) already created ecological impact, and further interventions will lead to significant changes in river flows and dynamics that will change the seasonal availability of freshwater in the coastal zone. Of specific concern is that the plan needs to

[7] Details related to the GBM explained in Whitehead PG, Barbour E, Futter MN, Sarkar S, Rodda H, Caesar J, Butterfield D, Jin L, Sinha R, Nicholls R, Salehin M. Impacts of climate change and socio-economic scenarios on flow and water quality of the Ganges, Brahmaputra and Meghna (GBM) river systems: low flow and flood statistics. Environ Sci Process Impacts 2015;17(6):1057—69.

revisited before executing the Jogighopa–Tista–Farakka Link Project or National River Linking Project in India (source: Delta Plan).[8]

3. IMPACT OF CLIMATE CHANGE ON WATER DEMAND AND SUPPLY

Water balance is the assessment of water resource and its use in the system. The computation includes all water receiving components within the systems as well as water losses from the system. The main principle of water balance is the difference between total incoming water and total losses equal to storage in the system. This study provides an understanding on consequences of climate change over time on water balance.

3.1 Water Balance

For water balance analysis of the study area, the calibrated soil and water assessment tool (SWAT) models[9] have been simulated for the time period of 1981–2012, and the hydrological components have been analyzed to compute average annual water balance [18] presented in Table 3.2. Input to the water balance is rainfall, while losses occur through evapotranspiration and percolation and as water contribute to stream flow through surface runoff and lateral flow. The average annual precipitation of the Ganges, Brahmaputra, and Meghna basins are 981, 1981, and 2816 mm, respectively, while the yearly actual evapotranspiration loss is 580, 652, and 910 mm, which is 59%, 33%, and 32% of annual precipitation, respectively.

In addition, evapotranspiration losses also occur through percolation. Annual percolation rate at the Ganges, Brahmaputra, and Meghna basins are 213, 767, and 1306 mm, respectively. Percolation loss is 22%, 39%, and 46% of annual rainfall for the three basins of the GBM basin, respectively.

After the losses of water through evapotranspiration and percolation, the remaining water contributes to stream flow as surface runoff and lateral flow. The annual average surface runoff for the Ganges, Brahmaputra, and Meghna basins is 144, 315, and 568 mm, respectively. The amount of surface runoff for the Meghna basin is high due to the high precipitation rate.[10]

[8] *Bangladesh Delta Plan 2010 Formulation Project: Coast and Polder Issues, Planning Commission.*

[9] SWAT model is a continuous-time, semi-distributed, process based river basin model. It was developed to evaluate the effects of alternative management decisions on water resources and nonpoint-source pollution in large river basins.

[10] Assessing the economic impact of climate change on agriculture, water resources and food security and adaptation measures using seasonal and medium range of forecasts.

Table 3.2 Annual Water Balance for the Ganges–Brahmaputra–Meghna Basins During the Period of 1981–2012

Basin	Precipitation (mm)	Evapotranspiration (mm)	Percolation (mm)	Surface Runoff (mm)	Lateral Flow (mm)	Groundwater Contribution to Flow (mm)	Snow Melt (mm)
Ganges	981	580	213	144	25	120	34
Brahmaputra	1981	652	767	315	134	552	212
Meghna	2816	910	1306	568	33	1211	0

Courtesy CEGIS. Assessing the economic impact of climate change on agriculture, water resources and food security and adaptation measures using seasonal and medium range of forecasts; 2013.

3.2 Water Availability for Base Conditions

ICIMOD [18] published research findings using SWAT simulation model for the basin of GBM to estimate the generated flow inside the basins. In this study, base year has been taken from 1981 to 2012. Basin-wide water availability of the base scenario has been analyzed. The average annual flow generated from the three basins is 1,217,930 Mm^3, from which 353,205 Mm^3 (29%) contributed from the Ganges basin, 713,681 Mm^3 (59%) contributed from the Brahmaputra basin, and 151,043 Mm^3 (12%) contributed from the Meghna basin. The flows are mainly concentrated in the monsoon period (June to October), and the maximum monthly flows are 91,804, 126,792, and 28,683 Mm^3, respectively, for the Ganges, Brahmaputra, and Meghna basins. For the Ganges basin, the peak flow occurs during August, while for the Brahmaputra and Meghna basin, it occurs during July.

3.3 Water Demand Data

Scenarios for water demands have been developed for baseline and future horizons corresponding to the climate scenarios. Data requirements for their calculation include human and livestock population, domestic, industrial, and irrigation water use. Detailed data on water use in countries upstream of Bangladesh are not available. Hence, for areas outside of Bangladesh, socioeconomic data from various publicly available sources were used. Methods to resolve the various formats and resolutions of data were developed similar to that used by Gaffin et al. [33], with an emphasis on providing a set of data that can be represented spatially. For Bangladesh, it was important to use locally sourced data and develop scenarios relevant to local water managers. Therefore, for the development of future water demands, the National Water Management Plan for Bangladesh was exploring the local data-based impacts of climate change using water resources regional impacts at a regional scale as explained in the paper of Fung et al. [34].

The socioeconomic scenarios were combined with the climate scenarios to provide an ensemble of future scenarios, giving an indication as to the possible impacts of climate change on water resources in Bangladesh. The main output of the work is a set of spatial—temporal distributed water availability indices, comparing water demands with supply, enabling areas of future scarcity to be identified. The index presented is the surface water availability index type 4, which is a ratio comparing the minimum of the 90% reliable monthly runoff during the whole run minus the impacts of climate change on water resources demand for that month and the overall 90% reliable monthly runoff plus the water demand for that month. Regions of water excess have a positive value and regions of water deficit a negative value. Despite the large increases in annual precipitation predicted by the RCMs,[11] the dry season is becoming drier for the more recent Providing Regional Climates for Impact Studies (PRECIS)[12] run. With the increases in water demand due mainly to population increases, there is a general increase in the water deficit, especially along the Ganges in India. The results suggest future changes could have major implications for trans-boundary water management. The large variation in magnitude of the climate change data from different climate models suggest that ensembles of runs are required to provide a better indication as to the likelihood of the modeled future water availability results. The Global Water Availability Assessment Model provides a useful tool to assess and highlight climate change impacts on water resources and will provide a more reliable picture of future water availability as more climate change data becomes available.[13]

[11] Regional Climate Models (RCMs) as explained in IPCC Home page at https://www.ipcc.ch/ipccreports/tar/wg1/380.htm.

[12] One of the UK's Met Office Hadley Center's key successes is the development of the regional climate modeling system PRECIS (Providing Regional Climates for Impact Studies) for details see: http://www.metoffice.gov.uk/media/pdf/i/6/Climate_impacts_model_capability.pdf.

[13] EU-funded project to examine the implications of changes in climate and sea level on water resources and coastal flooding in part of the Indian subcontinent with particular reference to Bangladesh. This project first used Regional Climate Change Model (PRECIS) by Met Office Hadley Center, combined with Global Water Avaiability Assessment Model (GWAVA) applied by CEH and Bangladesh partners, feeding Fluvial flooding model applied by Bangladesh partners and Sea Level and Storm Surge Model applied by Proundman Oceanographic Lab. For clear understanding see: http://www.wholesem.ac.uk/events/annual-conference/annual-conf-15/docs/annual-conf-15-presentations/helen-houghton-carr-presentation.

3.4 Water Availability Under Different Climate Change Scenarios

The monthly average flow from the GBM for base condition (1981–2012) and climate change scenarios for 2050 against three conditions (A2, high-emission scenario; B1, low emission, while A1B indicated for medium emission scenario) is shown in the following figures. The calibrated and validated SWAT model for the GBM basins has been simulated for three climate change scenarios (2050s). The simulated results have been compared with the base condition (1981–2012) to assess the impact on water availability due to climate change. The analysis results are presented here as monthly flow volume and yearly flow volume. A basin-wise comparison of the climate scenario results with the baseline scenario is presented in the following sections.[14]

Most of the climate model data suggested that annual runoff will increase by 7–12% by 2050, primarily due to accelerated melt in the upper Indus Basin, together with an increase in precipitation.[15] Laghari et al. in their study suggested that water availability might increase in the short term, but will decrease in the long term [45]. However, the projected future hydrology depends on the precipitation projections, which have a large uncertainty and large variation between annually averaged and seasonal projections among the general circulation models (GCMs).

3.5 The Ganges Basin

The monthly average flow for the three climate change scenarios has been compared with the base condition at the location of Harding Bridge of the Ganges basin. The climate change results for precipitation reflect on the basin outflow. The flow reduces during dry period and increase during monsoon and postmonsoon period. The annual flow increases for all of the three scenarios, but the flow decreases during the premonsoon period. Around 5–10% flow decreases during the premonsoon period and the impact of climate change is more significant.

[14] Assessing the economic impact of climate change on agriculture, water resources and food security and adaptation measures using seasonal and medium range of forecasts.

[15] Nepal S, Shrestha, AB. Impact of climate change on the hydrological regime of the Indus, Ganges and Brahmaputra river basins: a review of the literature. Int J Water Resour Develop 2015;31(2): 201–18.

Table 3.3 Annual Flow Volume Change Against the Climate Scenario in 2050 With Regards to Base Scenario for the Ganges–Brahmaputra–Meghna Basin

Basin Name	Base Annual Flow Volume (Mm^3)	Percentage Change of Annual Flow Volume		
		A2_2050s	A1B_2050s	B1_2050s
Ganges	353,205	8	17	9
Brahmaputra	713,681	5	9	5
Meghna	151,043	10	10	7

Courtesy CSIRO, WARPO, BWDB, IWM, BIDS, and CEGIS. Bangladesh integrated water resources assessment. Australia: CSIRO; 2014.

3.6 The Brahmaputra Basin

The monthly average flow of three climate change scenarios has been compared with the base condition of the Brahmaputra basin. The impact of climate change also follow almost the same trend like the Ganges basin with the exception that the dry season flow also increases for the Brahmaputra basin. The climate change impact in flow is highest for A1B (2050s) scenario.

It is found that there is a huge increase in flow during February and March, although there is a decrease in precipitation during that period. This is mainly due to the impact of temperature rise, which accelerates the rate of snowmelt early. From Table 3.3, it indicates changes in water flow apprehended for snowmelt contribution to surface runoff, which is about 11% more in the Ganges basin. This large amount of snowmelt water results in 24% increase of flow during March due to the influence of climate change.[16] The early snowmelt also results in the reduction of flow during June. There is a maximum of 4.3% reduction of flow that occurred for the A2 scenario during June. The peak flow increased by 5.9%, 7.8%, and 3.2% for A2, A1B, and B1 scenarios, respectively. The flow also increased during postmonsoon period (October–November), with a maximum increase of 17.2% [35].

[16] CSIRO, WARPO, BWDB, IWM, BIDS, and CEGIS. Bangladesh Integrated Water Resources Assessment. Australia: CSIRO; 2014.

3.7 The Meghna Basin

Here, the climate change impact of flow also reflects the trend of change in precipitation. The monthly flow during the dry season decreased and increased for monsoon period. The impact of 2050 (A2) scenario is highest for the Meghna basin, while the impact of A1B scenario is highest for the Ganges and Brahmaputra basin (Fig. 3.1).

Here, the A2 high emission scenario with 50-percentile ensemble 16 GCM results shows the highest flow bar. B1 low emission scenario and A1B medium emission scenario shows the lower *flow bar than A2 high scenario with respect to the base condition.*

3.8 Annual Flow Volume Changes of Three Basins of the Ganges–Brahmaputra–Meghna Basin

The annual percentage increases and decreases of the climate change scenarios of the GBM basin with respect to the base condition are presented in Table 3.3.

The previous table indicates that the A1B medium scenario shows the highest annual flow change percentage with respect to base condition for the Ganges (17%) and Brahmaputra (9%) basin. In the Meghna basin, A2 and A1B scenarios (Fig. 3.2) show the highest annual flow change (10%).

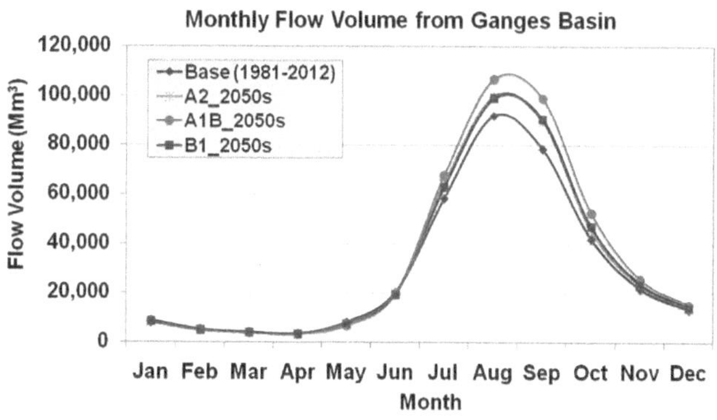

Figure 3.1 Comparison of climate change scenarios with base for the Ganges basin. *From CEGIS. Assessing the economic impact of climate change on agriculture, water resources and food security and adaptation measures using seasonal and medium range of forecasts; 2013.*

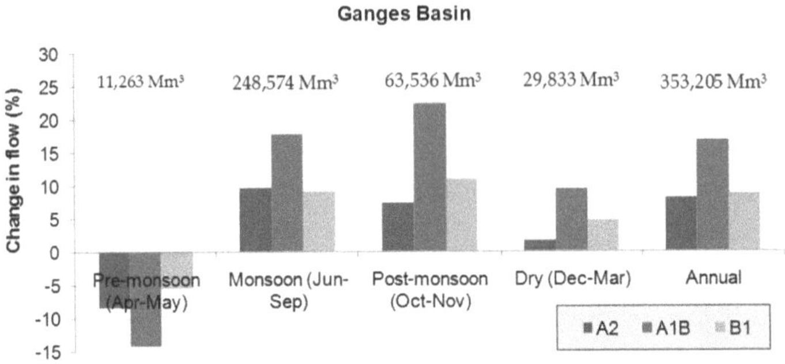

Figure 3.2 Change in seasonal and annual flow against different climate change scenario for the Ganges basin. *From CEGIS. Assessing the economic impact of climate change on agriculture, water resources and food security and adaptation measures using seasonal and medium range of forecasts; 2013.*

3.9 Reservoir and Irrigation Information

A number of dams and barrage have been constructed both in the Ganges and Brahmaputra basins for power production and irrigation purposes. Of special notice has been the Farakka Dam that was constructed with the purpose of diverting water from the Ganges into the dying Hooghly River in order to maintain the river navigable up to Kolkata. It was finalized in 1976, but has not been able to retain the further siltation of the Hooghly. Instead, an increasing amount of water has been diverted for irrigation over the years. As a consequence, the flow of the Ganges during the dry season has decreased, creating tensions between Bangladesh and India on sharing the available Ganges water during the dry season.[17]

Climate change has huge impacts on the regional level and at the country level with respect to water resource availability. As an upstream country, Nepal is mainly affected by changes in future climate, while Bangladesh, as a downstream country, is affected by both changes in climatic conditions and upstream developments. These problems are already visible during the dry season, when the available water resources are scarce. By the 2050s, these problems of the dry season water scarcity and monsoon flooding will further aggravate and create much larger sufferings for the people, their livelihoods, agriculture, and economy as a whole. Integrated water management can be a good option to solve these issues. It will incorporate changes and

[17] Wium Olesen K, Ammentorp HC, Hvam Petersen N. Mega deltas and the climate change challenges, http://www.irtces.org/zt/11isrs/paper/Kim_Wium_Olesen.pdf.

adjustments in water management structure operations, increase cooperation and coordination, and will be based on a sharing approach. A good example of basin-scale integrated water management is a water-sharing treaty (India and Bangladesh treaty on sharing of Ganges flow, India and Nepal Kosi agreement, etc.). These treaties can create a win—win situation for all countries in the basin by managing the water resources efficiently through upstream reservoir management and efficient irrigation systems by reducing losses from the system. This will also increase water availability during the dry season for the downstream countries.

4. NONAGRICULTURAL WATER DEMAND: A CASE STUDY OF BANGLADESH[18]

4.1 Domestic, Industrial, and Commercial Water Demands

The population of 64 districts for 2030 and 2050 has been projected based on the 2011 population of Bangladesh. The annual domestic water demand was then calculated for each of the districts for the years 2011, 2030, and 2050, considering the projected total population. The rate of water consumption might be varied in the future, depending on temperature and rainfall, but will be very insignificant in comparison with the other uncertainties of the estimation. Therefore, it has been considered constant for present and future conditions. The annual industrial and commercial water demand has been calculated considering the urban populations of 2011, 2030, and 2050. Region-wise annual urban and rural domestic water demand is shown in Table 3.4.

The annual domestic water demand for Bangladesh is estimated at 2716 Mm^3 in 2011, projected to increase by 50% and 99%, respectively, for 2030 and 2050. Industrial and commercial water demand is estimated as 79.3 Mm^3 for year 2011. By 2030 and 2050, industrial and commercial water demand will increase by 125% and 340%, respectively. Among the seven zones, annual domestic demand (661.2 Mm^3) is the highest in the North Central (NC) zone and the lowest (194.8 Mm^3) in the South Central (SC) zone, as domestic demand is directly related to total population and its living status. In addition, urban water demand will be much more than the rural domestic demand. For example, the North West (NW) zone was the

[18] Bangladesh Integrated Water Management Project: Technical Note on Non-agriculture Water Demand Estimation for Bangladesh [35].

Table 3.4 Annual Industrial and Commercial Water Demand by Hydrological Regions.

Hydrological Regions	Annual Industrial and Commercial Water Demand (Mm^3) 2011	2030	2050
North West (NW)	15	35	76
South West (SW)	8	21	45
North Central (NC)	29	57	96
North East (NE)	6	17	38
South East (SE)	6	17	38
South Central (SC)	3	10	23
Eastern Hills (EH)	9	18	31
Total	**76**	**175**	**347**

Courtesy Bangladesh Integrated Water Management Project: Technical Note on Non-Agricultural Water Demand Estimation for Bangladesh.
CSIRO, WARPO, BWDB, IWM, BIDS, and CEGIS. Bangladesh integrated water resources assessment. Australia: CSIRO; 2014.

largest region, with 37 million populations and had the highest water-demanding region in the base year 2011. In 2050, though, NW will contain the highest population, but will be the second highest water-demanding region after the NC region. The underlying factor for such changes is that the NC region will contain more urban population in 2050 than the NW region. In the case of annual industrial water demand, it is the highest in the NC zone because the urban population is high in this zone, and it is the lowest in the SC zone. Immediate river water pollution control measures should be the priority issue; otherwise, the quality of the surface water crisis will be a disaster case.

4.2 Forest Water Demand

Forest water demand for both natural forests and homestead forests has been calculated for every district with different climate change scenarios considering rainfall and evapotranspiration. Forest water demand has been calculated for 2030 and 2050, based on two forest cover scenarios: (1) considering present forest areas (about 12% of total area), and (2) considering 25% of the country area as forest area. Forest water demand in 2030 and 2050 by hydrological regions are presented in Table 3.5.

Total forest water demand for Bangladesh has been estimated to be 2928 Mm^3. The highest forest water demand (1091 Mm^3) is found in the EH region, and the lowest forest water demand (143 Mm^3) is observed in the NE region. Forest water demand is expected to change in response to

Table 3.5 Annual Forest Water Demand for Various Climate Change Scenarios

	Forest Water Demand (Mm^3) Considering Present Forest Area					Forest Water Demand (Mm^3) Considering 25% Forest Area			
Hydrological Region	**Base**	**2030s_avg**	**2030s_dry**	**2050s_avg**	**2050s_dry**	**2030s_avg**	**2030s_dry**	**2050s_avg**	**2050s_dry**
North West (NW)	109	136	152	147	154	159	177	172	180
South West (SC)	850	492	895	640	906	915	1678	1194	1698
North Central (NC)	243	146	190	157	196	241	313	259	324
North East (NE)	143	92	130	97	139	154	219	161	234
South East (SE)	181	177	255	213	265	324	466	392	484
South Central (SC)	311	212	296	247	310	393	549	458	574
Eastern Hills (EH)	1091	599	776	653	808	1159	1504	1264	1566
Total	**2928**	**1854**	**2694**	**2153**	**2779**	**3346**	**4905**	**3900**	**5061**

Courtesy CEGIS. Non-agricultural water use. Bangladesh Integrated Water Resources Assessment Project: Technical Note on Non-Agricultural Water Demand Estimation for Bangladesh; 2014.

case-specific climate change scenarios. It will decrease in 2030 and 2050 for the same forest area, as there is an increase in annual rainfall with climate change, while the demand will increase if the forest area increases to 25% of the total country area. About 37% and 26% of total demand will be reduced in 2030 and 2050, respectively, with average climate change condition if the forest area remains unchanged, while it will increase 14% and 33%, respectively if the forest area extends over 25% of the total country area. Forest water demand will be higher for dry conditions than average climate change conditions, as rainfall is expected to be higher in average conditions than in dry conditions.

4.3 Fisheries Water Demand

Total fishery water demand for Bangladesh has been estimated to be 5194 Mm^3 (open water capture fisheries water demand at 3070 Mm^3 and closed water culture fisheries water demand at 2123 Mm^3). The highest demand (both open and closed water fisheries) was found for the NW region (1800 Mm^3). Total fisheries water demand is the lowest for the EH region (46 Mm^3), which is mainly because of less closed water fisheries demand. Fisheries water demand has been calculated for various climate change scenarios considering rainfall and evapotranspiration. About 12% and 17% of total fisheries water demand will be reduced in 2030 and 2050, respectively (Fig. 3.3), because of average climate change conditions, while it will increase 8% and decrease 3%, respectively, in dry climate change conditions (Table 3.6).

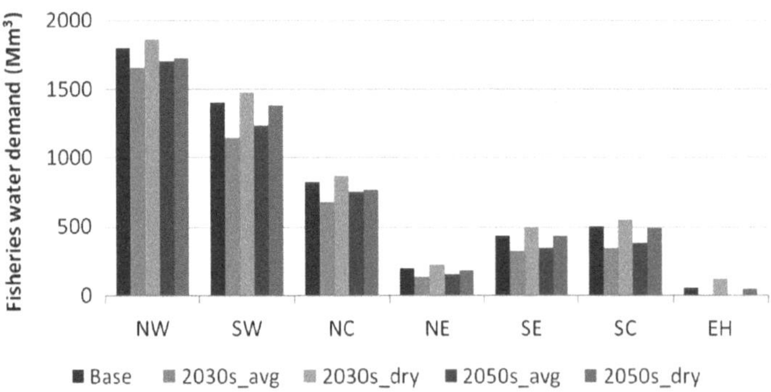

Figure 3.3 Annual fisheries water demand for different hydrological regions of Bangladesh. *CEGIS. Non-agricultural water use. Bangladesh Integrated Water Management Project: Technical Note on Non-Agricultural Water Demand Estimation for Bangladesh; 2014.*

Table 3.6 Water Demand of Fisheries for Various Climate Change Scenarios

	Annual Fisheries Water Demand for Climate Change (Mm^3)				
Hydrological Regions	**Base**	**2030s_avg**	**2030s_dry**	**2050s_avg**	**2050s_dry**
North West (NW)	1800	1656	1861	1704	1729
South West (SW)	1402	1145	1479	1234	1385
North Central (NC)	819	681	872	752	770
North East (NE)	195	134	222	155	180
South East (SE)	431	324	491	347	433
South Central (SC)	500	343	548	382	486
Eastern Hills (EH)	46	4	119	3	40
Total	**5194**	**4287**	**5594**	**4577**	**5023**

Courtesy CEGIS. Non-agricultural water use. Bangladesh Integrated Water Management Project. Technical Note on Non-Agricultural Water Demand Estimation for Bangladesh; 2014.

4.4 Environmental Flow

The concept of minimum flow in rivers came into practice during the 1970s, which explains a threshold flow value below which the health of the river degrades significantly if more and more water is withdrawn. Considering the importance, many scientists demonstrated the significance of river flow regime, such as high, medium, and low flows, and flood pulses for maintaining healthy river habitats and aquatic flora and fauna.[19] From the climate change impact on water perspective, it is critical to follow and apply the lessons learned from the past so-called development investment in the water sector. Water management in developed nations focused largely on maximizing flood protection, water supplies, and hydropower generation, the subsequent ecological and economic effects of these projects prompted scientists to seek ways to modify dam operations to maintain certain fish species and biodiversity. The initial focus was on determining the minimum flow necessary to preserve an endangered fish species in a river.[20] Environmental flows evolved from this concept of "minimum flows" and, later,

[19] Dhanya CT, Arun Kumar. Curr Sci 2015;109(6):1019–20. Central Inland Fisheries Research Institute, Assessment of environmental flows for Etalin hydro-electric project in the river Dri and Tangon, Dibang Valley District, Arunachal Pradesh, Barrackpore (West Bengal), Report, Kolkata, 2015, p. 93.

[20] https://en.wikipedia.org/wiki/Environmental_flow#Evolution_of_environmental_flow_concepts_and_recognition.

"instream flows," which emphasized the need to keep water within waterways [46]. Environmental flow[21] has been calculated at the locations where river flow data is available on the prominent rivers. In the southern part of Bangladesh, especially in the SE and SC regions, environmental flow is not calculated because of the unavailability of measured discharge data. Environmental flow in that region is not a major concern because of tidal influence. The impact of climate change on environmental flow has also been estimated for 2030 and 2050 and average and dry climate change conditions. The flow data of those particular locations have been provided by the IWM for the period of 1995–98. The percentage change of flow during that period has been calculated, and the change has been applied to estimate the climate change effect on environmental flow.

The maximum environmental flow (256 m^3/s) is found in the SW region in Kushtia District, and minimum environmental flow (2 m^3/s) is found in the SW region in Jessore District. The Gorai River flow pushes away the saline waterfront in the Passur River near Khulna. The average salinity at Khulna increases with the decrease of the Gorai flows. In the SW region of Bangladesh, salinity is increased significantly during the post-Farakka period (1975–92) than the pre-Farakka period (1965–74) [36]. Two salinity thresholds are considered in estimating the discharge requirements in the Gorai and Ganges rivers. The FAO [37] recommended a 750-$\mu\Omega$/cm salinity level for irrigation. However, the Master Plan Organization (MPO) [38] accepted a level of 2000 $\mu\Omega$/cm in the worst-case scenario. The former threshold is also permissible for human consumption. The discharge requirements resulting from the analysis indicate that keeping salinity below either of the threshold limits at Khulna in April requires the highest discharge of 240 and 158 m^3/s, as opposed to the current 52 m^3/s [36].

5. CLIMATE CHANGE AND WATER QUALITY ISSUES

5.1 Salinity

The GBM lower riparian deltaic regions are found to be agriculture-dominated economies, and in the face of emerging climate change impact,

[21] **Environmental flows** describe the quantity, timing, and quality of water flows required to sustain freshwater and estuarine ecosystems and the human livelihoods and well being that depend on these ecosystems.

surface water salinity is a growing concern, with reduced dry season flow in the rivers because of upstream withdrawals. The MPO Technical Report No. 10 [39], as one of the cases for evidence, looked at the salinity influence in coastal areas of Bangladesh and developed a relationship among river discharge and salinity at specific locations of Padma/Ganges River. The base data for the MPO report is from 1965 to 1983, and during this period, the average lowest flow was around 6000 m^3/s in February, and the corresponding salinity at Ilshaghat was around 715 $\mu\Omega$. For the present study, the base period spans from 1981 to 2012, with average February flow at Baruria of 4488 m^3/s. Based on the relationship, salinity corresponding to this value will be 1160 $\mu\Omega$, influenced by the flow reduction of Ganges and the Brahmaputra.

The future precipitation and temperature estimates that are due to climate change for the GBM basins have been collected from Climate Wizard of Washington University for three Special Report on Emissions[22] [40] scenarios (A1B, A2, and B1). The average ensemble result of 16 GCMs by 2050s has been utilized for the study. It has been found that the average annual temperature will increase by 1.3−1.8°C for the Ganges basin, while for the Brahmaputra and Meghna basin, the increase of temperature is 1.2−2.0°C and 1−1.4°C, respectively. It has been also found that the basin-wise annual precipitation will increase by 6−10%, 6−9%, and 7−8% for the Ganges, Brahmaputra, and Meghna basins, respectively, though the annual precipitation will increase in the future, but the dry season precipitation will decrease.

The model has been simulated for the years 1981−2012 to compute water availability for the GBM basin. Based on the annual water balance, it has been found that 981, 1981, and 3816 mm precipitation occurs for the Ganges, Brahmaputra, and Meghna basins, respectively, from which 144, 315, and 568 mm is converted to surface runoff, and 213, 767, and 1306 mm percolates to the shallow aquifer, respectively. The yearly actual evapotranspiration loss is 580, 652, and 910 mm for the Ganges, Brahmaputra, and Meghna basins, respectively. From long-term simulation results, it has been also found that annual average flow was 353,205 Mm^3 for the

[22] The Intergovernmental Panel on Climate Change (IPCC) Fourth Assessment Report (AR4) presents projections for global sea-level change of 0.21−0.48 m under the Special Report on Emission Scenarios (SRES) A1B scenario for the period 1980−99 to 2090−99, excluding carbon cycle feedbacks and excluding the recently observed dynamical changes in ice sheets.

Ganges basin, 713,681 Mm^3 for the Brahmaputra basin, and 151,043 Mm^3 for the Meghna basin.

The calibrated and validated SWAT model has been used to simulate climate change impacts. It has been found that there is an increase of flow during the wet season and a decrease in the dry season. For the Ganges basin, the average annual flow increases by 17%, 8%, and 9% for climate change scenarios A1B, A2, and B1, respectively, while for the Brahmaputra basin, the annual flow increased by 9%, 5%, and 5%, respectively, for the previous three scenarios. For the Meghna basin, the increment of average annual flow is maximum for the A2 and A1B scenarios (10%), while it is 7% for the B1 scenario. The impact of the A1B scenario is more significant than the other two climate change scenarios.[23]

5.2 Water Program Strategies and Response to Climate Change

5.2.1 Agriculture—Water and Climate Change Interface

Since 2010, 5-day flood forecasting and localized climate information products and services have been used in agriculture in the face of climate change adversaries. Local customized information provides a full range of advice regarding climate; its impacts on crops, livestock, fisheries;, and management practices to be followed to prevent, reduce, and/or manage risks. This tailored information assists farmers in making management decisions to reduce the risks and benefit from the opportunities of our variable and changing climate. An examination of weather patterns reveals that climate change is leading to a concentration of rain within a shorter period of days followed by periods of no rainfall or minimum rainfall not suitable for cultivation of paddies in drought-like conditions.

Since 2005, farmers growing monsoon season paddies in the GBM basin area often faced with the drought syndrome have started to cope with this adverse situation by changing the planting dates to derive maximum benefits, and introducing drought-tolerant crops. The traditional adaptation techniques sporadically attempted by farmers could be further strengthened through provision of weather forecasts of rainfall, along with supplementary data on solar radiation, temperature, soil moisture, etc. There is also a need for an interface with the country's Agricultural Extension Service providers

[23] Assessing the economic impact of climate change on agriculture, water resources and food security and adaptation measures using seasonal and medium range of forecasts.

who could translate the information on weather forecasts into practicable actions like land preparation, planting, and supplementary irrigation in a very user-friendly manner. Agriculture with customized weather service is thus likely to take a higher trajectory growth where the sector could be directly benefited by scientific knowledge, for example, in plant physiology, genetic engineering, etc.

5.3 Proposed Agrometeorological Forecasting System

The Regional Integrated Multi-Hazard Early Warning System for Asia and Africa (RIMES) is a regional cooperation of 26 countries in Africa and Asia on multihazard early warning. RIMES's objective to provide different needs and demands of member states in addressing gaps in the end-to-end multihazard early warning system could be instrumental in planning and management of water crises. Currently, RIMES is generating and providing localized and customized weather forecast results to national hydrometeorological organizations of the member states. In Nepal, the Department of Hydrology and Meteorology, and in Bangladesh, the Meteorological Department is regularly receiving RIMES forecasts. RIMES provides forecasts with three different lead times: short-term, medium-term, and long-term forecasts. Short descriptions of the three different lead times of RIMES forecast are given in the following:

Short-term weather forecast: The short-term weather forecast is the forecast range of up to 3 days (72 h), which can assist farmers in contingency planning. This forecast mainly contains the information about the rainfall (time and intensity) and temperature.

Medium-range weather forecast: The medium-range weather forecast is up to a week, which provides an idea about how the coming week looks. This forecast is similar to how the current BMD issues an agrometeorological forecast in Bangladesh, and such types of information help farmers with logistics planning, such as timing of fertilizer, pesticide selection, and irrigation application.

Seasonal weather forecast: This is a seasonal prediction of the weather. This forecast will better equip the farmers to longer-term resource planning and management like selection of crops and the sowing time of crops. A RIMES forecast contains information of rainfall and temperature for the present week, upcoming week, and current month. The RIMES result is a meteorological cum scientific format with numerical values of different weather parameters, which is understandable to experts and scientists but not easily understandable for the end-users, i.e., farmers. Therefore, it is very essential to convert this forecast for the farmers in an acceptable format.

6. WATER GOVERNANCE AND STRATEGIC OPTIONS AGAINST FUTURE CLIMATE SCENARIOS

Melting glaciers will cause an initial increase in river flows in winter and spring, but lead to a long-term reduction in water supply after 2050 as the ice resource depletes. Water stress will become a significant factor for agriculture and human consumption. In the short term, increasingly frequent flooding from increased glacial outflow and extreme rain events are anticipated.

At the local level, action plans and concrete implementation activities are required for irrigation systems, drinking water, and water-induced disasters. Training, awareness raising, and mobilization of resources are essential at the local level; this will require decentralized funds that are managed at the local level, for example, through community-based organizations.

The institutional frameworks governing irrigation and domestic use of groundwater in South Asia are not need-responsive and weak. It was observed that over the last three decades, irrigated areas have grown faster than the institutions needed to regulate this growth [41]. There is confusion and a lack of adequate scientific knowledge available to people about environment-friendly policy options, areas of responsibility, and failure to predict adverse impacts and consequences. Water rights remain unclear, and ad hoc decisions are taken by the regulatory bodies. In many cases, governments have subsidized electricity used to power water pumps for agricultural use. Groundwater depletion is now a serious threat for many regions in South Asia and can be addressed only through improved integrated resource management policies. Programs should be adopted by the government with due consideration of rational use of surface- and groundwater to facilitate sustainable recharge of underground aquifers [42].

6.1 National Water Policy

A case example of Bangladesh Water Policy, being a lower riparian member of the GBM basin, is discussed here to get an understanding of water management under stress, which is beyond the control of the country since 93% of the total GBM catchment lies outside of Bangladesh [43]. The National Water Policy was enacted in 1999 to address the objectives of improved water resources management and protection of the environment, recognizing that every public agency, community, village, and individual has an important role to play in ensuring that the water and associated natural resources of Bangladesh are used judiciously for the benefit of present and future generations. The policy recognizes that the process of planning and managing water resources requires a comprehensive and integrated analysis of relevant hydrological, topographical, social, political, economic, environmental, and institutional factors across all related water-using sectors.

The nature of drainage systems within the country requires that planning and management of river systems are undertaken within the context of hydrological regions. The principal river systems create natural boundaries for these regions. The hilly areas of the east form another hydrological region.

To address these issues through its responsible agencies, the policy of the government, among others, as reflected in the policy document, is the following:

- Undertake comprehensive development and management of the main rivers through a system of barrages and other structural and nonstructural measures.
- Develop water resources of the major rivers for multipurpose use, including irrigation, fisheries, navigation, forestry, and aquatic wildlife.
- Desilt water courses to maintain navigation channels and proper drainage.
- Delineate water stress areas based on land characteristics and water availability from all sources for managing dry season demand.
- Take steps to protect the water quality and to ensure efficiency of its use.
- Develop early warning and flood-proofing systems to manage natural disasters like flood and drought.
- Designate flood risk zones and take appropriate measures to provide desired levels of protection for life, property, vital infrastructure, agriculture, and wetlands.
- Undertake surveys and investigations of the problem of riverbank erosion, and develop and implement master plans for river training and

erosion control works for preservation of scarce land and prevention of landlessness and pauperization.

- Plan and implement schemes for reclamation of land from the sea and rivers, etc.[24]

7. SUMMARY

- Security of food, fiber, and fuel as well as growth and development associated with environmental health and poverty reduction are dependent on the conservation, quality, and availability of adequate flow of the GBM rivers. In the face of climate change impact, all member countries of GBM should come together to follow the "basin-wide approach" in policy, planning, and programming.
- Climate-resilient integrated water resource planning and management decisions in the basin should be consensus, based on scientific scenarios to tap potential opportunities against odds, and socioeconomic and ecological benefits in the face of changing climate and rising water pollution challenges.
- Priority should be given at the existing regional platforms, such as how SAARC may be instrumental in identifying GBM basin-wide trends and hotspots around the interface of water, agriculture, energy, and health; including other SAARC goals, those are extremely important for informed decisionmaking. This decision-making process needs paradigm shift from top—bottom to combined bottom-up—top-bottom processes with broad-based engagement of peoples in the hotspots, stakeholders including relevant institutions, civil society, researchers/scientists, media, and the government.
- The river systems are interchained, and hence, integrated and collaborative policies are required on a consensus basis for a sustainable solution, as one of the major challenges is that countries attach different valuation to water and do not approach the resource with a common understanding.
- About 90% of the natural hazards are caused by water-related issues, which will have serious consequences in changing scenarios by acting as a barrier to advancing economic development in the South Asia. It is predicted that this region will be threatened most by climate

[24] Bangladesh Delta Plan 2100 Formulation Project, Planning Commission, Government of Bangldesh, 2015.

change—related food security issues by the end of 2030. Regional hubs on comparative advantage basis logistics, food bank, knowledge, and data sharing should be in place in each country to extend support to each other in emergencies.

- On the eastern coast of India along the Bay of Bengal, recent study reveals the fact that SLR is comparatively high; and trends derived from altimeter data show 5 mm+/year, which is higher than those estimated from tide-gauge records over longer periods during the 20th century. On the northern coast of the Indian Ocean, this rate of SLR during 1993—2012 period reveals 3.2 mm/year as estimated.[25] The consequence of an SLR of 22 cm in 2050 and a 5% increase of cyclonic wind speed would be a call for an increase of the crest level for sea-facing embankments of 31 cm. Future planning and investment in water resource infrastructure needs to consider basin-wide integrated development approach.
- Most of the climate model data suggested that annual runoff will increase by 7—12% by 2050, primarily because of accelerated snowmelt in the upper basin, together with an increase in precipitation. However, another study suggested that water availability might increase in the short term, but will decrease in the long term, as the snow cover at the source of river water will be declined. Focus should be given on water conservation and appropriate technology in place to address this future disaster.
- The climate change scenario for 2050 reveals the fact that a large amount of snowmelt water results in a 24% increase of flow during March.[26] The early snowmelt also results in the reduction of flow during June. There is a maximum of 4.3% reduction of flow that occurred for a high-emission scenario during June. The peak flow increased by 5.9%, 7.8% and 3.2% for high-, medium-, and low-emission scenarios, respectively. The flow also increased during the postmonsoon period (October—November), with a maximum increase of 17.2%.
- It has been found that there is an increase of flow during wet season and a decrease in the dry season. For the Ganges basin, the average annual flow increases by 17%, 8%, and 9% for high-emission (A1B), medium-emission (A2) and low-emission (B1) climate change scenarios,

[25] http://www.currentscience.ac.in/Volumes/108/05/0966.pdf.

[26] CSIRO, WARPO, BWDB, IWM, BIDS, and CEGIS. Bangladesh integrated water resources assessment. Australia: CSIRO; 2014.

respectively, while for the Brahmaputra basin, the annual flow increased by 9%, 5%, and 5%, respectively, for the previous three scenarios. For the Meghna basin, the increment of average annual flow is maximum for A2 and A1B scenarios (10%), while it is 7% for the B1 scenario. The impact of the A1B scenario is more significant than other two climate change scenarios.

- It has been found that the average annual temperature will increase by 1.3–1.8°C for the Ganges basin, while for the Brahmaputra and Meghna basin, the increase of temperature is 1.2–2.0°C and 1–1.4°C, respectively. It has been also found that the basin-wise annual precipitation will increase by 6–10%, 6–9%, and 7–8% for the Ganges, Brahmaputra, and Meghna basins, respectively. Though the annual precipitation will increase in the future, in the dry season, precipitation will decrease.

CONCLUSIONS

The GBM basin holds the world's largest reserve of fresh and annually replenished ground and surface water and needs timely attention to maintain this resource and reap the benefit for people at large through an integrated basin-wide approach to follow water management principles. By the 2050s, this scenario will be changed, as the problem of dry season water scarcity and monsoon flooding will further aggravate and create much larger sufferings for people, their livelihoods, agriculture, and the economy as a whole. Climate-induced change in water regime and water pollution associated with the rapid urbanization and industrialization will affect the water availability and quality as well. In order to address these challenges common to the region, it is a critical decision issue to put top priority toward river conservation, effective governance, and mutually agreed regional policy to be implemented by the governments of the basin countries. A broad-based stakeholder's engagement of the basin region for an integrative approach is the critical need of the time to address the complex challenge of water management in this highly water-dependent environment. A free flow of technical and scientific data and information and sharing water-related decisions and practices would improve policymaking within the countries and be a confidence-building step in water dealing among regional countries.

Water-sharing treaties with effective assurance mechanisms can create a win–win situation for all countries in the basin by managing the water resources efficiently through environmentally sound and socially acceptable upstream–downstream reservoir management and efficient irrigation systems by reducing losses from the system. This will also increase water availability during the dry season for the downstream countries. The struggle against floods and droughts needs to be redefined, and the protection against saltwater intrusion and more frequent cyclones will require adaptation and new technologies. Climate change is expected to cause water stresses in new areas of the GBM basin and is likely to exacerbate the already critical water situation and quality as well. While the climate change projections derived from climate models are useful at large regional scales, the impacts need to be assessed at local scales with more precision. There is a considerable uncertainty associated with the exact nature of impacts on the water sector at local, river basin scales. The water management agencies and policymakers must take into account the uncertainties while planning for the future. Other than water availability, irrigation water demands and water quality parameters in a river are to be affected by climate change and frameworks to get the future projections considering climate change.

The immediate task at the existing regional platform, such as at SAARC, may be instrumental in identifying GBM basin-wide trends and hotspots around the interface of water, agriculture, energy, and health, and together with other SAARC goals, are extremely important for informed decisionmaking. This decision-making process needs a paradigm shift from top—bottom to a bottom-up process with broad-based engagement of people in the hotspots, stakeholders, including relevant institutions, civil society, researchers/scientists, media, and the government. This should be reflected in program design processes with built-in appropriate monitoring and reporting systems at regional levels, based on local- and country-level follow-up on water governance, water quality, and environmental health determinants, and the interplay of nonenvironmental determinants on health and productivity. Nevertheless, available insights do provide a basis for effective action.

REFERENCES

[1] UNESCO. The United Nations world water development report 4: managing water under uncertainty and risk. 2012. Chapter 4: Beyond demand: water's social and environmental benefits.

[2] UNESCO. The United Nations world water development report 4: managing water under uncertainty and risk. 2012. Chapter 7: Regional challenges, global impacts.

[3] Chowdhury R, Ward N. Hydro-Meteorological variability in the greater Ganges-Brahmaputra-Meghna basins. 2004.

[4] Biswas AK. Management of Ganges-Brahmaputra-Meghna system: way forward. Management of transboundary rivers and lakes. Springer Berlin Heidelberg; 2008. p. 143—64.

[5] JRCB (Joint Rivers Commission Bangladesh). Basin map of the Ganges, the Brahmaputra and the Meghna river; 2011.
[6] Malik R. Growth impacts of development and management of water resources in proceedings: annual IWMI-Tata partners meet: 2008. Hyderabad: International Water Management Institute; 2008.
[7] Iyer R. Water: charting a course for the future: II. Econ Polit Wkly 2001:1235−45.
[8] IPCC. The IPCC's fifth assessment report | What's in it for South Asia? In: Ahmed M, Suphachalasai S, editors. Assessing the costs of climate change and adaptation in South Asia; 2014.
[9] Wu X, Jeuland M, Sadoff C, Whittington D. Interdependence in water resource development in the Ganges: an economic analysis. Water Policy 2013;15(S1):89−108.
[10] Price G, Alam R, Hasan S, Humayun F, Kabir MH, Karki CS, Mittra S, Saad T, Saleem M, Saran S, Shakya PR. Attitudes to water in South Asia. 2014 [Chatham House Report].
[11] Rasul G. Food, water, and energy security in South Asia: a nexus perspective from the Hindu Kush Himalayan region. Environ Sci Policy 2014;39:35−48.
[12] Dore J. An agenda for deliberative water governance arenas in the Mekong. Water Policy 2014;16(S2):194−214.
[13] FAO. Irrigation in southern and eastern Asia in figures AQUASTAT survey − 2011. Rome: FAO; 2012. FAO Water Report, 37.
[14] Water Aid. In: Jones Elizabeth M, editor. An overview of the arsenic issues in Bangladesh. Dhaka: Water Aid Bangladesh; 2000.
[15] RDA (Rural Development Academy. Safe drinking water supply through RDA-developed low-cost deep tube well. 2001. A Report prepared by Matin MA, Khan MH, Khan MNI. Bogra: RDA, November 2001.
[16] Ahmed M. Integrated water resources management in Bangladesh. In: RSS Workshop on IWRM. Dhaka: Bangladesh University of Engineering & Technology; 2007.
[17] BADC (Bangladesh Agricultural Development Corporation). Groundwater zoning map and its application. In: A National seminar paper presented by Eftekharul Alam, April 9. Dhaka: Sech Bhaban; 2006.
[18] ICIMOD. In: Vidya RA, Sharma E, editors. Research insights on climate and water in the Hindu Kush Himalayas; 2014. One of the chapter contributors in this research is Hassan et al on Climate Change and Water Availability in the Ganges-Brahmaputra-Meghna Basin: impact on Local Crop Production and Policy Directives.
[19] CEGIS. Non-agricultural water use. Bangladesh integrated water resources assessment project report. Bangladesh: CEGIS; 2014.
[20] IPCC. Climate change 2014: impacts, adaptation, and vulnerability. Summary for Policymakers. 2014. p. 4.
[21] IPCC. Climate change 2013: the physical science basis. In: Stocker TF, Qin D, Plattner G-K, Tignor M, Allen SK, Boschung J, Nauels A, Xia Y, Bex V, Midgley PM, editors. Contribution of working group I to the fifth assessment report of the Intergovernmental Panel on climate change. Cambridge, United Kingdom and New York, NY, USA: Cambridge University Press; 2013. http://dx.doi.org/10.1017/CBO9781107415324. 1535 pp.
[22] Comprehensive Disaster Management Programme (CDMP). Spatial and temporal distribution of temperature, rainfall, sunshine and humidity in context of crop agriculture, prepared by the IWM and BUET; 2012.
[23] IWM. Surface water assessment of Bangladesh and impact of climate change. Bangladesh integrated water resources assessment project report. Bangladesh: IWM; 2014.
[24] Euroconsult Mott MacDonald, BETS, Socioconsult. Feasibility study of cluster of chars − integrated main report. Char development and settlement project Phase IV. Dhaka. April 2014.

[25] Gray WM. Global view of the origin of tropical disturbances and storms. Mon Weather Rev 1968;96(10):669–700.
[26] Ali A. Storm surges in the Bay of Bengal and their numerical modeling (No. 125/80). 1980 [SARC Report].
[27] Team CW, Pachauri RK, Reisinger A. Contribution of working groups I, II and III to the fourth assessment report of the intergovernmental panel on climate change. Geneva, Switzerland: IPCC; 2007.
[28] Unnikrishnan AS, Kumar KR, Fernandes SE, Michael GS, Patwardhan SK. Sea level changes along the Indian coast: observations and projections. Indian Academy of Sciences; 2006.
[29] Rabbani MG, Rahman AA, Islam N. Climate change and sea level rise: issues and challenges for coastal communities in the Indian Ocean region. Coast Zone Clim Change 2010:17–29.
[30] Asian Development Bank. Assessing the costs of climate change and adaptation in South Asia, Manila. 2014.
[31] Whitehead PG, Barbour E, Futter MN, Sarkar S, Rodda H, Caesar J, Butterfield D, Jin L, Sinha R, Nicholls R, Salehin M. Impacts of climate change and socio-economic scenarios on flow and water quality of the Ganges, Brahmaputra and Meghna (GBM) river systems: low flow and flood statistics. Environ Sci Process Impacts 2015;17(6):1057–69. Also quoted in *Bangladesh Delta Plan 2010 Formulation Project: Coast and Polder Issues, Planning Commission.*
[32] CEGIS. Assessing the economic impact of climate change on agriculture, water resources and food security and adaptation measures using seasonal and medium range of forecasts. 2013.
[33] Gaffin SR, Rosenzweig C, Xing X, Yetman G. Downscaling and geo-spatial gridding of socio-economic projections from the IPCC special report on emissions scenarios (SRES). Glob Environ Change 2004;14(2):105–23.
[34] Fung CF, Farquharson F, Chowdhury J. Exploring the impacts of climate change on water resources-regional impacts at a regional scale. Bangladesh: IAHS Publication; 2006. 308, p. 389.
[35] CSRIO, WARPO, BWDB, IWM, BIDS, CEGIS. Bangladesh integrated water resources assessment: final report. Australia: CSRIO; 2014.
[36] Mirza MMQ. Diversion of the Ganges water at Farakka and its effects on salinity in Bangladesh. Environ Manag 1998;22(5):711–22.
[37] FAO. Water quality for agriculture. FAO irrigation project. Morphological studies, vol. 3. Dhaka: Sir William Halcrow and Partners Ltd.; 1976.
[38] Master Plan Organization (MPO). Surface water availability. Dhaka: MPO; 1987. Technical Report No. 10.
[39] Master Plan Organization (MPO). National water plan, vol. 1. Dhaka: MPO; 1986.
[40] Meehl GA, Stocker TF, Collins W, Friedlingstein P, Gaye A, Gregory J, Kitoh A, Knutti R, Murphy J, Noda A, Raper S. Global climate projections. In: Solomon S, Qin D, Manning M, editors. Climate change 2007: the physical science basis. Contribution of working group 1 to the fourth assessment report of the Intergovernmental Panel on climate change; 2007.
[41] Barker R, Molle F. Evolution of irrigation in South and southeast Asia. Comprehensive assessment of water management in agriculture (program). Research Report 5. International Water Management Institute; 2007.
[42] Shah T. The groundwater economy of South-Asia: an assessment of size, significance and socioecological impacts. In: Giordano M, Villholth KG, editors. The agricultural groundwater revolution: opportunities and threat to development. Wallingford: CABI; 2007. p. 7–36.

[43] Ahmad QK, Ahmed AU. Regional cooperation in flood management in the Ganges-Brahmaputra-Meghna region: Bangladesh perspectives. In: Ahmad QK, Verghese BG, Iyer RR, Malla SK, editors. 1994. Converting water into Wealth: regional cooperation in Harnessing the eastern Himalayan rivers. Dhaka, Bangladesh: BUP/Academic Publishers; 2003. p. 192.
[44] Unnikrishnan AS, Nidheesh AG, Lengaigne M. Sea level rise trends off the indian coasts during the last two decades. Curr Sci March 10, 2015;108(5).
[45] Laghari AN, Vanham D, Rauch W. The Indus basin in the framework of current and future water resources management. Hydrol Earth Syst Sci 2012;16:1063–83. http://dx.doi.org/10.5194/hess-16-1063-2012.
[46] Flow: the essentials of environmental flows. In: Dyson M, Bergkamp GJJ, Scanlon J, editors. International union for conservation of nature and natural resources (IUCN). Switzerland, and Cambridge, UK: Gland; 2003.

CHAPTER FOUR

Forested Watersheds, Water Resources, and Ecosystem Services, with Examples from the United States, Panama, and Puerto Rico

M.C. Larsen
Smithsonian Tropical Research Institute, Balboa, Republic of Panama

1. INTRODUCTION

Forested watersheds offer access to vital natural resources, i.e., water supply for drinking, agriculture and industry, transport routes, and hydroelectric energy. Ecosystem services provided by these watersheds are numerous and, in addition to good quality water, include reduced peak river flow during storms; increased availability of groundwater and base flow in streams during seasonal dry periods and droughts; reduced soil erosion and landslide probability; enhanced resilience to wildfire, pathogens, and invasive species; biodiversity; genetic resources; and recreation [1−3].

These resource and esthetic benefits come with risks associated with floods, landslides, and wildfires, which are episodic natural disturbances in these settings. Furthermore, these disturbances compromise the provision of the ecosystem services listed previously. Nonetheless, natural landscape disturbance by floods has well-known benefits, i.e., delivery of nutrients to flood plains; landslides open forest gaps that create small-scale opportunities for successional vegetation growth, while hurricanes and wildfires similarly serve as large-scale mechanisms for resetting landscapes and creating new habitats.

More than half of the world population now lives in urban areas, which are expected to absorb all the population growth expected over the next four decades, mostly in the cities and towns of the less developed regions. The United Nations [4,5] has defined 23 megacities with at least 10 million inhabitants; all but 6 of these are in the developing world. These populations place large stresses on water and other resources.

Chemistry and Water
ISBN 978-0-12-809330-6
http://dx.doi.org/10.1016/B978-0-12-809330-6.00004-0

Societies that derive ecosystem services from forested watersheds face multiple challenges. Loss of forest cover during the 20th century has been well described [1,6]. This loss continues in the 21st century, and work by Hansen et al. [7] and Kim et al. [8] shows that forest cover in tropical America is in decline. For example, from 1990 to 2010 in Panama, forest cover decreased from 4.6 to 4.01 M ha [8]. Land cover change during this period is more complex in Puerto Rico, where, according to Gould et al. [9], forest increased in the eastern part of the island, regenerating on abandoned lands, while urbanization intensified in areas surrounding the Luquillo Experimental Forest (LEF; also located in eastern Puerto Rico). The reduction and fragmentation of forest cover compromises virtually all ecosystem services: water availability and quality, hydroelectric energy, wood products, carbon sequestration, maintenance of biodiversity, and reduction of natural hazard and vulnerability. Moreover, changing climate is already reducing the capacity for some forested watersheds to provide important services, as described in the following.

Societal use of forested watersheds and ecosystem services in the Americas, as elsewhere in the world, has increased substantially as global population has grown to the current level of 7.3 billion. The intensity of this use puts all ecosystem services at risk and requires attention at multiple societal and governmental levels so that these services are not severely compromised. This paper describes ecosystem services derived from new-world tropical and temperate forested watersheds and provides first-order examples of the value placed on these services.

In the US and other countries, public and private sector policies are increasingly placing values, often monetized, on ecosystem services so that these services can be better understood, and management decisions, including trade-offs, can be made collaboratively and transparently [10]. As these valuation and management approaches become more sophisticated, there is a concurrent need for more advanced and comprehensive monitoring and accounting systems, requiring investment from public and private entities. Some general US and Caribbean region examples are described in the following, with discussion of water resources and other ecosystem services. For Panama, the Panama Canal watershed is the focus, and in Puerto Rico, examples are drawn from the Luquillo Mountains.

2. CLIMATE CHANGE

Climate change is well documented and poses a number of direct challenges for our ability to continue to extract benefits from forested watersheds.

Additionally, population growth will require more arable land for agriculture, pushing managed forests onto areas of degraded and lower-quality soils [11]. One example of an observed change in climate is from long-term stream flow data in the northwestern US. According to Stewart et al. [12], peak snowmelt runoff is now approximately 2 weeks earlier than observed during the period from 1948 to 2000 in many western rivers and is predicted to be 30–40 days earlier as the 21st century progresses. Missouri River stream flow has also changed during the period from 1960 to 2011, with changes in climate and land use practices, according to a US Geological Survey (USGS) analysis of data at 227 streamgages?? [13]. Some regions (Kansas and southern Nebraska) of that watershed have reduced flows, while some northern tributaries, mainly in North and South Dakota, show increases in flow.

Water resource management challenges are likely to intensify over most land areas in the 21st century with the increases in the frequency, intensity, and/or amount of heavy precipitation that are expected as a result of a warmer atmosphere [14]. At the same time, warmer air temperatures and increases in intensity and duration of drought and heat stress are likely over many land areas [14,15], contributing to greater likelihood of major droughts, water stress, disease, and wildfires. Additionally, droughts and warmer temperatures stress forests, making them more susceptible to insect-borne diseases [15,16] (Fig. 4.1). Fluctuations between these

Figure 4.1 View of lodgepole pine forest in the northern Williams Range Mountains, Colorado, US. Pine bark beetles have killed more than 80% of the mature trees in this forest; mortality is visible in the red (dark gray in print versions)–brown (darker gray in print versions) color of the dead or dying trees. *Photo source: USGS, http://minerals.cr.usgs.gov/projects/colorado_assessment/.*

extremes of drought and heavy precipitation are likely to occur in spatial and temporal patterns that do not conform to past weather and climate patterns [17].

What have we already observed? Globally averaged air temperatures over land and ocean warmed by 0.85°C from 1880 to 2012. Moreover, the period from 1983 to 2012 was the warmest 30 years of the last 1400 years in the northern hemisphere [14]. Since 1901, an increase in average midlatitude northern hemisphere land area precipitation has been observed. From 1901 to 2010, global mean sea level rose by 0.19 m, and from 1979 to 2012, annual mean Arctic sea ice extent decreased 3.5–4.1% per decade [14]. For example, Arctic sea ice extent in January 2016 was the lowest January extent in the satellite record and was below the previous record January low in 2011 (Fig. 4.2). This is particularly relevant to Panama, as the Arctic Ocean is predicted to transition to a seasonally ice-free state during the middle of the 21st century [18]. The revenues derived from shipping through the Panama Canal are a major part of the economy of the nation. As navigational (bathymetric data are sparse) and security systems (coast guard resources in the region are limited) are developed, this will provide potential Arctic shipping routes as cost-effective alternatives to the Panama (and Suez) Canal because of distance reductions of 35–60%.

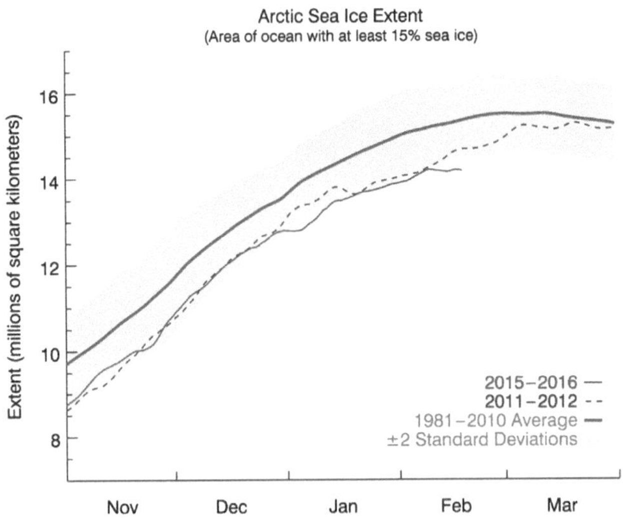

Figure 4.2 Winter Arctic sea ice cover for the period November through March. *Image courtesy of the National Snow and Ice Data Center, University of Colorado, Boulder, https://nsidc.org/arcticseaicenews/.*

Temporal and spatial patterns of precipitation distribution continue to change in North America and elsewhere, challenging water resource managers in their already difficult decision-making for allocation of water supplies. For example, water from the Lake Mead Reservoir is used for public supply and irrigation in Arizona, Nevada, California, and northern Mexico. The Las Vegas Valley is particularly dependent on the reservoir for its water supply. Barnett and Pierce [19] have estimated that there is a 50% chance that Lake Mead, a key source of water for the southwestern US, will be dry by 2021 if the climate warms as predicted and if regional water consumption is not reduced. There is evidence that we are seeing this change now. USGS and NASA Landsat satellite imagery from 2000 to 2015 shows that the lake, located at the Nevada/Arizona border, has been losing water volume for most of that period of record [20] (Fig. 4.3). The Southern Nevada Water Authority has been responding to this challenge by proactively implementing a wide range of water conservation strategies, such as lawn and golf course watering restrictions and incentives that include a rebate to customers of $2 per square foot (\$0.19 per m^2) of grass removed and replaced with desert landscaping up to the first 5000 square feet (465 m^2) converted per property. In addition, the Authority has extensive water reuse requirements and practices in place in public use areas.

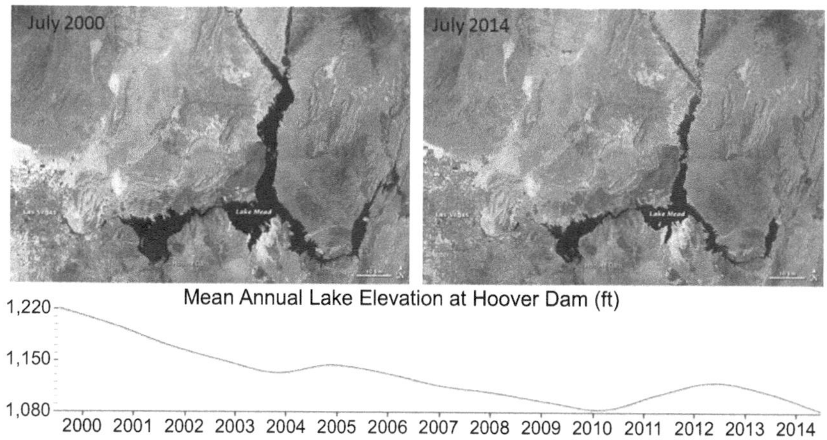

Figure 4.3 Landsat 7 and Landsat 8 images comparing Lake Mead areal extent in 2000 and 2014. Note substantial shrinkage. The mean annual lake elevation plot shows that water storage in the lake is in a multiyear decline. *Images from USGS/NASA, http://earthobservatory.nasa.gov/IOTD/view.php?id=86426.*

3. ECOSYSTEM SERVICES AND VALUATION

The Millennium Ecosystem Assessment established a benchmark for ecosystem services based on a 4-year United Nations assessment of the condition and trends of the world's ecosystems and the services we draw from them [21]. They further broadly defined ecosystem services as provisioning services, regulating, cultural, and supporting [22]. Although the term "ecosystem services" has become widely used and discussed, the concept is not new [23]; for example, von Thünen [24] discussed land use and landscape-derived services needed to sustain an agrarian-based self-sufficient state.

A regulating ecosystem service is easily understood, for example, as the improvement in water quality that is gained by protecting a watershed to enable the ecosystem to provide this service instead of depending on the construction and operation of water treatment facilities [21]. This difference has been described as green versus gray infrastructure.

Water sustains life on earth and is critical to nearly all other ecosystem services. Global estimates of ecosystem service values are provided by Coates et al. [25] and van der Ploeg et al. [26] in US dollars per hectare as $452 for water supply; $1966 for a set of regulatory services, which includes water flow regulation, waste treatment, and water purification; mitigation of extreme events (floods, droughts) and other regulatory services (air quality, climate regulation, erosion prevention, pollination, and biological control); and $398 for cultural services.

As has been stated by many, "*you can't manage what you don't measure.*" Measuring the effects ecosystem services valuation policies is complex because of the many forces involved, including climatic, local- to global-scale markets, and various intersecting, sometimes conflicting policies. In addition, long-term monitoring of forest cover, land use change, streamflow, agricultural practices, etc., is costly and in many cases not sustained by governmental agencies. For example, Sánchez-Azofeifa et al. [27] evaluated a payment for ecosystem services (PES) program that was implemented in Costa Rica in the 1990s to reduce deforestation. The authors did not observe significant effects of the PES program, but state that other policies, such as the creation of national parks and biological reserves, had already lowered deforestation rates and may have reduced the PES impact on land use practices.

3.1 United States: Ecosystem Service Payment Programs

PES is not new to the US. Starting in the 1980s, the US Conservation Reserve Program, which is administered by the US Department of

Agriculture, initiated payments to farmers to refrain from planting crops in environmentally sensitive areas. Wilson and Carpenter [28] synthesized the results of 30 different US studies from the period of 1971 to 1997 to describe the state of knowledge and gaps in understanding for the valuation of water resource ecosystem services. They stated that these valuation estimates tend to be specific to particular ecosystems and socioeconomic settings, and posit that physical and social scientists need to collaborate more effectively to improve future management and research.

The Obama administration, concerned over the impacts of climate change on the US private and public sector, has directed federal agencies with natural resource missions to develop ecosystem services assessment methods and approaches to devise innovative payment methods for ecosystem services. These approaches are designed to improve the management of these services [29]. According to Schaefer et al. [10], the federal agencies' ecosystem services approaches are grouped into enhancing investment in conservation and natural resource management, improving the cost-effectiveness of programs, making trade-offs transparent and avoiding unintended negative consequences of policy actions on ecosystems, enhancing resilience, and supporting public participation in the planning process. The agencies collaborated with the Duke University Nicholas Institute for Environmental Policy Solutions to produce a guidebook that describes and provides case study examples of ecosystem services approaches [30]. Because the guidebook was produced in 2014, it is not yet known how effective this strategy will be.

Carbon sequestration, water quality regulation, and biodiversity habitat protection, as well as suites of services such as wetland mitigation and conservation easements, have been assigned economic values in a number of states in the US [31]. Mercer et al. [31] describe three types of payments to landowners: payments directly from the government; voluntary payments from businesses, individuals, and nongovernmental organizations; and payments made to comply with government regulations, such as the Clean Water Act or the Endangered Species Act. The revenues for governmental and private sector PES derived from forested watersheds in the US were estimated at $1.9 billion in 2007 by Mercer et al. [31]. They note that because they lacked data on payments for some services, $1.9 billion is a conservative estimate. Their estimate includes $365 million from government sources (19%) and $1.5 billion (81%) from nongovernmental sources, including payments for wetland mitigation, conservation easements, and carbon offsets. Nongovernment payments come from conservation organizations such as the Nature Conservancy, the Trust for Public Land, the

Conservation Fund, and Ducks Unlimited, who provide funds to conserve land for ecosystem services such as water quality protection and biodiversity.

In another estimate of economic value obtained from an ecosystem, Batker et al. [32] estimated that the value of the natural infrastructure provided by the Mississippi River Delta ecosystem would be $330 billion to $1.3 trillion for a combination of hurricane and flood protection, water supply, water quality, recreation, and fisheries.

In a detailed report describing US federal and other programs that provide incentives for maintaining or enhancing ecosystem services, Scarlett and Boyd [33] include brief descriptions of various payment schemes from around the country:

- Florida pays farmers to maintain wetlands with a goal of improving water storage.
- In Seattle, new efforts to maintain natural landscapes have reduced storm water runoff at a cost that is 25% lower than traditional engineering solutions.
- Farmers in the Tualatin Basin, Oregon, were paid $6 million by water managers to plant trees along streams to meet water temperature requirements. The resulting shade along riparian corridors reduces water temperatures in the river channel and along the near-channel banks and floodplains. This allowed the water agencies to avoid a $60 million cost for refrigeration systems to cool wastewater effluent and storm water runoff. Riparian forests provide additional ecosystem benefits because they enhance bird and other wildlife habitats.
- New York City spent over $1.5 billion to protect and restore watersheds in the Catskill Mountains to maintain good water quality for the city. This avoided a $9 billion cost to construct and maintain water filtration and treatment facilities.
- The US Fish and Wildlife Service initiated a terrestrial carbon sequestration program to restore areas of the Lower Mississippi River Valley. They collaborated with a number of private and nonprivate entities, including energy companies and nonprofit organizations, to add 16,200 ha of restored habitat to their system of refuges and restored 32,400 ha to native habitats. In addition, they planted 22 million trees that, in the coming century, are estimated to sequester 33 million tons of carbon. This is roughly equivalent to the annual emissions from 5.5 million cars in the US.
- The state of Ohio started a Water Resource Restoration Sponsorship Program to provide loan rate reductions for wastewater treatment

projects in cases where the loan recipient commits to use a portion of their cost savings to protect watersheds and restore a land trust, park district, or other watershed protection project.

Avoiding hazardous areas such as floodplains reduces exposure for people and infrastructure. Many communities in the US use land management strategies, such as defining and zoning areas for conservation lands in floodplains, and other hazards, such as steep hillslopes. As discussed previously, this practice reduces exposure, thereby minimizing disaster costs associated with floods and landslides.

Kousky et al. [34] estimated benefits provided by floodplain conservation lands in an 11,330-ha area along the Meramec River, Missouri. In their analysis, approximately $13 million per year of flood damages were avoided by preventing development in the 500-year floodplain of the study area. They further estimated that this was a 38% reduction from average damages expected if these lands were not in a conservation area. In their simple benefit—cost analysis, they note that when considering this potential damage savings along with the recreational and esthetic benefits obtained, these conservation lands yield benefits for the region that exceed the opportunity costs of development.

It is widely recognized that vegetation, particularly forest, can stabilize steep slopes. Although not a US example, a study by Rickli and Graf [35] examined the effect of forest on shallow landslides. The authors showed that landslides were less frequent in forested terrain than in open land in six study areas in Switzerland. Their data also show that landslides mapped in forests occurred on steeper slopes than landslides mapped in open land.

Landslide losses are increasing in the US (and around the world) in association with growing population and development. Spiker and Gori [36] maintain that this trend will continue because of development in hazardous areas, expansion of transportation infrastructure, deforestation of landslide-prone areas, and climate change. These authors outlined a national landslide hazards mitigation strategy that would reduce the cost of landslide hazards and would require new partnerships between government, academia, and the private sector to sustain and expand a range of approaches, including research and development of mapping and other mitigation tools.

3.2 Ecosystem Services Obtained From the Panama Canal Watershed

The 3313 km^2 Panama Canal watershed is located at 9° north latitude with elevations that are generally 300 m or less above sea level, although several peaks reach 1000 m elevation [37,2]. Annual rainfall is variable across the watershed, from a low on the Pacific side of the isthmus of 1600 mm to more than 3000 mm on the Caribbean/Atlantic side. Approximately half of the watershed is in forest, mostly evergreen canopy, defined as tropical moist forest; however, forests near the Pacific coast are about 25% deciduous, while the wetter region near the Atlantic has few deciduous trees and includes wet forest and submontane?/forest [37,38].

Ecosystem services derived from the Panama Canal watershed provide a robust example of multiple high-value services with national, regional, and global significance [11]. Water is the most important control on virtually all canal watershed ecosystem services [39]. Annual precipitation in the canal watershed is reported as a volume of 8.9 km^3 for the period from 1993 to 2004 [40]. This translates to an annual stream flow volume of 4.4 km^3, with 2.6 km^3 (59%) used for lockages of vessels transiting the canal, 1.2 km^3 (27%) for hydroelectric power generation, and 0.27 km^3 (6%) for drinking water supply, according to an average canal watershed water budget published by Stallard et al. [2]. The balance, 7%, is mainly evaporation and groundwater infiltration [40].

Most of the nation's population of close to four million resides in or near the canal watershed, mainly along the canal route. Financial income is a major ecosystem service of the canal. A total of $1.91 billion in tolls were collected in 2014 for ships using the canal. About half of this is used for operations, and the balance goes into the general fund for the Republic of Panama. The Panama Canal Authority (ACP) has 9000 employees, but activities directly or indirectly related to canal operations generate some 200,000 jobs [41].

Shipping companies pay to use the canal because of major fuel and time savings, which prevents substantial burning of fossil fuel and consequent emission of greenhouse gases. For example, a ship traveling between New York and San Francisco saves about 13,000 km by using the Panama Canal instead of going around Cape Horn. About 14,000 ships use the canal every year [42]. Most of these are from the US, followed by those from China, Chile, Japan, Colombia, and South Korea. As such, the fuel savings and greenhouse gas emissions reductions achieved by the shipping companies from these countries (and others) are a valuable ecosystem service provided by the canal watershed but used globally. For example, shipping cargo from Shanghai to New York through the Suez Canal takes about 77 days for a round trip, but only 56 days per trip through the Panama Canal. When the Panama Canal starts use of its expanded set of locks in June 2016, which will allow greater capacity vessels to transit the system, it is expected that total reduced fuel consumption will decrease CO_2 emissions by an estimated 160 million tons in the first 10 years of operation.

Approximately 197,000 m^3 of water has been used for each vessel to transit the canal on average in recent years [43]. That totals 2.76 km^3 of water per year for shipping purposes, which, using the $1.91 billion in tolls, equals a value of 1.4 m^3 of water per dollar, or conversely, a value of $0.69 per m^3 of water. Using the 3313 km^2 area of the Panama Canal watershed, the annual shipping value per hectare of land is $5765. These are overly simplistic valuations of water and land, but provide a gauge of the value of the land used to support this particular water use in Panama. The approximate value does not include the important hydroelectric, esthetic, recreational, carbon sequestration, biodiversity maintenance, or overall ecosystem habitat values that are also provided by this water.

An additional complication of placing a specific economic value on water in the Panama Canal watershed is that each cubic meter of this water gains more importance during shortage periods. Average annual rainfall is 2659 mm at the Smithsonian Tropical Research Institute, administered Barro Colorado Island Nature Monument, located within the Panama Canal watershed (Fig. 4.4). Rainfall totals have a large interannual variation in accumulation, with a low of 1699 mm in 1997 and a high of 4487 mm in 1981. During drought years, the ACP has sometimes had to require that

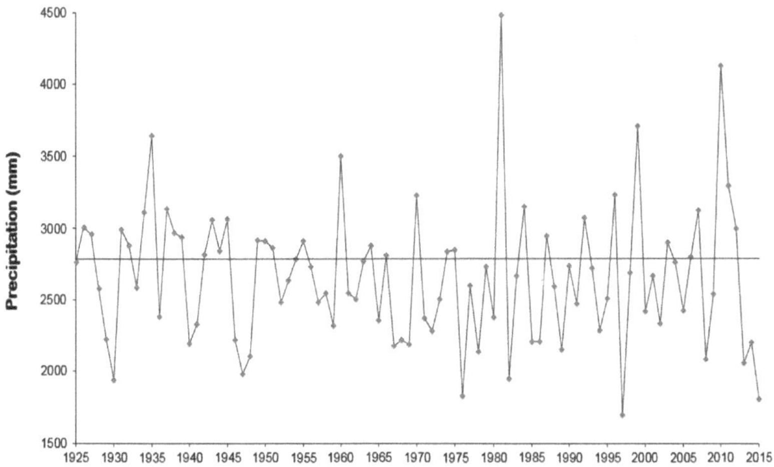

Figure 4.4 Annual rainfall totals, 1925 to 2015, Smithsonian Tropical Research Institute, Barro Colorado Island, Panama. Average for the period is 2659 mm. Note large interannual variation in accumulation. *Data source: Steven Paton, Smithsonian Tropical Research Institute, http://biogeodb.stri.si.edu/physical_monitoring/research/barrocolorado.*

vessels transiting the canal do so with reduced draft, which means that the fees paid by the shipping companies are also reduced.

Drinking water and energy production are the other major economically quantified ecosystem services of this watershed. Drinking water for more than half of the nation's population is obtained from the watershed; energy production for more than half of Panama's electrical energy supply is hydroelectric from dams in the canal watershed. In 2014, the canal generated $246 million in revenue from the sale of electric power and $29.4 million from the sale of potable water [43]. The Panama water authority (the Instituto de Acueductos y Alcantarillados Nacional) charges approximately $0.26 per m^3 to the consumer for potable water [44].

As noted previously, recreation, tourism, carbon sequestration, and maintenance of biodiversity are other important ecosystem services derived from the Panama Canal watershed. Total annual tourism revenue for the nation in 2015 from an estimated two million visitors was approximately $4 billion. Because data are limited, estimating economic values for carbon sequestration and maintenance of biodiversity is beyond the scope of this paper (see Ref. [11] for discussion of these services).

At 9° north latitude, Panama has the good fortune to be located just south of the Atlantic–Caribbean and Pacific hurricane zones. In the past 150 years of tracking of hurricanes, none have directly impacted the country. Nonetheless, floods caused by other weather systems, often convective disturbances associated with the location of the intertropical convergence zone (a dynamic band of convective moisture associated with the convergence of near-equatorial easterly trade winds from the northern and southern hemispheres), are not uncommon, and flood risk is the principal natural hazard faced by Panama, where many people live along or near riparian corridors. Storms with significant flooding in the canal watershed tend to occur at the end of the rainy season, for example: October 1923, November 1931, November 1932, November 1966, December 1985, December 2000, November 2004, and December 2010 [45]. A notable example of a major storm on this list with associated significant flooding is the event of December 2010. This storm, known as La Purisima, serves as a good illustration of flood and landslide hazard mitigation as an ecosystem service in the Panama Canal watershed [46]. The storm also illustrates what happens when hazard-related ecosystem services are at or beyond their limits when a rare, large-magnitude storm affects hillslopes and riparian corridors.

La Purisima, described as the largest 3-day storm in the Canal watershed's 100-year recorded history, was associated with the interaction of a frontal system and the intertropical convergence zone, and produced 760 mm of rainfall in 24 h. Mean stream flow for the principal canal watershed fluvial system, the Chagres River, was 908 m^3 per second, and a 3-day total steam flow volume of 235 million m^3 was calculated. This volume has a recurrence interval of approximately 300 years and was the largest flow recorded in the 78 years since recordkeeping began [46]. In a rare mitigation step, the ACP was forced to open the canal locks to discharge water, halting ship transit through the Canal for 17 h [46]. Additionally, the rainfall caused more than 500 landslides and temporarily closed the two roads that connect the two major cities of the country, Panama City and Colón. The landslides also introduced a massive pulse of sediment into river channels, raising water turbidity at a key public supply intake to 600 nephelometric turbidity units, closing water supply facilities and leaving parts of Panama City without normal water supply for 50 days. These aspects of the environmental response to this rare storm illustrate what happens when ecosystem services are fully or partially overwhelmed by the magnitude of the event.

About half of the canal watershed has been deforested, and the official policy in the canal watershed (Law 21) is to reforest in anticipation of regaining ecosystem services [2]. Canal watershed locks and dams were at their design limits during this flood, meaning that if there was much more stream flow, which would have been the case if more of the watershed had been deforested, the dam and the locks could have failed, a major disaster for Panama and world shipping. This averted disaster shows the high ecosystem service value of the forested areas of the Panama Canal watershed. Important services, including canal operations, were temporarily compromised, but canal infrastructure held up. Furthermore, an essential measure of the value of an ecosystem service with regard to hazard mitigation is loss of life. In spite of the large magnitude of this storm, few casualties were reported. The great importance of maintaining forest in this watershed, with extensive high-value infrastructure downstream, as well as critically important public water supplies, cannot be overemphasized. The environmental response provides a good example of when green (forested land) and gray (dams, locks) infrastructure is overwhelmed. The dams and locks (gray infrastructure) reached their design limits, and the green "infrastructure" (forests) was at capacity for mitigation.

With respect to ongoing management of flood hazard as an ecosystem service, the ACP has a flood control program that identifies, mitigates, and responds to conditions that pose a danger to communities and property located along riparian corridors (and on key ACP reservoirs and canal infrastructure) that could potentially interrupt canal operations [45]. The ACP, like many agencies that manage multiuse reservoirs (i.e., reservoirs used for a combination of flood control, hydroelectric energy production, drinking water supply, irrigation, and recreation) uses a complex set of metrics to control canal watershed reservoir levels to ensure water availability for human consumption, ship transit, and hydropower generation. One of the annual challenges faced by the ACP is associated with the timing and amount of rainfall delivered to the canal watershed by storms at the end of the wet season in December. The largest storms are often at the very end of the season, when reservoirs may be at, or close to, their maximum volume.

4. ECOSYSTEM SERVICES OBTAINED FROM THE LUQUILLO MOUNTAINS, PUERTO RICO

Puerto Rico, the smallest island (9000 km^2) of the Greater Antilles, is located in the northeastern Caribbean at 18° north latitude, about 1700 km southeast of Miami, US. It is an island of high relief with a maximum elevation in the central east–west trending mountain range of 1338 m. The rectilinear island measures 65 km north–south and 180 km east–west. Gradual forest removal began in the 1600s as land was cleared for agriculture by European settlers. After three centuries of extensive subsistence and plantation agricultural land use, most (94%) of Puerto Rico had been deforested by the late 1940s [9]. A shift away from agriculture toward industry began in the 1950s and resulted in much abandoned pasture and farmland that are now in secondary forest [9].

Topography in the Luquillo Mountains is rugged, stream channels are deeply incised, and annual rainfall averages more than 4000 mm in the upper elevations [47]. The mountains are largely within the boundaries of the El Yunque National Forest (EYNF), also known as the LEF, an intensely studied 11,300-ha preserve that is completely forested and under the administration of the US Forest Service. Because of the 1000-m elevational, temperature, and precipitation gradient, multiple forest types are present in the LEF, including subtropical moist forest and subtropical wet forest, with subtropical rainforest, lower montane wet forest, and lower montane rainforest at high elevations [9,48].

Prior to the 1898 US invasion, the Luquillo Mountains had been afforded some degree of forest protection during the 19th century by the Spanish crown because of the value of the hardwood there for shipbuilding and other purposes. This, along with localized cutting of wood to make charcoal, was one of the first described ecosystem services derived from the forest. During the 20th century, the mountains gained new uses as they were managed by the US Forest Service as a recreational area and as the Puerto Rico Water Authority (PRASA) began to use high-quality streamflow for drinking water supply in the region [49]. A first approximation of the value of public supply water from the LEF was estimated by Crook et al. [49] using stream flow from the nine rivers that drain the mountains. These rivers have modest water extraction sites, operated by PRASA, which is required to limit extraction in order to maintain minimum stream flow so as to sustain ecological function of the streams [50]. Water is extracted from 34 locations along these rivers, and on a typical day, 70% of stream flow from within the forest is diverted before reaching the ocean (Fig. 4.5). Two intakes draw particularly large amounts of water: the intake at Río Mameyes, which is permitted to extract 18,940 m^3/day, and the intake at Río Fajardo, permitted to extract 45,460 m^3/day [49].

In 2004, an approximate total of 0.252 million m^3/day of water was withdrawn from streams draining the LEF. PRASA charges $1.06 per m^3 for average residential customers. Using this price to the consumer for potable water in Puerto Rico, the daily volume of potable water withdrawn from the LEF has a total maximum possible value of approximately $267,000.

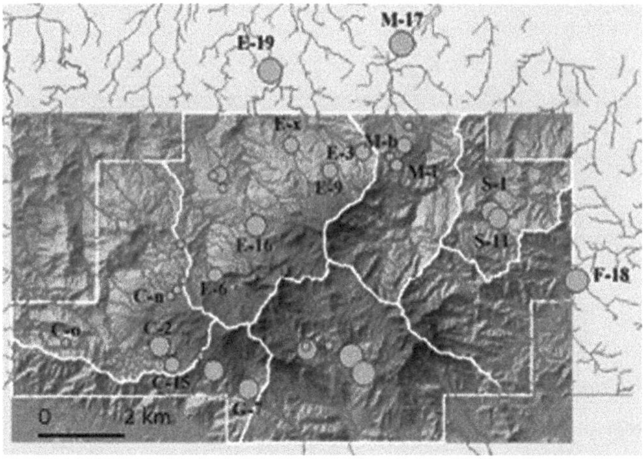

Figure 4.5 Map of public supply water intakes on streams draining the Luquillo Experimental Forest, eastern Puerto Rico. Forest boundary shown in white orthogonal lines; watershed boundaries are nonrectilinear white lines. Intakes shown with circles in which diameter is proportional to the intake withdrawal capacity, ranging from less than 10 m^3 per day to more than 10,000 m^3 per day. *Figure simplified from Crook KE, Scatena FN, Pringle CM. Water withdrawn from the Luquillo experimental forest, 2004. U.S. Department of Agriculture, Forest Service, General Technical Report GTR-IITF 34; 2007. 26 p.*

Hydropower represents only 1% of total electric energy for Puerto Rico, most of which (69%) is generated by oil-burning power plants [51]. Hydropower generation is severely limited because the 224 rivers in Puerto Rico are relatively short in length (a few tens of km), with only modest catchment size. This small watershed size, in combination with episodic droughts, makes hydroelectric energy unreliable [52]. A small hydroelectric facility on the south side of the Luquillo Mountains, on the Río Blanco, has a capacity to generate 5 MW, according to Liu et al. [51]. This is 12% of the 41.8 MW capacity from a total of 21 hydroelectric units on six rivers around the island. Puerto Rico's electricity costs are about 27 cents per kilowatt hour, approximately twice what they are in the US [53]. One megawatt equals 1000 kW, so at $0.27 per kilowatt, if the Río Blanco facility was operating at full 24 h/day capacity (it is reportedly not doing so), it would be producing electricity valued at $32,400 per day ($11.8 million/year).

The US Forest Service describes a "site visit" as the entry of one person to a national forest site or area to participate in recreational activities for an unspecified period of time. A "national forest visit" can be composed of multiple "site visits." In 2006, there were 1.336 million site visits to the EYNF, and in 2011, there were 1.123 million (written communication, Jose Ortega, Recreational Program Leader, EYNF, Puerto Rico, US Forest Service, September 8, 2015). The American Sportfishing Association [54] quantifies the economic value of visits to US Forest Service managed lands that are made for hunting, fishing, and wildlife viewing activities. Hunting and fishing are not permitted within the EYNF boundaries, so information for Puerto Rico was restricted to wildlife viewing activities. Bird watching is one the principal wildlife viewing activities as Puerto Rico, in combination with the US Virgin Islands, has approximately 270 species of birds [55]. Additionally, there is great interest in the dwindling populations of the once widely distributed Puerto Rican parrot. Between 2000 and 2003, an estimated annual average of $3.2 million was spent in Puerto Rico for wildlife viewing associated with the EYNF [54]. As the number of visitors to the forest has increased since 2003, it is likely that the economic contribution of wildlife viewing associated with the EYNF has also increased. US Forest Service data show an EYNF recreational visitor rate in excess of 1,000,000 per year.

Carbon sequestration and maintenance of biodiversity are other important services derived from the Luquillo Mountains and the forested 11,300 ha of the LEF, but these are beyond the scope of this paper.

Mitigation of flood and landslide hazards that not only threaten people and infrastructure, but compromise ecosystem services, is achieved largely through the practices of strong governance. An important part of the governance is minimization of forest removal in steeply sloping regions and zoning to prevent housing or other construction on or near the base of steep hillslopes [56]. Forested hillslopes provide a landslide hazard mitigation ecosystem service that also applies to flood hazard mitigation for people and structures located along riparian corridors. The presence of forest reduces storm runoff volume and reduces storm runoff peak stream flow in rivers, spreading the runoff volume over a larger time step than would occur if no forest were present [3].

In its recorded history, floods have caused the largest loss of life in Puerto Rico, which is the case for most countries around the world. Major floods during the 19th and 20th centuries were associated with rainfall delivered by tropical disturbances (depression, storms, hurricanes), and killed thousands [57]. Most of these flood deaths were prior to 1940 when zoning for housing location and construction standards were not well defined or regulated. Improved governance, including planning and zoning, has greatly reduced loss of life from flooding across the island. Effective governance is also evident in Puerto Rico, where a well-coordinated response system of governmental agencies is initiated each time that a tropical disturbance or other heavy rain threatens the island. Additionally, general education of the public for hazard preparation and a well-informed, decentralized civil defense network have combined to reduce loss of life to near zero during large storms.

CONCLUSIONS

The US, Panama, and Puerto Rico provide examples of a variety of payment programs for ecosystem services and for the services derived from forested watersheds, and offer insights into how we consider and take advantage of these services. The examples show the benefits and limitations of the ecosystem services provided by forested watersheds and how some of the services are valued. The examples also illustrate the importance of the maintenance and expansion of watershed forest cover as well as strong governance, which includes well-informed science- and engineering-based infrastructure zoning, planning, and design. Not surprisingly, because of its geographic size, large economy, and well-established natural resource regulatory policies, the US has the most developed set of PES.

In most countries, PES from forested watersheds is indirect. It is a value that is most commonly extracted for hydroelectric energy, water supply, and recreation through governmental or private sector charges to deliver these commodities to users. As the brief examples listed above demonstrate, most US programs that offer payments for ecosystem services don't focus on a single service. The programs are meant to incentivize general conservation practices, which support a range of benefits obtained from forested watersheds and other environments.

In spite of the numerous large investments made by federal, state, and local governments, as well as those made by nongovernmental organizations, many landowners do not participate in these PES programs. See Mercer et al. [31] for a discussion of data and statistics on this topic. Mercer et al. [31] further state that *"the economic and social forces that have led to forest fragmentation and loss in the US are so strong that PES payments have so far not had a significant impact on forest land use at the regional or national level"* and that *"changes in government and corporate policy will be critical for PES to result in large enough financial returns to effectively compete with development and other economic drivers of land use in the US in order to have a significant impact on the provision of forest-based ecosystem services."*

To best manage water resources and other ecosystem services, there is an increasing need for local land management actions and adaptation, which includes sustaining diverse forest cover, minimizing soil erosion and degradation, and assuring that road networks and essential infrastructure are well planned [58] and not built in areas subject to flood, landslide, and other hazards [59,60]. These actions assure that both the natural and built environments (green and gray infrastructure) are managed in coordination, improving and enhancing the benefits derived from each [61,62].

Additionally, mountains and rivers are often transboundary, crossing political and cultural divisions. As such, effective management of ecosystem services is highly dependent, not just on local strong governance, but also on the cooperation of local stakeholders, regional and national institutions, and in many cases, international institutions [63]. Additionally, timely access to governmental communication of accurate information associated with hazards, i.e., precipitation, streamflow, estimated fire probability, flood and landslide warnings, is key to effective response of at-risk communities so that loss of life is minimized.

Lastly, with changing climate, water resources management and flood (and landslide) hazard mitigation challenges are now increasing because the long-standing approach for estimating streamflow and flood probability is based on the principal of stationarity, which means that the present likelihood of streamflow and floods in a watershed can be well determined by examining the past 30 or more years of stream flow record. This approach has been weakened by changing rainfall and stream flow patterns observed in recent decades [17,64]. The intergovernmental panel on climate change (IPCC) [14] Fifth Assessment Report concluded that climate change has begun to affect the frequency, intensity, and length of many extreme events, thus increasing the need for additional timely and effective adaptation.

ACKNOWLEDGMENTS

Funding support for work on this paper was derived through the budget of the Smithsonian Institution. This paper was improved by comments from anonymous reviewers.

REFERENCES

[1] Noble IR, Dirzo R. Forests as human-dominated ecosystems. Science 1997;277: p.522–525.
[2] Stallard RF, Ogden FL, Elsenbeer H, Hall J. Panama canal watershed experiment: Agua Salud project. Water Resour Impact 2010;12(4):17–20.
[3] Ogden FL, Crouch TD, Stallard RF, Hall JS. Effect of land cover and use on dry season river runoff, runoff efficiency, and peak storm runoff in the seasonal tropics of Central Panama. Water Resour Res 2013;49(12):8443–62.
[4] United Nations. World urbanization prospects, the 2011 revision. UN Department of Economic and Social Affairs/Population Division; 2011. 318 p.
[5] United Nations, Department of economic and social affairs, population division. World Urbanization Prospects: The 2014 Revision, Highlights 2014 (ST/ESA/SER.A/352).
[6] Food and Agriculture Organization. State of the World's forests, FAO. Rome (Italy): United Nations; 1997.
[7] Hansen MC, Potapov PV, Moore R, Hancher M, Turubanova SA, Tyukavina A, Thau D, Stehman SV, Goetz SJ, Loveland TR, Kommareddy A, Egorov A, Chini L, Justice CO, Townshend JRG. High-resolution global maps of 21st-century forest cover change. Science 2013;342:850–3.
[8] Kim DH, Sexton JO, Townshend JR. Accelerated deforestation in the humid tropics from the 1990s to the 2000s. Geophys Res Lett 2015;42. http://dx.doi.org/10.1002/2014GL062777.
[9] Gould WA, Martinuzzi S, Parés-Ramos IK. Land use, population dynamics, and land-cover change in eastern Puerto Rico, Ch. B. In: Murphy SF, Stallard RF, editors. Water quality and landscape processes of four watersheds in eastern Puerto Rico. U.S. Geological Survey Professional Paper 1789; 2012. p. 25–42.
[10] Schaefer M, Goldman E, Bartuska AM, Sutton-Grier A, Lubchenco J. Nature as capital: advancing and incorporating ecosystem services in United States federal policies and programs. Proc Natl Acad Sci USA 2015;112(24):7383–9. http://dx.doi.org/10.1073/pnas.1420500112.
[11] Hall JS, Kirn V, Yanguas Fernández E, editors. Managing watersheds for ecosystem services in the steepland neotropics. Inter-American Development Bank. IDB Monograph, 340; 2015. 186 p.
[12] Stewart I, Cayan DR, Dettinger MD. Changes in snowmelt runoff timing in western North America under a 'business as usual' climate change scenario. Clim Change 2004; 62:217–32.

[13] Norton PA, Anderson MT, Stamm JF. Trends in annual, seasonal, and monthly streamflow characteristics at 227 streamgages in the Missouri river watershed, water years 1960–2011. U.S. Geological Survey Scientific Investigations Report 2014–5053. 2014. 128 p., http://dx.doi.org/10.3133/sir20145053.
[14] IPCC. Climate Change. In: Core Writing Team, Pachauri RK, Meyer LA, editors. Synthesis report. Contribution of Working Groups I, II and III to the fifth assessment report of the Intergovernmental Panel on climate change. Geneva (Switzerland): IPCC; 2014. 151 pp.
[15] Bentz BJ, Régnière J, Fettig CJ, Hansen EM, Hayes JL, Hicke JA, Kelsey RG, Negrón JF, Seybold SJ. Climate change and bark beetles of the western United States and Canada: direct and indirect effects. BioScience 2010;60:602–13.
[16] Clark JS, Iverson L, Woodall CW, Allen CD, Bell DM, Bragg DC, D'Amato AW, Davis FW, Hersh MH, Ibanez I, Jackson ST, Matthews S, Pederson N, Peters M, Schwartz MW, Waring KM, Zimmermann NE. The impacts of increasing drought on forest dynamics, structure, and biodiversity in the United States. Glob Change Biol 2016. http://dx.doi.org/10.1111/gcb.13160.
[17] Milly PCD, Betancourt J, Falkenmark M, Hirsch RM, Kundzewicz ZW, Lettenmaier DP, Stouffer RJ. Stationarity is dead: whither water management. Science 2008;319:573–4.
[18] Stephenson SR, Smith LC. Influence of climate model variability on projected Arctic shipping futures. Earth's Future 2015;3. http://dx.doi.org/10.1002/2015EF000317.
[19] Barnett TP, Pierce DW. When will Lake Mead go dry? Water Resour Res 2008;44: W03201. http://dx.doi.org/10.1029/2007WR006704.
[20] U.S. Department of the Interior, Bureau of Reclamation. Lake Mead at Hoover Dam, elevation. 2015. http://www.usbr.gov/lc/region/g4000/hourly/mead-elv.html.
[21] MEA. Millennium Ecosystem Assessment. Ecosystems and human well-being: a framework for assessment. Washington (DC): Island Press; 2003. 266 p. http://www.maweb.org.
[22] MEA. Millennium Ecosystem Assessment. Ecosystems and human well-being: synthesis. A report of the millennium ecosystem assessment. Washington (DC): World Resources Institute and Island Press; 2005. www.maweb.org/documents/document.356.aspx.pdf.
[23] SCEP, Study of Critical Environmental Problems. Man's impact on the global environment - assessment and recommendations for action. Cambridge (Massachusetts): The MIT Press; 1970. 319 p.
[24] von Thünen JH. Der isolierte Staat in Beziehung auf landwirtschaft und Nationalökonomie. 3rd ed. 1842. Berlin (Rostock).
[25] Coates D, Pert PL, Barron J, Muthuri C, Nguyen-Khoa S, Boelee E, Jarvis DI. Water-related ecosystem services and food security. In: Boelee E, editor. Managing water and agroecosystems for food security; 2013. p. 29–41.
[26] van der Ploeg S, de Groot RS, Wang Y. The TEEB valuation database: overview of structure, data and results. Foundation for sustainable development. Wageningen Neth 2010.
[27] Sánchez-Azofeifa GA, Pfaff A, Robalino JA, Boomhower JP. Costa Rica's payment for environmental services program: intention, implementation, and impact. Conserv Biol 2007;21:1165–73.
[28] Wilson MA, Carpenter SR. Economic valuation of freshwater ecosystem services in the United States: 1971–1997. Ecol Appl 1999;9(3):772–83.
[29] Executive Office of the President. Priority agenda for enhancing the climate resilience of America's natural resources (Executive Office of the President, Washington, DC). 2014. https://www.whitehouse.gov/sites/default/files/docs/enhancing_climate_resilience_of_americas_natural_resources.pdf.

[30] National Ecosystem Services Partnership. Federal resource management and ecosystem services guidebook (National Ecosystem Services Partnership, Duke University, Durham, NC). 2014. https://nespguidebook.com/.
[31] Mercer DE, Cooley D, Hamilton K. Taking stock: payments for forest ecosystem services in the United States; 2011.
[32] Batker D, de la Torre I, Costanza R, Swedeen P, Day J, Boumans R, Bagstad K. Gaining ground. Wetlands, hurricanes and the economy: the value of restoring the Mississippi River Delta. Tacoma (Washington): Earth Economics; 2010. http://www.eartheconomics.org/FileLibrary/file/Reports/Louisiana/Earth_Economics_Report_on_the_Mississippi_River_Delta_compressed.pdf.
[33] Scarlett L, Boyd J. Ecosystem services: quantification, policy applications, and current federal capabilities. Resources for the Future Discussion Paper No. 11−13. 2011. 78 p.
[34] Kousky C, Walls M, Chu Z. Measuring resilience to climate change: the benefits of forest conservation in the floodplain, p. 345−360. In: Sample VA, Bixler RP, editors. Forest conservation and management in the Anthropocene: conference proceedings. RMRS-P-71. Fort Collins, CO. US Department of Agriculture, Forest Service. Rocky Mountain Research Station; 2014. 494 p.
[35] Rickli C, Graf F. Effects of forests on shallow landslides − case studies in Switzerland. Snow Landsc Res 2009;82(1):33−44.
[36] Spiker EC, Gori PI. National landslide hazards mitigation strategy: a framework for loss reduction U.S. Geological Survey Open-file Report 00-450, 2000. 49 p.
[37] Condit RS, Robinson DW, Ibanez R, Aguilar S, Sanjur A, Martinez R, Stallard RF, Garcia T, Angehr GR, Petit LJ, Wright SJ, Robinson TR, Heckadon-Moreno S. The status of the Panama canal watershed and its biodiversity at the beginning of the 21st century. Bioscience 2001;51(5):135−44.
[38] Ibáñez R, Condit RS, Angehr GR, Aguilar S, Garcia T, Martinez R, Sanjur A, Stallard RF, Wright SJ, Rand AS, Heckadon-Moreno S. An ecosystem report on the Panama canal: monitoring the Status of the forest communities and the watershed. Environmental Monitoring and Assessment 1. 2002. p. 65−95.
[39] Stallard RF. Understanding natural capital, part a: geophysical context. In: Hall JS, Kirn V, Yanguas Fernández E, editors. Managing watersheds for ecosystem services in the steepland neotropics. Inter-american development bank. IDB Monograph, 340; 2015. p. 20−34.
[40] IDB. Panama canal expansion program. Inter-American Development Bank Environmental and Social Management Report PN-l1032. 2012. 38 p.
[41] Panama Canal Annual Report, 120 p., http://www.pancanal.com/eng/general/reporte-anual/index.html; 2014.
[42] Smith R. A hundred years old today, the Panama canal is about to get a lot bigger. Natl Geogr 2014. http://news.nationalgeographic.com/news/2014/08/140815-panama-canal-culebra-cut-lake-gatun-focus/?rptregcta=reg_free_np&rptregcampaign=2015012_invitation_ro_all#.
[43] Panama Canal, http://www.pancanal.com/eng/general/canal-faqs/physical.html; 2015.
[44] IDAAN, Instituto de Acueductos y Alcantarillados Nacional. http://www.idaan.gob.pa/; 2015.
[45] Cuevas JA. The Panama canal Authority's flood control program. In: Building knowledge bridges for a sustainable water future, proceedings of the second international symposium on building knowledge bridges for a sustainable water future, Panama, Republic of Panama, 21−24 November, 2011. Published by the Panama Canal Authority (ACP) and UNESCO; 2011.
[46] Espinosa JA. Water management in the Panama Canal during the December 2010 extreme flood. In: Tarte A, et al., editors. Second international symposium on building knowledge bridges for a sustainable water future, Panama, republic of Panama41−45. Panama (Republic of Panama): Panama Canal Authority and UNESCO; 2011. 301 p.

[47] Larsen MC. Landslides and sediment budgets in four watersheds in eastern Puerto Rico, Ch. F. In: Murphy SF, Stallard RF, editors. Water quality and landscape processes of four watersheds in eastern Puerto Rico; 2012. p. 153–78. U.S. Geological Survey Professional Paper 1789.
[48] Harris NL, Lugo AE, Brown S, Heartsill Scalley T, editors. Luquillo experimental forest: research history and opportunities. EFR-1. Washington (DC): U.S. Department of Agriculture; 2012. 152 p.
[49] Crook KE, Scatena FN, Pringle CM. Water withdrawn from the Luquillo experimental forest, 2004. U.S. Department of Agriculture, Forest Service, General Technical Report GTR-IITF 34. 2007. 26 p.
[50] PRASA. Puerto Rico Aqueduct and Sewer Authority Report. 500 p., http://www.gdb-pur.com/investors_resources/documents/praqueductsewerauth02a-fin-295mm.pdf; 2012.
[51] Liu H, Masera D, Esser L, editors. World small hydropower development report 2013. United Nations Industrial Development Organization; 2013. International Center on Small Hydro Power, www.smallhydroworld.org.
[52] Larsen MC. Analysis of 20th century rainfall and streamflow to characterize drought and water resources in Puerto Rico. Phys Geogr 2000;21:494–521.
[53] Gross D. Why is Puerto Rico burning oil to generate electricity? Slate. 2014. http://www.slate.com/articles/business/the_juice/2014/05/puerto_rico_is_burning_oil_to_generate_electricity_it_s_completely_insane.html.
[54] American Sportfishing Association (ASA). State and national economic effects of fishing, hunting and wildlife-related recreation on U.S. Forest Service-Managed ands. 2007. Report prepared for the Wildlife, Fish and Rare Plants U.S. Forest Service U.S. Department of Agriculture. 11.15.13, http://www.fs.fed.us/biology/resources/pubs/wildlife/usfs_wildlife_based_recreation_economic_contributions_1_03_07.pdf.
[55] Raffaele HA. A guide to the birds of Puerto Rico and the Virgin Islands. Princeton (New Jersey): Princeton University Press; 1989.
[56] Keefer DK, Larsen MC. Assessing landslide hazards. Science 2007;316:1136–8.
[57] Ramos-Ginés O. Estimation of magnitude and frequency of floods for streams in Puerto Rico: new empirical models. U.S. Geological Survey Water-Resources Investigations Report 99-4142, 1999. 41 p.
[58] Larsen MC, Parks JE. How wide is a road? The association of roads and mass-wasting disturbance in a forested montane environment. Earth Surf Process Landforms 1997;22:835–48.
[59] Larsen MC, Torres Sanchez AJ. The frequency and distribution of recent landslides in three montane tropical regions of Puerto Rico. Geomorphology 1998;24:309–31.
[60] Larsen MC, Wieczorek GF. Geomorphic effects of large debris flows and flash floods, northern Venezuela, 1999. In: Latrubesse E, editor. Tropical geomorphology with special reference to South America. Zeitschrift für Geomorphologie Suppl, vol. 145; 2006. p. 147–75.
[61] Green PA, Vörösmarty CJ, Harrison I, Farrell T, Sáenz L, Feketea BM. Freshwater ecosystem services supporting humans: pivoting from water crisis to water solutions. Glob Environ Change September 2015;34:p.108–118.
[62] Sáenz L, Mulligan M, Arjonac F, Gutierrez T. The role of cloud forest restoration on energy security. Ecosyst Serv 2014;9:180–90.
[63] ISDR. Hyogo framework for action 2005–2015: building the resilience of nations and communities to disasters. World Conference on disaster reduction 18–22 January 2005, Kobe, Hyogo, Japan. International strategy for disaster reduction. 2005. Extract from the final report (A/CONF.206/6), www.unisdr.org/wcdr. 25 p.
[64] Larsen MC. Global change and water resources, where are we headed? Water Resour Impact Am Water Resour Assoc 2012;14(5):3–7.

CHAPTER FIVE

Water Quality and Sustainability in India: Challenges and Opportunities

R.K. Sharma, M. Yadav, R. Gupta
Green Chemistry Network Centre, University of Delhi, Delhi, India

1. INTRODUCTION

Water is a vital resource among other critical substances on this planet for sustaining all forms of life, food production, and economic development. It is a unique gift to mankind from nature and is considered as one of the most manageable natural resources. All these properties of water impart great utility to human beings. Over the millennia, available water has been used in a sustainable and equitable manner through water-conservation technologies, different cropping patterns, agricultural systems, and conservation-based style of living.

However, the rapid pace of industrialization, population expansion, and urbanization have significantly contributed to severe deterioration of water quality by interfering with the natural hydrological cycle. Furthermore, growing global demand for water and poor management has disturbed its quality and has caused significant damages to the environment.

According to the World Bank, 21% of communicable diseases in India are associated with unsafe water, and according to the WHO, 97 million Indians lack access to safe water today [1]. Even though, the per-capita availability of water in the capital New Delhi is greater than that of Paris, the city does not provide reliable water. Despite investments in the infrastructure of water sanitation, many low-income communities in India continue to be deprived of safe drinking water. Problems such as endemic fluorosis have emerged as one of the most alarming public health issues of the country. In fact, contaminated water often carries toxic and carcinogenic impurities, which can cause serious health problems in humans. Not only has the water quality in the country deteriorated, but its excessive utilization by domestic, agriculture, and industrial sectors has led to depletion of water resources in many parts of the country. There is no single cause for the

Chemistry and Water
ISBN 978-0-12-809330-6
http://dx.doi.org/10.1016/B978-0-12-809330-6.00005-2

country's water crisis, but a combination of factors, including population growth, unaccounted for water, dwindling groundwater supplies from overextraction by farmers, inadequate pricing, and insufficient investment in treatment facilities at the federal, state, and local levels have contributed to the degradation of water quality [2].

Water quality is now being recognized as a major crisis in India, and the users and communities at large have to play a key role in maintaining hygiene near water sources in order to improve water quality. Consumers also have to improve the ways in which they collect and store water as to avoid contamination while collecting, storing, and using water resources. For the effective management of a sustainable water-quality plan, it is essential to have a policy that addresses all components and prioritizes control strategies according to the needs of the country [3]. The present chapter focuses on various challenges and opportunities associated with water quality and sustainability, which can help in improving existing water systems.

2. WATER RESOURCES IN INDIA

The world's oceans cover about three-fourths of the earth's surface. According to UN estimations, about 2.7% of the total available water on the earth is fresh water of which 75.2% is available in frozen form in polar regions and 22.6% is present as groundwater. Thus only a very small fraction is accessible for human consumption. Fortunately, India is enviably endowed with water resources. The country is blessed with highly crisscrossed rivers, streams, and snowcapped mountains that can meet an assortment of water requirements of the nation. The major water resources in India are discussed in the following sections.

2.1 Surface Water

India is a land of numerous mountains and rivers. Snowcapped Himalayas in the north provide water to many rivers. There are around 1500 glaciers in the Himalayan region and several natural lakes. India is also blessed with a river system comprising more than 20 major rivers with several tributaries. Many of these rivers are perennial and others are seasonal. Additionally, there are 13 major river basins in the country, which supply 85% of the country's total surface water and fulfill the needs of 80% of the nation's population [4]. This surface-water resource also plays a major role in agriculture, hydroelectricity generation, livestock production, industrial activities, forestry, fisheries, recreational activities, etc.

Box 5.1 India's Rainfall

- Occurs mainly in the months of June to September
- Is unevenly distributed across the country
- It is probable that some regions experience both flood and drought
- Average annual rainfall is about 125 cm

2.2 Groundwater

In addition to surface water, groundwater is also an important source of water. Around 2.8 million km^2 of Indian territory is filled with utilizable groundwater. Approximately 80% of domestic needs in rural India and 50% of urban and industrial uses are met from groundwater. In urban areas, groundwater serves as a source of relatively low cost and high-quality water supply.

2.3 Rainfall

Furthermore, India receives annual precipitation of about 4000 km^3, including snowfall. Out of this, monsoon rainfall is of the order of 3000 km^3. But there is great regional variation in the distribution of rainfall. India receives around 80% of its annual rainfall in the months of June to September [4,5] (Box 5.1).

Regardless of the country's abundant resources, many parts of the country have started to experience the effects of water scarcity. It is reported that the annual per capita water availability has decreased from 6042 cm^3 (in 1947) to 1545 cm^3 (in 2011). According to the estimates of the Ministry of Water Resources, this number will further reduce to 1340 cm^3 by 2025 and to 1140 cm^3 by 2050. Since the population and economic growth is increasing constantly, the Ministry of Water Resources has estimated that around half of the demand for water will be unmet by 2030.

3. OVEREXPLOITATION OF WATER IN INDIA

According to a 2010 report, agriculture is the largest consumer of water in India (85%), followed by industry and energy (9%) and domestic users (6%).

3.1 Agriculture

Agriculture is a highly inefficient consumer of water. For instance, the water productivity of rice in India is 0.2–0.26 kg per cubic meter of water, which is half that of China. Water productivity for other cereals is 0.32–0.42 kg per cubic meter of water, just one-third of that of China [6].

3.2 Industry

The industry and energy sectors of India also use water inefficiently. These sectors pay three types of charges related to water: a water cess, tariff to municipalities or other suppliers of water, and cost of extraction of water from rivers or groundwater. But the level of these charges is low in comparison to their industrial output due to which their thirst for water is increasing. Recently released figures from the Ministry of Environment and Forests show that the total amount, collected as water cess from industries in different states, has increased by Rs 40 crore in the past 3 years [6]. Though the amount collected as water cess is not enormous, the expanding pattern is an indication of how the use of water by industries is expanding over time. With the government concentrating on its "Make in India" objective, the share of water usage by industries or manufacturing sectors will undoubtedly increase. It will then make the task of water conservation significantly more vital.

3.3 Domestic

In India, the approximate usage of water/home/day is 900 liters, and in Delhi, for example, loss of 30–40% drinking water has been observed. Consumable water is wasted in several ways, e.g., from unnecessary running of tap during kitchen work, during toothbrushing/shaving and long showers, excessive flushing of toilets, too much water used to wash laundry, for home garden/lawn and car washing, etc. In addition to this, leaky taps contribute 20–100 gallons/day [7].

In spite of multiple benefits of water, we are responsible for its shortage and quality deterioration. Thus water, being a prime resource, needs special attention.

4. PARAMETERS OF WATER QUALITY

The extent of water pollution in the nation can be determined by the status of water quality. Various pollutants such as trace metals, pesticides, inorganic ions, and other micropollutants are present in water bodies. Water-quality criteria are based on the different parameters that characterize the water quality. In some cases, the criteria set a maximum level for the concentration of a substance in a particular medium beyond which it is harmful, and in other cases, water-quality criteria are set at minimum admissible concentrations. These criteria are usually developed for raw water and its relative use in drinking, domestic usage, irrigation, aquatic ecosystem, or as a habitat for biological species.

In order to set water-quality objectives, the central pollution control board (CPCB) has classified water resources into five different classes (A-E) according to their purpose of utilization (Table 5.1) [8]. These five classes have been applied over stretches of Yamuna and Ganga to set objectives and conduct surveys so as to compare their quality with that required for its best use. With the evaluation in hand, one can find the deficiencies and plan suitable measures to control pollution from the major sources. Table 5.1 depicts various water-quality variables like pH, DO, BOD, TC, etc., for a period of 5−7 days. Organic pollution is measured in terms of biochemical oxygen demand (BOD) and dissolved oxygen (DO), whereas the presence of pathogens in water is determined by total coliform (TC) and fecal coliform (FC) [9].

5. FACTORS RESPONSIBLE FOR WATER-QUALITY DETERIORATION

Water pollution in India is a serious issue. Almost 80% of India's surface water is polluted, and an alarming percentage of groundwater reserves are contaminated by various organic and inorganic pollutants. Water quality is affected by various sources, which are classified as point, nonpoint, and transboundary sources (Fig. 5.1). Thus these sources have been rendered

Table 5.1 Water-Quality Variables

		Parameters			
Use of Water	**Class**	**pH**	**DO (mg/L)**	**TC (MPN/100 mL)**	**BOD**
Drinking water source without conventional treatment but after disinfection	A	6.5−8.5	≥6	≤50	≤2
Outdoor bathing	B	6.5−8.5	≥5	≤500	≤3
Drinking water source with conventional treatment and disinfection	C	6.5−8.5	≥4	≤5000	≤3
Propagation of wildlife and fisheries	D	6.5−8.5	≥4	Free ammonia = ≤ 1.2 mg/L	
Irrigation and industrial purposes	E	6.0−8.5	Electrical conductivity = ≤ 2250μ℧/cm Boron content = ≤ 2.0 mg/L Sodium absorption ratio = ≤ 26		

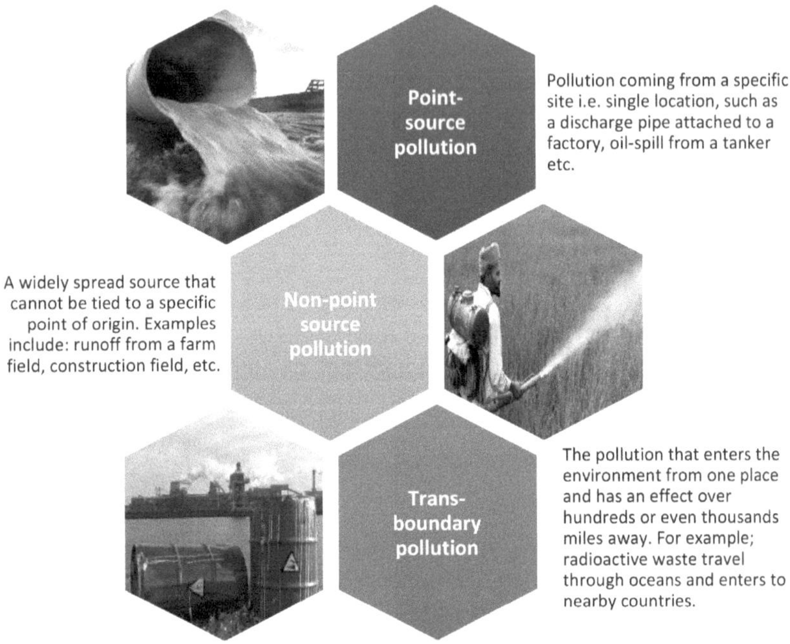

Figure 5.1 Classification for sources of pollution.

unsafe for various activities such as agricultural, industrial, and for domestic use [10]. Furthermore, degraded water quality can add to water scarcity as it limits its accessibility for both human use and for the ecosystem.

Various activities responsible for water-quality degradation are discussed in the following sections (Fig. 5.2):

5.1 Industrial Activities

Untreated wastewater discharge from industries is a major cause of water pollution, which reduces the accessibility of clean water. Effluents from various pharmaceutical and drug industries are discharged into rivers and other water bodies without being treated. It is estimated that around 26,900 million liters per day (MLD) of wastewater is generated in the country, out of which only 26% is treated [11]. Untreated water accumulates in surface water bodies and eventually filters into underground water aquifers. According to a study of Centre for Science and Environment (2004), every liter of released wastewater further pollutes 5–8 liters of water [6]. This lack of wastewater-treatment capacity in cities results in degradation of water quality. Thus it is imperative to measure the contamination from industrial effluents and to screen groundwater consistently.

Figure 5.2 Various activities responsible for water-quality deterioration.

5.2 Modern Agricultural Practices

Agricultural activities generally degrade the quality of soil and groundwater due to extensive seepage and percolation of water containing pesticides and fertilizer residues. They increase the nutritional content of water, thereby allowing organisms to proliferate such as algae and other disease-causing microorganisms.

5.3 Disposal of Domestic Sewage

Household-borne effluents contribute a considerable extent of water contamination in India. It is estimated that 75–80% of water contamination by volume is from residential sewerage. This situation is further exasperated by the transfer of untreated city sewage into these water bodies. Between 1991 and 2008, flow of untreated sewerage multiplied from around 12,000 MLD to 24,000 MLD in Class I and II towns.

5.4 Inadequate Sanitation Facilities

Inadequate sanitation facilities is one of the primary reasons for groundwater and surface water pollution in the nation. The results of a recent report by WaterAid, an international organization working for water sanitation and hygiene, show that around 20% of urban households have inadequate sanitation facility. The majority of them rely on onsite sanitation systems such as septic tanks and pit toilets, which have become major contributors to water pollution.

5.5 Cultural Practices

Various religious practices such as immersion of idols in surface water bodies contribute to deteriorating water quality. In fact, water bodies have been considered as dumping grounds for various offerings, which has degraded the potability of surface water.

5.6 Improper Water Supply

In many areas of the country, the pipelines supplying water have been identified in a shocking state. At many places, they pass through drains. These worn out channels conveying consumable water have developed microcracks. Therefore the risk of waste and sewage water entering from drains into the potable water cannot be ruled out in many cases.

6. HEALTH AND ECONOMIC BURDENS DUE TO WATER POLLUTION

High rates of mortality and morbidity due to water-borne diseases are well recognized in India. In fact, World Bank has estimated that 21% of diseases in India such as cholera, dysentery, jaundice, and diarrhea are water related [1]. As reported, 1.5 million children under age 5 die each year due to water-borne diseases and the country experiences huge economic losses each year due to these diseases. Water pollution is also responsible for poor nutritional standards and development in children. Table 5.2 shows the impact on human health due to the presence of various constituents in water beyond their permissible limits.

Poor water quality not only spreads disease and causes death but also hampers socioeconomic progress. Economic loss is encountered at the household level in terms of cost of treatment and wage loss during illness, and loss of working days and amount of time and money spent by the government on the treatment of the sick affects national productivity [12].

Table 5.2 Sources of Water Contaminants and Their Effects on Human Health

Element	Source/Cause	Health Impact
Fluoride	Due to contamination of groundwater with fluorine rich rocks	High concentration can cause skeletal fluorosis (severe bone damage), dental fluorosis (staining of tooth enamel), digestive disorders, skin diseases, anemia, stiffness of back, difficulty in performing natural movements
Arsenic	Due to deep digging for groundwater, geogenic contaminant	Acute effects: vomiting, esophageal and abdominal pain, diarrhea Chronic effects: skin, lungs, urinary bladder and kidney cancer, arsenical skin lesions, pigmentation changes, lead to arsenicosis
Iron	Due to leaching of iron from soil into groundwater (in high rainfall areas)	Digestive disorders, skin diseases, dental problems, damage blood tissues
Nitrate	Application of nitrogen rich fertilizer and animal manure, airborne nitrogen compounds given off by industry and automobiles, septic and sewage discharge	Cause of Blue Baby disease, increase risk of cancer
Salinity	Overextraction of groundwater and seepage of seawater	Offensive taste to water, affect osmotic flow and movement of fluids
Heavy metals	Due to industrial effluents	Cause damage to central nervous system, kidney, and other metabolic disruptions
Persistent organic pollutants (POPs)	From fertilizers in agriculture and from industrial sources	Hypertension, hormonal dysfunction, growth retardation
Pesticides	From fertilizers in agriculture and from industrial sources	Weakened immunity, tumor formation, reproductive and endocrinal damage
Coliform bacteria such as *Escherichia coli*	Due to sewage contamination in water	Cause gastrointestinal diseases such as typhoid, cholera, gastroenteritis, and jaundice Cause fever, nausea, vomiting, gastrointestinal inflammation

7. RESPONSIBILITIES OF VARIOUS AGENCIES

The various agencies created for assessing water quality are listed in Table 5.3 with their respective responsibilities. These agencies include central boards, state public health engineering departments, state boards, district engineering agencies, and other outside governmental agencies [13].

8. CHALLENGES AND OPPORTUNITIES

Water has an ability to absorb pollution up to a certain level without influencing its quality. But since the pollution loads from various anthropogenic activities have exceeded the natural regenerative capacity of water, we need to identify the challenges faced by the present system and to come up with holistic and people-centered solutions for water quality and sustainability [14].

8.1 Need for Water-Resource Management: Conservation and Utilization

Water is a resource for the whole community, and its desired quality and quantity should be maintained so that it can be shared equitably by present and future generations [15].

8.1.1 Rainwater Harvesting

India has a strong tradition of water harvesting. Since antiquity, our ancestors have met their minimum water requirements by collecting rainwater in various structures, which have served as lifelines for local people are continued till today and (Table 5.4). However, these conventional methods need to be combined with modern technologies and scientific tools. Rainwater harvesting should be encouraged by the government of India by providing subsidies for the establishment of water harvesting infrastructure. The benefits of harvesting rainwater are manifold if it will be implemented to domesticated areas. The filtered and stored rainwater can be used for various household uses, for washing, heating and cooling applications in industries, and also for small-scale irrigation purposes in agriculture. Therefore the Central Ground Water Authority (CGWA) has directed all residential groups including institutions, schools, hotels, and industries to set up roof top rainwater-harvesting systems on their premises (Box 5.2).

Table 5.3 Responsibilities of Various Agencies Involved in Water-Quality Assessment

Concerned Ministry	Name of Agency	Responsibility
Ministry of Water Resources (MoWR)	Central Water Commission (CWC)	• Deals with all matters related to surface water • Regulates the use of surface water for irrigation, industry, and drinking water purposes • Manages interstate water allocation disputes
Ministry of Water Resources (MoWR)	Central Groundwater Board (CGWB)	• Provides scientific inputs for management, investigation, periodic assessment, and regulation of groundwater resources • Implements schemes for recharge amplification
Ministry of Environment and Forests (MoEF)	Central Pollution Control Board (CPCB)	• Conducts surveys about status of pollution and promotes pollution control strategies • Implements national environmental laws and sets standards
Ministry of Environment and Forests (MoEF)	State Pollution Control Board (SPCB)	• Demands information from any industry about their compliance with the acts formed by CPCB • Owns the power to punish noncompliant companies
Ministry of Environment and Forests (MoEF)	National Rivers Conservation Directorate (NRCD)	• Implements plans to improve the quality of rivers in India
Ministry of Rural Development (MoRD)	Rajiv Gandhi National Drinking Water Mission (RGNDWM) (under the Department of Drinking Water Supply)	• Formulates policies, sets standards • Provides funds and technical assistance to the states for rural water supply and sanitation activities
Ministry of Agriculture (MoA)		• Involved in planning, formulation, monitoring, and reviewing of watershed-based developmental project activities

(Continued)

Table 5.3 Responsibilities of Various Agencies Involved in Water-Quality Assessment—cont'd

Concerned Ministry	Name of Agency	Responsibility
Ministry of Urban Development (MoUD)		• Formulates policies, provides guidance for the urban water supply and sanitation sector
Ministry of Health and Family Welfare	Central Bureau of Health Intelligence (CBHI)	• Collects, compiles, analyzes, and disseminates the information on health conditions caused by waterborne diseases
—	Bureau of Indian Standards (BIS)	• Drafts standards pertaining to drinking water quality
—	Nongovernmental Organizations (NGOs)	• Acts as a mediator between the community and various water-related agencies to facilitate water project activities • Has the ability to reach remote rural areas and promotes local participation

Table 5.4 Traditional Water-Harvesting Systems

Traditional Water-Harvesting Systems	Description	Found in
Zing	Tanks for collecting glacier water	Ladakh
Naula	Wells or ponds	Uttranchal
Kuhl	Surface channels that divert water from natural flowing streams for irrigation	Himachal Pradesh
Apatani	Terraced plots connected by inlet and outlet channels	Arunachal Pradesh
Zabo	Impounding runoff	Nagaland
Bamboo drip irrigation	Water from streams in the hills is brought to the plains via bamboo pipes for drip irrigation	Meghalaya
Dongs	Ponds for irrigation	Assam
Ahar-pynes	Embanked catchment basin and channels	Bihar
Bengal's inundation channels	Inundation canals	West Bengal
Baolis	Stepwells	Delhi, Gujarat, Rajasthan
Kunds	Underground storage	West Rajasthan
Jhalaras	Tank	Gujarat, Rajasthan
Nadi	Village ponds	Jodhpur, Rajasthan
Tankas	Underground tank	Bikaner, Rajasthan
Khadins	Embankment across lower hill slopes	Jaisalmer. Rajasthan
Talab	Reservoirs	Bundelkhand, Madhya Pradesh
Bhandaras	Check dams	Maharashtra
Kere	Series of tanks	Central Karnataka
Ramtek model	Intricate network of groundwater and surface waterbodies, connected through surface and groundwater canals	Ramtek, Maharashtra
Surangam	Horizontal well	Kasargode, Kerela
Ooranis	Pond	Tamilnadu
Jackwells	Bamboo pipes are used to lead water into shallow pits	Great Nicobar Island

Box 5.2 Advantages of Rainwater Harvesting

1. Standing water percolates into the ground and increases groundwater level
2. Expands greenery in surrounding areas
3. Reduces flood and soil erosion
4. Fulfills the requirement of nonpotable uses such as watering gardens, flushing, cleaning, and bathing
5. Low maintenance cost

8.1.2 Wastewater Treatment

In addition to conservation, wastewater recycling is equally crucial for sustainable water usage (Fig. 5.3). About 76% of generated wastewater is disposed of untreated into water bodies. Even a major part of industrial sector lack adequate effluent treatment plants (ETPs). The major reason behind this scarcity of ETPs is the huge capital costs that come with them. Also, it is not economically viable for the small-scale industries to set up their own recycling plants. However, due to the importance of wastewater recycling, these small-scale industrial units are opting for common effluent treatment plants (CETPs). Therefore setting up of ETPs must be made compulsory for all large-scale industries [16].

The sewage generated by households is also not treated. Although it is the municipalities who are responsible for sewage treatment, only 19% of the total sewage generated is actually treated [6]. Since the cost for establishing a sewage network is high, the CPCB is encouraging the use of decentralized sewage treatment plants (STPs). For instance, in Chennai, all new housing colonies are directed to set up their own STPs. Tariff structures should be devised in such a way that freshwater costs more than sewage treated water, and it should be provided at a subsidized rate. All these regulations need to be supported by awareness programs in order to incentivize use of recycled water.

8.1.3 Dual Water Supply

Dual water systems are currently being used in several parts of country in order to reduce the burden on fresh-water resources. These systems appear to be successful in terms of using purified water exclusively for drinking purposes and using raw water for other purposes. In addition, this measure is cost-effective since it saves on resources used to provide clean water.

It is also advisable to use low-cost, simple technologies instead of modern water-purification techniques for rural areas since they are easy to operate and do not require any technical knowledge.

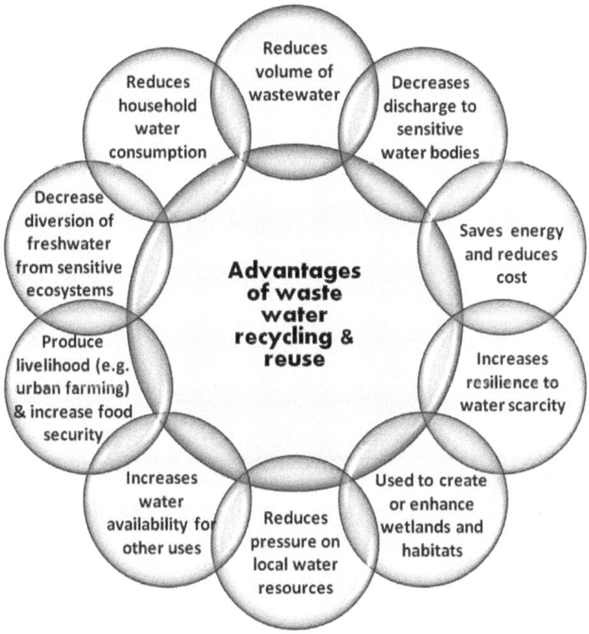

Figure 5.3 Benefits of recycling and reusing wastewater.

8.1.4 Use of Desalination Technologies

Desalination refers to a process by which purified water is retrieved from saline water using different forms of energy. Depending on the source and salinity, saline water is classified as either brackish water or seawater. Desalination produces two streams—fresh water and brine (a more concentrated stream) [17]. The growing need for freshwater has led to a dramatic increase in the demand for desalination facilities and is becoming less of a "nonconventional" water resource. A recent report suggests that there are around 1000 functional desalination plants in India [18]. But most of them run on technologies such as thermal distillation, electrodialysis, reverse osmosis, or a combination of thermal and membrane distillation [19]. These technologies are associated with two major constraints—one is high-energy consumption and the other is high plant cost. The energy required to run these plants is generally supplied in the form of either steam or electricity, which is generated using conventional fossil fuels. However, increasing environmental concerns due to the nonrenewable nature of fossil fuels has created concerns about the use of desalinating technologies, which must be addressed. Utilization of renewable sources such as solar, wind, and geothermal can potentially address these concerns, thereby providing a clean supply of energy for desalination. Thus a governmental push for utilizing such sources for desalination can increase participation in this market, which will augment fresh-water requirements of India and contribute to its sustainable development.

8.2 Surveillance of Water Quality

To ensure that the quality of water is maintained, it is crucial to monitor water resources on a regular basis. Accurate information is required to answer a number of questions such as what the problem is, where it is occurring, how serious it is, and what are the factors responsible for it. One of the weakest links in controlling water pollution in India is an insufficient number of monitoring stations and frequency of monitoring pollution. Presently, water-quality monitoring is done by the CPCB at 2500 monitoring stations under a three-tier program: Global Environmental Monitoring System (GEMS), Monitoring of Indian National Aquatic Resources System (MINARS), and Yamuna Action Plan (YAP) [20]. Regarding frequency, it has been found that 68% of the stations monitor water quality on a monthly basis while 32% do so on a half yearly basis. Thus there is an urgent need to increase the number of monitoring stations as well as their frequency [21]. However, remote villages still remain untouched by monitoring agencies. Part of the reason for this is that collecting water samples and transporting them to distant laboratories is a tedious task due to transportation and communication problems. Thus there is a need to establish monitoring stations with adequately equipped laboratories, both in terms of manpower and infrastructure, in such remote areas.

Furthermore, biomonitoring (analyzing water quality in terms of toxic metals, pesticides, inorganic ions, and POPs) should be done every month to keep track of water quality and the results should be made available online to ensure transparency. Also, it should be ensured that there is no duplication of efforts or overlapping of functional areas of the monitoring agencies.

8.3 Data Management and Information Flow

Poor management of data is another important challenge to be addressed. Although plenty of data is available on chemical contamination, there is no such system to consolidate it. Also, there is no scientific monitoring that relates the data of water quality to the cause and effect of disease burden in the community. Thus due to the mismanagement of information flow, the desired results have not been obtained.

Web-based geological information systems (GISs) can be an alternative to assessing water-quality data over the Web, since they require little user knowledge and experience. The data from various sources can be obtained and stored in a relational database management system (RDBMS) with water-quality assessment tools [1]. These assessment tools can help in the identification of different management options, and accessibility of information and assessment tools over the Web can increase public participation and hence help in controlling water degradation.

8.4 Quick and Initial Assessment of Water Quality at Community Level

Guzzling water in summer is a common way to stay hydrated so that lost liquids can be renewed and heat-related illness is kept under control. But drinking tap water, which often carries contaminants, can cause various water-related diseases. Thus field-level diagnostic kits should be developed for the quick assessment of water quality. Various state departments have acquired these field-testing kits under the Rajiv Gandhi Drinking Water Mission [22]. But the lack of regular supply of reagents for performing water-quality tests has limited the effectiveness of these tests.

8.5 Interagency Coordination and Public-Private Partnership

Issues related to access to water resources has led to water conflicts both between and within nations. This is due to the fragmented approach at both the state and central levels. Thus better coordination among ministries and departments must be ensured for the effective implementation of management strategies. Corporatization or public-private partnership would help increase the efficiency of water-quality monitoring and decrease the burden of sample collection and evaluation. Government laboratories could be leased out to private agencies on a contractual basis. In this manner, the government can segregate its work by giving the responsibility of operation and maintenance of their laboratories to private agencies and focusing on quality control. Additionally, there should be involvement of local villages for the surveillance of water quality. There should also be willingness at community level to pay for the operation and maintenance of treatment plants.

8.6 Balancing Sanitation, Hygiene, and Water Quality

Sanitation and hygiene are often seen as social and behavioral issues, whereas water quality as a scientific and technical affair. There is a need to stop open defecation and to promote toilet coverage for the prevention of water-quality degradation. Also, water emanating from these facilities should not lead to gross water contamination of water resources. Thus technological methods to treat water pollution should go hand in hand with issues related to sanitation and hygiene.

8.7 Build-Neglect-Rebuild Paradigm

Although noteworthy advancements and projects have improved water quality and infrastructure for water supply, many of them remain incomplete. This has given birth to the so-called "build-neglect-rebuild" syndrome. The major factor that contributes to this syndrome is insufficient financial resources for proper operation and maintenance.

8.8 Awareness and Accountability

India is characterized by disparity in the levels of awareness, education, poverty, practices, and rituals due to which water quality and quantity conservation has become an extensive challenge. Water-conservation awareness is an understanding of the need to utilize water efficiently at all stages in order to promote change in state of mind and behavior with respect to water management and use. But even today, many traditional methods for water-quality monitoring have been publicized without understanding their effectiveness through appropriate research. Thus there is a need to create awareness among the rural and urban masses at a grassroots level about water quality and waterborne diseases. Of course, it will take time to change established habits, but if all systems, institutions, and societies work toward a common goal, the overexploitation of water can be reduced.

Water conservation should begin with the efficient usage of water in household chores and by educating people about how diminishing the measure of the water they utilize each day can significantly conserve water. There are several water-saving tips, e.g., brushing and shaving using a mug rather than running the tap can save up to 78 liters of water. Similarly, using a modern dual-flush system in toilets, 80 liters of water can be conserved (Fig. 5.4). Thus by adopting simple water-conservation methodology, 40% of the water can be saved/family [7].

In addition to conservation, various central and state agencies are working on improving water quality by promoting the use of chlorine bleach for disinfecting water sources like wells. But due to lack of appropriate guidelines, there is no regulation of dosage and frequency of chlorination. Thus bleaching powder is often seen dumped into these wells, leading to excess residual chlorine, which increases the risk of cancer. This creates the need to develop IEC (information, education, and communication) material for

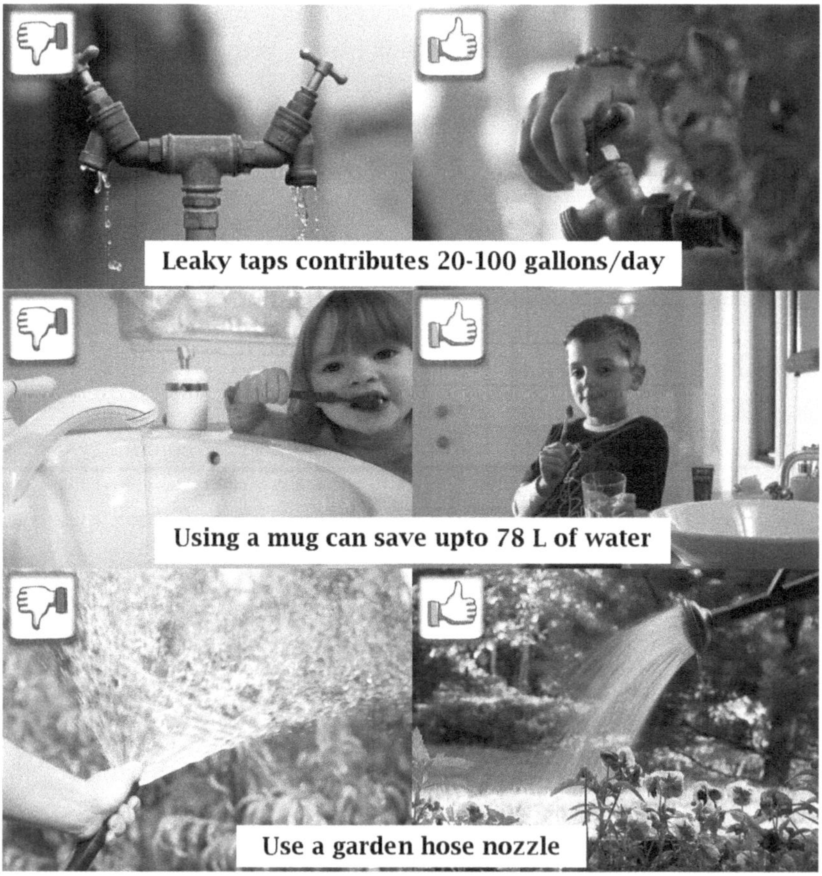

Figure 5.4 Tips for the efficient usage of water.

the benefit of public health and safety [23]. But it has been observed that IEC material often contains many technical elements that are beyond the understanding of common people. Therefore developing IEC material in different regional languages should be a stepping stone to control degrading water quality.

8.9 Revision of Drinking Water-Quality Standards

Recent WHO guidelines state that the standards for drinking water should be developed by taking into account a nation's geography and climatic conditions. Therefore there is a major challenge in front of the agencies to establish realistic drinking water-quality standards based on scientific research.

But it has been observed that the Bureau of Indian Standards (BIS) has pretty much embraced WHO rules for drinking water without making the vital adjustments for Indian conditions. For instance, as per WHO, the permissible level of fluoride and arsenic in drinking water is 1.5 mg/L and 10 μg/L, respectively. Because the Indian climate is tropical, the consumption of water is high. Consequently, the probability of body burden due to fluoride and arsenic also increases. Malnutrition also aggravates the disorders caused by them. Therefore such factors should also be considered when determining a realistic national drinking water standard.

8.10 Decentralization of Water Regulatory Authorities

There is a need for decentralization of water regulatory authorities and restructuring of governance frameworks. Local institutions often have better knowledge of water issues and thus should be allowed to negotiate and jointly decide water-management strategies for taking effective measures. Since large-scale training and capacity building are needed for effective monitoring, managing, and maintaining of systems in India, communities and local institutions like panchayats must actively participate in the planning and execution of the water-supply programs. Furthermore, they must undertake the monitoring of water sources and be aware of simple remedial measures. The rural community should also be involved in every stage of the process of community based water-quality management [24].

8.11 Amendments of Water Laws

Since water is a limited resource, its sharing and distribution requires an administrative system. This can be accomplished by forming water laws. Laws are required (1) to ensure that the distribution of good-quality water is equitable, (2) to protect the water resources from exploitation, and (3) to assure sustainability.

There are laws that focus on aspects such as limitations on depth of wells, declaration of groundwater protection zones, etc. But legislation should also be created in order to avoid excessive pumping of groundwater by farmers, which prevents the accessibility of groundwater resources by others. Also, the implementation of these laws should be checked to substantiate the success of them.

Recently, new legislation called the National Green Tribunal Act 2010 was established for effective and expeditious treatment of cases related to environmental protection. It allows the victims of environmental damage and pollution, including water pollution, to claim damages and compensation. This law is a critical takeoff from the current enactment where apart from shutting down a polluting industry, cutting down the water and power supply, and criminal punishment for those responsible for running it, legislations are made available to the victims of water pollution.

Thus the need for stringent regulations is urgent.

8.12 Need for Human Resources

Resources in terms of manpower are inadequate. Many laboratories are understaffed and hence sample collection, storage, and analysis are affected. Thus there should be training programs for data management, analysis, and validation to increase human resources. In many parts of rural India, various treatment plants are defunct due to lack of operation and maintenance. Millions of rupees can be saved if local workers are given training for primitive maintenance of these plants. Another drawback in this division is that there are only a couple of organizations/colleges in India that offer specialized courses in environmental science. Thus there is a lack of experts who deal in managing water and sanitation issues. Establishments managing water and sanitation are regularly headed by chairmen and administrations who have minimal experience in addressing these issues.

CONCLUSIONS

The impending water crisis and degrading water quality are now apparent in several parts of the country. We need to act decisively to secure our water assets and use them sensibly. In this regard, a new approach that can deal with water administration and its use is required, which must be understood at all levels, national, state, and also by general society. In order to accomplish this aim, major efforts are needed both by government and communities to promote rainwater-harvesting systems and watershed-management programs at the local level all over the country. There should be interagency coordination among central and state governments and they should empower local groups with knowledge, understanding, and real-time information on the status of groundwater so as to manage extraction in a cooperative way. Additionally, stringent regulations should be made to improve the quality of water as well. Of course, wastewater recycling needs more focus. The government should mandate all industries or a cluster of industries to establish common effluent treatment plants to recycle all their wastewater and strengthen state pollution control boards to enforce effluent standards. For this, extensive awareness and education programs to adopt safe and conservative water adaptation policy should be embraced in parallel.

Implementation of low-cost sanitation systems with lower subsidies, proper price charging for water in cities, greater household involvement, effective data collection, planning and monitoring activities, variety of technology choices, rural drainage systems, awareness building, engaging NGOs and local groups, and human resource development are some chief areas to be considered. Appropriate forms of public-private partnerships, evolution of sound policies, and emphasis on sustainability with political commitment are also prerequisites to bring change.

REFERENCES

[1] Srikanth R. Challenges of sustainable water quality management in rural India. Curr Sci 2009;97:317–25.
[2] Sharma RK, Gulati S, Puri A. Green chemistry solutions to water pollution. In: Ahuja S, editor. Water reclamation and sustainability. Elsevier Publications; 2014. p. 57–75.
[3] Kathpalia G, Kapoor R. Water policy and action plan for India 2020: an alternative. Water Policy Action Plan 2002;2020:14–7.
[4] Kumar R, Singh R, Sharma K. Water resources of India. Curr Sci 2005;89:794–811.
[5] Chatterjee K, Water resources of India, in 2014.
[6] India Infrastructure Report. Water: policy and performance for sustainable development. Oxford University Press; 2011.
[7] Kurunthachalam SK. Water conservation and sustainability: an utmost importance. Hydrol Curr Res 2014;5:1.
[8] Sharma RK, Gulati S. Water quality issues and solutions in India. In: Ahuja S, editor. Comprehensive water quality and purification. Elsevier Publications; 2014. p. 21–39.
[9] Von Sperling M. Wastewater characteristics, treatment and disposal. IWA Publishing; 2007.
[10] Rao SM, Mamatha P. Water quality in sustainable water management. Curr Sci 2004; 87:942–7.
[11] Kaur R, Wani S, Singh A, Lal K. Wastewater production, treatment and use in India, in: national report presented at the 2 nd regional workshop on safe use of wastewater in agriculture; 2012.
[12] Daumerie D, Savioli L, Crompton DWT, Peters P. Working to overcome the global impact of neglected tropical diseases: first WHO report on neglected tropical diseases. World Health Organization; 2010.
[13] Khurana I, Sen R. Drinking water quality in rural India: issues and approaches. Water-Aid India 2008.
[14] Kenel P, Witherspoon JT. Practices to promote sustainable water sources. Proc Water Environ Fed 2005;2005:5650–6.
[15] Jayyousi A. Protection of the quality and supply of freshwater resources: application of integrated approaches to the development, management and use of water resources, localizing agenda 21 in Palestine. 2001. p. 161–205.
[16] Jhansi SC, Mishra SK. Wastewater treatment and reuse: sustainability options. Consilience J Sustain Dev 2013;10:1–15.
[17] Sinha R. Desalination & water purification technologies. Mumbai: Goverment of India; 2010.

[18] Abraham T, Luthra A. Socio-economic & technical assessment of photovoltaic powered membrane desalination processes for India. Desalination 2011;268:238–48.
[19] Beltrán JM, Koo-Oshima S. Water desalination for agricultural applications. FAO Land water Discuss Pap 2006;5:48.
[20] Bhardwaj R. Water quality monitoring in India—achievements and constraints. In: Iwg-env, International work Session on water Statistics, Vienna; 2005. p. 1–12.
[21] Kumar S, Murty M. Water pollution in India: an economic appraisal, in. Oxford University Press; 2011.
[22] Gandhi R. National drinking water mission. New Delhi, India: Ministry of Rural Development; 1993.
[23] Kumar SG, Kar SS, Jain A. Health and environmental sanitation in India: issues for prioritizing control strategies. Indian J Occup Environ Med 2011;15:93.
[24] Coussens C. Global environmental health: research gaps and barriers for providing sustainable water, sanitation, and hygiene services: workshop summary. National Academies Press; 2009.

FURTHER READING

[1] http://wrmin.nic.in/forms/list.aspx?lid=297.
[2] http://www.arlingtoninstitute.org/wbp/global-water-crisis/606.
[3] http://www.worldbank.org/en/country/india.
[4] http://articles.economictimes.indiatimes.com/2014-12-28/news/57462656_1_water-cess-water-resources-ministry-water-usage.
[5] http://wrmin.nic.in/writereaddata/Guidelines_for_improving_water_use_efficiency.pdf.
[6] http://timesofindia.indiatimes.com/home/environment/pollution/80-of-Indias-surface-water-may-be-polluted-report-by-international-body-says/articleshow/47848532.cms.
[7] http://timesofindia.indiatimes.com/home/environment/pollution/Around-80-of-sewage-in-Indian-cities-flows-into-water-systems/articleshow/18804660.cms.
[8] http://cpcbenvis.nic.in/cpcb_newsletter/sewagepollution.pdf.
[9] http://cpcb.nic.in/.
[10] http://wrmin.nic.in/.
[11] http://www.nbr.org/.
[12] http://www.mdws.gov.in/.
[13] http://www.who.int/water_sanitation_health/dwq/gdwq0506.pdf.

CHAPTER SIX

Challenges and Solutions to Water Problems in the Middle East

Y. Shevah
H.G.M. Consulting Engineers & Planners Ltd., Netanya, Israel

1. INTRODUCTION

1.1 The Middle East Region

The Middle East (ME) is a transcontinental region centered in Western Asia. It extends over an area of over 5.0 million square miles from the Black Sea in the north to the Arabian Sea in the south, and about 1000 miles from the Mediterranean Sea in the west to the mountains of Iran. The ME covers the eastern Mediterranean coast of Egypt, Palestinian Authority (Palestine), Israel, Lebanon Syria, and Turkey through the desert to Iraq and Arabia, and to the East through Iran to the Caspian, the Caucasus, and the Black Sea. In the southeast, the Arabian Peninsula along the Red Sea, Arabian Sea, and Persian Gulf, the region includes the countries of Bahrain, Kuwait, Oman, Qatar, Saudi Arabia, the United Arab Emirates, and Yemen. The ME is rather a geographical concept termed also as Near East, Fertile Crescent, and the Levant. By other definitions, the Middle East is stretched to include the North African countries and termed MENA, forming the alliance—the Arab World, or the Arab League (see Map).

The physical geography of the Middle East and North Africa (MENA) is varied. It comprises sea coasts, mountains with peaks rising as high as 19,000 ft., deserts, fertile plains, dissected by major rivers including the Nile Delta in Egypt, the Tigris and Euphrates watersheds of Mesopotamia, and the Jordan River. Vast deserts are common in the region including the Sahara Desert, which runs across North Africa, essentially limiting settlement along the Mediterranean coastline.

The MENA constitutes various ethnicity, religions, and national identity, amounting to about 300 million inhabitants in 2007 (about 5% of the world's population) and is projected to grow to about 500 million by 2030 [1]. In general, the region is endowed with natural wealth, although not evenly distributed, and is oil dependent [2]. This wealth was translated

Chemistry and Water
ISBN 978-0-12-809330-6
http://dx.doi.org/10.1016/B978-0-12-809330-6.00006-4

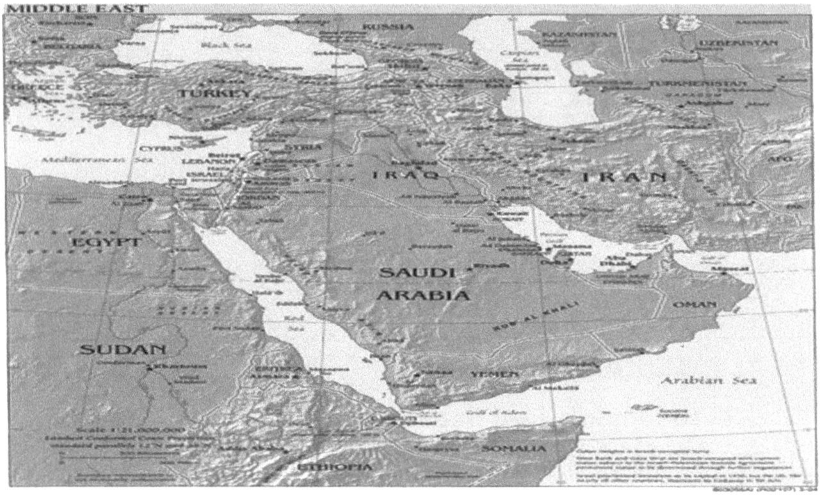

Map Middle East physical map 2004. *Based on US Central Intelligence Agency.*

into regional and international influence and prestige, but was insufficient to implant a sense of political, economic, and social progress and equality. The current sectarian divides and the entrenched rift between Sunnis and Shiites Muslims has immensely complicated the regional politics and stability, although the solidarity with Arabs engaged in conflict with non-Arab countries is still strong [3].

2. CLIMATE AND TRENDS IN CLIMATIC CHANGE IN THE MIDDLE EAST

2.1 Regional Climate

Climatically, the ME is arid and semiarid, ranging from the temperate Mediterranean coast, to the extreme heat of the arid desert areas, to snowy mountains. Generally, the climate ranges from the warm summers and cold winters of highland Turkey and Iran, through hotter summers and cool winters of northern Mesopotamia and the Mediterranean coast, to the extreme temperatures of the Arabian Desert. Modulated by complex topography, inland water bodies, and proximity to the Mediterranean Sea, the annual average precipitation ranges between 700 mm and above, in the northern states to about 60 mm in the Arab Peninsula and nearly zero on the west sides of continents. Most rainfall falls in winter; the summer is almost completely dry, with occasional rainy events triggered by anomalous synoptic conditions [4].

The regional climate was extensively studied, analyzing century-long observed trends in precipitation, temperature, and sea-level pressure, supported by climate models [4]. The results strongly suggest that the prevailing harsh climatic conditions will worsen over time as a consequence of human interference and anthropogenic impacts. This would be reflected by a reduction in precipitation and increase in temperature and evaporation, while the soil moisture, groundwater replenishment, and stream flows would be reduced [5]. In figures, the probability of severe and persistent droughts in this region would be reflected in a decline in rainfall (−10% to −25%), soil moisture (−5% to −10%), and runoff (−10% to −40%) as against an increase in evaporation (+5% to +20%) [5].

This trend was already apparent in the years 2006−09, which were the worst three years of drought on record and more recent temperatures in 2014 and 2015, which were the hottest years since 1880. The average temperature in 2015 was 58.62°F (14.79°C), passing by a record margin of 0.29°C the 2014 record [6] showing a statistically significant rising trend in temperatures ($p < .01$) during the 21st century and 13% drop in the rainfall ($p < .05$) since 1931 [6]. This trend indicates that Israel could see a 10−20% decrease in overall precipitation, while rainfall in Jordan and Lebanon could be 20% and 30% lower, respectively [7]. A 10% decrease in precipitation, coupled with a 2°C rise in surface temperatures, could reduce the amount of water in the Jordan River basin by 45−60% [7]. In Turkey, winter precipitation could drop by 20−30%, with mountain snow packs decreasing by up to 70−90%, resulting in reduced flow of the Tigris and Euphrates rivers, the vital water sources for large portions of the ME [8].

The apparent extreme climate in recent years cannot be explained by natural variability alone. It is attributed to a combination of El Nino, which releases immense amounts of stored heat into the atmosphere, and to man-made anthropogenic activity [9,10].

2.2 Climate Social and Environmental Impacts

With 15 out of the 18 countries around the world that are at "extreme risk," the ME, as a whole, is classified as an extremely high-risk water scarcity region, in which water resources are limited and unequally shared in time and space [11]. As a consequence, many countries have very limited water resources including two countries, Bahrain and Kuwait, which effectively have no domestic water supplies, using desalination to meet most of their water needs [12].

It is almost certain that the predicted climate change would seriously damage the region, affecting water, public health, agriculture, energy, biodiversity, coastal infrastructure, natural disasters, national security, and more. The region may face significant negative economic and social impacts, giving rise to increasing tension and conflict among political and economic players in a region that is already experiencing natural and anthropogenic threats, resulting in intense floods, desertification, and droughts.

In the short term, increased groundwater extraction will probably be used to make up some of the shortfall, increasing sharply the excessive withdrawals since the onset of drought in 2007 [12]. Hot, dry conditions will increase the risk of wildfires of forest, while warmer and longer growing seasons could lead to an explosion of pest populations. For some crops, grains in particular, warmer temperature will mean accelerated vegetative growth and reduced yields [13,14]. Similarly, rising sea levels will pose an increased risk of saltwater intrusion into adjacent aquifers, especially in areas undergoing high levels of groundwater extraction [15]. Rising sea levels will be no less serious for the Nile Delta, in which a sea-level rise of 1 m would flood a large area, forcing out about 10.5% of Egypt's population from their homes and hitting half of Egypt Delta's crops.

On the Tigris and Euphrates rivers, which are vital for agriculture, hydropower generation, and domestic consumption, the unilateral construction of dams by Turkey, holding a larger portion of the flows needed by Syria, Iraq, and Iran, in the downstream, could increase tensions between the riparian states [12]. The Jordan River basin, which draws water from Lebanon, Syria, Jordan, Israel, and the West Bank, is another potential source of international tension [7].

In general, climate trends play a role in demography, causing outmigration, such as that seen in the movement of refugees fleeing conflict in Iraq and Syria. Intense conflicts over scarce resources could spark a massive wave of environmental refugees, forcing people to move out due to sudden or gradual alterations in the natural environment. In the case of northeastern Syria, just as Syria was exploding with immigrants from the Iraq war, the persistent drought of 2007–10 devastated the farming communities forcing people to flee their homes to migrate internally or internationally, into the neighboring countries and beyond, greatly contributing to the Syrian civil war [16,17].

2.3 Adaptation to Climate Change and Global Warming

These impacts could greatly disrupt the livelihood of the affected populations. Thus substantial mitigation and adaptation efforts, combined with stronger cooperation in transboundary water management, are essential for the region to successfully navigate the effects of climate change and to reduce potential conflicts over increasingly scarce water resources. As a way forward, the development of renewable energy and decarbonizing of the economies have to be accelerated, in line with the UN treaties—Kyoto Protocol, 1997, Copenhagen Climate Summit, 2009, and the most recent Paris Summit, 2015, aiming to capita the earth's warming to 2°C, or 3.6°F, by the end of the 21st century [18].

During the Paris Summit (2015), 180 countries committed themselves to be part of the new deal, among them, the United States, which has pledged to reduce emissions by 26% below 2005 levels, the European Union by 40% below 1990 levels, and Israel, with about 0.1% of the world's population and contribution of about 0.2% of global greenhouse gas emissions, has pledged to reduce greenhouse gas emissions by 26% of the current levels. China pledged to increase by 20% its energy production from renewable sources by 2030. Advanced monitoring systems such as space-borne sensors, satellites, commercial jets, smokestacks, and communications towers could be deployed to closely and effectively track the emissions of greenhouse gases [18].

3. WATER RESOURCES SCARCITY IN THE MIDDLE EAST

The MENA region is relatively water poor, with an average of approximately 1000 m^3/capita per year, far below the world's annual average of 8000 m^3/capita [19]. The ranking of 180 countries on the basis of water availability revealed that the majority of MENA countries, including Bahrain, Gaza, Israel, Jordan, Kuwait, Libya, Oman, Qatar, Saudi Arabia, Tunisia, UAE, and Yemen, were in the bottom 10% [20]. Consequently, as of 2015, almost all the countries are assumed to be below the level of severe water scarcity, which is defined as less than 500 m^3 per capita per year, with nine countries below 200 m^3 and six below 100 m^3. Close to the Mediterranean Sea Coast (Israel, Palestine, and Jordan), the annual volume of renewable water resources is less than 200 m^3/capita compared to about 800–1000 m^3/capita in neighboring countries, 2500 in the UK, 3500 m^3/capita in France, and 10,000+ m^3 in the United States [20]. By 2025, only Lebanon and Iraq are expected to remain above the water-scarcity level [21].

3.1 Surface Water

The region depends on the precipitation that maintains the rivers and recharges the groundwater aquifers. The main rivers such as the Euphrates and Tigris rivers and their numerous tributaries are the principal freshwater supply, but the importance of groundwater is increasing as surface supplies become less reliable and predictable [22]. Surface-water resources are estimated to be about 224 billion m^3/year of which 60% is transboundary water, originating in other countries [23]. Egypt is the most extreme case, with some 95% of its water coming from the Nile. Euphrates and Tigris rivers originate in Turkey and the Jordan River and tributaries are shared by Israel, Palestine, and Jordan.

The variable rainfall in quantity, intensity, and distribution and the continuous droughts of recent years have drastically reduced water availability, affecting the ecosystems and the hydrological patterns, placing aquatic and rain-fed agriculture and consequently rural livelihoods at risk. The increase in temperature and consequently the rates of evaporation reduce the water volume in lakes and reservoirs, while intense and short storms require expanded storage to offset runoff variability with high economic and environmental costs.

Diminished snowmelt in Lebanon and Syria and extensive diversion of the flow led to the drying of rivers such as the Khabur River in northeast Syria and the decline in discharge at Ras Al Ain from an average of 60 m^3 per second to nil after 2001, and other rivers that once flowed to the Mediterranean Sea and the Jordan River, supporting rich aquatic and wetland ecosystems are now seriously depleted and polluted. Political tensions and regional conflicts also significantly affect the shared water resources. Investments in hydraulic infrastructure by upstream riparian states changed the water availability of downstream users. The Greater Anatolia Project of Turkey, for example, affects water availability in Syria and Iraq and dams constructed in Ethiopia and Sudan affect Egyptian water [1,24]. The changing circumstances also affect Egypt, which depends on the Nile for more than 95% of its water needs. In other cases, such Israel and its neighbors, water is an additional factor in the ongoing political tensions. The frightful situation of surface water in this part of MENA can be best illustrated by the case of the Jordan River.

The case of the Upper and Lower Jordan River, the Sea of Galilee, and the Dead Sea. Water of the Jordan River and its tributaries was diverted by Lebanon, Syria, Israel, and Jordan for use by agriculture, industry, and drinking water purposes. The diversion of the flow has had an adverse impact on the aquatic systems, altered the hydrological cycle, eliminated the wetlands, and reduced the biodiversity and the natural functioning of the aquatic ecological systems. Due to the continuous drought of recent years, the Sea of Galilee water level has drastically dropped by almost 2 m below the lowest permissible red line, to −214.87 m bmsl [24], severely threatening the lake's ecosystems and water quality. The diversion of the flow has also had a drastic effect on the Dead Sea, which has lost much of its surface area from 950 km^2 in the beginning of the 20th century to the current 392 km^2, while the water level dropped by about 1 m a year, from 394 m below sea level in the 1960s to 429.04 m in 2015 [25]. As the Dead Sea level continues to drop, rapid changes in the ecological balance of this fragile ecosystem and the surrounding oases and wetlands are observed [26]. These changes are also evident from the hundreds of sinkholes which occurred along the shoreline, ripping up buildings, beaches, and road sections of the north-south artery. To restore a bit of the flow of the river, the Israeli Water Authority is discharging water to sustain a basic flow in the Jordan River downstream of the Sea of Galilee.

3.2 Groundwater

Groundwater sources in the region are quite limited, not exceeding a total of 50 billion m^3/year [27]. The erratic rainfall and consequently the low recharge of the aquifers and excessive abstraction pattern have greatly altered the natural state of groundwater, leading to serious depletion of underground aquifers and to seawater intrusion, in coastal aquifers, with consequences for both short- and long-term water quality. Underground aquifers that recharge during wet periods are abstracted much faster than they can naturally replenish and the reserves have been drafted at unprecedented rates.

In many parts of the region, accentuated by weak regulations and subsidized power, groundwater aquifers are already significantly exploited, exceeding the rate of natural recharge and becoming unsustainable. It is becoming more difficult and expensive to draw on these sources in which the water level gets deeper and requires greater pumping power. In addition, as deeper wells are drilled, greater concentrations of salt and naturally occurring toxic elements are found that need more treatment to make water potable. This further increases costs and puts water supplies at risk. Groundwater depletion will likely be worsened by reduced rainfall and natural recharge in addition to the increase in the intrusion of salt water to coastal aquifers.

3.3 Water Quality

Reduced inflow to lakes and streams induces eutrophication of the water bodies, a process where water bodies receive excess nutrients that stimulate excessive growth of plankton and zooplankton [28]. Thermal pollution increases evaporation and warmer waters affecting salt balance and harming habitat-forming species such as coral, oysters, and mussels [29]. The projected frequent droughts and the lower stream flows will further adversely affect the quality of the water bodies due to pesticides, pathogens, sediments, dissolved organic carbon as well as emerging pollutants found in the enriched runoff. The water bodies are subjected to these hazardous substances, harming aquatic life and the water quality [30]. The anthropogenic sources of pollutants also contribute to the presence of bioavailable metals causing metal toxicity [31].

The higher concentrations of biodegradable organic material, in terms of biological oxygen demand (BOD) and chemical oxygen demand (COD), may lead to the decomposition of the organic material and excess release of nutrients, such as N and P, to produce dense algal bloom that decreases the dissolve oxygen (DO), causing hypoxia in the water layers.

Eutrophication is a worldwide problem and a large number of lakes are exposed to it, hindering many of their functions including the supply of drinking water, recreation, and as cultural and bird sanctuaries. Excessive production of planktonic algae leads to oxygen deficiency, fish death, reduced biodiversity, and the formation of the periodical nitrogen-fixing bacteria, cyanobacteria. The harmful algal blooms (Cyano-HABs)—blue green algae—emit foul odor and produce toxins that harm the ecological balance, indicating that the lake's natural "buffering capacity" may be in danger [28].

Similarly, groundwater pollution is caused either from saline water that flows laterally into the depleted zone (especially near coasts) or from the deeper layers of water, which are more saline. A good example is the coastal aquifer shared by Israel and the Gaza Strip in which inadequately treated sewage effluent, runoff, fertilizers, and pesticides residues and leachate from landfills and seawater intrusion contribute to very high concentrations of nitrates and chlorides exceeding standard limits for drinking water and causing irreversible damage to some parts of the aquifers [32]. Being a shared aquifer, the discharge of raw sewage from the Gaza Strip flows north easterly along the Mediterranean Sea to the beaches of Israel, endangering the health of Gazans and Israelis alike. Weak regulations and overall low capacity to regulate and enforce those regulations further exacerbates the issue [33].

4. REGIONAL WATER DEMAND

4.1 General

The regional water demand for agriculture and domestic and industrial uses exceeds the available resources. Even if in the coming years the precipitation was much higher than normal, this would not be enough to relieve the region from the gripping drought conditions of recent years.

Of the total demand, the domestic water consumption was about 16 billion m^3 in 2002 [34], reflecting an average per capita domestic water consumption of about 200 L/day with wide variation among the countries, ranging from 300 to 750 L/capita/day in the Gulf Cooperation Council (GCC) countries and much less in the rest of the region. As of now, many people in the region, especially in those countries with scarce resources, lower income, and or riddled by war and conflicts, do not have regular access to safe drinking water and many more lack access to sanitation services [35]. Demographic growth, economic and social development as well as rapid urbanization and industrialization, higher per capita demand, and the preference for high-quality domestic water will increase the demand for water and competition for less and less water.

Irrigated agriculture is a major consumer utilizing 80–90% of available water in most MENA countries [1] and will continue to be the predominant water user to keep pace with the growing demand for food and poverty alleviation in the rural areas, adding to the instrumental value of irrigation in maintaining environmental services and ecosystem resilience. Conservation of agriculture is also of interest because of its potential for climate change mitigation in the form of increased carbon sequestration. However, the available water resources are not able to meet current and future demands, and the anticipated climate change that will lead to more frequent and longer droughts is expected to decrease the supply of water, while the demand for water for irrigation of crops will increase with temperature [22]. In addition, potential crop yields tend to fall at high temperatures, so the productivity of water in agriculture will fall [13,14]. Also, keeping a basic flow in rivers and streams will be necessary for aquatic ecology restoration purposes. Therefore, farmers have to recognize that water is a scarce commodity for which the demand by the other sectors of economy is on the increase. Water used by the other sectors yields a much higher value to the national GDP, as compared to the marginal contribution of agriculture, adding to the very low irrigation water use efficiency of about 30%. For example, in Jordan the agricultural sector accounts for 75% of all water consumption and produces only 2% of the GDP [36].

Due to low availability and continued reduction in available freshwater, water will become a rare commodity unable to satisfy the aggregated demand of all sectors and the issue would become more and more alarming each year. These drivers are likely to continue to intensify, magnifying the inevitable need for rational water management and the development of additional sources to satisfy the need for water demand for all, including irrigated agriculture and the environment.

Water agencies have to recognize the competition among the various water users and to set fair rules for intersectoral transfer of water, avoiding conflict, moderating and planning for the structural transition of water allocations. In particular, readjusting the incentive system that can reduce agricultural water use while equitably promoting water conservation, misuse of water and pollution, and unaccounted for water, as discussed in Section 5 of this chapter and considering the specific demographic and economic conditions and water situation in the various states comprising the MENA region, as briefly described in the following.

4.2 State of Water in Selected States of the Middle East and North Africa

4.2.1 Syria

Given its stronger dependence on the annual rainfall and inadequate groundwater, Syria is far more vulnerable to drought. Variability in rainfall patterns, retreat of mountain glaciers, and higher evaporation rates in reservoirs bring new climatic risks that need immediate and long-term mitigation. The persistent drought of 2007—10, coupled with natural resource mismanagement, contributed to the broad collapse of farming in Syria and particularly northeastern Syria, the "breadbasket" region of Syria. For example, the multipurpose Tishrin Dam in northern Aleppo east of the Euphrates River, which is captured by the Islamic State, and the Khabour Dam, which supply a large command irrigation area, are drying.

In addition to the drought and the political unrest, water-resource mismanagement led to a massive depletion of groundwater and inadequate water supplies, adding to a highly inefficient urban water infrastructure. Cities are already experiencing serious water shortage, e.g., the Damascus water network leaks up to 60% of the water in circulation. Agriculture is characterized by unsustainable farming practices due to inefficient irrigation techniques: e.g., flood irrigation in which 60% of water used is wasted, as well as the irrigation of heavily subsidized water-intensive crops.

The inefficient management of available water resources contributed to land degradation and desertification. The resulting low agricultural yields and water and food insecurity has devastated the farming communities, forcing people to flee their homes [6]. The displaced population added to the popular uprising against the regime in March 2011, which triggered an armed uprising across Syria, demanding democratic reform and end of repression, dragging the country into a full-scale civil war. To date, the civil war has claimed close to 500,000 lives and more than 2 million people were displaced and as many were forced to flee across the border [16]. Along with other political, economic, and social grievances, which erupted in the wake of the "Arab Spring" revolutions, Syria will become drier in the future as the predicted global warming may bring more severe climate events and hardship.

4.2.2 Jordan

Jordan is one of the fourth driest countries in the world unable to meet the water demand. With a population of about 7 million and close to 3 million refugees, Jordan has become overwhelmed by the demand on its water supplies, amounting to an annual amount of about 750 million m^3. The renewable freshwater resources include about 300 million m^3 (37%) of surface water and the rest are groundwater (54%), which are overly exploited through hundreds of illegal wells contributing to the depletion of aquifers.

Currently, most of the aquifers, like Jafr Dhulia, are depleted and others, like the Azraq and Agib basins, are showing signs of fast depletion and increasing groundwater salinity [37]. Other aquifers (Sirhan, Hamad, and Azraq basins) in the northeast are saline while high concentrations, above permissible levels, of nitrates, phosphates, and salinity were observed in the Mafraq, Agaba, and Ramtha aquifers in the Amman-Zarka area. The available water reflects 145 m^3/capita/year, far below the international poverty line of 500 m^3/year. Most of the available water (71%) is used to irrigate land in the Jordan Valley (some 33,000 ha) and in the Plateau (some 44,100 ha). Irrigation practices are poor and inefficient, contributing 2.1% the GDP [1].

Jordan faces the challenges of a constantly growing water demand and a weak, unstable economy. There are places in Amman where water is supplied just once a week, and people collect water in pails. To ease some of the water shortage, water is transferred from Israel to Jordan.

To alleviate the water-stress situation, the irrigated agriculture will need to be capped and regulated, increasing the use of drip-irrigation methods and discourage the irrigation of water-intensive crops (banana and citrus). The extremely high water losses and unaccounted-for-water (50%) should be reduced to a normal 10–15% losses [38] and water reuse has to be increased. Wastewater collection and treatment services serve about 4 million people (62% of the population), generating about 100 million m^3 of wastewater per year, of which the Amman's effluent is mixed with storm water in the King Talal Dam and used for irrigation in the Jordan Valley.

To address the increase in water demand and the growing competitive use, Jordan has drafted a Master Plan aimed at equitable allocation of water, intra- and intersectoral transfers of surface and groundwater rights, and introduction of advanced technologies including metering systems, drip irrigation, water reuse, brackish, and seawater desalination. Jordan also aims for electronic data acquisition, decision support systems, and private sector participation to complete the Disi-Amman Water Conveyance, the Red Sea–Dead Sea Project, and expansion of water reuse [39].

4.2.3 Israel

Israel is among the countries in the Mediterranean basin with about 200 m^3 per inhabitant per year, an amount that will decline further with the growth of the population [21]. Israel recognized that water management is central to development and made it a national plan, developing a national water system that allows conjunctive use of surface and groundwater combining the Sea of Galilee and the aquifers to secure a steady supply in dry and wet years. To balance between surface and groundwater stocks and flows within and between years, the major freshwater lake—the Sea of Galilee—is linked to the system and artificial recharge of aquifers is practiced, using check dams, quarries, and dual wells, as well as recharge with treated wastewater in a secluded aquifer.

Water and sanitation services are provided at their true cost; investment in infrastructure and irrigation systems, smart metering, and leak detection make Israel water independent, despite being one of the world's driest countries. Israel has reached most of its water-saving potential. Homes and businesses are equipped with low-flow and low-pressure fixtures, avoiding water-thirsty urban landscaping, as well as cutting agricultural water use through extremely improved efficiency of the irrigation systems and reuse of the wastewater. To comply with stringent drinking water-quality standards, water-treatment plants have been built to reduce the turbidity level to less than one nephelometric turbidity unit (NTU) [21].

Improved water management and the extended development of nonconventional water resource, up to 70% of the total water demand by the year 2030, is coupled with other advances in agricultural production and energy. Turning the desert into a farm land and tapping renewable energy, moving away from coal fired power plants to burning natural gas. Photo voltaic power generation is also on the increase aiming to reach a target of 17% of the national total energy requirements by 2030.

For a population of 7.5 million in 2010, the total water demand has reached 1800 million m^3, used by the urban sector (40%) and the agricultural sector (60%). For the projected population of 15.5 million in 2050, an increment of about 1000 million m^3/year will be required to reach a total amount of about 2800 million m^3/year. Of this amount only 28% will derive from natural resources, the remaining 72% will be supplied from nonconventional sources: treated effluent (26%) and brackish and seawater desalination (46%) [40]. The desalination of seawater, which began in 2005, has evolved to the construction of five major desalination plants in 2015, supplying 600 million m^3 (75%) of the domestic consumption at a production costs of $0.5–0.7/$m^3$. The vision is for Israel to use its relative advantage to assist the neighboring states to secure water supply and to release adequate flow for the restoration of drying streams and replenishment of the aquifers that were overexploited during drought years.

4.2.4 Palestine

The Palestinian Territory covers a total area of 6020 km^2, split between the Gaza Strip and portions of the Jordan River West Bank. The West Bank is a landlocked territory on the west bank of the Jordan River with a total area of 5655 km^2, surrounded by Jordan to the east and Israel to the south, west, and north. The Gaza Strip is a narrow coastal strip of land along the Mediterranean Sea with a total area of 365 km^2, bordering with Egypt to the south and Israel to the north and east. Under existing arrangements, the Palestinian Authority has control over civil administration and the Israeli authorities have control over the security [41].

With a population of about 1.5 million people, the water demand in **Gaza Strip** is met by groundwater pumped locally from the coastal aquifer shared with Israel and supplemented by water transfer from Israel, providing 10 million m^3 of a total consumption of 172.4 million m^3 (2010). Urban consumption accounts to 96 million m^3 or 55% of the total supply. Water-supply efficiency is very low; nonrevenue water ranges between 35% and 52% in the various localities. Furthermore, the local aquifer is overexploited, high above the safe pumping rate (50–60 million m^3/year), causing intrusion of seawater and salinization of the abstracted water, adding to pollution from sewage and fertilizers. Nitrate and chloride concentrations exceed the recommended WHO drinking water standards of 45 mg/L for NO_3 and 250 mg/L for chlorides. Consumption of contaminated water without treatment may attract cholera typhoid, dysentery diarrhea, salmonellosis, and hepatitis [42]. To protect their health, the population turns to decentralized point-of-entry and point-of-use water supply systems comprising reverse osmosis, ion exchange, and distillation systems to reduce organic and inorganic contaminants and to bottled water [43].

West Bank Water Resources. The West Bank's water supply is obtained from wells, springs, and from the Israeli water supply system, providing 55.5 million m^3 (87%) of the 85 million m^3 used for domestic consumption (2010). The consumption rate amounts to an average of 100 l/c/d, with a wide variation between urban and rural settlements, of which only 73 l/c/d are actually consumed. Sanitation is confined to the urban area and no improved sanitation services are provided in the rural areas [44]. Agriculture is a leading sector of the Palestinian economy, contributing greatly to the national GDP and a major source of employment, both in irrigated farming and rain-fed farming. Potable water is used to the amount of 69 million m^3 in the West Bank and 81 million m^3 in Gaza Strip for the irrigation of 11,822 ha out of the total arable area of 15,482 ha of arable land [45].

The Palestinian Water Authority suffers from inadequate institutional arrangements, insufficient cost-recovery, poor operation, and maintenance, and overall lack of sound management practices. The nonrevenue water amounts to 28%. The guiding vision of the Palestinian Water Authority (PWA) is to achieve equitable and sustainable management and development of water resources, to break the circle of poverty and ill health [45]. Also, to improve the level of service, rising the consumption rate, pricing and cost recovery, to reduce the unaccounted for water and to enforce the law against unauthorized (unlicensed) abstraction and illegal connections [46].

4.2.5 Egypt

The Nile–Lake–Nasser system is Egypt's only renewable supply source for surface water, and constitutes 95% of the country's total water resources. The rest of Egypt's water resources is mainly fossil (nonrenewable) groundwater found in the coastal zones, deserts, and Sinai and is estimated at 3 to 4 BCM/year. Egypt has no significant rainfall and is dependent almost entirely on 55.5 billion m^3/year of water from the Nile River. The available water per capita amounts to about 900 m^3, just below the scarcity level of 1000 m^3/capita, and is expected to fall to 670 m^3 by 2017 [47].

Urban water supply coverage is almost 100%, amounting to 130 L/day to 275 L/day in 2000. But the unaccounted for water ranges from 15 to 56%, with a weighted average of 30%. Sanitation is confined to the urban area and no improved sanitation services are provided in the rural areas. Agriculture is the major water user; approximately 85% of the Nile water is used for irrigation. Agriculture accounts for 20% of the GDP and 35% of the labor force. The responsibilities for water-resources management and regulation are carried out through multiple and segregated agencies and departments working and budgeted in parallel. The agencies have very little incentive to be accountable to the service characterized by low tariff, poor cost recovery, poor service, and low consumer expectations. The inadequate funds to sustain investment and operation at a proper standard level contribute to the widening gap between demand and supply.

Egypt faces the challenge of improving the productivity and sustainability of the water-supply system to be able to meet the water requirements of the growing population, the rising living standard, and the needs of industries and agriculture. Due to its dependence on irrigation, the Egyptian agriculture is vulnerable to potential climate change, which could affect irrigation, crop yields, and livestock water supply [48]. Also, the likely rising sea level may disrupt the stable and sustainable supplies [15]. In this context, the initiation of pilot projects aimed at improving the management and efficiency of the irrigation and drainage systems as well as water system and establishing holding companies that attract private sector participation are the right way forward.

4.2.6 Yemen

In Yemen, water is derived from rainfall, springs, seasonal flows, runoff, and groundwater, but the per capita supply is one of the lowest in the world (135 m^3/capita/year), compared to the regional average of 1000 m^3/capita/year [6]. The extreme water scarcity has created competing demands between the various sectors of the economy [49], although only about 55% of the urban population has access to piped water supply and only 31% have access to a sewerage system. In rural areas, these percentages drop to 45 and 21%, respectively [50]. The water shortfall has caused a massive and persistent problem of unsustainable groundwater extraction in both highland and coastal areas, threatening the public health, the agricultural sector, and the economy. Many wells have dried up, and groundwater salinization and pollution are worsening as aquifers are exhausted. Many farmers who lived in the Sa'ada Basin, where groundwater levels have fallen more than 50 m, have abandoned their farms or converted most of these areas to rain-fed agriculture [51].

The current situation greatly complicates the challenges of governance for Yemen to avoid the critical point at which it might not only constrain economic development but also threaten social stability, as is presently evident. The capacity of the Yemeni government to control extraction of groundwater of the depleted aquifers and the outcomes in the water sector is restricted by its weak technical, administrative, and enforcement capacity. The government inability has led to the economic marginalization of those unable to compete in power and money. The delegation of some official responsibilities to the participatory basin committees and to the local authorities is being introduced, assuming that decentralized structures will be more efficient to control abstraction and drilling and less politically driven than centralized ones [51].

4.2.7 Tunisia

Tunisia is among the countries in the Mediterranean Basin in which water availability per person is less than the annual amount of 500 m^3. This ratio will decline to 360 m^3 by 2030, when the population will have grown to approximately 13 million. The major water-using sector is agriculture, which consumes almost 80% of Tunisia's available water resources. Tunisia must expect a future of water shortages exacerbated by more frequent droughts and climate change and the increasing water demand. The anticipated climate change will further reduce the amount of renewable freshwater resources [52].

The unavoidable reality facing Tunisia has led to conserving and deriving maximum benefit from limited water resources and to crafting policy and strategy instruments that balance sustainable water management against the various political and social conflicts. In this context, Tunisia to balance between the annual and inter-annual availability of surface and groundwater resources, dams and artificial water recharge are implemented to store water surplus as available in rainy years. Tunisia has promoted the use of nonconventional water including reuse of treated wastewater. Most of Tunisia's 194 million m^3/year urban wastewater is treated and used for irrigation. Desalination of brackish water, to the amount of about 60,000 m^3/day and about 45,000 m^3/day by the public and private sectors, is part of the integrated management of the water cycle [52].

The Tunisia national water strategy needs to be further developed to address degradation of natural ecosystems and irreversible long-term damage to aquifers, avoiding excessive use of groundwater and poorly managed surface water as well as the operation of water infrastructure. In recognition that conservation of resources is far more productive than looking for ways to secure new reserves, Tunisia plans to implement a demand-based water management strategy, incorporating participatory management, helping local organizations and beneficiaries/stakeholders to take control of operating and maintaining their water distribution facilities, all aiming at the sustainable management and use of available water resources, as further elaborated on in the next section.

5. WATER MANAGEMENT

Given the increasing political recognition of the importance of water for the region and going beyond merely finding engineering solutions, it seems that advanced management systems that include regulatory framework and public–private partnerships are most essential for sustainable use of water resources. Macroeconomic, trade liberalization, agricultural, and urban policies that relate to hydrology, political science, law, institutional economics, and engineering should all be wholly integrated in water management, as comprehensively reviewed by Jagannathan et al. [53] and highlighted in the following.

5.1 Water-Management Principles

Good water-management principles include: **transparency**, **accountability**, and **participation** [54].

- **Transparency** implies openness and visibility in which accountability and participation are built. Effective and efficient institutions are required for transparency mechanisms to facilitate good governance in the form of regulatory bodies that create mutual accountability between the various stakeholders.
- **Accountability** implies that water utilities have the capacity to respond to the population water demand, while accounting to the financiers and policy makers stakeholders regarding the expenditure of public funds and safeguarding the environment, and,
- **Participation** represents the involvement of civil society organizations, consumer groups, project beneficiaries, and affected communities in all

stages of the management and administration processes in urban and agricultural water sectors, encouraging public involvement in water planning and decision making [54].

Successful water-management systems follow a clear pattern that integrates the following components:

- Allocation of available water resources (Bargaining).
- Regulations, defining the water services to various use groups (water laws and codification);
- Participation of the agencies (government, private, cooperative, etc.) responsible for delivering water services (Delegation);
- The contractors of the infrastructure necessary to deliver water to consumers (Engineering); and
- Monitoring and evaluation of adaptive measures (Feedback) [54].

5.2 Integrated Water-Resources Management

The above management principles are incorporated in integrated water resources management (IWRM), the new management approach that substitutes the conventional supply-driven approach and integrates investments, financial, economic, and institutional instruments in routine water management [55]. Integrated water resources management balances between the growing water demand from the various users including domestic, environment, industry, commerce, and agriculture in relation to cost recovery and other drivers, notably energy, virtual water, political reluctance, tariffs, etc. Integrated water resources management also emphasizes conservation and sustainable use prior to accelerated development and use of nonconventional water sources, water reuse, and seawater desalination.

Several countries in the region have already adopted the IWRM approach as the preferable platform for improved water-sector management, reflecting commitment of the water agencies to employ IWRM principles [56]. However, as described in Section 4.2 above, the region as a whole is still characterized by poor service delivery exacerbated by high operation and maintenance (O&M) costs and low end-user tariffs, resulting in poor O&M and unreliable service and deferred maintenance. With few exceptions, unaccounted for water amounts ranging between 40% and 50% [1], while the existing financing arrangements are not conducive to efficiency in the delivery of water services [48].

To address the issues of water scarcity and better management, intensive efforts are still required for the region to make a breakthrough in the adoption of advanced IWRM principles, covering legislative and institutional aspects, technology, investment, public awareness, and capacity building, as briefly reviewed in the following.

5.2.1 Water Laws, Institutional Capacity, and Regulation

Countries in the region are in the process of adapting or revising existing water laws, aiming to align Islamic and customary laws with the realities of modern water laws and policy [57]. But many have to enact drought decrees that foresee water-use restrictions and allocation of scarce water resources to the most prioritized use and enforcing and updating policies and regulations related to pollution control, water-resources protection, and conservation of ecosystems. Key issues such as groundwater management, water rights, and wastewater reuse are also to be addressed together with codification, financing, subsidies, and bidding rules, aiming at creating incentives for private sector participation in water supply infrastructure.

In this context, the challenge is to develop water agencies that are capable of developing policies that are specially related to water scarcity, incorporating climate change perspective and exposed to the changing realities. The new agencies should be able to observe and accomplish the objectives and goals of the enterprise, combining sound institutional regulations, strong work ethic, and the use of technology that could improve the performance of the water utility. A supporting pillar is the formation of water-user associations that help the management of water utilities in the delivery of a service that responds to the demands of water users, in terms of timely and equitable water supply that meets their requirements.

5.2.2 Demand Management

Demand management means efficiency, economic and environmental sustainability, and conservation to secure long-term water supplies while meeting strict criteria for socioeconomic, financial, and environmental sustainability and public health requirements.

Specific challenges to be met are:

- Regulation of water allocation among and within sectors to regulate permit systems for abstraction.
- Employment of reliable metering, data and exchange of information; monitoring of progress against a set of objectives; awareness; capacity building; and meaningful stakeholder participation.
- Expanding water services to vulnerable communities, encouraging local initiatives in building and managing the services.

- Reforming agricultural policies and food crop productivity, increasing irrigation efficiency to increase irrigated areas, and significantly reducing the region's water deficit and food imports.
- Securing environmental and ecological sustainability to promote the use of treated wastewater and drainage water.
- Considering water's real value, including social and financial costs, to identify the most effective strategy to address the escalating water crisis.
- Building partnerships with beneficiaries and the private sector, thus encouraging participation in modern, timely, and well-monitored water-delivery services.

5.2.3 Agriculture Water Policy

Agricultural policy is prominent among the many factors that would shape the region's vulnerability to water scarcity. Adaptation of appropriate administrative and technical measures will allow maintaining agricultural production, while satisfying the needs of the other sectors. Farmers have to change water-usage patterns at a time when crop water requirements are on the increase, facing the challenge of managing alternating water abundance and scarcity [53]. Therefore mitigation measures to improve water-use efficiency have to be employed, gaining farmers' appreciation of the value of water.

Advanced irrigation technology, which allow precise delivery of water and nutrients to the plants, saving 50% in water and 30% in nutrients as well as increasing yields by 20–30%, while reducing labor requirements and ultimately increase farmers income should be employed, substituting the highly inefficient flood-irrigation method, which is used on 80% of all irrigated land [58]. New smart water devices are available to monitor water-usage patterns, to detect any unusual usage such as a burst pipe or a leak, and to alert the consumer and or turning off the water flow remotely until repairs are made [59]. Also, new techniques are available to minimize nonbeneficial consumption (such as evaporation from reservoirs, open canals, etc.). Economic incentives may be required to encourage farmers to switch to improved irrigation techniques, which currently are limited to only 4–5% of all irrigated land, and to adjust the cropping systems to irrigation with alternative water sources to substitute scarce potable water.

Virtual water. Virtual water is an integral component of the water cycle of the arid zones, being the water that is imported or exported through intensive commodities—flowers, fruits and vegetables, legumes, and cereals.

Import/export of virtual water is a strategic choice to address the growing gap between supply and the growing demand [60]. Under international free trade markets, scarce water resources can be alleviated by importing rice, wheat, dairy, and beef imports that result in a significant import of virtual water, which could amount to a significant portion of the water cycle, e.g., Algeria (87%), Egypt (31%), Jordan (398%), Libya (530%), and Saudi Arabia (580%) [61].

5.2.4 Water Charges

In general, water prices for consumers do not reflect the real value of water, especially in the agriculture sector, though in some countries pricing is now tiered, in which higher levels of usage bear higher unit charge and paying the full price. But in most cases, attempts to introduce sensible pricing and tradable water rights are constrained and get circumscribed because of fear of political opposition and the widely held belief that Islamic traditions require water to be a free commodity [62]. Low tariff and price subsidies, irrespective of water cost, encourage misuse of water and lead to utilization of resources at unsustainable rates. Cost recovery and water price that closely reflect the scarcity value are therefore necessary for water-use efficiency and financial sustainability without compromising the principle of water as a "public good." Targeted subsidies are recommended for low-income users, waiving the charge for individuals below a certain income limit.

5.2.5 Public–Private Partnership

Public–private partnership (PPP) is enabling an environment in which a scale up of private-sector participation in water investments and management is facilitated and opened for a new financing and costing system, involving cost sharing with the private sector and the community. Under the PPP, the private investor builds and operates the system and earns a commercial return by collecting water fees from consumers. The PPP is founded on full cost recovery, volumetric pricing, formal water entitlements, and participation in financing and management. A key feature is the involvement of the community in which the consumers understand their needs regarding the service and performance standards that are to be achieved and their ability to absorb the cost of service and willingness to pay for water. The PPP creates a sense of ownership of the utility, since consumers become more proactive in dealing with emerging problems and in resolving social and technical problems that the top-down approach fails to resolve.

5.2.6 Data Management

In recognition of the irreversible adverse consequences of global warming, the first challenge concerns the systematic collection of data and information, as required to reduce the negative climatic impact on vulnerable population groups. Valid data are essential in order to coordinate among sectors, to communicate the associated risks for water security, and to propose solutions to reduce gaps and contribute to a sustainable water-resources management. However, given today's trends on climate change, passed data series are not always relevant for the assessments of current and future water availability, as apparent by the precipitation in recent years, which has been significantly lower than historical trends. Therefore advanced tools should be incorporated in gathering systematic, well-planned, and accurate hydrological data including geophysical, land, and water productivity data.

Monitoring devices installed at various points in a water-supply system may provide the water utility with online water quantity and quality data, remotely obtained from gauges, meters, and valves, regulating water pressure and flow. These devices can detect abnormality in the flow, pressure, and water quality above set thresholds, sending warning signals to the operators at the command center [59]. Similarly, a combination of geographic information systems (GIS), remote sensing, and modeling techniques helps to collate both dynamic and static characteristics of the land, vegetation, and climate. Space-based remote sensing can identify potential changes well before the conditions on the ground are evident. Remote sensing technology may provide historical time series as well as near-real-time data about the major components of the hydrological cycle: air temperature, rainfall, land use, irrigated areas, evapotranspiration, soil, moisture, and aquifer storage, while GIS is a helpful tool in creating easily accessible maps, figures, and tables.

Remote sensing and Web-GIS interfaces enable the possibilities for assessing land and water productivity and for mapping water-management systems, such as irrigation channels and dams, providing decision-makers with real-time information [63]. These techniques are also useful for the management of country/basin level as well as shared water resources, allowing optimal allocation of limited water among various competing needs/sectors. On a regional basis, these techniques may strengthen integration and interregional collaboration on the management of shared resources, avoiding multiple agencies collecting similar data.

Currently, many of the MENA countries are still lacking the capacity and resources to analyze and work with advanced GIS and remote-sensing monitoring devices. Thus an initiative was launched in Egypt, Gaza, Morocco, West Bank, and Yemen by the Arab Water Council, NASA, USAID, and the World Bank to obtain and analyze the data, based on the Gravity Recovery and Climate Experiment (GRACE) [12,64,65].

5.2.7 Training and Capacity Building

The water sector touches on water supply, food production, health, industry, energy, ecosystem, and recreation. As such, it requires skills and knowledge drawn from scientific, technological, economic, health, legal, and social disciplines. Effective training is essential to engage competent human resources able to integrate and interpret data, to identify gaps and to learn from past experiences, and to reflect on the future [66]. In line with the new development, capacity is required to harness and integrate advanced monitoring tolls and to blend nonconventional water resources (desalination, wastewater reuse, cloud seeding, utilization of saline water, etc.) into the hydrological cycle in a sustainable manner and without creating adverse negative consequences.

In this context, the MENA region, through the leadership of the Arab Water Council, has established several institutions including the Ameba Water Academy (AWA) [67] and MEDREC [68], an international organization working on solutions to fresh water scarcity, with the objective to establish regional centers of excellence. These institutions aim to promote innovative perspectives in order to make the most of water in the MENA region. The centers train and create new water managers able to absorb various disciplines, representing all aspects of water management, codification, and regulation and promote and improve water innovation, governance, leadership, management, and technological changes for sustainable growth. These institutes create a link between R&D of incubators and science parks and innovations, pilot schemes, and end products.

6. INCORPORATION OF NONCONVENTIONAL WATER SOURCES IN THE WATER CYCLE

Nonconventional water sources need to be integrated in the water cycle to supplement fresh water resources as required to satisfy the growing water demand of the urban population and the consumption of the economic sectors. As such, nonconventional water resources, namely, water reuse and seawater desalination, will have to be developed, on a large scale, to close the water supply and demand gap. Generally, water reuse is intended for agricultural irrigation and landscaping, while desalination is principally intended for domestic consumption. Both resources are playing a greater role as a cost-effective solution and are regulated by specific law and quality standards for the intended use (agriculture, urban, industrial).

6.1 Wastewater Treatment and Reuse

6.1.1 General

Water allocated for urban water supply is discharged back to the environment as a return wastewater flow. Of each cubic meter used by the household 75–85% is a return flow that has to be adequately treated and safely disposed. In a region in which fresh water is scarce it is necessary to take one step further and to treat and reuse the effluent for suitable use [69].

Available wastewater treatment technologies can convert wastewater into renewable valuable products including irrigation water, industrial process water, biogas, heat, electricity, and nutrient-rich biosolids for soil conditioning and even to drinking water. However, municipal wastewater contains human fecal waste and biological and chemical contaminants that are discharged by commercial and industrial facilities. The waste contains excessive loads of nutrients, a wide variety of contaminants and pathogens, and inorganic micro-pollutants, having human health and ecological impacts. Also, potentially harmful pharmaceuticals that can alter the biology of some organisms that live in the recipient water bodies may be detected.

Environmental safety and health impacts of these wastewater contaminants have long been a matter of concern regarding their potential threat to landscape, agricultural workers and to crops and soils irrigated with wastewater effluents, and to the public at large [70,71]. Therefore these harmful contaminants must be removed from sewage before widespread use of the effluent to meet stringent quality standards as well as public health regulations for reuse and disposal of effluents to land and or water bodies.

6.1.2 Wastewater Constituents

The contaminants and pathogens found in wastewater include a variety of organic and inorganic matters and micropollutants that may be classified as:

- Physical: suspended solids
- Biological: pathogens, microbial agents, and antibiotics compounds
- Chemicals: pH, alkalinity, ions, metals, fats, oils and grease (FOG), nutrients, and micronutrients

Microbial contaminants. Pathogenic microorganisms end up in sewage in large quantities not just during disease outbreaks, but also routinely in smaller amounts from people who are not overtly infected. These include bacteria: *E. coli*, *Salmonella*, and *campylobacter*; protozoa: Cryptosporidium, Giardia and various amebae; viruses: hepatitis, many types of enteroviruses, norovirus, rotavirus, and adenovirus; and other pathogens.

Chemical contaminants. The number of potential chemicals in sewage is almost limitless, but most are present in small and minute quantities. The principal nutrients are organic and inorganic nitrogen and phosphorus. The nitrogenous products can be found in the forms of ammonia, amines, nitrate, nitrite, or more complex molecules. The various classes of inorganic chemicals associated with wastewater and thought to be of importance to human and ecosystem health were listed by the National Research Council [72].

Among the diversified chemicals found in wastewater, there are other traces of biologically active chemicals that are found in human and veterinary medicinal and personal care products called endocrine compounds (ECs) or pharmaceuticals and personal care products (PPCPs). These compounds and their degraded derivatives have potential impacts on aquatic and terrestrial ecosystems and human health [73,74]. Other byproducts of chemicals used for the disinfection of drinking water may also be found.

The occurrence, fate, and effects of these constituents in regards to human health and ecological impacts have yet to be fully understood and are the focus of environmental research [32]. All of which, including ECs, PPCPs, and disinfection by-products, have to be removed or transformed to harmless compound to reach the permissible levels of total suspended solids (TSS), BOD, fecal coliforms, and residual chloride.

6.1.3 Wastewater-Treatment Process

Reliable treatments that meet strict water-quality requirements for intended reuse application have been developed and applied to protect public health and to gain public acceptance [75]. The wastewater-treatment process is generally categorized as preliminary and primary followed by secondary and tertiary treatment steps. The process includes a physical step intended for the separation and removal of suspended solids and a biological step for the decomposition of particles and the digestion of residual solids. The biological step may be followed by a physicochemical step that is intended for the removal of the remaining impurities and disinfection of the final effluent.

Each physical/biological treatment step removes contaminants to a degree, through a combination of biological and physical steps, aiming for the degradation of the organic compounds using suspended microbes, followed by a sedimentation stage designed to separate the solids from the liquid fraction. Disinfection can be applied at multiple stages to control septic odors, kill microorganisms, or provide oxidative effects.

The multiple technologies and configurations of wastewater treatment as well as the engineering and chemical aspects are well documented [75] and highlighted in the following.

Preliminary treatment. The pretreatment includes grease separation and grit and screenings removal using bar screens. The grit is discharged into a container and the fat is skimmed and conveyed to a fat-collection pit and sent for treatment and recycling. The screenings are conveyed by conveyors and compacted before discharge into a container for disposal. An odor-control system sucks and treats the contaminated air from the pretreatment building.

Primary treatment. From the bar screens, the wastewater is conveyed into an anaerobic cell equipped with mixers, in which the raw wastewater is mixed with a stream of a return activated sludge (RAS) emerging from the sludge clarifiers to generate the mixed liquor, rich with bacteria able to remove phosphates, that is delivered through a selector that distributes the mixed liquor to the aeration tanks for secondary treatment.

Secondary treatment. In this step, air or oxygen is injected to degrade organic compounds using suspended microbial organisms, followed by sedimentation to separate the solids from the liquid fraction. The configuration of the aeration tank includes aerated zones, in which carbonaceous and ammonia oxidation (nitrification) take place, and anoxic zones in which denitrification take place. The process removes organic and microbial contaminants as well as nitrogen and phosphorous.

Secondary sedimentation tanks. The aeration process is followed by settling in circular sedimentation tanks in which the biomass is separated from the secondary effluent. The effluent is diverted into a storage reservoir for further treatment and the settling material (the sludge) is partly returned to the aeration tank, as RAS, and the excess sludge (WAS) is sent for treatment before disposal.

Sludge treatment. The WAS is thickened to 5% solids and then digested in aerobic or anaerobic digesters (to produce biogas) and dewatered. The dried digested solids are transferred to composting plants in which the sludge is converted to organic fertilizer or sent to a landfill or incinerated.

Tertiary treatment. The secondary effluent undergoes a purification step to remove the remaining suspended solids and microorganisms such as Cryptosporidium and Giardia and fecal coliform. The process includes flocculation, filtration, and disinfection to produce high-quality effluent, suitable for unrestricted irrigation and other industrial applications. A variation of this process is the soil-aquifer treatment (SAT) process in which the secondary effluent is infiltrated through sand layers into a confined aquifer, which serves as an underground storage.

Membrane biological reactor (MBR). Semipermeable microfiltration or ultrafiltration membranes may be installed in the aeration tanks to become a MBR, aiming to stimulate the degradation process and to reduce the process resident time, the footprint, and energy consumption [76]. The MBR is a newer method that combines biological treatment with membrane separation to replace conventional coagulation, sedimentation, and filtration. The process results in a higher quality effluent with a smaller footprint and lower cost. The MBR may be followed by reverse osmosis to remove almost all residual organic and inorganic chemicals and any residual trace organic chemicals to subparts-per-billion levels.

Direct potable reuse (DPR). Treatment of municipal wastewater and reuse of the effluent to produce potable water supplies including direct potable reuse (DPR) is being explored, especially in the United States. Direct potable reuse is a form of potable reuse (unfinished) in which treated wastewater is introduced into an environmental buffer (e.g., a groundwater aquifer or surface-water reservoir, lake, or river) for a period of time before the blended water is introduced into a water-supply system, as detailed in the Framework for Direct Potable Reuse [77]. The procedures cover regulatory considerations (public health risks) and technical issues related to the production of advanced treated water and public support and outreach. The DPR involves the use of state-of-the art advanced water-treatment technologies including membranes, reverse osmosis, and advanced oxidation to remove viruses, bacteria, chemicals, and other contaminants that may be present in the effluent.

Underground and surface storage. To adjust between the almost even generation of the effluent and the seasonal requirements of the irrigation systems, a storage component in the form of underground (artificial recharge) [78] or surface reservoirs [79] may be added. The reservoir, in addition to storing water, acts as a large biological reactor in which physical, biological and chemical processes add to improved quality of the impounded water. Adsorption and precipitation, ion exchange, and die-off processes, among others, act to improve the quality of the incoming effluent.

6.1.4 Treatment-Quality Performance

Raw wastewater quality and quality improvement over the various treatment stages compared to the applied standards for unrestricted reuse for irrigation and other nonpotable uses and disposal to water bodies are given in Table 6.1.

The resulting contents of solids, microbial pathogens, and chemicals indicate that the applied treatment processes produce an effluent that is suitable for unrestricted irrigation and other industrial applications. Heavy metals concentration were found to be below the standards by one or two orders and the concentration of fecal coliforms after disinfection and a contact time of 30 min yielded on average 1.8 units in 100 mL. Regarding the estrogenic contaminants, which may transform to potential hormonally active byproducts [82], they may be found in the original or in transformed forms in minute quantities, not easily detectable [83].

Energy consumption and cost of treatment. Energy consumption and cost of treatment varies depending on the type of treatment and the intended final use. Typical energy requirements are about 0.35 kWh/m^3 for secondary effluent and 0.5 kWh/m^3 for tertiary effluent, respectively, compared to 0.9 kWh for advanced potable reuse and 3.5 kWh for seawater desalination. The cost of treatment varies widely between US\$0.45 and US\$0.75/m^3 and US\$0.55/m^3, on average, of which the capital cost is US\$0.12/m^3, operation US\$0.31/m^3, maintenance US\$0.10/m^3, and miscellaneous US\$0.02/m^3.

Municipalities are responsible for financing and operating the treatment plants at their own cost and may sell the effluent to farmers, usually much below cost. Seen by users as inferior to freshwater, the price charged to farmers is only US\$0.02/m^3, in Tunisia [84], while in Syria and Yemen it is provided free of charge [85]. In Israel, the price is 20% lower from that of the national water system [40].

6.1.5 Water-Reuse Benefits

Water reuse is emerging as an established water-management practice in several water-stressed countries and regions because of human health and agronomic and environmental advantages that provide the following benefits:

- Substitution of fresh water that can be released for other uses
- Securing water for land irrigation under scarce and arid conditions
- Guaranteed continuously supply distinct of the seasonality of the natural flow

Table 6.1 Wastewater-Treatment Systems and Quality Standards

Parameter	Unit	Raw Sewage[1]	Secondary Effluent[1]	Tertiary Treatment		Acceptable Standard[4] for	
				Artificial Recharge[2]	Surface Reservoir[3]	Irrigation	Estuaries
Temperature	°C	na	19.4	23.5	22		
pH		7.38	7.45	7.3	8.1	6.5–8.5	7–8.5
BOD	mg/L	385	6		8	10	10
COD	mg/L	878	40	0.56	59	100	70
SS105	mg/L	484	6	1.26	4.2	10	10
Turbidity	NTU	4.5	2.9	0.2	17.2		
NH4 as N	mg/L	48.1	2.9	0.11	21.9	20	1.5
Nitrate as N	mg/L	0.72	0.72	4.5	28.2		
Total N	mg/L	68.7	7.3	4.9		25	10
Ptot	mg/L	5.7	0.8		3.7	5	
PO4 as P	mg/L	5.5	0.6	0.15	8.5		
Chloride	mg/L	214	223	232	403	250	400
Sodium	mg/L	162	153	174	277	150	
Conductivity	Micromohs	1568	1410	1400	2212	1400	
SAR		4.3	4.8	4.1	6.32	5	
Detergents	mg/L	14.3	<0.11	0.006	<0.11	2	0.5
Fecal coli	N0/100 mL	na	7.7 E + 04	0	2.9	10	200

(Continued)

Table 6.1 Wastewater-Treatment Systems and Quality Standards—cont'd

Parameter	Unit	Raw Sewage(1)	Secondary Effluent(1)	Tertiary Treatment: Artificial Recharge (2)	Tertiary Treatment: Surface Reservoir(3)	Acceptable Standard (4) for: Irrigation	Acceptable Standard (4) for: Estuaries
Trace Elements							
Boron	μg/L	0.27	0.26	0.21	300	400	
Cadmium	μg/L	<25	<0.2	0	0.2	10	5
Cyanide	μg/L	<1	<5	0.5	3	100	5
Cobalt	μg/L	<25	<3	1.7	3	50	
Chromium	μg/L	<26	<2	0	4	100	50
Copper	μg/L	118	19	1.3	6.7	200	20
Lead	μg/L	<25	<2	0	2	100	8
Mercury	μg/L	<0.44	<0.1	0	na	2	0.5
Molybdenum	μg/L	<25	<3	0.6	5.3	10	
Nickel	μg/L	<25	<4	5.4	26	200	50
Selenium	μg/L	<25	<2	0	2	20	

Based on Mor R, Kraitzer T, Michail M, Elkayam R, Sherer D, Shoham G, et al. Soreq mechanical biological wastewater treatment plant operation report, 2014. Tel-Aviv: Mekorot National Water Co., for Mey Ezor Dan, Agricultural Cooperative Water Society Ltd.; 2015; Halperin R, Asor R, Smulevitz S, Aharoni A, Kishon A. Project complex monitoring report, 2011. Tel-Aviv: Mekorot National Water Co. publication; 2014; Halprein R, Aharoni A. Monitoring report of the third line project, 2013. Mekorot National Water Co & Technion, Israel Institute of Technology; 2014; Ministry of the environment. http://www.sviva.gov.il/subjectsEnv/Streams/SewageStandards/Pages/Milestones.aspx.

- Provision of fertilizer, N, P, and K and other essential nutrients
- Contribution to the restoration of rivers, wetlands, and the aquatic ecology

Adequately treated wastewater used for irrigation can substitute natural water that can be diverted for domestic consumption and to reduce the pressure on scarce water resources, while contributing to the preservation of the green space, avoiding the discharge of effluent into water courses and pollution of streams, aquifers, and the sea. These and other environmental, public health, and economic benefits of water reuse are widely documented [86,87], indicating that water reuse helps to secure potable water supply for the population and postponement of development of costly water-supply systems including import, storage, transfer, and or seawater desalination schemes, having considerable environmental toll and capital expenditure.

6.1.6 Environmental Impact

Stringent regulations govern the quality of effluent used for reuse and disposal of effluents into water bodies. Strict compliance to these regulations secures the protection of the environment, avoiding the pollution of streams, aquifers, and the sea. Likewise, the use of effluents for agricultural irrigation yields tangible benefits to the municipalities, the farmers, and the environment since it is the most convenient way and ecologically sound method of disposal of wastewater. However, for a successful operation, the perception and the negative attitude toward water reuse has to be openly addressed and safety issues associated with water reuse on human and environment have to be fully clarified, including possible soil salinity and built up of salts in irrigated soils [88] as well as the impact of new chemicals and accumulation of persistent endocrine substances in the ecosystem that can act in an additive manner and impact on the environment [89]. Such substances were detected in fish exposed to wastewater-polluted streams [90] and in drinking water-distribution systems [91], but the evidence about their accumulation in soils and the likely effect on the environment and human health are not conclusive [92]. Nevertheless, taking into consideration the numerous compounds that are discharged into the environment and the in situ attenuation processes, the environmental impacts of water reuse require further and in-depth investigations.

6.1.7 Water Reuse in MENA

6.1.7.1 General

Basically, much of the wastewater collected in MENA is discharged untreated into the sea, the streams, or on land. Wastewater-treatment plants, if they exist, are often overloaded, underdesigned, and plagued by poor operation and maintenance and consistently produce effluent quality that cannot be safely reused. Despite its ranking as the most arid and water scarce in the world, water reuse for agriculture and irrigation purposes has been surprisingly uneven and slow across the region because of economic and institutional constraints, restricting the sustainable and safe reuse of wastewater. Currently, only 55% of wastewater is treated and 15% is reported to be reused in agriculture, landscape irrigation, industrial cooling, and environmental protection [93].

Major hindrances and causes for low intake have been reported [94] to include:

- **Lack of political commitment** and of national policies/strategies.
- **Technical constraints**, including insufficient infrastructure and improper maintenance of infrastructure where available.
- **Financial constraints**. High construction costs of sewerage networks and treatment systems as well as the cost of construction of water-reuse systems, where the seasonality of the irrigation does not match the year-round supply of wastewater.
- **Public acceptance and social awareness**. Low demand for reclaimed water and unwillingness to pay by the farmers where low pricing of fresh water does not adequately reflect the true cost. Further, farmers prefer to use alternatives water sources in spite of higher costs, because of the social stigma and crop restrictions associated with reuse.
- **Standards and regulations** are currently too strict to be achievable, coupled with weak regulatory compliance and enforcement.

Generally, the lack of consumer awareness of water scarcity remains a major obstacle, although the pervasive water scarcity, urbanization, and the increasing impacts of climate change will undoubtedly cause a shift in local perceptions of the importance of properly capturing and using reclaimed water. Significant progress with water reuse has been already achieved by Israel, Jordan, Tunisia, and others, setting the basic conditions for wastewater treatment and safe reuse, as demonstrated by the following case studies.

6.1.7.2 Selected Case Studies

6.1.7.2.1 Israel Israel is one of the best cases of wastewater reuse where much innovation has taken place. The country's long history of reuse has enabled it to develop extensive expertise since 1953 when Israel drafted the world's first set of standards for wastewater reuse, which have continued to evolve to reflect the latest scientific findings on microbiological and chemical risks [95] and the perfection of the technologies [96]. These achievements have led to the realization of the treatment of 470 million m^3 (90%) of 510 million m^3 of the generated wastewater and utilization of 400 million m^3 in 2012 [97]. Water reuse provides 38% of the water allocated to agriculture and is projected to increase to 585 million m^3 (51%) in 2025 and 900 million m^3 (67%) in 2050 [40]. Typically, wastewater treatment and reuse follow two different systems presented by the Dan region and the Kishon schemes, as follows.

The Dan district wastewater-treatment and reclamation facility is a major scheme of its kind. The facility serves the Tel-Aviv metropolitan area and treats on average 357,260 m^3/day or about 130 million m^3/year (25% of the countrywide wastewater generation in 2014) [78]. The treatment process includes the infiltration of secondary effluent into infiltration basins (artificial recharge) surrounded by production wells for soil aquifer treatment (SAT). After a long detention period, the reclaimed water is recovered and pumped through a 70″, 54 km long, branching into two pipelines (48″ 26 km and 24″ 20 km long) ending in seasonal reservoirs, from which the reclaimed water, after chlorination, is distributed to consumers for the irrigation of 55,000 ha [80].

Kishon complex wastewater treatment and reuse scheme—surface storage. The scheme serves the Haifa metropolitan area and reuse of the treated effluent amounts to about 35 million m^3/year for irrigation at the Izreel Valley, about 30 km away via a network of surface reservoirs with a total storage volume of 25.7 million m^3 and expanding on an area of 330 ha [79]. After a detention period of up to 270 days, the effluent is disinfected to reduce the fecal coliforms to less than 10 in 100 mL, before being delivered to the irrigated land according to the irrigation schedule.

Similar wastewater treatment and reuse schemes were constructed to serve small and medium towns to produce secondary effluents that are stored in about 325 surface reservoirs, ranging in volume between 8 and 0.5 million m^3 with a total volume of about 200 million m^3 and used for the irrigation of 75,500 ha [97].

6.1.7.2.2 Jordan Jordan's venture into the development of wastewater reuse capacity has been motivated by the freshwater scarcity (approximately 150 m^3/capita/year) [98] combined with a degradation in the quality of groundwater sources. The government position regards wastewater as a resource rather than waste and all wastewater-treatment projects are required to include a water-reuse component. In accordance, the government has enacted well-defined policy, laws, and standards governing water reuse in agriculture, which has developed to reach 70 million m^3/year, mostly in the Jordan Valley, or about 10% of the total national water supply. Other water-reuse projects promoted by private enterprises are used to irrigate palm plantations near the Aqaba wastewater-treatment plant. Other unregulated water-reuse projects are found where users have no alternative water supply. The unplanned water reuse has become an important and reliable source to augment the base flows of water streams [99] for irrigation along Wadi Zarqa near Amman.

These case studies demonstrate that wastewater reuse can be integrated coherently with conventional water resources encompassing environmental management, water demand, water conservation, and financing arrangements. Such initiatives can be replicated, leading to increased demand and willingness to use and pay for reclaimed water by users.

6.2 Desalination

6.2.1 General

Desalination of seawater and brackish groundwater is universally recognized as an attractive option to produce fresh water for drinking purposes, especially where there is no alternative water supply or where other alternatives to augment conventional water supplies have grown more expensive than the cost of desalination. At present, thousands of desalination plants have been installed to supply drinking water for the rapidly growing population in water-scarce countries including the MENA region, which produces 15 million m^3/day of desalinated water out of the global capacity of 24 million m^3/day, a capacity that is projected to double by 2020 [100]. Boosted by the persistent drought of recent years, seawater desalination has evolved into a long-term solution and is expected to play a significant role in the growing water demand in the MENA region.

The desalination is carried out using thermal, electrical, and pressure systems that greatly differ in terms in technology, energy consumption, and cost. The most common desalination technologies in the region are the distillation and the reverse osmosis (RO) based plants. The distillation technology such as multistage flash (MSF), multieffect distillation (MED), and vapor compression (VC) include a change of phase to separate the distilled water from the feed water, requiring substantial energy and discharge of a heavily concentrated brine back into the sea [101]. The RO desalination consists of feeding seawater through a semipermeable membrane under a pressure level higher than the solution osmotic pressure, typically 50–80 bar for seawater [102,103].

Historically, in the ME, thermal distillation technologies have dominated the seawater desalination market, partly because of cogeneration of water and power by the oil-rich countries neighboring the Persian Gulf where energy prices are low, especially Saudi Arabia where most of its water needs are satisfied by distillation desalination. Over the years, alternative processes have been developed, most notably the RO process, a membrane technology, which is less energy intensive. The RO technology, which dominates some sectors of the desalination market, is the preferred technology for the construction of flexible and modular plants for the desalination of brackish and seawater to satisfy current and future drinking water demand [104].

6.2.2 RO Desalination Technology

The critical step in the RO process is the pretreatment stage, which is performed to remove the impurities that could cause fouling and damage to the semipermeable membranes. Multimedia filtration, ion-exchange systems, and scale inhibitors are used for the removal of suspended particles and colloids before feeding the raw water into the membrane vessel. The pressurized feed water flows across the membrane surface where it is separated into a soft permeate water, containing 2 to 4% dissolved solids, and a concentrated stream—the brine—containing the removed ions and organics that emerge from the membrane module as a concentrated solution under high pressure and energy that can be recovered using advanced energy-recovery devices. The permeate recovery is typically limited to about 45% for seawater and 75% for brackish water.

The whole process is monitored to control the suspended particles and colloids using silt density index (SDI), differential pressure, the permeate flow, solids rejection, and pressure drop coefficient. These parameters indicate the fouling potential and extent of fouling and/or scaling of membranes, the major causes of premature membrane element replacement. The SDI reflects on the performance of the filtration process, indicating the remaining concentration of suspended solids particles and colloids in the filtered water. An SDI value of less than 3 is required in the raw water fed to the RO membranes. When the RO membranes are scaled or blocked, the permeate conductivity begins to increase, indicating that the percent rejection is decreasing.

6.2.3 Typical Desalination Facility

A typical sea desalination facility is a sizable installation with extended infrastructure sited close to a seashore and includes a deep-sea inlet intake, pumping feed water with total dissolved solids of up to 50,000 mg/L and at a varying temperature of 10–32°C. The RO process includes a pretreatment stage consisting of multimedia and cartridge filters before feeding the raw water into the RO pressure vessels [105]. The vessels contain semipermeable membranes with high flux, higher surface area per unit volume, and high salt rejection to produce permeate (45% of the inlet water) containing 100–500 ppm of total dissolved solid (TDS). Energy-recovery devices (a turbine, energy exchanger, or pressure exchanger) are incorporated to recover 80–85% of the waste energy released during the depressurization of the residual brine before discharge back to the sea. The brine (55% of the inlet water) is returned back to the sea while the antiscaling chemicals and anticorrosion products are treated before disposal. The product water is posttreated with lime to comply with WHO guidelines regarding drinking water quality.

6.2.4 RO Performance, Energy Consumption, and Cost

The energy consumption (SEC) of the RO process is 2.91 kWh/m^3, out of a total plant average energy consumption of 3.5 kWh/m^3, and the emission of CO_2 amounts to 286 g/m^3. The cost of desalinated water is sensitive to the costs of both energy and financing. But advances in desalination technologies have already reduced the cost of desalination to approximately US\$0.52/m^3 [106] and the cost may reduce further, driven by improvements in the separation membranes and decreases in energy consumption. Carbon nanotubes, radial deionization, ceramic and polymeric membranes, and biomimetic aquaporin membranes are some of the new approaches that will drive desalination to become an indispensable element in water management, paving the way for the acceptance of desalination as a practical solution to tackle the shortfall in fresh water. Further advances are expected in the RO process including the use of nanomembranes, forward osmosis, and ion-concentrate polarization, while hydraulic turbocharger and positive-displacement pistons will increase the waste-energy recovery to 90–95% [107,108]. In most MENA countries, the private sector has been a key player in promoting desalination, providing the necessary capital and efficient management and operation.

6.2.5 Environmental Impact of Desalination

On the positive side, desalination on a large scale would contribute to the conservation of conventional water resources, to preserve the aquatic ecosystems in rivers, and to halt groundwater depletion and seawater intrusion due to overly exploited aquifers. The RO technology is well established and clearly the least energy-intensive system; it is expected to become the most convenient system to produce drinking water for the population and also for irrigation of high-value crops, especially in coastal areas. Technology advances are likely to continue, reducing the cost of production, while the costs of exploitation of conventional surface water and groundwater are on the increase. Such development is considered in the plan to expand the desalination of seawater from the current 650 million m^3/year in 2015 to 1.6 billion m^3/year, as required to satisfy the water demand of Jordan, Palestine, and Israel in 2040 [109].

Desalination, on the other hand, affects the environment in many different ways, including the occupation of sensitive coastal land and the marine environment due to the installation of deep-sea intake and discharge of return flows back to the sea. The use of cleaning chemicals, energy consumption and the associated greenhouse gas emissions, and the disposal of the concentrated saline waste are all threats to the environment. Other local impacts are associated with the siting of the plant and seawater intake, which can affect the biodiversity of ecologically sensitive coastal areas and entrapment of aquatic creatures in the seawater intake, while disposed chemicals can damage the fragile marine ecosystem [110].

Mitigation measures are employed to minimize any damage to the coastal and marine environment and a stringent long-term monitoring system is employed to avoid as much a possible any adverse impact on the sea environment. Of major impact is the offshore seawater intake, which is built as a submerged structure with slow intake velocity to avoid interference with aquatic life. Similarly, the filter backwash stream is treated to remove suspended solids before being mixed with the rejected concentrated brine form the RO vessels. The combined stream is blended with returning cooling water from a nearby power plant before being discharged to sea. Alternatively, the mixed stream is discharged to the deep sea using diffusers to achieve a good mixing with sea water, as required to avoid rising temperature at sea and sinking of the brine at the bottom.

7. REGIONAL COOPERATION

Water remains a significant cause of potential conflict between riparian states as water scarcity can intensify latent conflicts, magnifying the differences between the upstream state and the downstream state whose ability to retaliate and reciprocate to water policies created by the upstream state is minimal [110]. Even though feasible practical solutions to alleviate water scarcity are available, as outlined in the above sections, it is quite obvious that achieving adequate supply and equal management of shared water under scarcity cannot be materialized without genuine cooperation between the neighboring states. Furthermore, because of the relative political stability and asymmetry in the economic strength of the riparian countries, there is a difference in the way they view how to reach sustainable management of a particular transboundary source. The riparian states may also differ in how to preserve and protect the source from pollution, based on their capacity to accept an inferior water supply, which is higher in a poor country than in a prosperous country, whereas the ability to improve the situation is the opposite, as evident from the accumulated regional experience.

The main transboundary rivers in the region, the Nile, the Jordan, the Tigris, the Euphrates and Shatt Al—Arab, among others, have already been affected by the prolonged droughts and climate-induced risks. Israel, Jordan, Palestine, and Syria are already affected and their water bodies are far from filling to design capacity. In the southern part of the region, Egypt runs the risk of losing a considerable part of its Nile Delta agriculture due to the expected reduced flow from the Ethiopian highlands and the rise in temperatures and evaporation rates in Lake Nasser, which will reduce the water stock available for downstream users. Thus irrespective of the upstream-downstream inherited conflicts, the current severe water scarcity problem provides a strong impetus for riparian states to combine forces to sensibly confront the continuous drop of water that may destabilize the whole region in the ensuing decades.

7.1 International Water Law

The general platform for protecting water rights and interests in shared surface and groundwater is the International Water Law [111], which protects the rights of all riparian states, and the "Helsinki Rules," 1996 [112], which promote the principle of reasonable and equitable sharing of the international rivers, promoting the understanding that water resources are critical to societies and ecosystems. The treaties are binding upon national governments in their bilateral or multilateral arrangements for managing transboundary water resources. The core obligations are to prevent, control, and reduce transboundary environmental impacts and their socioeconomic implications to ensure reasonable and equitable use of transboundary waters through good cooperation.

However, these pacts do not provide comprehensive international agreements to adequately accommodate the specific conditions of each of the river basins in the world and in this case of the water courses of the MENA region. Additional conventions are required as indicated by Heddadin and the League of Arab Nations [113,114] to provide the basis for the riparian states to recognize their joint and individual vulnerability to prolonged droughts, floods, or both and therefore to cooperate in the search for mitigation measures that would minimize the impacts of extreme events and adaptation to climatic change.

Seeing beyond conflicting interpretations of legitimate water rights and recognizing the interdependence between the states within the region, the way forward is to institutionalize rational water allocation to benefit the entire ecosystem, allowing water to be utilized to its highest economic and environmental values [115]. A positive approach that overcomes political and social opposition would permit improved watershed management irrespective of political boundaries as well as water trade and delivery of wastewater effluent to where they can be safely used. Such understanding could lead to cooperation in other interrelated sectors, and energy in particular, to enhance peace, stability, and economic growth [116,117].

7.2 Regional Cooperation Initiatives

Instances of positive cooperation on the management of transboundary water are already apparent in the region and significant initiatives have been made, although much remains to be accomplished. Acceptable institutional mechanisms for cooperation among affected stakeholders are required to embark on ongoing cooperation such as the Nile River Basin and the early initiatives related to the Jordan River, as summarized below.

7.2.1 Nile Basin

The Nile River basin extends over 10 nations, 8 upstream riparian (Ethiopia, Eritrea, Uganda, Rwanda, Burundi, Congo, Tanzania, and Kenya) and 2 downstream (Sudan and Egypt). Egypt, Ethiopia, and Sudan signed a declaration regarding the impact of the construction of the 750 MW—the Grand Ethiopian Renaissance hydroelectric project—on the Blue Nile River, a new project that might have a tremendous impact on water availability in the downstream [118]. The new agreement, signed by the three parties, includes the study of the dam's impacts on the downstream states using an advanced satellite-based water-management and forecasting system [119] to evaluate water resources, irrigation, and actions against impending floods or droughts and also to improve the reliability of the stakeholder decision-support system regarding flood warning, reservoir management, and irrigation planning [120].

7.2.2 Israel, Jordan, and Palestine Cooperation

The Israeli-Palestinian Oslo Agreement 1991 [121] and the Peace Treaty between Israel and Jordan in 1994 [122] reflected a shift from hydroconflict to hydrocooperation and led to several joint efforts between the three countries. These agreements addressed water-allocation issues between the two states, as well as making a provision for the storage of excess water from the Yarmouk River for use by Jordan. The joint efforts that derived from the interstate agreements see water as an element for cooperation rather than a cause for conflict and confrontation as well as a topic of building up a positive interdependence and trust and joint resilience to the predicted climate change, as elaborated on below.

The Lower Jordan River. The normal flow in the Jordan River, which threads through Jordan, Israel, and the Israeli-occupied West Bank, ceased back in 1964, when more than 90% of its natural water flow was diverted to satisfy the needs of Syria, Jordan, and Israel. The remaining flow, which ends in the Dead Sea, consists mainly of raw sewage, fish pond, and agricultural drainage and excess storm water. However, the river, being rich in natural ecology, the route of migrant birds, pilgrimage, and of economic potential induced the three counterparts to reach across geopolitical borders in search of equitable and sustainable means for the rehabilitation of the river. A master plan prepared by EcoPeace (a cross-border organization devoted to environmental conservation) [123] outlines 127 projects, creating thousands of tourism jobs, boosting agricultural output, and revenue to the impoverished region.

The Red Sea–Dead Sea Project. The Dead Sea, shared between Israel, Jordan, and Palestine, has economic, cultural, touristic, and environmental values. But because of diversion of the Jordan River water upstream and huge losses to evaporation, the seawater level is dropping at an accelerated rate of about 1 m/year to the current 428 m below sea level. To restore the seawater level, the three parties agreed on the Red Sea–Dead Sea Project that will transfer, annually, 2 billion cubic meters of water from the Red Sea to the Dead Sea, feeding on the way a desalination facility to produce 800 million cubic meters of potable water and a hydroelectric power plant. The bidding for the first phase of this project—Phase—I is underway [124]. It includes the construction of a desalination facility near the Aqaba Airport to generate 100 million m^3/year of brine, which will be pumped for the first 60 km and then flow by gravity along 120 km to the Dead Sea, allowing for hydropower generation. The desalinated water amounting to 85 million m^3/year will be used to supply Jordanian and Israeli consumers in the south in exchange for water to be supplied by Israel to Jordanian and Palestinian consumers in the north.

Gaza Strip Water and Power Supply. Gaza Strip suffers from inadequate water supply and improper treatment of sewage because of lack of power. The untreated sewage pollutes the streams and tributaries shared by the two states as well as the sea coast. The discharge of raw sewage to the Mediterranean Sea flows north easterly to the beaches of Israel, endangering the public health and the sea desalination works [33]. Of mutual interest, Israel supplies a certain amount of drinking water to Gaza Strip and plans to increase the amount as well as provide the power to operate the wastewater treatment plants, an act that will benefit the environment and public health.

Solar Energy—Water Nexus. An innovative plan is intended to promote the exchange of water for energy in which solar fields installed in the Jordanian desert would generate electrical power to supply Jordan, Israel, and the Palestinian Authority, in exchange for water to be supplied by Israel to its neighbors [125]. Israel uses advanced desalination plants on the Mediterranean Sea shore and thus is able to produce adequate desalinated water for all, while Jordan with its high solar radiation and large tracts of desert land is ideal for the establishment of large-scale solar power fields to generate substantial power for the three states at a very competitive cost of $0.058/kWh [126]. Considering that energy consumption accounts for more than 30% of the cost of desalinated water [106], the low cost of the clean energy would reduce the total cost of the desalinated water, and in addition, the photovoltaic power will cut down on greenhouse gases in line with the pledge made in the Paris Summit 2015 to reduce greenhouse gases emission.

Interlinking water, land, energy, and environment interventions as envisioned would contribute to optimizing the use of vital natural resources while addressing environmental needs and climate change impacts [127]. In this particular case, it will provide tangible solutions and promote innovation and strong cooperation in the ME region.

8. DISCUSSION

The World Economic Forum [128] placed water crises among the top three global issues that would have the highest impact on the world economy in the next 10 years. Adding that the search for solutions to solve the problem of water scarcity is still urgent, as has ever been. The forum concluded that leverages are needed to motivate public–private cooperation that would promote water-resilient economy in a warming world.

The MENA region situated in an arid and semiarid climate is among the most water-stressed countries and by 2040, 14 states are expected to top the list of the world's most deprived water-scarce states in the world, including Israel, the Palestinian Territories, Jordan, the Arab peninsula, and the North African countries, which have already reached a point of severe water stress lacking sufficient freshwater resources to support their respective populations. The climate change and continuous droughts in recent years have added to the devastating consequences of water scarcity and the unsustainable state of water resources and land-use practices, where water is not typically priced to reflect the true cost and scarcity value.

The chronic stress of water as already apparent is associated with both the quantity and quality of the available resources. The reduced quantity and the deteriorated biological and chemical quality affect every aspect of life, the public health, the ecosystems and the environment. The lack of water is reflected in deficient water supply and poor sanitation services as well as poverty and food insecurity due to inefficient irrigation. The extreme droughts that have become more frequent and persistent will render the water situation more acute and amplify the "Arab Spring" phenomena of endured economic crisis, sociopolitical instability, armed conflicts, and large-scale environmental migration.

Today, it is becoming even clearer that the observed long-term trends in declined precipitation and increased temperature are due to human interference and anthropogenic pollution and less consistent with patterns of the natural variability. Thus, the effects of the anticipated climate change are likely to exacerbate the situation in which the region is likely to experience more competition and conflict among countries and within sectors, communities, and individuals over water. The allocation of water for the growing number of competing water consumers—agriculture, domestic consumption, industry, hydroenergy, and ecosystems—under the changing climate would place a heavy burden on the water sector, as a whole. This situation negates the UN Millennium Development Goals (MDG), which addressed the elimination of extreme poverty and hunger and the new UN Post-2015 Development Agenda for a world where everyone has access to safe, sufficient water, and nutritious food, without compromising the needs of future generations.

Under such circumstances, which are complicated by changes in the hydrological cycle, in the form of reduced quantities and timing of precipitation and high rates of evapotranspiration, sustainable measures responding well to water-scarcity challenges meeting current and future needs are highly essential if a drastic water crisis is to be avoided. However, satisfying water needs cannot be met simply by the construction of costly and environmentally undesirable dams, conveyance systems, and other infrastructure. Instead, a mix of demand management, innovation, conservation, and regulation, coupled with water reuse, limited desalination, and water trade is suggested. Such measures will accelerate and pave the way for improved water management and water-use efficiency, making the most of the available resources.

The focus should be on the application of cost-effective solutions that reduce the demand for water based on available technologies. For the urban sector, such solutions may include application of water treatment technology to improve the quality of unfit drinking water and improved maintenance of water supply systems to avoid excessive losses of water in the distribution networks. In agriculture, introduction of low volume irrigation technology, improvement of the cropping systems adjusted to the prevailing climate are suggested. Appropriate economic charge for water used for irrigation is also essential to reflect the water scarcity value, balancing between the need for food production, the national food security policy and the recovery of the increasing cost of water. These measures may be supplemented by augmentation of the supply side, harnessing nonconventional sources, taking advantage of the drastic drop in renewal energy prices to extend water reuse and the desalination of brackish water and seawater, which emerge as the optimal solutions to filling in the widening gap between demand and supply, providing a realistic approach to the severe regional water scarcity.

Great progress has been made in water reuse with the introduction of advanced wastewater treatment systems that generate effluents highly suitable for unrestricted irrigation and other nonpotable uses, incorporating energy-recovery units able to convert the organic content of municipal sewage to energy. Similarly, desalination provides an important tool to solve water shortage, providing unlimited manufactured water for domestic use. As applicable solutions for the provision of additional water for the world arid regions facing water shortage, water reuse and desalination are a "game changer" for the water sector, although their environmental implications, regarding energy, health, and other impacts should be carefully considered while defining the potentials and limitations of such options. The issues of operational (timely delivery of water), financial (cost recovery), and environmental (salinity risks, chemicals, and pharmaceuticals, including the exposure to chronic low dose of complex mixtures of chemicals—water reuse) obstacles that hinder the widespread use of water reuse and desalination should be resolved, as well as the social aspects (perception and attitude toward water reuse and desalination), which avert the demand for these resources.

The progress in advance water management, as discussed, opens the way for riparian states in conflict to pursue positive transboundary interactions and to adopt collaborative policies and bilateral solutions that were not practical in the past. A large percentage of the region's water resources are transboundary and many states depend on shared rivers and/or aquifers to secure their water supply. But only a few formal agreements for joint water management of shared resources have been achieved so far. Making matters difficult, negotiations at the *trans*-regional scale are politically contentious.

Nevertheless, taking into consideration the large disparity in economic and technical capability of the various states, the shared water resources provide a unique opportunity to reduce conflicts by adopting advance cooperation toward shared benefits in the form of optimized allocation and tackling transboundary water pollution and securing the ecosystem's integrity.Therefore, it is positive for riparian states to explore relative advantages that will lead to allocation and sharing of transboundary water resources based on economic terms instead of established water rights. Such approach would overcome suppressing sentiments and political barriers, in favor of national interests, creating a mutual constructive cooperation to solve inter states water issues.

A mechanism for sustainable regional water management should be created to pave the way for the evolution of a regional water community able to create new opportunities for resolving prolonged water-related conflicts. A breakthrough in this direction is the possibility to view water and energy as interdependent sectors and to contemplate on how water and energy sectors meet customers' requirements through the innovative "water and energy nexus" approach. Forward-thinking business models for pricing energy and valuing water are being developed to facilitate the exchange of water for energy between parties, leading to cross-sector and end-use efficiency of water and energy. Further coordination of action through public—private partnerships would crystalize partnerships helping to meet the MENA water-energy challenges.

CONCLUSIONS

Important progress has been made in the past decade to deal with water scarcity through water cooperation, yet much more remains to be done, as highlighted in this chapter in which past achievements and opportunities and challenges were assessed, presenting ideas for the way forward. Issues were analyzed from the practical perspectives, furthering the understanding of how to fill capacity gaps to ensure effective water cooperation. In the absence of transboundary cooperation, the ME may be at risk of war for water, a struggle that can be avoided by mutual agreements that can be translated into programs and actions, assuming that a vigorous political commitment can be achieved. The ongoing water-sharing initiatives may set an important precedent for the region, inspiring confidence that cooperation rather than conflict is preferred to relieve tensions and resolve long-standing water problems.

ACKNOWLEDGMENT

This work was supported by the Division of the Chemistry and Environment, International Union of Pure and Applied Chemistry (IUPAC) and Malta Conferences Foundation (MCF) and is gratefully acknowledged.

REFERENCES

[1] Bucknall JA, Kremer T, Allan J, Berkoff N, Abu-Ata M, Jarosewich-Holder Y, et al. Making the most of water scarcity accountability for better water management results in the Middle East and North Africa. MENA Development Report on Water Information. Washington (DC): World Bank Publication; 2007.
[2] Regional Economic Outlook. Middle East and Central Asia. Washington (DC): International Monetary Fund; 2015. https://www.imf.org/external/pubs/ft/reo/2015/mcd/eng/pdf/menap1015.pdf.
[3] Heller MA. The Arab world 2007. In: Heller MA, editor. Middle East strategic balance 2007–2008. Tel-Aviv: Institute for Nation Security Studies; 2008. p. 49–56.
[4] Givati A, Rosenfeld D. The Arctic oscillation, climate change and effects on precipitation in Israel. Atmos Res 2013;132–133:114–24.
[5] IPCC (Intergovernmental Panel on Climate Change, Impacts, Adaptation, and Vulnerability. Contribution of the working Group II to the fifth assessment report of the Intergovernmental Panel on climate change. Cambridge (UK) and New York (NY). Cambridge: Cambridge University Press; 2013.
[6] Kelley CP, Mohtadi S, Cane MA, Seagerc R, Kushnir Y. Climate change in the Fertile Crescent and implications for the recent Syrian drought. Proc Natl Acad Sci 2015;112(11):3241–6. http://dx.doi.org/10.1073/pnas.1421533112.
[7] Chenoweth J, Hadjiniclaou P, Bruggeman A, Lelieveld J, Levin Z, Lange MA, et al. Impact of climate change on the water resources of the eastern Mediterranean and Middle East region: modeled 21st century changes and implications. Water Resour Res 2011;47(6):1–62.
[8] Bozkurt D, Sen OL. Climate change impacts in the Euphrates-Tigris Basin based on different model and scenario simulations. J Hydrology 2013;480:149–61.
[9] NOAA (National Oceanic and Atmospheric Administration). Climate change and variability. 2016. http://www.ncdc.noaa.gov/climate-information/climate-change-and-variability.
[10] Hansen J, Sato M, Ruedy R, Schmidt GA, Lo K. Global temperature in 2014 and 2015, http://data.giss.nasa.gov/gistemp/; http://www.columbia.edu/-mhs119.

[11] Maddocks A, Young RS, Reig P. Ranking the world's most water-stressed countries in 2040, http://www.wri.org/blog/2015/08/ranking-world%E2%80%99s-most-water-stressed-countries-2040.
[12] Voss KA, Famiglietti JS, de Linage LM, Rodell C, Swenson M. Groundwater depletion in the Middle East from GRACE with implications for trans-boundary water management in the Tigris-Euphrates-Western Iran region. Water Resour Res 2013;49(2):904–14.
[13] Kundzewicz ZW, Mata LJ, Arnell NW, Doll P, Jimenez B, Miller K, et al. The implications of projected climate change for freshwater resources and their management. Hydrol Sci J 2008;53(1):3–10. http://dx.doi.org/10.1623/hysj.53.1.3.
[14] Cline WR. Global warming and agriculture: impact estimates by country. Washington (DC): Center for Global Development and Peterson Institute for international Economics; 2007.
[15] Susnik J, Vamvakeridou-Lyroudia LS, Baumert N, Kloos J, Renaud FG, La Jeunesse J, et al. Interdisciplinary assessment of sea-level rise and climate change impacts on the lower Nile delta, Egypt. Sci Total Environ 2015;503–504:279–88.
[16] Gleick PH. Water, drought, climate change and conflict in Syria. Weather Clim Soc 2014;6:331–40. http://dx.doi.org/10.1175/WCAS-D-13-00059.1.
[17] de Châtel F. The role of drought and climate change in the Syrian uprising: untangling the triggers of the revolution. Middle East Stud 2014. http://dx.doi.org/10.1080/00263206.2013.850076.
[18] Cornwall W. Carbon trackers could help bolster climate vows. Science 2015; 350(6267):1450–1. http://dx.doi.org/10.1126/science.350.6267.1450.
[19] FAO (Food and Agriculture Organization of the United Nation). Review of world water resources by country. Rome: FAO; 2008. Water Reports 23, http://www.fao.org/docrep/005/y4473e/y4473e00.HTM.
[20] Shevah Y. Adaptation to water scarcity and regional cooperation in the Middle East. In: Ahuja S, editor. Comprehensive water quality and purification, vol. 1. Amsterdam: Elsevier; 2014. p. 40–70.
[21] UNESCO (United Nations Educational, Scientific, and Cultural Organization). Water availability per person per year. 2009. www.unesco.org/bpi/wwdr/WWDR_chart1_eng.pdf.
[22] Famiglietti JS. The global groundwater crisis. Nat Clim Change 2014;4(11):945–8. http://dx.doi.org/10.1038/nclimate2425.
[23] FAO (Food and Agriculture Organization of the United Nations). Near East agriculture towards 2050: prospects and challenges. In: In the 29th FAO regional conference for the near East, NERC/08/3, Cairo; 2008. K1699E.
[24] Morris ME. Water and conflict in the Middle East: threats and opportunities. Stud Confl Terror 1997;20(1):1–13. http://dx.doi.org/10.1080/10576109708436023.
[25] Israel Water Authority, Water resources data, http://www.water.gov.il/Hebrew/ProfessionalInfoAndData/Data-Hidrologeime/Pages/default.aspx.
[26] Malkawi A, Husein I. Water in Jordan. In: Lerman Z, editor. Proceedings of the frontiers of chemical sciences V: research and education in the Middle East. Paris (France): Malta Conferences Foundation; November 15–19, 2011. p. 20–1.
[27] UNDP (United Nations Development Program), Technical report on groundwater management in the Mediterranean and the Water Framework Directive. The mediterranean groundwater Working Group (Med-Euwi WG on groundwater), http://www.semide.net/initiatives/medeuwi/JP/GroundWater.
[28] Merkel D, Zohari T, Sukenik A. Sea of Galilee management under low water level conditions and impact on water quality. Israel Water authority, Ministry of Infrastructure, Tel-Aviv. 2008 [In Hebrew)].

[29] Kintisch E. Earth's lakes are warming faster than its air. Science 2015;350(6267):1449. http://dx.doi.org/10.1126/science.350.6267.1449.

[30] Tal A, Al Khateeb N, Nagouker N, Akerman H, Diabat M, Nassar A, et al. Israeli/Palestinian trans-boundary stream restoration and management: lessons for the future. Int J River Basin Manag 2010;8(2):207–13. http://dx.doi.org/10.1080/15715124.2010.491797.

[31] Rüdel H, Muñiz CD, Garelick H, Kandile NG, Miller BW, Munoz LP, et al. Consideration of the bioavailability of metal/metalloid species in freshwaters: experiences regarding the implementation of biotic ligand model-based approaches in risk assessment frameworks. Environ Sci Pollut Res 2015;22:7405–21. http://dx.doi.org/10.1007/s11356-015-4257-5.

[32] Larsen MC. Global change and water availability and quality, challenges ahead. In: Ahuja S, editor. Comprehensive water quality and purification, vol. 1. Amsterdam: Elsevier; 2014. p. 11–20.

[33] Herut B, Shefer E, Gordon N, Galil B, Lubinevsky H, Tibor G, et al. The national monitoring program of Israel's Mediterranean coastal waters – scientific report for 2012. IOLR Report H62/2013. 2014.

[34] FAO (Food and Agriculture Organization of the United Nations). AQUASTAT Database. Global information system on water and agriculture. 2008. http://www.fao.org/nr/water/aquastat/main/index.stm.

[35] Abu-Zeid K, Wagdy A, Elbadawi O, Meguid AA. State of the water report in the Arab world. The Centre for Environment and Development for the Arab Region and Europe (CEDARE) and Arab Water Council (AWC); 2004. p. 1–64.

[36] Schyns JF, Hamaideh A, Hoekstra AY, Mekonnen MM, Schyns M. Mitigating the risk of extreme water scarcity and dependency: the case of Jordan. Water 2015;7:5705–30. http://dx.doi.org/10.3390/w7105705.

[37] Salameh E, Bannayan H. Water resources of Jordan: present status and future potentials. Amman (Jordan): Friedrich Ebert Stiftung and Royal Society for the Conservation of Nature; 1993.

[38] Quna S, Magiera L, Nolte S, Taha S, Bohnet D. Planning Jordan's water future. lessons learnt from the water sector planning support project. Amman: Ministry of Water and Irrigation; Deutsche Gesellschaft fur Technische Zusammenarbeit (GTZ); 2006.

[39] Ministry of Water and Irrigation (MWI). Water for life: Jordan's water strategy 2008–2022. Amman (Jordan): Ministry of Water and Irrigation; 2009.

[40] OECD. Environmental performance reviews, Israel. Part II, Chapter 4, Water. Paris: OECD; 2011.

[41] FAO AQUASTAT, http://www.fao.org/nr/water/aquastat/countries_regions/PSE/PSE-map_detailed.pdf; 2008.

[42] Eran O, Bromberg G, Milner M. Water, sanitation and energy crises in Gaza. Humanitarian, environmental and geopolitical implications with recommendations for immediate measures. Tel-Aviv: EcoPeace Middle East and INSS; 2014.

[43] Tamimi AR, Shbak NA. Qualitative use of home reverse osmosis filters of some areas in Gaza Strip. Palestinian Hydrology Group for Water and Environmental Resources Development Report. Jerusalem. 2011.

[44] Palestinian National Authority. Millennium development goals: progress report. Ramallah: Palestinian National Authority; 2010. http://www.undp.ps/en/newsroom/publications/pdf/other/mdg10.pdf.

[45] UNDP (United Nations Development Program). Climate change adaptation strategy and program of action for the Palestinian Authority. Program of assistance to the Palestinian people, Jerusalem. 2010.

[46] Attili S. Water supply report 2010. Rammalla: Palestinian Water Authority; 2011.

[47] Simas J, Morelli J, El Sadani H. Egypt: irrigation innovations in the nile delta. In: Jagannathan NV, Shawky AM, Kremer A, editors. Water in the Arab world. Washington (DC): The World Bank, Middle East and North Africa (MNA) Region; 2009. p. 433–46.
[48] Shawky A, Jagannathan Egypt NV. Water sector public expenditure review. In: Jagannathan NV, Mohamed AS, Kremer A, editors. Water in the Arab world. Washington (DC): The World Bank, Middle East and North Africa (MNA) Region; 2009. p. 37–58.
[49] Warner K, van de Logt P, Brouwer M, van Bodegom AJ, satijn B, Galema F. Climate change profile in Yemen, Netherlands Commission for environment assessment. Dutch Sustainability Unit (MER); July 2015. http://api.commissiemer.nl/docs/os/i71/i7152/climate_change_profile_yemen.pdf.
[50] Ministry of Water and Environment. National water resources authority. Groundwater management and agricultural development in Yemen. 2009. http://www.un.org/esa/dsd/dsd_aofw_wat/wat_pdfs/meetings/ws0109/2_Yemen_Salem.pdf.
[51] Ward C. Water conflict in Yemen: the case for strengthening local resolution mechanisms. In: Jagannathan NV, Mohamed AS, Kremer A, editors. Water in the Arab world. Washington (DC): The World Bank, Middle East and North Africa (MNA) Region; 2009. p. 233–67.
[52] El Hedi Louati M, Bucknall J. Tunisia's experience in water resource mobilization and management. Development and climate change. Background Note, World Development Report (WDR). 2010. http://siteresources.worldbank.org/INTWDRS/Resources/477365-1327504426766/8389626-1327509184256/WDR2010_BG_Note_Louati.pdf.
[53] Jagannathan NV, Mohamed AS, Kremer A, editors. Water in the Arab world. Washington (DC): The World Bank, Middle East and North Africa (MNA) Region, the World Bank; 2009. p. 562.
[54] Perry CJ, Bucknall J. Water resource assessment in the Arab world: new analytical tools for new challenges. In: Jagannathan NV, Mohamed AS, Kremer A, editors. Water in the Arab world: management perspectives and innovations. Washington (DC): The World Bank Middle East and North Africa (MNA) Region; 2009. p. 97–118.
[55] Global Water Partnership (GWP). Integrated water resources management (IWRM)-(TAC background paper no. 4). Stockholm (Sweden): GWP; 2000.
[56] Salman MA, Bradlow D. Regulatory frameworks for water resources management: a comparative study. Washington (DC): World Bank; 2008. https://openknowledge.worldbank.org/handle/10986/11757.
[57] Morill J, Simas J. Comparative analysis of water laws in MNA countries. In: Water in the Arab World: management perspectives and innovations, [Chapter 16]. pp. 285–334, http://www.ce.utexas.edu/prof/mckinney/ce397/Topics/MENA/World_Bank_2009.pdf.
[58] Perlman H. Irrigation: drip/micro-irrigation. 2015. http://water.usgs.gov/contact/gsanswers?pemail=hperlman,kimmartz&subject=USGS+Water+Science.
[59] DeLay HJ, Smart water technology: benefits, challenges and three action steps for utilities. Black & Veatch Holding Co., http://bv.com/Home/news/solutions/water/smart-water-technology-benefits-challenges-and-three-action-steps-for-utilities.
[60] Allan JA. Virtual water - the water, food, and trade nexus useful concept or misleading metaphor? Water Int 2003;28(1):106–13. http://dx.doi.org/10.1080/02508060.2003.9724812.
[61] Orlowsky B, Hoekstra AY, Gudmundsson L, Seneviratne SI. Today's virtual water consumption and trade under future water scarcity. Environ Res Lett 2014;9:074007.

[62] Water governance in the Arab region managing scarcity and securing the future united nations development program. New York, 10017, USA: Regional Bureau for Arab States (RBAS); 2013.

[63] Adeel Z, Aslov S, Unver O. Water cooperation: views on progress and the way forward. Hamilton (Canada): UN University, Inst. For water, environment and Health; 2015. http://www.scidev.net/global/water/news/share-tech-risk-water-conflict-warns-un-think-tank.html#sthash.KGqJodOb.dpuf.

[64] Richey AS, Thomas BF, Min-Hui L, Reager JT, Famiglietti JS, et al. Quantifying renewable groundwater stress with GRACE. Water Resour Res 2015;51(7): 5217–38. http://dx.doi.org/10.1002/2015WR017349/full.

[65] Reager JT, Gardner AS, Famiglietti JS, Wiese DN, Eicker A, Lo MH. A decade of sea level rise slowed by climate-driven hydrology. Science 2016;351(6274):699–703. http://dx.doi.org/10.1126/science.aad8386.

[66] Al-Roubaie A. Building knowledge capacity for sustainable development in the Arab world. Int J Innovation Knowl Manag Middle East & North Afr 2013;2(1):7–20.

[67] AWA (Amoeba Water Academy - Inspiring Arab Water Leaders. 2009–2015 Amoeba water Academy. Abu Dhabi: United Arab Emirates; 2009. http://www.awacademy.ae/.

[68] MEDRC (Middle East Desalination Research Center) International Organization Working on Solutions to Fresh Water Scarcity, Oman, www.medrc.org.

[69] NRC (National Research Council). Issues in potable reuse: the viability of augmenting drinking water supplies with reclaimed water, committee to evaluate the viability of augmenting potable water supplies with reclaimed water. National Research Council; 1998. p. 280. http://www.nap.edu/catalog/6022.html.

[70] Shuval HI. The development of health guidelines for wastewater reclamation. Water Sci Technol 1991;35:131–5.

[71] Shuval HI, Lampert Y, Fattal B. Development of a risk assessment approach for evaluating wastewater reuse standards for agriculture. Water Sci Technol 1997;35:15–20.

[72] NRC (National Research Council). Water reuse: expanding the Nation's water supply through reuse of municipal wastewater. Washington (DC): The National Academy Press; 2012.

[73] Ternes TA, Joss A, Siegrist H. Scrutinizing pharmaceuticals and personal care products in wastewater treatment. Environ Sci Technol 2004;38:392A–9A.

[74] Ying GG, Kookana RS, Kumar A, Mortimer M. Occurrence and implications of estrogens and xenoestrogens in sewage effluents and receiving waters from South East Queensland. Sci Total Environ 2009;407:5147–55.

[75] Tchobanouglous G, Burton FL, Stensel HD. Wastewater engineering: treatment and reuse. Metcalf & Eddy Inc.. 4th ed. Boston: MeGgraw Hill; 2003. p. 1771.

[76] Hai FI, Yamamoto K, Lee CH. Membrane biological reactors (MBR), theory, modeling, design, management and applications to wastewater reuse. IWS Publishing; 2013. p. 504.

[77] Mosher JJ, Vartanian GM, editors. Framework for direct potable reuse. Alexandria (VA): Water Reuse Research Foundation, WEF, AWWA, NWRI; 2015.

[78] Mor R, Kraitzer T, Michail M, Elkayam R, Sherer D, Shoham G, et al. Soreq mechanical biological wastewater treatment plant operation report, 2014. Tel-Aviv: Mekorot National Water Co., for Mey Ezor Dan, Agricultural Cooperative Water Society Ltd.; 2015.

[79] Halperin R, Asor R, Smulevitz S, Aharoni A, Kishon A. Project complex monitoring report, 2011. Tel-Aviv: Mekorot National Water Co. publication; 2014.

[80] Halprein R, Aharoni A. Monitoring report of the third line project, 2013. Mekorot National Water Co & Technion, Israel Institute of Technology; 2014.

[81] Ministry of the environment. Public health standards for sewage treatment and reuse of effluent. 2010. http://www.sviva.gov.il/subjectsEnv/Streams/SewageStandards/Pages/Milestones.aspx.
[82] Barber LB. Emerging contaminants. In: Ahuja S, editor. Comprehensive water quality and purification, vol. 1. Amsterdam: Elsevier; 2014. p. 245–66.
[83] Lapworth DJ, Baran N, Stuart M, Ward RS. Emerging organic contaminants in groundwater: a review of sources, fate and occurrence. Environ Pollut 2012;163: 287–303.
[84] Lahlou AA. Wastewater reuse. In: Baroudy E, Lahlou A, Attia B, editors. Managing water demand: policies, practices and lessons from the Middle East and North Africa. London: IWA Publishing; 2005. p. 11–24.
[85] Baquhaizel S, Mlkat S. Wastewater management and reuse in the Republic of Yemen. In: Proceedings of the regional workshop on health aspects of wastewater reuse in agriculture. Amman (Jordan): WHO; 30 October – 2 November, 2006.
[86] Mantovani P, Asano T, Chang A, Okun D. Management practices for nonpotable water reuse. Alexandria (VA): Water Environment Research Foundation; 2001.
[87] Asano T, Burton F, Leverenz H, Tsuchihashi R, Tchobanoglous G. In: Eddy M, editor. Water reuse: issues, technologies, and applications. New York: McGraw Hill; 2007.
[88] Levi G, Fine P, Goldstein D, Eizencot A, Zilberman A, Hazan A, et al. Long term effluents irrigation and sub-soil sodification. Vineyards irrigation - case study. Water Eng 2013;86:38–44 [in Hebrew].
[89] Focazio MJ, Kolpin DW, Barnes KJK, Furlong KK, Meyers MY, Zaug SD, et al. A national reconnaissance for pharmaceuticals and other organic wastewater contaminants in the United States - in untreated drinking water sources. Sci Total Environ 2008;402(2–3):201–16.
[90] Barber LB, Vajda AM, Douville C, Norris DO, Writer JH. Fish endocrine disruption responses to a major wastewater treatment facility upgrade. Environ Sci Technol 2012;46(4):2121–31.
[91] Benotti MJ, Trenholm RA, Vanderford BJ, Holady JC, Stanford BD, Snyder SA. Pharmaceuticals and endocrine disrupting compounds in U.S. drinking water. Environ Sci Technol 2009;43(3):597–603.
[92] Shevah Y. Water scarcity, water reuse, and environmental safety. Pure Appl Chem 2014;86(7):1205–14. http://dx.doi.org/10.1515/pac-2014-0202.
[93] World Bank and the Arab World Council. Water reuse in the Arab world, from principle to practice. In: A summary of proceedings, expert consultation wastewater management in the Arab world; May 2011. p. 22–4. Egypt, Dubai-UAW, 2011.
[94] WHO (World Health Organization). A regional overview of wastewater management and reuse in the Eastern Mediterranean Region. Geneva: WHO Regional Office for the Eastern Mediterranean, CEHA; 2005.
[95] Tal A. Seeking sustainability: Israel's evolving water management strategy. Science 2006;313(5790):1081.
[96] Juanico M, Milstein A. Semiintensive treatment plants for wastewater reuse in irrigation. Water Sci Technol 2004;50(2):55–60.
[97] Cohen A, Feiman D, Koller N. National survey of wastewater collection, treatment and reuse for irrigation, 2012. Nature and Reserves Authority for the Water Authority; 2014 [in Hebrew].
[98] FAO (Food and Agricultural Organization), AQUASTAT Information System on Water and Agriculture Database, U.N., Rome, http://www.fao.org/nr/water/aquastat/water_use/index.stm; 2003.
[99] McCornick PG, Hijazi A, Sheikh B. From wastewater reuse to water reclamation: progression of water reuse standards in Jordan. In: Scott CA, Faruqui NI, Raschid Sally L, editors. Wastewater use in irrigated agriculture. UK: CABI Publishing; 2004.

[100] Frost & Sullivan News Release, Thursday, October 29, 2015, http://www.prnewswire.com/news-releases/desalination-plants-to-double-by-2020-300166838.html.
[101] Gebel J. Thermal desalination processes. In: Kucera J, editor. Desalination: water from water. Hoboken, NJ (USA): John Wiley & Sons, Inc.; 2014. http://dx.doi.org/10.1002/9781118904855.ch2.
[102] Mark W. The Guide Book to membrane desalination technology. D'Aquila (Italy): Balaban Desalination Publications; 2007.
[103] Burn S, Gray S, editors. Efficient desalination by reverse osmosis. A guide to RO practice. IWA publishing; 2015. p. 272.
[104] Schorr M, editor. Desalination, trends and technologies. InTech; 2011. http://dx.doi.org/10.5772/583. http://www.intechopen.com/.
[105] Sauvet-Guichon B. Ashkelon desalination plant, a successful challenge. Desalination 2007;203:75–81.
[106] Yechiel A, Shevah Y. Optimization of energy costs for SWRO desalination plants. Desalination Water Treat 2012;46(1–3):304–11.
[107] Elimelech M, Phillip WA. The future of seawater desalination: energy, technology, and the environment. Science 2011;333(6043):712–7.
[108] Namboodiri V, Rajagopalan N. Desalination. In: Ahuja S, editor. Comprehensive water quality and purification, vol. 2. Amsterdam: Elsevier; 2014. p. 98–119.
[109] Shevah Y. Water resources, water scarcity challenges, and perspectives. In: Ahuja S, de Andrade JB, Dionysiou DD, Hristovski KD, Loganathan, BG, editors. Water challenges and solutions on a global scale. ACS Symp Ser, vol. 1206; 2015, [chapter 10], pp. 185–219. http://dx.doi.org/10.1021/bk-2015-1206.ch010.
[110] Lokiec F. Sustainable desalination: environmental approaches. In: The proceedings of the international desalination association world congress on desalination and water reuse 2013/Tianjin, China; October 20–25, 2013.
[111] International Water Law Project, addressing the future of water law and policy in the 21st century, http://www.Internationalwaterlaw.org.
[112] Convention on the protection and use of trans boundary watercourses and international lakes, Helsinki. March 17, 1992. http://www.unece.org/fileadmin/DAM/env/water/pdf/watercon.pdf.
[113] Haddadin MJ. Water: triggering cooperation between former enemies. Water Int 2011;36:178–85.
[114] League of Arab States (LAS), Institutional and legal issues in managing shared water resources: the Arab region's experience. LAS, General Secretariat, Economic Affairs, Department of Environment, Housing and Sustainable Development, Cairo, http://portal.worldwaterforum5.org/wwf5/en-us/Pages/Home-Page.aspx, 2008.
[115] Sadoff C, Greiber T, Smith M, Bergkamp G. Share - managing water across boundaries. Gland (Switzerland): IUCN (International Union for Conservation of Nature and Natural resources); 2008.
[116] Zeitoun M, Mirumachi N. Transboundary water interaction I: reconsidering conflict and cooperation. Int Environ Agreements 2008. http://dx.doi.org/10.1007/s10784-008-9083-5.
[117] The Blue Peace. Rethinking Middle East water. Strategic Foresight group. 2011. http://www.strategicforesight.com/publication_pdf/40595Blue%20Peace_Middle%20East.pdf.
[118] Stokstad E. Power play on the Nile. Science 2016;351(6276):904–7. http://dx.doi.org/10.1126/science.351.6276.904.
[119] NASA. Monitoring the water cycle in Arab region using NASA satellite and data assimilating technology. USA: a proposal to USAID; 2008.

[120] Zaitchik BF, Simane B, Habib S, Anderson MC, Ozdogan M, Foltz JD. Building climate resilience in the blue Nile/Abay highlands—a role for earth system sciences. Int J Env Res Pub Health 2012;9(2):435–61. http://dx.doi.org/10.3390/ijerph9020435.
[121] Israel Ministry of Foreign Affairs. Israeli-Palestinian Interim agreement on the West Bank and the Gaza strip, Washington, DC. September 28, 1995. Anne III, Article 40, http://www.mfa.gov.il/mfa/foreignpolicy/peace/guide/pages/the%20israeli-palestinian%20interim%20agreement.aspx.
[122] Israel Ministry of Foreign Affairs. Treaty of peace between the state of Israel and the Hashemite Kingdom of Jordan. October 26, 1994. Article 6, Water, http://www.mfa.gov.il/mfa/foreignpolicy/peace/guide/pages/israel-jordan%20peace%20treaty.aspx.
[123] EcoPeace Middle East. Regional NGO master plan for sustainable development and regional investment strategy for the Jordan river valley in the Jordan valley. Final Report. June 2015. p. 190.
[124] Jordan valley authority, project concept for RSDS/initial phase - red sea desalination project at Aqaba. February 2014.
[125] EcoPeace Middle East. A water and energy nexus as a catalyst for Middle East peace energy. Amman, Bethlehem, and Tel Aviv; August 2014. p. 28. http://foeme.org/uploads/14379199451~%5E$%5E~Water_Energy_Nexus_Web4.pdf.
[126] Bkayrat R. Desalination is a thirsty business, but does solar provide an answer? December 20, 2015 Updated: December 21, 2015, http://www.thenational.ae/opinion/desalination-is-a-thirsty-business-but-does-solar-provide-an-answer.
[127] Rodriguez DJ, Delgado A, DeLaquil P, Sohns A. Thirsty energy. Washington (DC) 20433: The World Bank; 2013. http://www.gwp.org/Global/ToolBox/References/ThirstyEnergy_WorldBank_2013.pdf.
[128] World Economic Forum. Global risks 2016 report, http://www.weforum.org/press/2016/01/united-nations-world-bank-group-launch-high-level-panel-on-water-in-davos.

CHAPTER SEVEN

Challenges and Solutions to Water Problems in Africa

S.O. Wandiga, S. Ngigi Mbugua, J.W. Macharia, M.A. Otieno
University of Nairobi, Nairobi, Kenya

1. INTRODUCTION

Africa is a continent well endowed with natural resources including minerals, wild animals, forests, lakes, rivers, and grasslands. Despite this richness, it has one of the biggest, driest, arid and semiarid lands, occupying roughly 2 billion hectares of the continent, or 65% of Africa's total land area [1]. The population of Sub-Saharan Africa stood at 853,231,271 in mid-2010 and is growing at 2.3% per year, compared to the world average of 1.2% per year. It is projected to reach 1245 million in 2025 and 2069 million in 2050 [2]. About 900 million people live in arid and semiarid lands globally. Africa hosts 400 million people in its arid and semiarid lands. They are dependent on nature for livelihoods, as the majority are hunters and gatherers; few practice agriculture (rain-fed or irrigated) and sustain pastoralist lives. The pastoralists' way of living depends on cattle, goats, sheep, and camels. The majority use milk and blood for everyday food. Some eat meat of the same animals. However, they feel poor and deprived of dignity if their animals die or are stolen. For this reason, the morans (young warriors) are ready to die for the defense of their cattle.

Besides the people living in arid and semiarid lands, Africa has the fastest growing urban population of about 320 million (37%, growing at 3.9% per year) and is forecasted to reach 654 million people by 2030 by UN-Habitat 2011 [3,4]. The majority of the new urban dwellers find residence in the peri-urban areas with no access to centralized or market supplies of potable water.

Africa has experienced and continues to undergo dynamic changes in the use of its natural resources. There are instances of underexploitation of some resources as well as overexploitation of some resources, such as forest products, oils, and minerals that are exported without value addition. Water is one critical resource that has seen abuse through industrial waste disposal

Chemistry and Water
ISBN 978-0-12-809330-6
http://dx.doi.org/10.1016/B978-0-12-809330-6.00007-6

activities or overwithdrawals. Water scarcity is increasing in the continent due to climate change and climate variability. The impact of the climatic changes is also altering the pasture areas and sources of water for animals and humans. The reduced water availability has become one of the major reasons for the use by 1.9 billion people globally of either unimproved or improved (unimproved sources of drinking water include surface water, unprotected springs, and unprotected dug wells) water sources. It is estimated [5] that 502,000 diarrheal deaths in low- and middle-income countries can be attributed to insufficient and unsafe drinking water [5]. Africa and Southeast Asia contribute the vast majority of the deaths due to having the highest numbers of vulnerable groups of children, malnourishment, and people living with HIV.

Chemical sciences can play an important role in alleviating the water issues in Africa. Chemical sciences have ability to analyze and quantify the offending water impurities that cause diseases. Furthermore, chemical sciences have the ability to find remedies to the illnesses that pollutants cause. Africa has relied heavily on engineering technologies for reticulating and purifying water for home use. The lack of access to piped water has sparked interventions using alternative point-of-use (POU) technologies [6]. One of the most common POU water purification systems is the ceramic water filter, which has gained widespread use around the world as an inexpensive method to treat microbial contaminated water for potable use [7]. Ceramic filtration involves the use of porous ceramic (fired clay) to filter microbes or other contaminants from drinking water. The ceramic filters are in the range of microfilters and are therefore able to trap particles that are in the micro range, as is the case with the suspended particles, particles responsible for turbidity, and some bacteria [8]. When natural coagulants like *Moringa oleifera* are used as preprocessors to ceramic filtration, the efficiency of ceramic filters is greatly improved.

Several water filtration technologies have been developed by researchers' initiatives and nongovernmental organizations to resolve potable water scarcity [9]. Some of the most studied and surveyed techniques used for water purification around the globe are: chlorination, aluminum sulfate coagulation, combined coagulant—chlorine disinfection systems, sunlight exposure techniques such as solar disinfection (SODIS) and UV radiation, filtration systems such as reverse osmosis, ceramic filtration, and biosand filtration [10,11]. Concerns have, however, arisen regarding the cost of such methods and the use of such water treatment techniques, mainly due to the formation of potentially harmful disinfection by-products, such as trihalomethanes and haloacetic acids, upon reaction with the naturally occurring organic matter and halide ions [12]. Additionally, the efficiency of the chlorination process is affected by the presence of nitrite (nitrification) and suspended solid particles, which also require high investments for scrubbing safety equipment (i.e., filter systems) and the need for dechlorination to meet stringent Environmental Protection Agency regulatory limits [13]. On the other hand, aluminum sulfate has been linked to Alzheimer's disease [14]. Further, the cost of purchasing synthetic coagulants and disinfectants is also high, leading to increased water treatment costs that translate further to high pricing for treated water in Africa [13].

Ceramic filters are made of clay mixed with water and a combustible material, like sawdust, and then fired [15]. Firing burns out the sawdust, leaving out tiny pores that allow water molecules to pass through while particles are retained. The pore size and flow rate of these filters varies, depending on the size of combustible material used in the mixture [7]. The pores can be made small enough to remove virtually all bacteria and protozoa by size exclusion, down to 0.2 μm, in the range referred to as microfiltration. Small-scale ceramic filtration has a long history, having been used in various forms since antiquity [4]. Millions of these porous clay ceramic filters are in use in several countries in the African, Asian, and South American continents [9].

1.1 Photocatalysis Using TiO_2

The mechanism for the photoactivation of electrons to jump from valence band to conduction band of TiO_2 has been well studied. The energy required for this transition is in the UV region (−3.2 eV). In order to use sunlight for catalysis, it is necessary to reduce the energy band to the visible range. The two principal modes by which the activity of TiO_2-based photocatalysts can be extended into the visible are doping and surface modifications [14]. Doping generally means the insertion of atoms, or group of atoms or ions, into a host's crystal lattice. The main objective of doping is to induce a bathochromic shift, i.e., a decrease of the band gap or introduction of intraband gap states, which results in the transition in the visible light [15]. Doping of TiO_2 with suitable ions [16], such as chromium, iron, or tin, or with nonmetallic atoms like sulfur [17], nitrogen [18,19], or carbon [20,21] induces interband states. Hence, additional absorptions of lower energy (i.e., longer wavelengths) lead to the promotion of electrons into the conduction band of TiO_2 [15,22].

The study of POU technologies that can be easily adopted by rural communities to produce portable water has been the focus of many theses at the University of Nairobi laboratory. The technology design is simple enough, affordable and accessible. The studies have involved two systems, namely, ceramic filters combined with *M. oleifera* and the synthesis of nanoparticles of titanium oxide: tungsten oxide and titanium oxide—nitrate composites. This paper describes the studies.

1.2 Experiments

1.2.1 Manufacturing of Ceramic Pots

The flow rate of each filtering element is used as a primary form of quality control once a filter mixture formulation has been established and effectiveness confirmed with microbiological testing [23]. There are many variables in the manufacturing process, which are considered to influence the flow rate and/or effectiveness of the filtering element, including the type of clay (particle size, distribution, sand content, and plasticity), the burn-out material (type, size, and humidity of burn-out material), the clay-to-burn-out ratio, the amount of water added to the mixture, the manufacturing method (molded by hand, pressed, wheel-thrown), drying time and conditions, firing temperature, time and location in kiln, size of the filtering element, and capacity and the thickness of the filter [24].

There were two types of ceramic filters used in the study. The first type of fired ceramic pots was molds of clay prepared at Fine Arts Department, Kenyatta University, on a clay—sawdust powder volume ratio basis by mixing clay and appropriate amount of sawdust powder with water [24]. Four rations, i.e., 50:50, 55:45, 60:40, and 65:35 of clay-to-sawdust ratio were selected and frustum-shaped filters prepared from them in triplicates. The filters were dried under shade for 5 days and then fired by first preheating at 100°C for 2 h, followed by sintering at 850°C for 8 h (Fig. 7.1). Another set of filters was prepared in the same way, but fired at 650°C. Water of known levels of contamination was passed through the filters and the filtrate collected for analysis. For the determination of flow rates, distilled water was passed through.

The second type of pots used in this research was molded by hand at Oriang' Pottery, Homa Bay County, Nyanza, Kenya (Fig. 7.2). The ratio of clay to sand was 3:1. The pots were fired openly using dried grass and cow dung [24]. J. Rayner [15] recommends that the clay should have an acceptable level of plasticity, dry rate of shrinkage, and after firing, acceptable rates of total shrinkage and absorption. Clay, *M. oleifera* seeds, and tree branches and sawdust were obtained and dried under shade until completely dry by touch, then ground separately to obtain powder. The synthesized ceramics with 60:40 clay—sawdust and 3:1 clay-to-sand ratios were used for the study of flow rates, turbidity reduction, and *Escherichia coli* filtration [23—25].

Figure 7.1 Samples of molded clay ceramics.

Figure 7.2 Ceramic clay pots handmade by Oriang' pottery.

1.2.2 Determination of Flow Rates, Turbidity, and Escherichia coli *Counts*

Major water sources were Lake Victoria, pans, and dams [26,27]. Other sources were piped water, the Nairobi River, and a few boreholes. The main sources of water pollution include agrochemicals, defecation in bushes due to lack of pit latrines, wastewater from industries, and surface water runoff that end into water resources [27]. The water obtained from the pans is also contaminated by livestock urine and feces, since the pans are not protected. Water of known levels of contamination was passed through the laboratory-molded filters and the filtrate collected for analysis. For the determination of flow rates, distilled water was passed through, and the flow ranged between 20 mL/h and 104.5 mL/h.

The Oriang' pottery filters were initially soaked under water for 1 h to allow saturation. The filters were then filled to the brim with river water. The intent was to roughly mimic a typical residential usage pattern by filtering water from an actual source. The filtrate was collected and volumes determined after every 30 min for 3 h. An average was then done to obtain the flow rate per hour. There was a trend of reduced flow rate over time during the course of the study of the filters, which can be attributed to clogging. This finding relates to Van Halem's study of the Potters For Peace (PFP) filters, which showed that flow rate decreased over the 12-week testing period with a 70–80% decrease in filter discharge [28]. The initial flow rates of the pots ranged from 60 mL/h to 130 mL/h; this reduced to between 20 mL/h and 50 mL/h as a result of clogging. Lower flow rates were recorded for the pots with greater thickness of the walls. It has been shown that the original flow rate of PFP filters can be restored by scrubbing the filter [29].

Turbidity was determined [23] using turbidity meter model LaMotte TC-3000e Tri-Meter. The spectrophotometric turbidity meter was calibrated using standard solutions of turbidity 1 and 10 Nephelauxetic Unit (NTU) with zeroing of the meter done with distilled water. Water samples were placed in the glass bottles provided and the bottles placed into the sample holders. The turbidity readings were taken and recorded.

E. coli count was determined using the 3M *E. coli* Petrifilm plates. The count plate was placed on a flat surface, top film lifted, and 1 mL of sample dispensed onto the center of the bottom film. Slowly, the top film was rolled down onto the sample to prevent the entrapment of air bubbles. The plate was then left undisturbed for 1 min to permit even distribution of the sample and solidification of the gel. The plates were then incubated in a horizontal position with the clear side up at 37.5°C for 24 ± 2 h. After the incubation, the plates were removed and enumeration of *E. coli* carried out. Blue colonies associated with entrapped gas were counted as *E. coli*, while other coliform colonies were red and associated with gas bubbles.

1.2.3 Synthesis of TiO_2 Composites as Photocatalysts

Solid powder precursors of TiO_2 and WO_3 (Drug Houses Limited purity of 99.4% and 99.5%, respectively) were utilized and hydrolysis was carried out in acidified isopropanol (J.T. Baker, Inc.) [25]. The addition of nitric acid, together with 12-h stirring, followed by drying and calcinations at 575°C for 2 h yielded the desired product. The nanoparticles were characterized using a Philips MiniPal2 X-ray fluorescence (XRF) instrument (Shimadzu) and an X-ray diffraction (XRD)-600 analyzer (Shimadzu), at two theta (2θ) angles between 0–80°. The mean crystallite size was determined from the broadening of the most intense line at full width at half maximum (FWHM), based on the Debye–Scherrer equation [30]. Our result showed a particle size diameter of 18.99 nm.

Experiments were carried out with the water samples collected from the Nairobi Dam next to the Kibera slums South of Nairobi, 6 km from the Department of Chemistry where analysis was done. Samples were transported in 2-L amber bottles and stored at 4°C prior to analysis. All experiments were performed within 3 days of sample collection at the natural pH, 7.3, of the samples without any pH adjustment. The water contained 2600 colony-forming units (CFUs)/mL *E. coli*. A portion of the water solution (10 mL) was diluted by a factor of 10, which gave a count of 260 CFU/mL after 24 h incubation.

N-doped TiO_2 was prepared by sol–gel method. Titanium (IV) isopropoxide (97%, Sigma Aldrich) as a titania precursor, 25% ammonia solution, and isopropanol in the ratio 2:1:10 was measured and the mixture put in a conical flask. This was then stirred at 250 rpm for 1 h at room temperature and pressure (rtp). Initially, there was a rapid hydrolysis and condensation reaction, causing formation of a colloid. The resulting white gel was dried in the oven at 120°C for 1 h to remove any excess NH_3 and other by-products of hydrolysis. The dried material was taken for calcination at 673°C overnight. Calcination was carried out in porcelain crucibles. Similarly, the mean crystalline size as determined previously was found to be 18.372 nm.

1.3 Characterization of the TiO_2 Composites

1.3.1 Characterization of TiO_2–WO_6 Photocatalyst

The optimum working photocatalyst from the previously mentioned results was subjected to characterization using XRF and XRD data. A Philips MiniPal 2 instrument was used for XRF, and a Shimadzu 600 XRD analyzer for XRD was used. The data for the pure titanium (IV) oxides and that of the mixed product is as shown in Figs. 7.3 and 7.4.

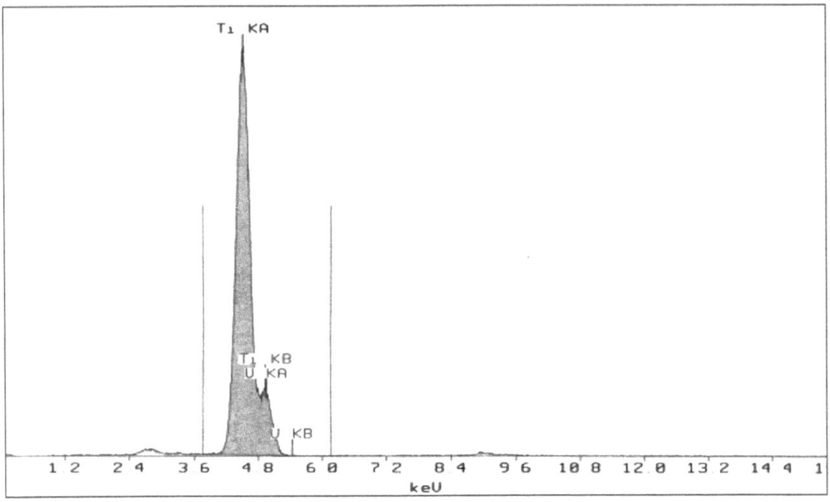

Figure 7.3 XRF data for pure TiO_2 showed a single titanium (IV) oxide peak at 99.4% purity.

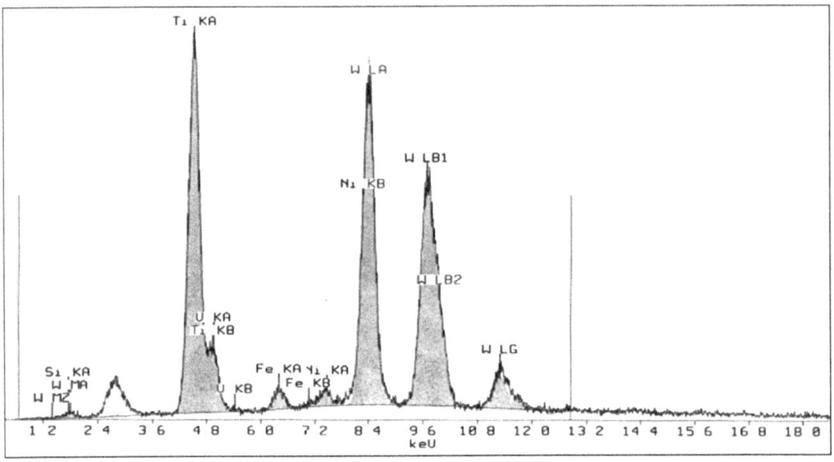

Figure 7.4 1:1 TiO_2:WO_3 composite.

The XRD data for the pure TiO_2 and titanium (IV) oxides—tungsten (VI) oxide in Fig. 7.4 are: 2θ values of diffraction peaks observed in XRD spectra of pure TiO_2 were at 25.6, 37.8, 48.0, 53.9, 55.0, 62.7, 68.90, 70.36, and 75.2. These peaks were assigned to reflections from (101), (004), (200), (105), (211), (204), (116), (220), and (215) crystal planes, which mainly corresponds to anatase phase of TiO_2 [31]. The base peak broadening was an FWHM of 0.42500°.

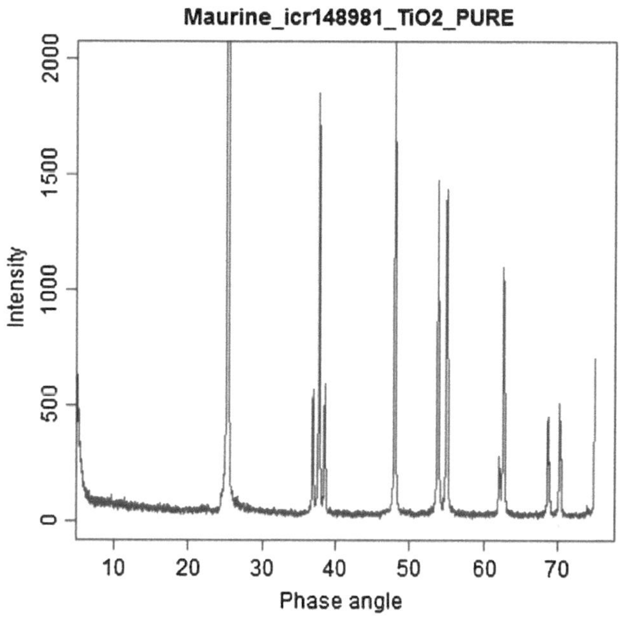

Figure 7.5 X-ray diffraction pattern for pure TiO_2.

The data in Fig. 7.5 was then used to determine the particle size of the synthesized TiO_2—WO_3 compound using the Debye—Scherrer equation [30].

1.3.2 Synthesis of TiO_2—NO_3 Photocatalyst

N-doped TiO_2 was prepared by sol—gel method. Titanium (IV) isopropoxide (97%, Sigma Aldrich) as a titania precursor, 25% ammonia solution, and isopropanol in the ratio 2:1:10 were measured and the mixture put in a conical flask. This was then stirred at 250 rpm for 1 h at rtp. Initially, there was a rapid hydrolysis and condensation reaction, causing formation of a colloid. The resulting white gel was dried in the oven at 120°C for 1 h to remove any excess NH_3 and other by-products of hydrolysis. The dried material was taken for calcination at 673°C overnight. Calcination was carried out in porcelain crucibles.

1.3.3 Characterization of TiO_2—NO_3 Photocatalyst

1.3.3.1 X-Ray Diffraction Analysis

The phase composition and the crystal size of the N-doped TiO_2 were evaluated using XRD analysis. The products were characterized by XRD at a scanning rate of 5 °/min in the 2θ range of 10—70° using a Philips X-ray diffractometer (7602 EA Almelo, the Netherlands) with Cu Ka radiation ($\lambda = 0.1542$ nm).

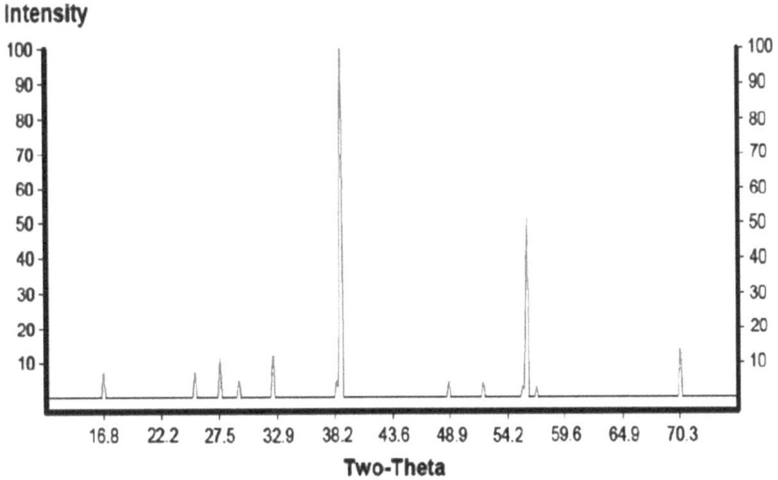

Figure 7.6 X-ray diffraction spectra for 1:1 Ti/W–doped photocatalyst.

In Fig. 7.6, 2θ values of diffraction peaks were observed in XRD spectra of pure TiO_2 at 25.5, 38.0, 48.2, 54.5, 55.3, 63.1, 68.90, 70.36, and 75.2. These peaks were assigned to reflections from (101), (004), (200), (105), (211), (204), (116), (220), and (215) crystal planes, which mainly correspond to anatase phase (Veluru et al.) [33]). This shows that the anatase is the dominant phase, and this is beneficial since anatase is reported to be more photocatalytically active than rutile [31].

From the XRD results (Fig. 7.8), the particle size for the N-doped TiO_2 was calculated using the Debye–Scherrer formula [30].

N-doped TiO_2 SEM Micrograph TiO_2 SEM Micrograph

Figure 7.7 SEM micrographs for N-doped TiO_2 and TiO_2 photocatalysts.

1.3.3.2 Scanning Electron Microscope Analysis

Fig. 7.7 shows the scanning electron microscope (SEM) micrographs for N-doped TiO2 and the original TiO2 photocatalysts. The results of SEM revealed that the photocatalysts consist of irregular agglomerated nanoparticles. The images revealed that the morphology and size of the particles were not changed on incorporation of N into TiO2 by treatment with ammonia solution. The morphology of the TiO2 particles was not altered on N-doping due to the comparable atomic size of nitrogen with that of oxygen (PalaezPelaez et al., 2012) [32]).

1.4 Inactivation of *Escherichia coli*

1.4.1 TiO_2—NO_3 Photocatalytic Inactivation of Escherichia coli

The optimal weight of the N-doped photocatalyst for inactivation of *E. coli* was found to be 0.10 g. The weights above 0.10 g did not cause additional inactivation of *E. coli*, whereas the mass below 0.10 g had lower inactivation [24] Fig. 7.9.

The fact that the mass of the photocatalyst above 0.10 g did not translate into enhancing inactivation of *E. coli* could be attributed to the fact that although the larger catalyst concentration provides more catalytically active centers for the absorption of photons that translate into more electrons and

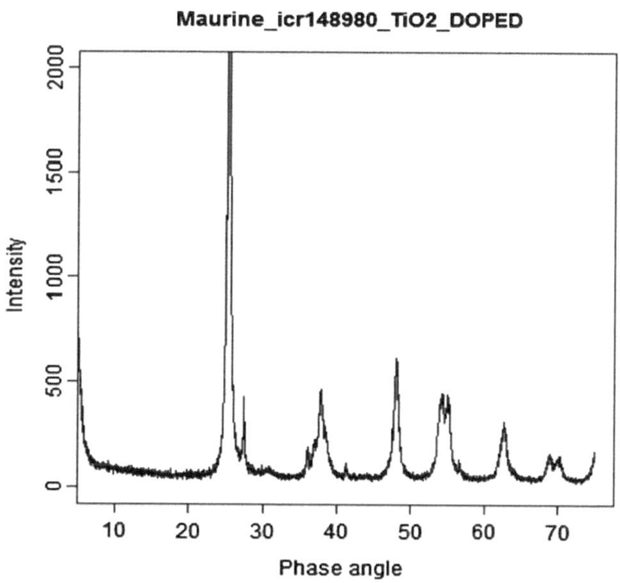

Figure 7.8 X-ray diffraction pattern for N-doped TiO_2

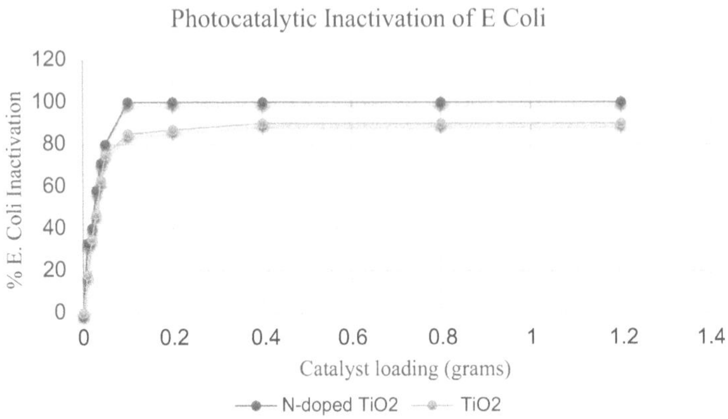

Figure 7.9 Percent of *E. coli* inactivation against catalyst loading.

holes are generated, the excess photocatalyst may also act as an optical filter and impede the further penetration of incident light into the medium. This effect has been also reported by Zhang et al. [34]. For other experiments on *E. coli* inactivation, 0.10 g of the photocatalyst was thus taken as the optimal amount.

The doped $TiO_2{-}NO_3$ was used in one experiment to inactivate *E. coli* [24]. An initial 1 mL of the raw water was transferred directly onto the 3M Petrifilm. This was then placed in an incubator at 37.5°C for 24 h. The resultant *E. coli* and coliform were noted by counting the blue and red colonies formed, respectively. One hundred mL of the raw water with 0.1 g of the catalyst added to it was then transferred into three clear polypropylene bottles and exposed to sunlight for 60 min. In three other polypropylene bottles, 100 mL of the water was added with 0.10 g of TiO_2 and exposed to sunlight for 50 min. In another set of three polypropylene bottles, 100 mL of the water was added without addition of the catalyst and also placed in sunlight for the same duration of time. One mL of each solution was pipetted out after every 10 min and placed onto the 3M Petrifilm, then incubated for 24 h, after which the resultant *E. coli* and coliform were noted by counting the blue and red colonies formed (Fig. 7.9).

A second experiment [25] with $TiO_2{-}WO_6$ doped catalyst was similarly used to inactivate *E. coli*. Tests using 50 mL of the water samples and 0.75 g/L of the photocatalyst were carried out in triplicates and 1 mL aliquots sampled every 15 min for a total illumination time of 75 min. A TUV 8W G8T5 lamp from PHILIPS was used as energy source. The intensity reaching the solution was 0.3520 watts (350–560 nm) as measured using an optical power meter radiometer (USB2000 spectrometer from Ocean Optics). The test sample was positioned approximately 3 cm from the blue light bulb lamp. The sample was continuously stirred using a magnetic stirrer (400 rpm). Control tests without the photocatalyst were carried out to establish inactivation by photocatalysis only. Dark control experiments were also carried out to exclude UV light. This was done by tightly covering the reaction vessel with aluminum foil to exclude all light. All experiments were performed in triplicates.

1.5 Inactivation of *Escherichia coli* Using *Moringa oleifera* and Clay Filtration

1.5.1 Coagulation and Disinfection

Two hundred mL of the contaminated water was put in a 250 mL conical flask. Various amounts of ground *M. oleifera* seeds powder, ranging from 5 to 200 mg, were added to the conical flasks. The flasks were shaken in an orbital shaker at 150 rpm for 4 min, followed by slow mixing at 50 rpm for 10 min. A blank sample was treated in a similar way to act as a control. The conical flasks were then removed from the orbital shaker and left to stand on the bench for 1 h. At this point, the measurements were taken for the various target contaminants to determine the coagulative and disinfection effects.

1.5.2 Extraction of Hexane-Soluble Oil from Moringa oleifera *Seeds*

Hexane-soluble oil was extracted from the seeds by soaking around 5 g of dry *M. oleifera* seed powder in 50 mL high performance liquid chromatography-grade hexane in 100-mL beaker. The contents of the beaker were shaken for 15 min in an orbital shaker before being allowed to stand overnight. Whatman filter paper was used to filter out the seed cake residue. This seed cake residue was dried in the air for 1 h before drying in an oven at 105°C for another 2 h and cooling it to room temperature. The weight difference of the seed cake was used to compute the percentage composition of the oil.

1.6 Bleaching of Methylene Blue Dye with N-Doped $TiO_2{-}NO_3$

A stock solution of methylene blue (MB) dye was prepared in distilled water. It was further diluted as required.

1.6.1 Effect of Catalyst Weight on Methylene Blue Degradation

MB was used as the model organic pollutant to evaluate the photocatalytic activity of the N-doped TiO_2. 40 mL of aqueous dye solution (0.05 gL^{-1}, pH 7) was put into different vials, and 0.05 g, 0.10 g, 0.20 g, 0.40 g, 0.60 g, 0.80 g, and 1.00 g of the catalyst were added to the vials and stirred in triplicate. Before illumination, the suspensions were mixed using a magnetic stirrer for 2 h in the dark to ensure that the mixture reached adsorption equilibrium. These were then exposed to sunlight for 2 h. The pH was adjusted with 0.1-M NaOH solution. The photocatalyst was separated from the solution by centrifugation and the absorbance of the remaining clear liquid measured by UV-visible spectroscopy at 616.8 nm.

2. RESULTS AND DISCUSSION

2.1 Determination of Flow Rates

The filtration rate of the ceramic pots depends on the thickness of the walls, the ratio of the clay to the sawdust, the method used to make it, and the expertise in pottery. The ceramic pots made by mold in the laboratory had flow rates ranging between 20 mL/h and 104.5 mL/h. The filtration rate increased with increases in the amount of sawdust for the first two filters, which could be attributed to the fact that as the sawdust increased, more pores were available for the passage of water. However, a decrease is observed with an addition of more sawdust to the clay. A high amount of sawdust added caused the loss of elasticity of the clay. For this reason, the 55:45 and 50:50 ceramic filters were made from an inelastic mold of clay that implied making filters with thicker walls to compensate for the inelasticity; see Fig. 7.10.

As a result, their flow rates decreased, with the 50:50 filters having the lowest flow rate. The flow rates can be enhanced without compromising the efficiency of the filters by reducing the wall thickness of the filters and making larger filters that can accommodate more water. The filtration rate is lost with the clogging of the pores by particulates. On average, the best performing filter could filter an average of 17.75 mL/h. The low flow rates could be attributed to the fact that the *M. oleifera* powder was so fine, implying that the pores formed after firing the filter were also tiny and could therefore not allow much water to pass through at a time.

By varying the grain size of the sawdust (300, 600, and 900 μm) and using a clay:sawdust ratio of 1:2, Varkey and Dlamini [35] obtained flow rates

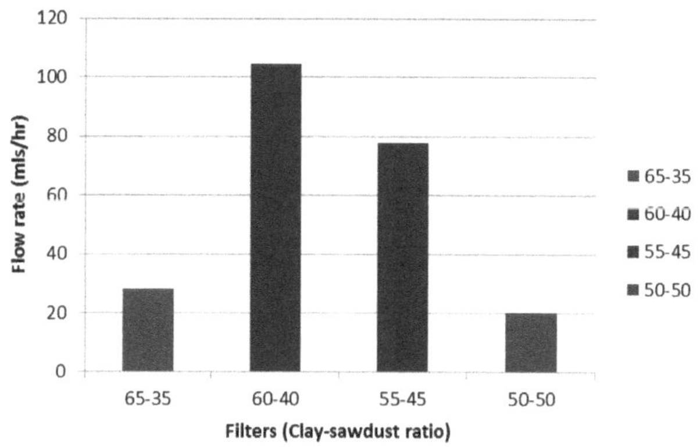

Figure 7.10 Illustration of the filters' flow rates of molded pots.

ranging between 50 and 140 mL/h, which compares to flow rates recorded for some of the filters prepared in this work.

2.1.1 Reduction of Turbidity

Sample water with different initial turbidity was used in testing the filters' efficiency in turbidity removal. For all the water samples, the residue turbidity was below 3 NTU. It was observed that the higher the turbidity of the raw water, the lower the turbidity of the treated water recorded, as illustrated in Table 7.1 Naturally occurring turbid water obtained from Nairobi River at the Chiromo Bridge was used in the experiment, and therefore, turbidities higher than 65 NTU could not be achieved.

The improved efficiency with an increase in initial turbidity has also been reported by Sagara [36]. He suggested that as the particles in water get filtered, the effective pore size of the filter gets smaller, due to the clogging of the pores by the filtered particles. This therefore implies that the highly turbid water would reduce the pore size more and thus the higher efficiency.

Table 7.1 Efficacy of the Filters in Turbidity Removal for Water of Different Initial Turbidities

Initial Turbidity (NTU)	Residue Turbidity (NTU)
5	2.366 ± 0.059
10	2.060 ± 0.046
20	1.446 ± 0.047
50	1.213 ± 0.032
65	0.896 ± 0.031

However, very turbid water would result in total clogging, thus preventing any filtration to take place.

2.1.2 *Inactivation of* Escherichia coli

The molded ceramics were used with sample water obtained from Nairobi River containing 390 CFU/mL *E. coli* and 530 CFU/mL other coliforms. Nairobi River has generally been reported to contain high levels of the indicator organism, *E. coli*. Musyoki et al. [37] recorded an average of 980 ± 130 CFU/mL for the river before the Dandora Sewage Treatment Plant (DSTP) and an average of 1000 ± 110 CFU/mL after the DSTP point. The high levels of the fecal coliform contamination could be attributed to the surface runoffs from the city as well as wastewater pollution from the informal settlements located alongside its course, in addition to effluent from the sewage treatment plant [38]. The 3M Petrifilm incubation results are shown in Fig. 7.11.

In Fig. 7.11, the M2 and S1 are plates incubated with water filtrate. The absence of the blue colonies in these plates indicates that all the *E. coli* have been filtered out. The enumeration of the *E. coli* after filtration and incubation was done and the results presented in a tabular form in Table 7.2. The

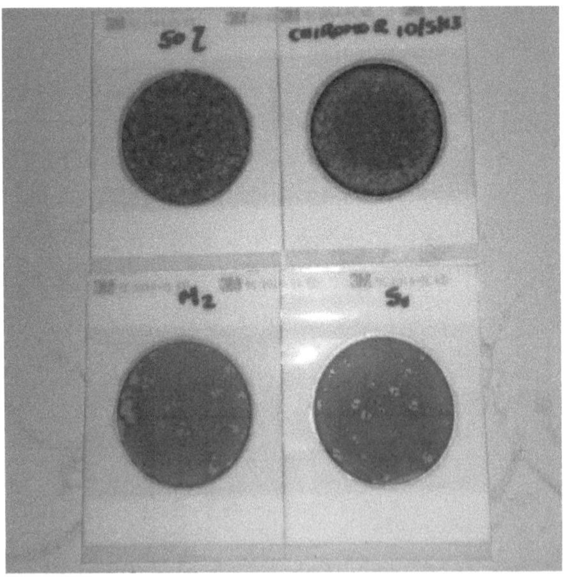

Figure 7.11 Pictorial presentation of the *Escherichia coli* 3M Petrifilm results after filtration.

Table 7.2 Residue *Escherichia coli* Count After Filtration

	***E. coli* (CFU/mL)**
Raw water	390.00 ± 10.00
Filter 1	0.00 ± 0.00
Filter 2	0.00 ± 0.00
Filter 3	0.33 ± 0.58
Filter 4	0.00 ± 0.00

results indicate that three of the four sets of ceramic filters prepared eliminated all the *E. coli*.

The ceramic filters recorded an average efficiency of 99.98% *E. coli* reduction. From literature, ceramic filters have been reported to remove *E. coli*, with an efficiency ranging between 98% and 100% (Simonis and Basson [8]). By using water spiked with $6.0 \star 10^6$ CFU/mL *E. coli*, the same authors obtained an average reduction efficiency of 99.999% after filtration through ceramic filters.

The inactivation of *E. coli* study using ceramics made by Oriang' pottery initially had 1 mL of the raw water transferred directly onto the 3M Petrifilm. This was then placed in an incubator at 37.5°C for 24 h. The resultant *E. coli* and coliform were noted by counting the blue and red colonies formed, respectively. A second 100 mL of the raw water with 0.10 g of the TiO_2-NO_3 catalyst added to it was then put into three clear polypropylene bottles and exposed to sunlight for 60 min. In three other polypropylene bottles, 100 mL of the water was added to 0.10 g of TiO_2 and exposed to sunlight for 60 min. In another set of three polypropylene bottles, 100 mL of the water was added without addition of the catalyst or TiO_2 and also placed in sunlight for the same duration of time. One mL of solution from each propylene bottle exposed to the sun was pipetted out after every 10 min and placed onto the 3M Petrifilm and then incubated for 24 h, after which the resultant *E. coli* and coliform noted by counting the blue and red colonies formed. The results are given in Table 7.3 and show the level of inactivation being reduced from 1001.667, 1028.333, and 1006.667 CFU/mL to 103.333, 278.333, and 281.667 CFU/mL for TiO_2-NO_3, TiO_2, and SODIS, respectively.

The use of an N-doped TiO_2 photocatalyst achieved 98.5% inactivation of *E. coli* in raw river water within 50 min compared to TiO_2, which achieved 81.1% and SODIS 76.7% inactivation of the *E. coli* under similar conditions, from the initial *E. coli* concentration of 1011.67 CFU/mL.

Table 7.3 Comparison of *Escherichia coli* Inactivation Using N-doped TiO_2, TiO_2, and Solar Disinfection (SODIS)

Time (min)	N-doped TiO_2 Count 1, Count 2, Count 3			Mean	Standard Deviation	TiO_2 Count 1, Count 2, Count 3			Mean	Standard Deviation	SODIS Count 1, Count 2, Count 3			Mean	Standard Deviation
10	1005	1000	1000	1001.67	2.887	1030	1025	1030	1028.33	2.887	1005	1010	1005	1006.67	2.889
20	790	795	795	791.67	2.89	810	810	800	806.67	5.773	850	846	840	845.33	5.03
30	490	490	495	491.67	2.89	630	625	628	627.67	2.52	655	650	660	655	5.000
40	250	255	260	255.00	5.00	450	460	460	456.67	5.77	460	460	450	456.67	5.77
50	110	100	100	103.33	5.77	280	280	275	278.33	2.89	285	280	280	281.67	2.89

The results revealed that the fastest inactivation is for N-doped TiO_2 rather than undoped TiO_2 photocatalyst and SODIS alone. No *E. coli* inactivation was observed in any control bottles.

2.1.3 Effect of Firing Temperature

To investigate the effect of firing temperatures, one set of the molded ceramic filters was fired at 650°C while the other one was fired at 850°C. Both set of ceramic filters were similar in appearance after the firing and were equally effective in removing fecal coliforms and turbidity. This is graphically represented in Fig. 7.12.

Since firing at 650°C was equally as effective as firing at 850°C, firing for this type of clay used should be done at the lower temperature to save energy. The temperature should be high enough to vitrify clay and make it hard and resistant to stress. This ensures that it does not change shape when water is added to it and neither will it contaminate the water filtered through.

2.1.4 Coagulation and Disinfection Using *Moringa oleifera* Seeds

Both synthetic and naturally turbid water was used in testing the effects of the seeds and the optimal dose on turbidity. Various amounts of the seeds were applied to water at this turbidity and the residue turbidities recorded after 0, 4, and 18 h. The results are presented in Fig. 7.13. The optimal *M. oleifera* dosage was found to be 0.2 g/L, corresponding to a reduction from 100 to below 5 NTU after 1 h. This dose further reduced the turbidity to 2.74 and 0.83 NTU after 4 and 18 h, respectively. Optimal *M. oleifera*

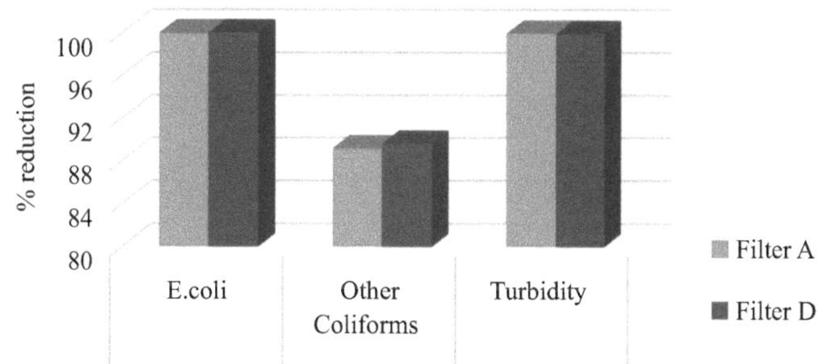

	E.coli	Other Coliforms	Turbidity
Filter A	100	89.18238994	99.94983022
Filter D	100	89.62264151	99.93248835

Parameters

Figure 7.12 Efficiency of two filters: Filter A fired at 850℃ and Filter D fired at 650℃.

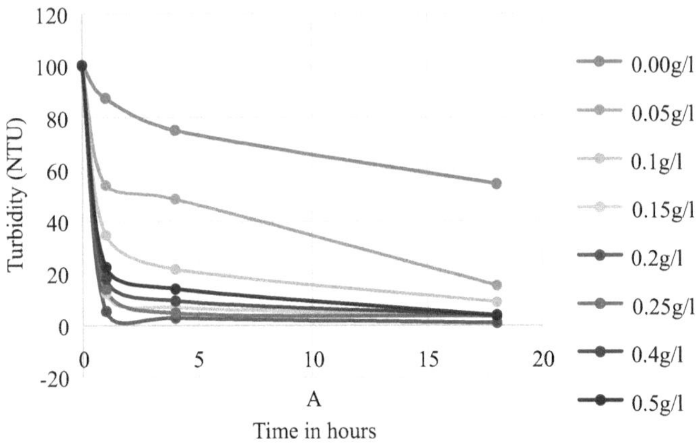

Figure 7.13 *Moringa oleifera* optimal dosage for synthetic turbid water at 100 NTU initial turbidity.

dosage as low as 0.1 g/L has also been reported initially. Yongabi et al. [38], in his findings, reported that 2 g of crushed *M. oleifera* seeds were used to treat 20 L of water in Malawi, resulting in residue turbidity below 5 NTU.

2.1.5 Synthetic Turbid Water at 16.3 NTU

The synthetic water at initial turbidity of 16.3 NTU required a lesser amount of the seeds, as indicated in Fig. 7.14. The optimal *M. Oleifera* dosage for water with an initial turbidity of 16.3 NTU was 0.025 g/L. The turbidity was reduced to 2.88 NTU and 1.44 NTU after 1 and 4 h, respectively, corresponding to a reduction efficiency of over 91.2%.

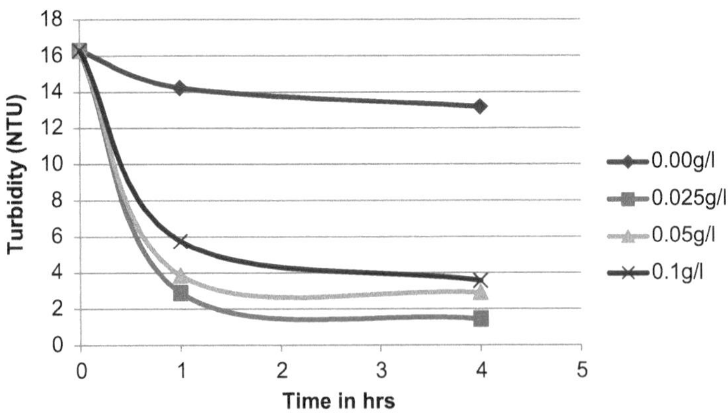

Figure 7.14 Optimal *Moringa oleifera* dosage for synthetic turbid water at initial turbidity of 16.3 NTU.

The lowest turbidity value achieved for synthetic water sample with an initial turbidity of 16.3 NTU was 1.44 NTU, which is high compared to the 0.82 NTU value achieved by using water at initial turbidity of 100 NTU. The behavior has been reported by other scholars and could be attributed to the fact that low turbidity waters contain limited colloidal matter, hence, a very limited interparticle contact system for the polyelectrolyte [39,40].

2.1.6 Natural Turbid Water

The reduction of turbidity of natural water at an initial turbidity of 22.5 NTU was tested to obtain the optimal *M. Oleifera* dosage. Fig. 7.15 shows the results of the test.

The optimal dose was found to be 0.2 g/L, which lowered the turbidity from 22.5 to 4.02 NTU. This corresponds to a 59% reduction. Mathematically, the optimal dosage for this natural water is around 10 times greater than that of the synthetic turbid water of the same initial turbidity range. However, this lowest residue turbidity achieved is higher than that of the synthetic turbid water of 16.3 NTU, which gave a residue turbidity of 1.44 NTU.

Pritchard et al. [40] had previously reported that the seeds were not as effective with the natural turbid water as they were with artificial turbid water. They reported that natural water requires far higher doses (15 times) of *M. oleifera* seeds than the model turbid water to successfully produce coagulation. It has been proposed that the reduced efficiency with water of lower and natural turbidity could be as a result of formation of small and less dense flocs that could not settle effectively [41].

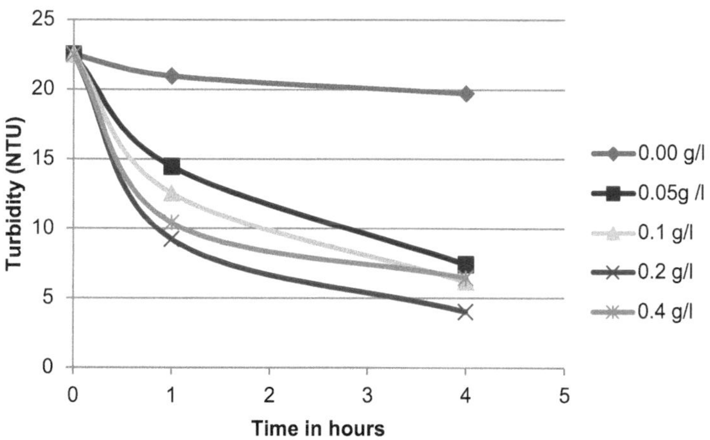

Figure 7.15 Optimal dosage of *Moringa oleifera* for natural turbid water.

For all the cases studied, the optimum dosage that reduced turbidity to acceptable limits for drinking water was found to be equal or less than 0.2 g/L. Increasing the dosage beyond the optimal dosage resulted in slight increase in the residue turbidity. Sutherland et al. [42], in their work, have mentioned that once the optimal dosage is achieved, the excess polyelectrolyte proteins repel each other due to their charged nature, leading to the flocs floating or suspending in the water. Such floating flocs could be filtered to achieve lower turbidity [43].

2.1.7 *Inactivation of* Escherichia coli *Using N-Doped TiO_2 Photocatalyst*

Synthesized N-doped TiO_2 amounts of 0.10 g, 0.20 g, 0.40 g, 0.60 g, 0.80 g, and 1.00 g were weighed and used to prepare separate pastes (see Table 7.4). The pastes were then applied in the different Oriang' ceramic clay pots and put in an oven at 90°C for 3 h. One hundred mL of raw water was transferred into each pot and allowed to percolate through the pores. From the filtrate, 1 mL was pipetted and placed directly onto the 3M Petrifilm. This was incubated at 37.5°C for 24 h and the resultant number of *E. coli* and total coliform colonies determined.

The raw water used had *E. coli* concentration of 1600 CFU/mL. Increasing the amount of N-doped catalyst from 0.10 to 1.20 g used as paste on the pots resulted in 93.75% removal of *E. coli*. On the other hand, the use of the clay pot without addition of the photocatalyst resulted in 85% inactivation of the *E. coli*. The highest concentration of the catalyst gave a result that was below the recommended standard for *E. coli* concentration for potable water according to the World Health Organization [44], which requires *E. coli* concentration to be nil.

Table 7.4 Effect of Filtering Raw Water With *Escherichia coli* Using Clay Pots Pasted With Various Amounts of N-doped Photocatalyst

Water Sample	*E. coli* (CFU/mL)
Raw	1600
Filtered with just the clay pot	240
Filtered with 0.10-g photocatalyst	180
Filtered with 0.20-g photocatalyst	160
Filtered with 0.40-g photocatalyst	100
Filtered with 0.80-g photocatalyst	100
Filtered with 1.20-g photocatalyst	100

2.1.8 *Effect of TiO_2—WO_6 Catalyst Load Versus Efficiency of* Escherichia coli *Removal From Water*

E. coli and related bacteria constitute about 0.1% of gut flora, and fecal—oral transmission is the major route through which pathogenic strains of the bacterium cause disease. Cells are able to survive outside the body for a limited amount of time, which makes them ideal indicator organisms to test environmental samples for fecal contamination [45]. The second tests were carried out using the *E. coli* plates on the raw water, as collected from Nairobi Dam. The test returned "Too Numerous To Count" (TNTC) colonies, as shown in Fig. 7.16.

The same raw water was tested against different amounts of the synthesized TiO_2—WO_6 photocatalyst under UV light from an 8W tube from Phillips. One-mL aliquots were taken at 15-min intervals and cultured using the 3M Petrifilm (*E. coli*) plates for 24 h. It was found that at TiO_2—WO_3 photocatalyst amounts of 0.125 and 0.25 g/L, the CFUs remained TNTC even after 1 h photocatalysis (see Table 7.5). However, at a catalyst amount of 0.5 g/L, the number dropped and could be counted to about 80 CFUs/mL after 1 h.

Experiments were carried out, starting with the water as collected from the field. This water was estimated to carry about 2600 CFU/mL. This was done by using a water sample at a dilution of 10, which gave a count of an average of 260 CFU/mL after 24 h incubation. The catalyst amount in all

Figure 7.16 Coliform culture plate showing *Escherichia coli* in the freshly sampled raw water.

Table 7.5 Effect of TiO_2–WO_6 Catalyst Load Versus Efficiency of *Escherichia coli* Removal From Water

Efficiency Versus Catalyst Amount

Amount of Catalyst in L	Mean N_t	Mean N_i	Efficiency: $[(N_i - N_t)/N_i]*100$
Plain water sample	2600	2600	0
UV only	233	260	10
Dark reaction (catalyst only)	180	260	31
0.125	180	260	31
0.25	113	260	56
0.5	100	2600	96
0.75	1	2600	100
1	2	2600	100

N_i, initial number of *E. coli* colony forming units; N_t, number of E. coli colony forming units at a time, t.

test solutions was adjusted to simulate a test environment with a 2600 CFU/mL *E. coli* amount. Tests were carried out in triplicates and aliquots sampled every 15 min for 75 min. The data at different catalyst loads are as in Fig. 7.17. Control tests were carried out to establish photokilling by photocatalysis only.

A graphical trend of the residual fraction with time in minutes using different photo catalyst amounts was then generated, as shown in Fig. 7.17. Photocatalytic action is evident, as the highest residual fractions were obtained during photolysis and dark reactions. It can be seen that photolysis gradually reduced the residual fraction consistently, while during

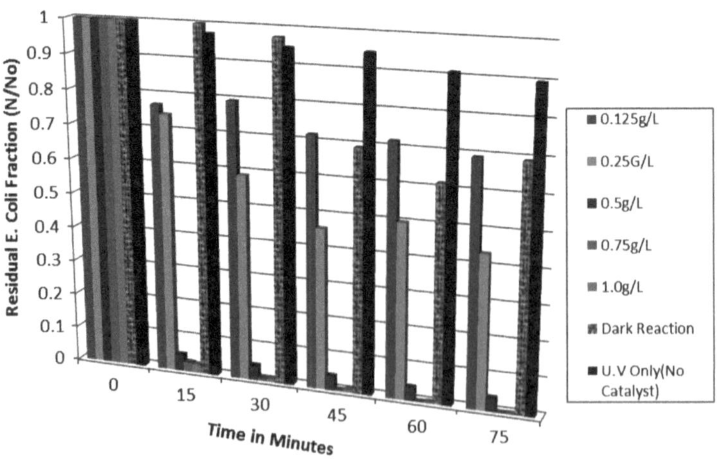

Figure 7.17 Residual *Escherichia coli* fraction over time at different catalyst amounts.

the dark reaction, only a slight decrease of the residual fraction was observed at 75 min. This can be explained that the reduction during dark reaction was as a result of adsorption of the *E. coli* onto the surface of the $TiO_2{-}WO_3$ photocatalyst. With time, the *E. coli* got desorbed back into the solution, thereby raising the number of *E. coli* detected after 75 min.

The photokilling effectiveness of the catalyst increased with catalyst amount up to 0.75 g/L optimum amount. Increasing the photocatalyst amount resulted in no further increase in photokilling effectiveness. A plot of degradation efficiency versus amount of catalyst was plotted, which showed that 0.75 g/L catalyst amount as the optimum. Consequently, other experiments were conducted at this catalyst amount. The effectiveness of photocatalytic degradation of *E. coli* with the amount of catalyst is as shown in Fig. 7.18. Percent removal efficiency rises with photocatalyst amount up to 0.75 g/L and then decreases at a photocatalyst amount of 1.0 g/L.

2.2 Use of *Moringa oleifera* Seeds in Water Purification

2.2.1 Effect of De-oiling the Seeds on Turbidity Removal

The *M. oleifera* seeds were de-oiled and the hexane soluble oil content found to be 26.40 ± 1.04% of the total mass of the seeds. From literature, the seeds have been reported to contain over 40% edible oil [46]. The de-oiled seeds

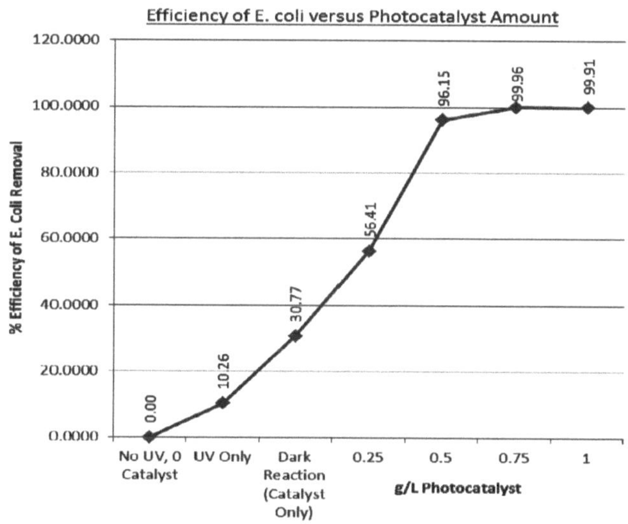

Figure 7.18 Effect of $TiO_2{-}WO_3$ catalyst concentration on *Escherichia coli* removal N = CFU/mL. Irradiation Time = 75 min, pH = 7.3.

Table 7.6 Comparison of the Shelled and De-oiled Seeds in Turbidity Removal

Amount of *Moringa oleifera* Seeds (g/L)	Residue Turbidity (NTU)	
	Shelled Seed Powder	De-oiled Seed powder
0.05	4.925 ± 0.078	4.85 ± 0.071
0.1	3.725 ± 0.078	3.66 ± 0.057
0.2	2.45 ± 0.212	2.45 ± 0.106
0.25	3.665 ± 0.078	4.83 ± 0.028
0.5	5.44 ± 0.042	6.62 ± 0.056

were as effective as the shelled seeds in turbidity removal, as illustrated by the residue turbidities in Table 7.6.

2.2.2 Disinfection Properties of the Seeds

An experiment with raw water at an initial concentration of 260 CFU/mL and 300 CFU/mL *E. coli* and other coliforms, respectively, was used to establish the disinfection properties of the seeds. Fig. 7.19 is a plot of the residue number of the fecal coliforms after 1 h against the various amounts of the seed powder used.

The application of various dosages resulted in the reduction of *E. coli* residue fraction to a range of 15–60 CFU/mL, corresponding to reduction efficacy range of 55.67–92.33%. One gram per liter dosage recorded the best efficiency of 92.33% reduction. The results were within the range reported by other scholars. Yongabi et al. [38] reported a 95% efficiency reduction of the total aerobic mesophilic bacterial counts in *E. coli* as well as coliforms counts. They reported an optimal dosage of between 4 and 5 g/L. Pritchard et al. [40]

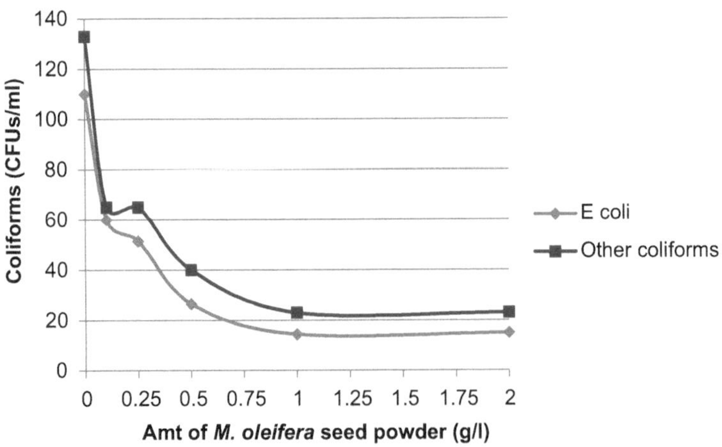

Figure 7.19 Optimal dose disinfection results of *Moringa oleifera* seeds.

reported different efficiencies at different working conditions. By changing the initial microbial load and initial turbidity of the test water, they observed efficiency that ranged between 84% and 88% for artificial model water and 77% and 88% for natural water. Their model water was spiked with 100–300 CFU/mL *E. coli*, while their river was reported to contain an average of 26.5 CFU/mL. For their work, they reported an optimal dose ranging between 0.75 and 1.25 g/L.

2.2.3 *Effect of Time on* Escherichia coli *Disinfection with* Moringa oleifera

The minimum amount of *M. oleifera* seeds that could result in maximum disinfection was determined when the treated water samples were left to stand for up to 6 h. The jar tests were carried out as previously outlined with the sampling of water for inoculation conducted after 1, 3, and 6 h. Fig. 7.20 is a pictorial presentation of the effect of time on the disinfection properties of the *M. oleifera* seeds.

The blank sample had a high residue number of coliforms, as evidenced from the blue and red dots, even after leaving the waters to stand for 6 h. With a dose of 0.1 g/L, the residue coliforms considerably reduced after 1 h with an almost 100% reduction after 3 and 6 h. Enumeration of residue coliforms was carried out and the results presented in Table 7.7.

The *E. coli* counts were found to reduce for the first 3 h for all the dosages. After 6 h, the *E. coli* residue counts were increasing for dosages lower than the optimal dose. For instance, at the lowest dose of 0.1 g/L, the residue count

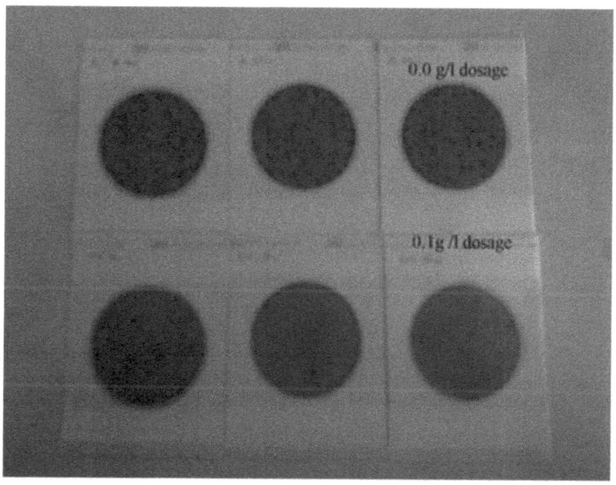

Figure 7.20 Pictorial illustration of the effect of time for the 0.0 and 0.1 g/L dosages after 1, 3, and 6 hours, starting from left.

Table 7.7 Effect of Time on *Escherichia coli* Disinfection With Various Amounts of *Moringa oleifera*

Amount of *M. oleifera* (g/L)	Residue *E. coli* Count With Time			
	0 h	**1 h**	**3 h**	**6 h**
0	260 ± 10	125 ± 5.0	92.3 ± 4.2	96.5 ± 7.8
0.1	260 ± 10	115.3 ± 4.7	76.3 ± 4.9	92.3 ± 2.6
0.2	260 ± 10	86.7 ± 2.5	67.3 ± 5.9	69 ± 2.8
0.25	260 ± 10	69 ± 2.6	36 ± 3.6	59.7 ± 1.5
0.5	260 ± 10	43.3 ± 3.8	26 ± 5.3	39.7 ± 8.5
1	260 ± 10	28 ± 2.6	3.7 ± 2.5	3.0 ± 1.0
2	260 ± 10	28.7 ± 4.2	2.0 ± 1.0	2.7 ± 1.5

after 3 h is 76.3 ± 4.9 CFU/mL. After 6 h, the same sample recorded a residue *E. coli* count of 92.3 ± 2.6 CFU/mL. The increase would be as a result of the residue *E. coli* feeding on the nutrients present in the water and multiplying, thus resulting in the increase. At a dose of 1 and 2 g/L, the residue number of *E. coli* was 3.7 ± 2.5 and 2.0 ± 1.0 CFU/mL, respectively, after 3 h. After 6 h, the average residue counts were 3.0 ± 1.0 for the 1 g/L dosage and 2.7 ± 1.5 CFU/mL for the 2 g/L dosage. Since almost all the *E. coli* were eliminated after the first 3 h, there was no significance increase in the number of coliforms after leaving the samples to stand for an extra 3 h.

2.2.4 Effect of pH

A study was conducted to understand the effect of pH on the effectiveness of *M. Oleifera* on *E. coli*. A dosage of 1 g/L of N-doped TiO_2 catalyst was chosen and the samples pH adjusted to 5.0, 7.0, and 9.0 conditions. The disinfectant tests were carried out and the enumeration of *E. coli* after 1, 3, and 6 h undertaken.

Fig. 7.21 gives the results for the effect of pH on *E. coli* disinfection. For all the pH values chosen, the residue fraction after 1 h was between 0.02 and 0.06 mol ratio with the highest residue figure recorded at pH 5 and the lowest figure at pH 9. The same trend was observed after the third and fourth hours, but with reduced discrepancy between the residue fractions values.

A basic media of pH 9 was found to be a better working pH for the *M. oleifera* seeds, while acidic conditions were found to lower the efficiency of the seeds. Generally, natural water pH ranges between 6.5 and 8.5; therefore, it was concluded that the seeds would work well in neutral water. Previous work [47] had reported that the seeds were more effective in neutral

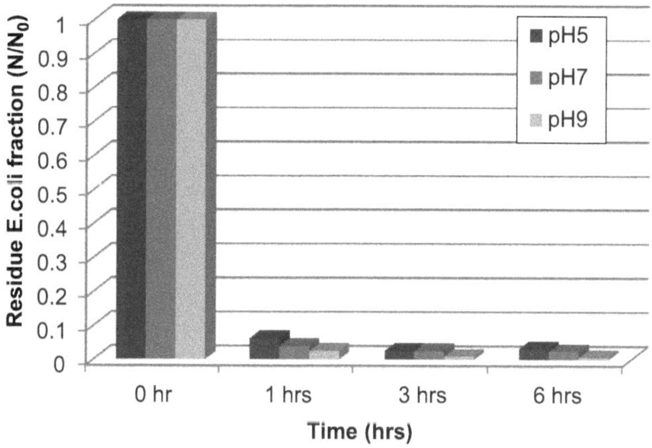

Figure 7.21 The effect of pH on the effectiveness of *Moringa oleifera* on *Escherichia coli.*

pH, but also observed that alkaline conditions were overall more favorable than acidic conditions.

The seed kernels of *M. oleifera* contain significant quantities of low molecular weight (water-soluble proteins) that carry a positive charge. When the crushed seeds are added to raw water, the proteins produce positive charges acting like magnets and attracting the predominantly negatively charged particles (such as clay, silt, bacteria, and other toxic particles in water) (Sutherland et al. [48]). Since bacteria in water are generally attached to solid particles, treatment with *M. oleifera* powder can leave water clear with 90–99% of the bacteria removed (Bukar et al. [49]).

M. oleifera is negatively charged and attacks the membrane of the microorganism.

2.3 Use of N-Doped TiO_2 Catalyst in Degradation of Dyes

2.3.1 Methylene Blue Degradation

The results of degradation of MB under dark conditions showed a decrease in MB concentration during the first hour of the test, but thereafter concentration remained unchanged after the second hour. The results suggested that the adsorption equilibrium of dye on photocatalyst surface had reached within the first 1 h. Discoloration of MB in the presence of N-doped TiO_2 surface was consistent with the widely reported photocatalytic properties of TiO_2.

2.3.2 Effect of Irradiation Time on Methylene Blue Degradation

Fig. 7.22 shows a linear plot of C/C_0 for the photodegradation of MB under visible light after 100 min of illumination. In the absence of light irradiation,

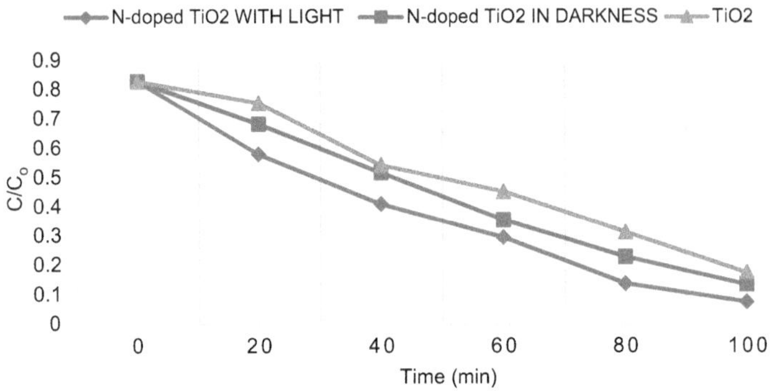

Figure 7.22 Plot of C/C_o of methylene blue with given time.

the use of N-doped TiO_2 photocatalyst achieved only 83.7% degradation. However, in the presence of the photocatalyst and light irradiation, 91.01% degradation of MB was achieved under similar conditions. The results suggested that the N-doped TiO_2 catalyst had higher photocatalytic activity under the light irradiation. Increased irradiation of the solution that contained catalyst up to 120 min resulted in 99.6% MB degradation. Under similar conditions, the undoped TiO_2 was able to degrade 78.7% of MB.

In order to evaluate the rate of photodegradation of MB under visible light, the natural log of the ratio of the remaining concentration of MB was plotted against exposure time to the photocatalyst under irradiation. Fig. 7.23 shows the findings of the plot of the concentration of MB $[\ln(C_t/C_o)]$ against exposure time.

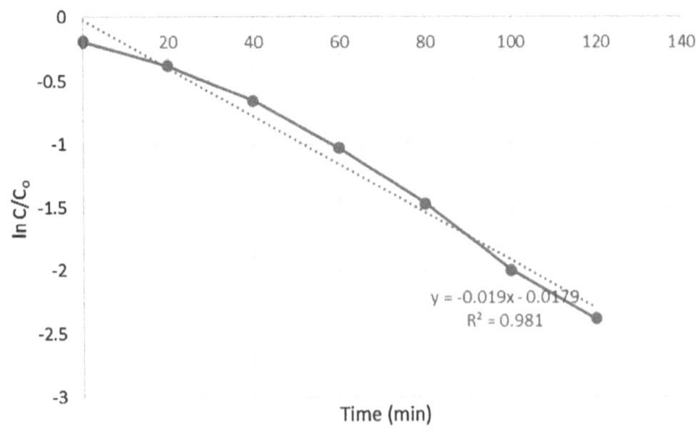

Figure 7.23 Linear plot of ln (C/C_o) for degradation of methylene blue under light irradiation after 120 min.

The data agreed best with the Langmuir–Hinshelwood kinetic model for reaction rate dependence on initial reactant concentration.

$r = dC/dt = kKC/(1 + KC)$ where r is the oxidation rate of the blue dye in (mg/L min), C the concentration of the dye (mg/L), t the illumination time, k the reaction rate constant (mg/L min), and K the adsorption coefficient of the reactant (L/mg). When the initial concentration, C_o, the equation can be simplified to an apparent first-order equation:

$$dc/_{dt} = kKc \tag{7.1}$$

$$\int dC/C = kK \int dt \tag{7.2}$$

Integrating the equation and taking boundary conditions of $C = C_o$ at $t = 0$ gives

$$\ln(C_o/C_t) = kKt = k_{opp}t \tag{7.3}$$

Or

$$C_t = C_o e^{-kt} \tag{7.4}$$

which is the first order rate equation.

Rearranging the equation, one gets:

$$\ln C_t = -kt + \ln C_o \tag{7.5}$$

where

C_t = dye concentration at time, t

K = the apparent first order reaction rate constant t = the time in minutes

C_o = the original dye concentration in ppm

The equation obtained is in the form of $y = mx + c$. A plot of $\ln(C_t)$ versus t gives a straight line, the slope of which upon linear regression equals the apparent first-order rate constant $k$app. $= 1.9 \times 10^{-2}$ min^{-1}.

2.3.3 Effect of Dosage of the N-Doped TiO_2 Photocatalyst on Degradation of Methylene Blue

The results of degradation of MB after application of different amounts of the photocatalyst are shown in Fig. 7.24. The data revealed that photodegradation increased initially with increasing amount of the photocatalyst up to 0.60 g. This suggests that with increasing the amount of the photocatalyst initially, both the number of dye molecules adsorbed and the number of

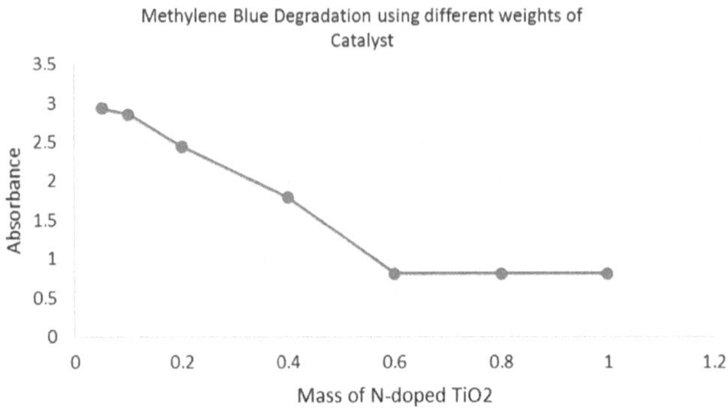

Figure 7.24 Plot of absorbance of methylene blue against catalyst loading.

photons absorbed increased promoting degradation. However, for very high concentrations of catalyst in the solution, turbidity increases, decreasing light penetration into the solution, and consequently decreasing photodegradation. Similar findings were reported by Franco et al. [47].

The optimal amount of the photocatalyst on degradation of MB was 0.6 g. The N-doped TiO_2 photocatalyst has good adsorption capability because of large surface area, which enables the diffusion of the MB molecules freely into the N-doped TiO_2. However, many pairs of hole electrons produced by irradiation are quickly recombined, resulting in reduced photocatalytic degradation of MB molecules (Mohapatra and Parida [50]).

2.3.4 Effect of pH on Degradation of Methylene Blue

Increasing pH of the reaction mixture from 2 to 7 was found to increase photodegradation of MB from 23% to 97.1%. On the other hand, the rate of degradation decreased from 97.1% to 77% as the pH increased from 7 to 11 (Fig. 7.25).

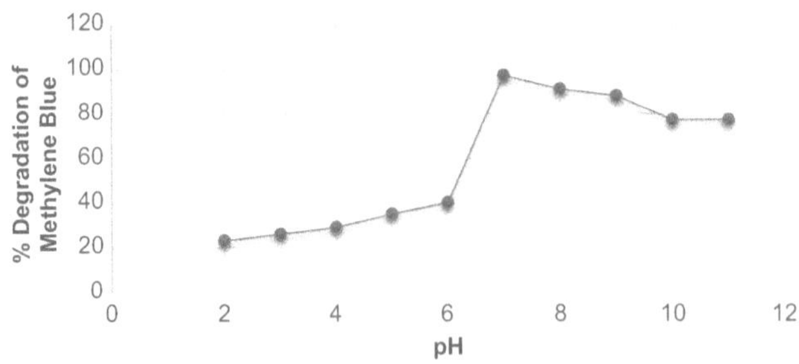

Figure 7.25 Effect of pH on methylene blue degradation by N-doped TiO_2 catalyst.

The highest rate of photodegradation (97.1%) was recorded at pH of 7.0. Consequently, pH 7 was selected as the optimal pH condition for methylene blue photodegradation experiments. The reduction in photodegradation rate at higher pH could be attributed to the formation of carbonate ions, which are effective scavengers of OH^- ions, thus reducing the degradation rate [25,51]. At higher pH values, the hydroxyl radicals are rapidly scavenged so that they do not have the opportunity to react with the dye molecules [25,52].

2.4 Pesticide Degradation

Photocatalytic experiments were carried out in duplicate. Table 7.8 shows the average concentration of each pesticide studied in the initial sample and the final concentration after the photocatalytic treatments. The irradiation time was 120 min. In addition to the photocatalytic treatments, blank tests were carried out. These blank tests were solutions of pesticides with N-doped TiO_2 without irradiation and with irradiation without catalyst. Only for some pesticides were degradations different from zero achieved.

The pesticides most electively removed by the photocatalytic treatment, reaching yields higher than 50%, are heptachlor, Endosulfan I, Endosulfan II, p,p′-DDE, dieldrin, endrin, endrin aldehyde, p,p′-DDT, endosulfan

Table 7.8 Concentrations of Pesticides in Sample Before and After Irradiation With Photocatalyst

	Pesticide Concentration (ng/mL)		
	Initial	After Irradiation With N-doped TiO_2	% Degradation
α-HCH	1834.579	1824.643	0.5
β-HCH	840.825	530.353	36
δ-HCH	942.845	942.245	0.06
ϒ-HCH	2083.952	1441.093	30
Aldrin	1435.774	932.506	35
Heptachlor	1407.028	674.902	52
Heptachlor epoxide	847.595	355.005	58
Endosulfan I	1670.898	770.164	53.9
P,p′-DDE	1190.777	449.848	62
Dieldrin	1607.398	719.145	55.2
Endrin	1610.083	612.002	61.9
Endrin aldehyde	366.892	117.568	67.9
Endosulfan II	1692.598	594.442	64.8
P,p′-DDT	121.857	44.385	63.6
Endosulfan sulfate	590.955	289.601	50.9

sulfate, and heptachlor epoxide. The degradation percentages obtained after the photocatalytic treatment were lowest for α-hexachlorocyclohexane (α-HCH) and δ-hexachlorocyclohexane (δ-HCH), that is, 0.5 and 0.06, respectively.

In the cases of heptachlor, p,p′-DDE, p,p′-DDT, and aldrin, blank experiments show an important degradation of pesticides by N-doped TiO_2 without irradiation and irradiation without a catalyst. Thus photocatalytic treatment and irradiation are both responsible for their degradation. This is illustrated on Table 7.9.

For example, the degradation of methyl red gives the degradation of methyl red with increase in catalyst amount up to 0.6 g and does not increase further (Fig. 7.26). Similarly, the degradation is complete after 200 min (Fig. 7.27). The bleaching of methyl red dye by light reduces the absorbance with time but does not completely clear the dye (Fig. 7.28).

2.5 Use of TiO_2-WO_3 Degradation of Pesticides and Dyes

2.5.1 Pesticides Mixture

A mixture of α-HCH, β-lindane, benzene hexachloride, and δ-HCH were treated with TiO_2-WO_3 catalyst with one sample wrapped with aluminum

Table 7.9 Blank Experiments of Pesticides With N-doped TiO_2 Without Irradiation and With Irradiation Without Catalyst

Pesticide	% Degradation by N-doped TiO_2	% Degradation by Visible Light
Heptachlor	43	18
P,p′-DDE	40	27
P,P′DDT	24	29
Aldrin	50	21

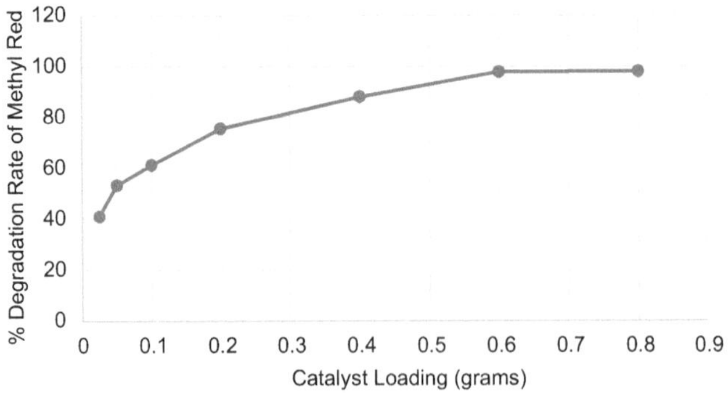

Figure 7.26 Percentage degradation of methyl red with varying catalyst loading.

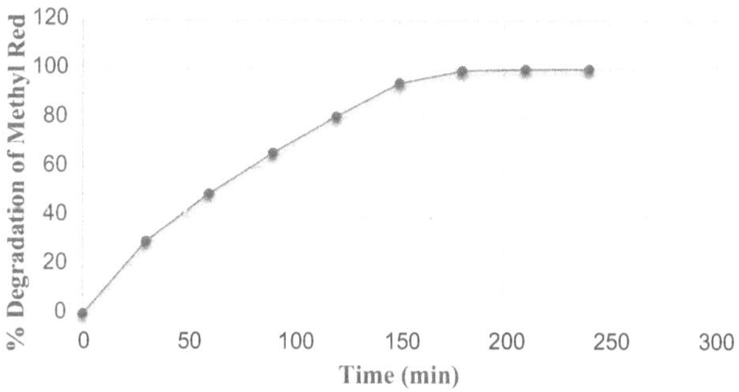

Figure 7.27 Percentage photodegradation of methyl red exposed to light mixed with 0.7 g/L over time.

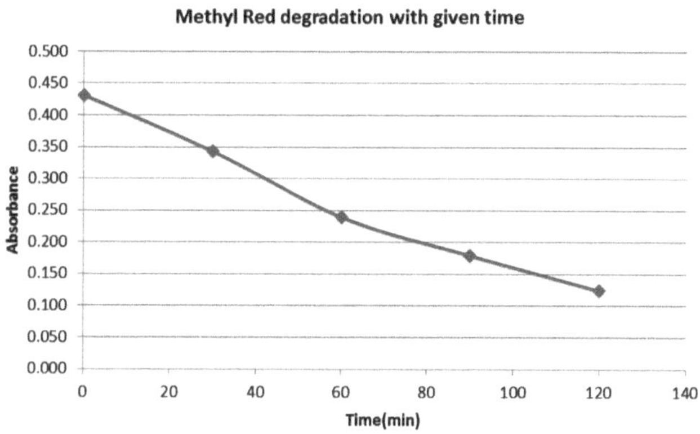

Figure 7.28 Absorbance of methyl red dye in the solution exposed to light irradiation alone without catalyst.

and the other sample exposed to irradiation over 1 to 4 h. Fig. 7.29 shows the residual fraction of the four pesticides in the mixture, taking the equilibrium adsorption in the dark as reference. It was found that β-lindane had the lowest degradation rate with a residual fraction of 0.09 remaining after 4 h photocatalytic degradation. δ-HCH had the highest degradation rate going below detection limit after 4 h.

2.5.2 *Effect of pH on* Escherichia coli *Photokilling Efficiency with TiO_2-WO_3 Added*

The effect of pH on photokilling efficacy of 0.75 g/L catalyst added was investigated. The data on the residual fraction with time at different pH values starting with 2600 CFU/mL count of *E. coli* in the test water is in

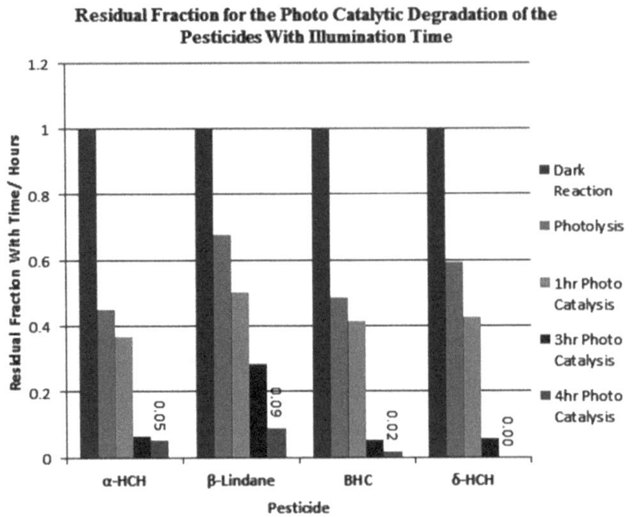

Figure 7.29 Degradation fractions for pesticides solutions treated with TiO_2–WO_3 with and without irradiation over time.

Fig. 7.30. It was found that varying pH did not significantly vary the *E. coli* inactivation rate after 15 min.

It can be seen that pH 7.3 was best for the photokilling of the *E. coli*. Efficiency was lowest at pH 4, but followed almost a similar trend for all

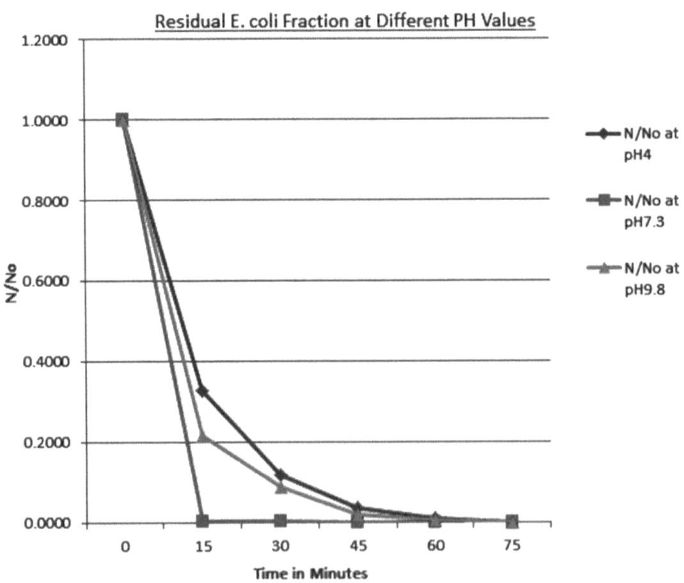

Figure 7.30 Effect of pH on photokilling of *Escherichia coli*. 0.75 g/L TiO_2–WO_3 catalyst added.

pH from 45 min of illumination. These agreed with the findings by Angela-Guiovana Rincón and Cesar Pulgarin [53], who found that the photolytic inactivation of *E. coli* was independent of initial pH between 4.0 and 9.0 [54]. Heyde and Portalier [55] explained that the negligible *E. coli* reaction to acid conditions was due to the presence of an acid tolerance response in the bacterium itself, which secreted the acid-induced proteins for bacterial acid shock protection.

2.5.3 Disinfection Kinetics for the Photokilling of Escherichia coli *Using the* TiO_2–WO_3 *Photocatalyst*

A disinfection kinetics model is usually used to compare results obtained under different experimental conditions. One of the most fundamental and simple models reflecting the principle of disinfection kinetics is the Chick-Watson model [52], $\log (N/N_0) = -kCT$. This model assumes that the rate of inactivation of the microorganisms by the disinfectants is primarily affected by the combination of the concentration of the disinfectants/catalyst (C) and the contact time (T) ($C \times T$, which is expressed as the CT value). The CT value is widely used as a measure of compliance with disinfection requirements in drinking water treatment [53].

Fig. 7.31 shows the log reduction of the *E. coli* with time according to the Chick-Watson model. The plot was then used to determine the *E. coli* disinfection kinetics using the Chick-Watson model.

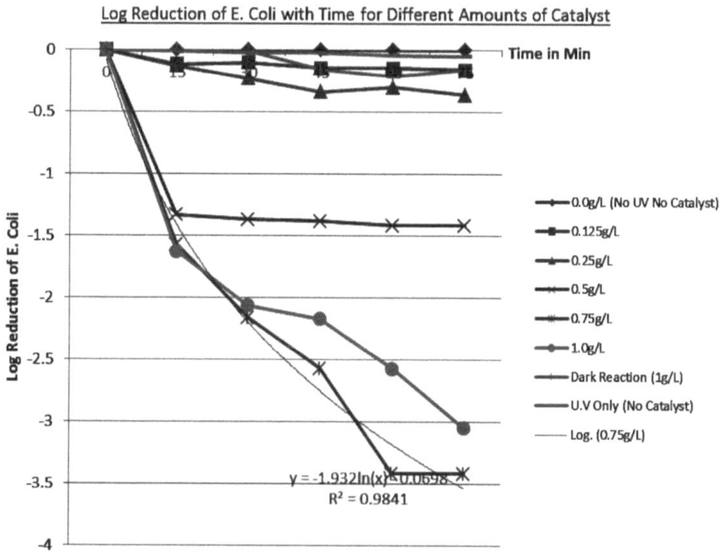

Figure 7.31 Log inactivation of *Escherichia coli* with time, pH 7.3.

Using the equation for log inactivation, the rate constant for the photo-killing was determined as follows:

$$LogN/N0 = -kCT \tag{7.6}$$

where k = the rate constant.

C = the amount of catalyst, 0.75 g/L

T = the illumination time

The reaction rate constant at different time intervals was determined for the 0.75 g/L photocatalyst amount and the data entered, as shown in Fig. 7.32. It can be seen that the rate constant is highest within the first 15 min, decreasing gradually as time progresses. This can be explained that the bacteria develop resistance as they spend more time in the reaction conditions.

After 15 min photocatalysis, more than 2-log reduction (90%) has already been achieved at 0.75 g/L catalyst amount and pH 7, the log reduction rising to 3415-log reduction after 45 min of photo inactivation.

The following tests were conducted in triplicates using the raw water and 0.75 g/L catalyst to investigate the photokilling efficiency on total coliforms using solar radiation. Water samples were exposed to the midday sun, first without the catalyst (solar only), and then on addition of the catalyst (solar - + catalyst) for 75 min. The water was then sampled for incubation using the 3M *E. coli* plates for 24 h at 37.2°C. The residual fraction of *E. coli* with photocatalytic time was analyzed, as shown in Fig. 7.33.

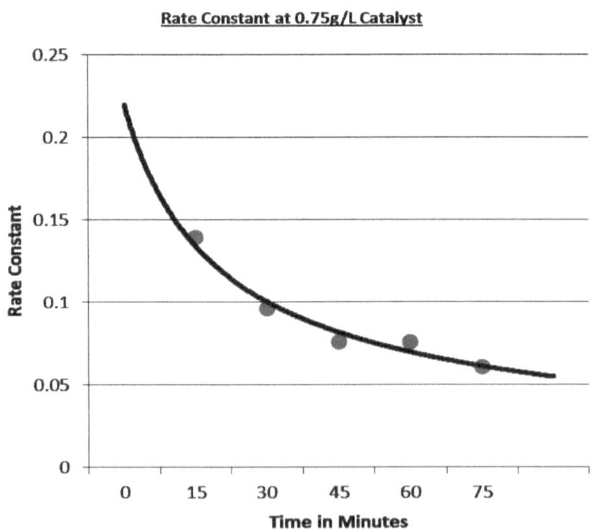

Figure 7.32 Reaction rate constant variation with time.

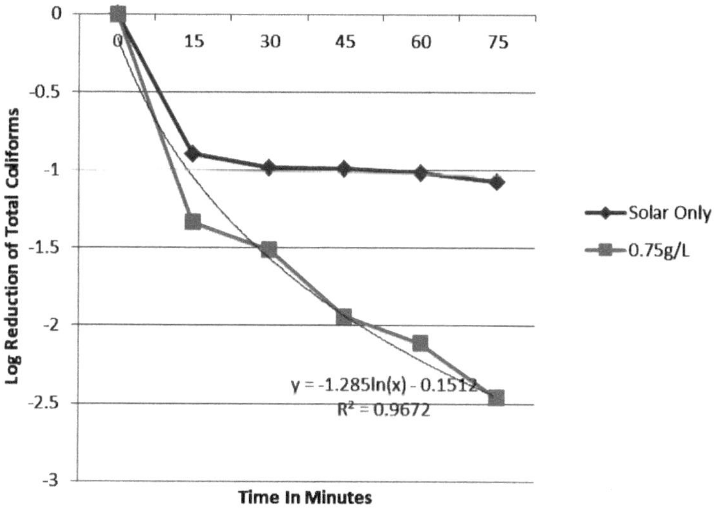

Figure 7.33 Total coliform log reduction with solar illumination time.

It compares the efficiency of SODIS only with that of photocatalysis. It shows that using the photocatalyst enhanced the reduction of total coliforms by more than a factor of 20 as compared to using solar irradiation alone, reaching an efficiency of coliform removal of 99.65% or log 2.5 *E. coli* inactivation.

Though the conditions for this experiment were mild and maintained in their natural state, the results produced proved that though sunlight alone can be used for disinfection as in SODIS, the combination of the effects of both SODIS and photocatalysis is a far more effective procedure for degradation of *E. coli*. In this experiment, photocatalytic removal of *E. coli* was 24.44 times higher than using solar alone.

The comparison between the new catalysts synthesized and studied, that is, TiO_2—NO_3 and TiO_2—WO_3, show both effective against disinfection of *E. coli*. However, the titanium—tungsten composite is more effective than the N-doped titanium catalyst. Both are not very effective in mineralization of pesticides. On the other hand, both are able to degrade dyes very well. The catalysts used in combination with clay ceramics give total elimination of *E. coli*. We are investigating incorporation of catalysts in local ceramics made by community women.

ACKNOWLEDGMENT

The authors wish to acknowledge the USAID/National Academy of Science PEER Project No. PGA-2000001967 for the support grant, Dr. Vincent Madadi for providing the 3M

Escherichia coli plates, Ms. Dorice Situma for providing SPE cartridges, the Department of Chemistry of the University of Nairobi, ANCAP, and the Jaramogi Oginga Odinga University of Science and Technology.

REFERENCES

[1] UNEP. Status of desertification and implementation of the United Nations plan of action to Combat Desertification. Report of the Executive Director. UNEP/GCSS.III/3. Nairobi. 1991.

[2] Allen A, DÁVILA JD, Hofmann P. In: The peri-urban water poor: citizens or consumers?, Vol. 18. International Institute for Environment and Development (IIED); 2006. p. 333–51. http://dx.doi.org/10.1177/0956247806069608. 333, http://water.worldbank.org/sites/water.worldbank.org/files/iuwm_ch1.pdf www.sagepublications.com.

[3] UN Habitat. Sustainable urbanization. 2014. Available http://www.un.org/en/ecosoc/integration/pdf/economiccommissionforafrica.pdf.

[4] United Nations, Department of economic and social affairs, population division. World population prospects: The 2015 revision. (Medium variant). https://populationpyramid.net/sub-saharan-africa/2015/.

[5] WHO. Preventing diarrhea through better water, sanitation and hygiene: exposures and impacts in low-and middle-income countries. Geneva: World Health Organization; 2014a. http://www.who.int/water_sanitation_health/gbd_poor_water/en/2014/.

[6] Sobsey MD. Managing water in the home: accelerated health gains from improved water supply. Geneva: WHO; 2002.

[7] Bielefeldt AR, Kowalski K, Scott Summers R. Bacterial treatment effectiveness of point-of-use ceramic water filters. Water Res 2009;43(14):3559–65. http://dx.doi.org/10.1016/j.watres.2009.04.047.

[8] Simonis JJ, Basson AK. Evaluation of a low-cost ceramic micro-porous filter for elimination of common disease microorganisms. Phys Chem Earth 2012;36(14):1129–34.

[9] Sobsey MD, Stauber CE, Casanova LM, Brown JM, Elliott MA. Point of use household drinking water filtration: a practical, effective solution for providing sustained access to safe drinking water in the developing world. Environ Sci Technol 2008;42:4261–7.

[10] Naddafi K, Mahvi AM, Nasseri S, Kokhtari M, Zeraati H. Evaluation of the efficiency of clay pots in-removal of water impurities. Iran J Environ Health Sci Eng 2005;2(2):12–6.

[11] Plappally A, Chen H, Ayinde W, Alayande S, Usoro A, Friedman Kc, Are E, Ogunyale T, Yakub I, Leftwich M, Malatesta K, Rivera R, Brown L, Soboyejo A, Soboyejo W. A field study of the use of clay ceramic water filters and influences on the general health in Nigeria. J Health Behav Public Health 2011;1(1):1–14.

[12] Grieken R, Marugan J, Sordo C, Martinez P, Parblos C. Photocatalytic inactivation of bacteria in water using suspended and immobilized silver-TiO_2. Appl Catal B 2009;93:112–8.

[13] Zhang J, Zhang F, Luo Y, Yang H. A preliminary study on cactus as coagulant in water treatment. Process Biochem 2006;41(3):730–3.

[14] Ghebremichael KA, Gunaratna KR, Hongbin H, Harry B, Gunnel D. A simple purification and activity assay of the coagulant protein from *Moringa oleifera* seed. Water Res 2005;39:2338–44.

[15] Rayner J. Current practices in manufacturing of ceramic pot filters for water treatment. MSc thesis. Loughborough University; 2009.

[16] Carp O, Huisman CL, Reller A. Photoinduced reactivity of titanium dioxide. Prog Solid State Chem 2004;32:33–117.

[17] Gupta SM, Tripathi M. A review of TiO_2 nanoparticles. Chin Sci Bull 2011;56:1639–57. http://dx.doi.org/10.1007/s11434-011-4476-1.

[18] Chatterjee D, Dasgupta S. Visible light induced photocatalytic degradation of organic pollutants. J Photochem Photobiol C 2005;(6):186–205.
[19] Ohno T, Mitsui T, Matsumura M. Photocatalytic activity of S-doped TiO_2 photocatalyst under visible light. Chem Lett 2003;32:364–5.
[20] Asahi R, Morikawa T, Owaki T, Aoki K, Taga Y. Visible-light photocatalysis in nitrogen-doped titanium oxides. Science 2001;293:269–71.
[21] Sakthivel S, Kisch H. Daylight photocatalysis by carbon-modified titanium dioxide. Angew Chem Int Ed 2003;42:4908–11.
[22] Winzenburg A, Faust R. Efficient photocatalysis in the visible with TiO_2/phthalocyanine-hybrid particles. Kassel University Press; 2012. p. 177–84.
[23] Macharia JW. Development of a low cost water purification system- A case study ceramic filters and *Maringa oleifera* seeds. M/Sc thesis. University of Nairobi; 2015.
[24] Otieno MA. Water purification using clay impregnated with N-Doped titanium oxide-case study lake Victoria basin, Homa Bay county. M/Sc. thesis. University of Nairobi; 2016.
[25] Mbugua SN, Wandiga SO, Kamau GN. Photo catalytic inactivation of *Escherichia coli* using titanium (IV) oxide-tungsten (VI) oxide nanoparticles composite. Int J Photocatal Phot 2015;119:204–11.
[26] Bakibinga-Ibembe JD, Vuai SA, Mungai NW. Environmental laws and policies related to periodic flooding and sedimentation in the Lake Victoria Basin (LVB) of East Africa. Afr J Environ Sci Technol 2011;5(5):367–80.
[27] UNESCO. Kenya National Water Development Report. 2006. unesdoc.unesco.org/images/0014/001488/00148866E.pdf.
[28] Van Halem D. Ceramic silver impregnated pot filters for house-hold drinking water treatment in developing countries. M.S. thesis-2006. Delft, Netherlands: Delft Univ. of Technology; 2006.
[29] Lantagne DS. Investigation of the potters for peace colloidal silver impregnated ceramic filter report 1: intrinsic effectiveness; 2001.
[30] Habibi MH, Nasr-Esfahani M, Giti E, Baharak H. Nanostructure thin films of Titanium-Dioxide coated on glass and its anti UV effect for living organisms. Curr Nanosci 2010;6:324–9.
[31] Ocwelwang AR, Tichagwa L. Synthesis and characterization of Ag and nitrogen doped TiO_2 nanoparticles supported on a Chitosan-PvaeNanaofibre support. IJARCS 2014;1:28–37.
[32] Pelaez M, Nolan N, Pillai S, Seery M, Falaras P, Kontos AG, Dunlop PSM, Hamiltone JWJ, Byrne JA, O'Shea K, Entezari MH, Dionysiou DD. A review on the visible light active titanium dioxide photocatalysts for environmental applications. Appl Catal B Environ 2012;125:331–49. http://dx.doi.org/10.1016/j.apcatb.2012.05.036.
[33] Veluru JB, Appukuttan SN, Zhu P, Seeram R. Synthesis and characterization of rice grains like Nitrogen-doped TiO_2 nanostructures by electrospinning photocatalysis. Material Letters 2011;65:3064–8.
[34] Zhang L, Yan J, Zhou M, Liu Y. Photocatalytic inactivation of bacteria by TiO_2-based compounds under simulated sunlight irradiation. Trans Nonferrous Met Soc China 2012;24:743–9
[35] Varkey AJ, Dlamini MD. Point-of-use water purification using clay pot water filters and copper mesh. Water SA 2012;38:721–6.
[36] MSc Thesis Sagara Junko study of filtration for point-of-use drinking water treatment in Nepal. Massachusets, USA: Massachusetts Institute of Technology; 2000.
[37] Musyoki AM, Mbaruk SA, Mbithi JN, Maingi JM. Water-borne bacterial pathogens in surface waters of Nairobi river and health implication to communities downstream Athi river. Life Sci Pharma Res 2013;3:1–7.

[38] Yongabi KA, Lewis DM, Harris PL. Application of phytodisinfectants in water purification in rural Cameroon. Afr J Microbiol Res 2011;5(6):628–35.

[39] Weber Jr WJ. Physicochemical processes for water quality control. New York: John Wiley & Sons; 1972.

[40] Pritchard M, Craven T, Mkandawire T, Edmondson SA, O'Neill JG. A study of the parameters affecting the effectiveness of *Moringa oleifera* in drinking water purification. Phys Chem Earth 2010-b;35:791–7.

[41] Arnoldsson E, Bergman M, Matsinhe N, M.Persson K. Assessment of drinking water treatment using *Moringa Oleifera* natural coagulant. Vatten 2008;64:137–50.

[42] Sutherland JP, Folkard GK, Mtawali MA, Grant WD. *Moringa oleifera* at pilot/full scale. In: Pickford, editor. Water, Sanitation, Environment & Development: Proceedings of the 19th WEDC Conference, Accra, Ghana, 6th–10th September 1993; 1994.

[43] Arama P, Wagai S, Ogur J, Walter A, Owido S, Mahagayu C. Harvesting surface rainwater—purification using Moringa-oleifera seed extracts and aluminumsulfate). Agric Ext Rural Dev 2011;3(6):102–12.

[44] WHO. In: Guidelines for drinking-water quality. 3rd ed., Vol. 1. Geneva: World Health Organization; 2004.

[45] Feng, P, DW, Stephen (ret), MA, Grant (dec), W. Burkhard. Enumeration of *Escherichia coli* and the Coliform Bacteria. Bacteriological analytical manual chapter 4 2002; Food and Drug Administration. http://www.fda.gov/Food/FoodScienceResearch/LaboratoryMethods/ucm064948.htm.

[46] Lalas S, Tsaknis J. Characterization of *Moringa oleifera* seed oil variety "Periyakulam 1". J Food Compos Anal 2002;15:65–77.

[47] Franco A, Neves MC, Carrott MMLR, Mendonça MH, Pereira MI, Monteiro OC. Photocatalytic decolorization of methylene blue in the presence of TiO_2/ZnS nanocomposites. J Hazard Mater 2009;161:545–50.

[48] Sutherland JP, Folkard GK, Grant WD. Natural coagulants for appropriate water treatment: a novel approach. Waterlines 1990;8:30–2.

[49] Bukar A, Uba A, Oyeyi T. Antimicrobial profile of *Moringa oleifera* Lam. Extracts against some food-borne microorganisms. Bayero J Pure Appl Sci 2010;3(1):43–8.

[50] Mohapatra P, Parida KM. Photocatalytic activity of sulfate modified titania 3: decolorization of methylene blue in aqueous solution. J Mol Catal A Chem 2006;258:118–23.

[51] Feng W, Nansheng D, Helin H. Degradation mechanism of azo dye C.I. reactive red 2 by iron powder reduction and photooxidation in aqueous solutions. Chemosphere 2000;41:1233–8.

[52] Karkmaz M, Puzenat E, Guillard C, Herrmann JM. Photocatalytic degradation of the alimentary azo dye amaranth. Mineralization of the azo group to nitrogen. Appl Catal B 2004;51:183–94.

[53] Angela-Guiovana-Rincón, Cesar Pulgarin. Effect of pH, inorganic ions, organic matter and H_2O_2 on *E. coli* K12 photocatalytic inactivation by TiO_2 Implications in solar water disinfection. Appl Catal B Environ 2004;51:283–302.

[54] Heyde M, Portalier R. Acid shock proteins of *Escherichia coli*. Federation of European microbiological societies. Microbiol Lett 1990;69:19.

[55] Cho M, Chung H, Yoon J. Disinfection of water containing natural organic matter by using ozone-initiated radical reactions. Appl Environ Microbiol 2003;69(4):2284–91. http://dx.doi.org/10.1128/AEM.69.4.2284-2291.2003.

[56] Deleted in review.

CHAPTER EIGHT

Comparative Analysis of Existing Water Resources Data in the Western Balkan States of Bosnia and Herzegovina, Macedonia, Montenegro, and Serbia

J. Markovski[1,2], K.D. Hristovski[2], L.W. Olson[2]
[1]University of Belgrade, Belgrade, Serbia
[2]Arizona State University, Mesa, AZ, United States

1. INTRODUCTION

Establishing a sustainable water management infrastructure is an essential component of developing a multidimensional socioeconomic, cultural, and political system of a country [1,2]. For Bosnia and Herzegovina, Macedonia, Montenegro, and Serbia, building such systems has proven particularly challenging because of the turbulent political past. During the period of 1952–1990, these countries operated under the principles of communist ideology and socialist economic systems as part of the former Yugoslav federation [3,4]. Exiting the Second World War era as generally agriculture-dominated countries and with infrastructure not able to support rapid growth, the main economic development emphasis of this socialist federation was placed on accelerated industrialization without any concerns for protection or management of the environment. These circumstances hindered any coordinated efforts to adequately manage the environmental infrastructure, including the water management system. Any scarce data that was collected to characterize the environmental system in the Yugoslav federation was frequently considered either strategically unimportant or confidential, and generally excluded from public record.

With the fall of communism and Yugoslavia, a transition from socialist-based system to a free market–oriented economy under the auspice of democracy was initiated in the newly independent republics. Driven by an influx of new knowledge and ideas, this transition induced dramatic socioeconomic and political changes in the former Yugoslav republics, which

Chemistry and Water
ISBN: 978-0-12-809330-6
http://dx.doi.org/10.1016/B978-0-12-809330-6.00008-8

necessitated deep reforms of the environmental principles and the overall water management systems [3,4]. However, scarcity of data made it virtually impossible to analyze or describe the state of the water management systems in these republics. This problem became amplified in light of the recent global warming trends and the climate change impacts on local water resources [2,5]. Considering that the fragile economies of these countries could be significantly impacted by fluctuations of the national water resources, it becomes imperative to better understand the climate change–induced vulnerabilities of these resources [6,7]. The initial step in understanding the vulnerabilities of any water system necessitates analysis and generation of a clear picture that describes the state of the existing water resources. However, in absence of comprehensive data sets, it becomes difficult to initiate this first step. Therefore, a critical need exists to assess the pool of available information, which is required to analyze the existing water resources in these former Yugoslav countries. This study examines available data to better understand the condition of the existing water resources in four non-EU former Yugoslav republics: Bosnia and Herzegovina, Macedonia, Montenegro, and Serbia. The ultimate goal of this study is to identify the existing data gaps and inconsistencies that hinder the initiation of in-depth water resource analyses needed for developing a sustainable integrated water management system. With this aim, this study: (1) compiles and compares the available data related to water resources, quality, supply, and sanitation; (2) investigates the existing temperature and precipitation change trends over the last decades; and (3) identifies potential impacts of the climate change on the national water resources of these four former Yugoslav republics.

2. GEODEMOGRAPHIC AND SOCIOECONOMIC OVERVIEW OF BOSNIA AND HERZEGOVINA, MACEDONIA, MONTENEGRO, AND SERBIA

Considering that socioeconomic, cultural, demographic, and geopolitical determinants may have a strong influence on an environmental system, for example, a water system of a country, it becomes imperative to contextualize these factors and better understand their implications and influences.

Located in the western part of the Balkan Peninsula, Bosnia and Herzegovina, Macedonia, Montenegro, and Serbia (without Kosovo) encompass about 168,232 km^2 in southeastern Europe (Fig. 8.1) [8]. Individually Serbia

Figure 8.1 Location of former Yugoslav republics in the Balkans Peninsula.

occupies 46.07% of total area (77,498 km^2), followed by 30.44% of Bosnia and Herzegovina (51,209 km^2), 15.28% of Macedonia (25,713 km^2), and 8.21% of Montenegro (13,812 km^2). Arable land covers the northern area of Serbia as part of the Pannonian Plains and smaller parts of Bosnia and Herzegovina and southern Macedonia. Topographically, however, the rest of this region mainly consists of hills and mountainous regions with peaks frequently exceeding 2000 m. Terrain elevation declines toward the coastal area of Adriatic Sea extended in Montenegro and partly in Bosnia and Herzegovina.

The total population of these four republics is estimated to be about 13.73 million, of which slightly more than half live in urban areas [9]. The Republic of Serbia has the largest population of about 7.21 million, and it is followed by Bosnia and Herzegovina with an estimated 3.84 million, Macedonia with an estimated 2.06 million, and Montenegro with about 0.62 million. Respectively, dominant ethic groups are Serbs, Bosnians, Croats, Macedonians, and Montenegrins. Orthodox Christianity is the most widespread in Macedonia, Montenegro, and Serbia, while Islam is dominant also in communities of the Bosnia and Herzegovina Federation.

Since the beginning of the 1990s, this region has exhibited a general demographic decline induced by the civil wars, dwindling economies, and the so-called "brain drain" [10,11]. The disintegration of the Yugoslav market and introduction of sanctions to some of the republics in former Yugoslavia facilitated the collapses of their already weak economies, which were built during the 40 years of Yugoslavia. These economic regressions caused rapid decreases in the GDPs of the former Yugoslav republics, increased unemployment, and emigration of the highly skilled population to North America, Australia, and the EU [10]. To illustrate this economic decline, the nominal 2015 per capita GDP ranged from $4030 in Bosnia and Herzegovina, $4867 in Macedonia, and $5102 in Serbia to $6373 in Montenegro [12]. Although the economy is partly recovered after the 1990s, GDP is not even close to the $17,764 of nominal GDP per capita of former Yugoslavia from the 1980s [13]. This downward trend was fueled by the high unemployment rates, which were among the highest in Europe. For example, in 2014, the estimated unemployment rates were 27.6% in Macedonia, 27.5% in Bosnia and Herzegovina, 18.6% in Montenegro, and 16.8% in Serbia [14]. Furthermore, the drop in GDP during this period was mirrored by population decline trends [10]. During 1991–2010, the population of Bosnia and Herzegovina and Serbia decreased by about 14.3% and 8.3%, respectively. Based on unreliable estimates, a much smaller population decrease of 0.6% was estimated for Macedonia, while in Montenegro, an increase of 0.8% was recorded [10].

3. OVERVIEW OF AVAILABLE WATER RESOURCES IN BOSNIA AND HERZEGOVINA, MACEDONIA, MONTENEGRO, AND SERBIA

3.1 Available Annual Water Resource Potential in Bosnia and Herzegovina, Macedonia, Montenegro, and Serbia

Fig. 8.2 summarizes the estimated national water balances with respect to available water resource potential that could be utilized by Bosnia and Herzegovina, Macedonia, Montenegro, and Serbia. Depending on the origin, the available annual water resource potential in these former Yugoslav republics could be divided into: (1) in-country–generated water, which represents a difference between precipitation and evaporation; and (2) surface water influx, which is comprised of the inflow from rivers originating outside country boundaries. Considering that no available data exists to accurately describe the existing groundwater resource potentials in these republics, available water resource potentials were based on conservative

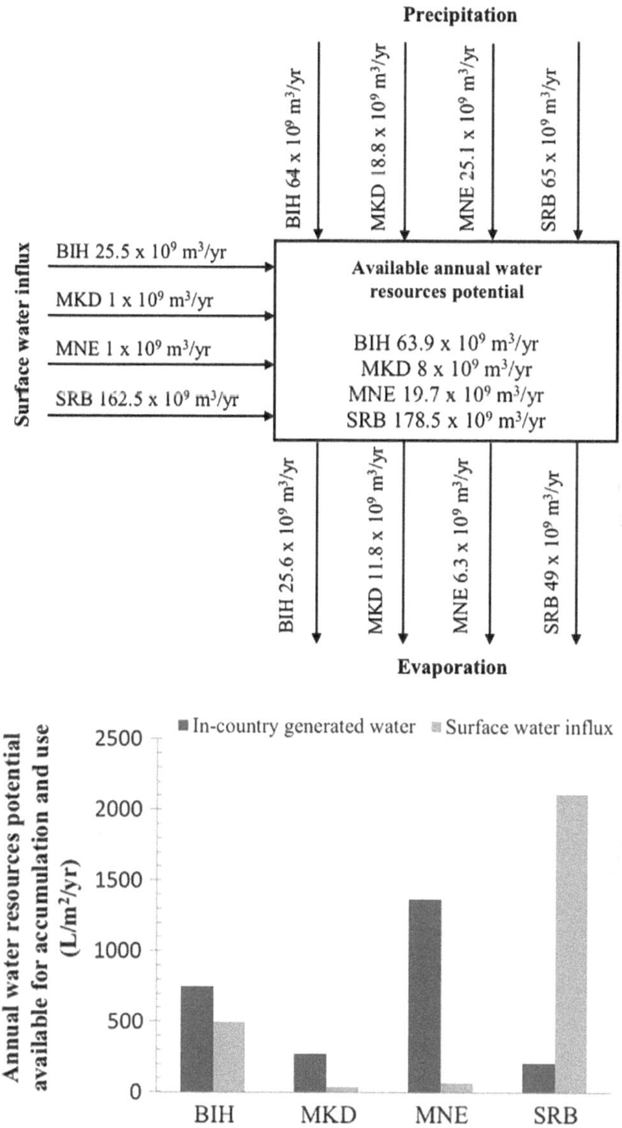

Figure 8.2 The balance and origin of national water resources potential in Bosnia and Herzegovina (BIH), Macedonia (MKD), Montenegro (MNE), and Serbia (SRB).

estimates that include only the surface water influx and the in-country–generated water.

For Bosnia and Herzegovina, the water input from precipitation is about 1250 $L/m^2/yr$ [15]. An estimated portion of about 500 $L/m^2/yr$ is lost to evaporation, which leaves about 750 L/m^2/yr to be available for accumulation and use [15]. This input and the estimated surface water influx of 25.5×10^9 m^3/yr [15], or 500 L/m^2/yr when normalized against Bosnia and Herzegovina's territory, create an available annual water resource accumulation potential of 1250 L/m^2/yr (64×10^9 m^3/yr).

The water input from precipitation in Macedonia is slightly lower. Here, about 730 L/m^2/yr enter the country via precipitation, of which an estimated 460 L/m^2/yr are lost to evaporation [16]. The remaining 270 L/m^2/yr comprise the majority of available water with only additional 40 L/m^2/yr of surface water influx (when normalized against Macedonia's territory) [16,17], generating available water resource accumulation potential estimated at about 310 L/m^2/yr (8×10^9 m^3/yr).

In Montenegro, the water input from precipitation is highest in the region, with an estimated 1820 L/m^2/yr [18]. This is not surprising, considering that European maximum rainfall is in the mountainous areas of Montenegro, where precipitation averages about 5000 L/m^2/yr and frequently reaches 7000 L/m^2/yr during rainy years [19]. An estimated portion of about 460 L/m^2/yr of the precipitated water returns to the atmosphere via evaporation, leaving about 1360 L/m^2/yr of water available for accumulation and use. With about 10^9 m^3/yr (or normalized 65 L/m^2/yr), the surface water influx [18] contributes only a small portion to the available annual water resources potential of Montenegro, which is estimated at about 1430 L/m^2/yr (20×10^9 m^3/yr).

Of the estimated 840 L/m^2/yr, which enters Serbia as precipitation, about 640 L/m^2/yr returns into the atmosphere via evaporation, while around 210 L/m^2/yr are available for accumulation and use [20]. The surface water influx by several large rivers is an order of magnitude greater than the available water from precipitation. It is estimated that about 162.5×10^9 m^3/yr enters Serbia, or 2110 L/m^2/yr when normalized against Serbian territory [17,20]. With available annual water resource accumulation potential estimated to 2320 L/m^2/yr (178.5×10^9 m^3/yr), Serbia is first among four federal Yugoslav republics, primarily due to high surface water influx as a result of its favorable geographical location.

Absence of reliable data related to surface water outflow, groundwater recharge, and existing groundwater estimates seriously impedes any efforts to develop complete water balances for all of these former Yugoslav republics. Significant data gaps exist that prevent complete quantification of water losses to evaporation upon withdrawal and use. Existing reports infer that an estimated 98% of water from Serbia and a majority of water from Bosnia and

Herzegovina are discharged into the Black Sea via the Danube watershed [15,20]. Montenegro surface waters are discharged mainly into the Adriatic Sea [18], while Macedonian water mainly enter the Aegean Sea via the Vardar watershed [21]. Furthermore, inconsistencies and gaps exist in the reported values. For example, the presented water resource potential balances in Fig. 8.2 were estimated of reports that do not always indicate how or when (e.g., the exact year) the data was generated. Although they address data collected during same decade, data variability is not often presented, which makes it difficult to assess any fluctuations in available water resources in these countries, which is imperative for assessing vulnerability of these resources in context of climate changes.

3.2 Water Withdrawals for Potable and Nonpotable Supply in Bosnia and Herzegovina, Macedonia, Montenegro, and Serbia

To support demands for potable and nonpotable uses, water is withdrawn from both groundwater and surface water accumulations. According to the currently available data, estimated water withdrawals are estimated at 4900 m^3/capita/yr in Bosnia and Herzegovina [15], 680 m^3/capita/yr in Macedonia [22], 5580 m^3/capita/yr in Montenegro for 2014 [23], and 580 m^3/capita/yr in Serbia [24] for different years during the period of 2010–2014. Although these estimates could be considered representative for the given period, it is important to emphasize that they are reported for different years in this period because of data scarcity.

As illustrated in Fig. 8.3, water withdrawn to support potable water needs are about 90 m^3/capita/yr in Bosnia and Herzegovina [15], 110 m^3/capita/yr in Macedonia [22], 180 m^3/capita/yr in Montenegro [23], and 90 m^3/capita/yr in Serbia [24]. Groundwater serves as the primary source for potable water supplies, while the rest of the supply is provided from surface water accumulations. Groundwater supplies about 80 m^3/capita/yr for potable uses in Bosnia and Herzegovina [15], 80 m^3/capita/yr in Macedonia [25], 175 m^3/capita/yr in Montenegro [23], and 60 m^3/capita/yr in Serbia [24].

When these estimates are compared with data reported by the European Environment Agency (EEA) for the 2000–2011 period, as illustrated in Fig. 8.4 [17], Serbia, Bosnia and Herzegovina, and Macedonia could be characterized as countries with medium water withdrawal rates for public supply. The withdrawal rates are similar to Turkey, Luxemburg, or Slovenia, which is another former Yugoslav republic but a full member of the EU.

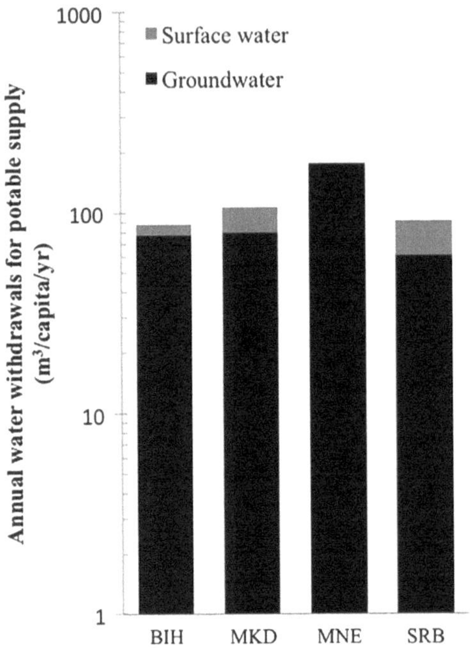

Figure 8.3 Annual water withdrawals for potable supply in Bosnia and Herzegovina (BIH), Macedonia (MKD), Montenegro (MNE), and Serbia (SRB).

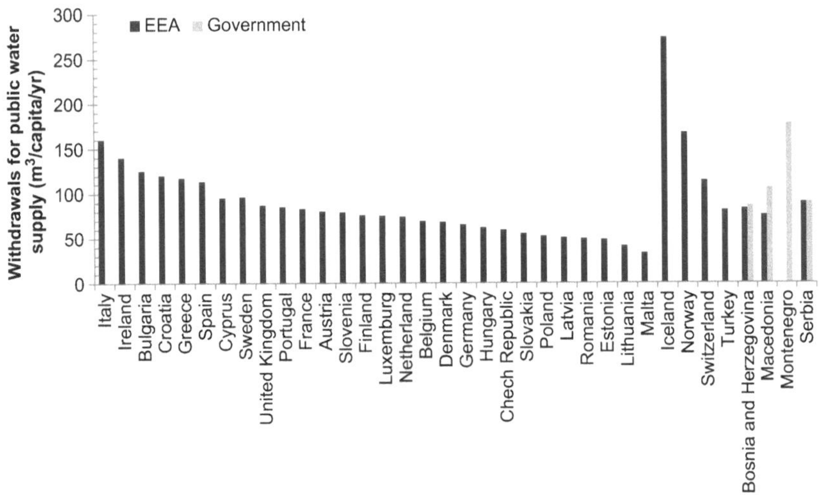

Figure 8.4 Annual water withdrawals for potable supply according to the European Environment Agency (EEA) for the period 2000–2011 and the latest national government reports.

The previously given estimates are in agreement with estimates provided by Serbia and Bosnia and Herzegovina to the EEA [15,24]. Interestingly, however, the previous estimated values were about 30% higher than the values reported by EEA for Macedonia, and they ranked Montenegro second highest in Europe for water withdrawals rates, right after Iceland [22,23].

When nonpotable water withdrawals are examined, as illustrated in Fig. 8.5, it could be noted that they are about 5–25 times greater than the ones reported for potable water. Montenegro and Bosnia and Herzegovina have the highest nonpotable water withdrawals with an estimated 4810 m^3/capita/yr and 5400 m^3/capita/yr [15,23], respectively. Surprisingly, Macedonia and Serbia exhibit an order of magnitude lower withdrawal rates for nonpotable water uses than Montenegro and Bosnia, with an estimated 570 m^3/capita/yr for Macedonia and 485 m^3/capita/yr for Serbia [22,24]. This divergence between Montenegro/Bosnia and Herzegovina and Macedonia/Serbia nonpotable water withdrawals could possibly be attributed to the more mountainous terrain of the first two republics and greater surface water accumulation potential then the latter two republics. Specifically, almost all nonpotable water in Montenegro and Bosnia and

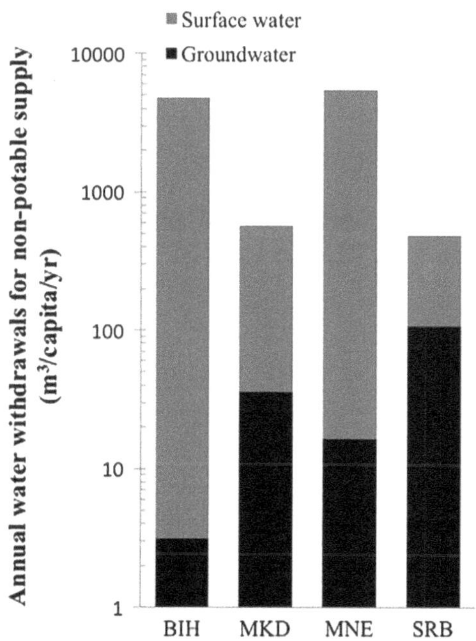

Figure 8.5 Annual water withdrawals for nonpotable supply in Bosnia and Herzegovina (BIH), Macedonia (MKD), Montenegro (MNE), and Serbia (SRB).

Herzegovina is acquired from surface water accumulations, with annual withdrawals estimated to 4800 m^3/capita/yr in Bosnia and Herzegovina [15] and 5380 m^3/capita/yr in Montenegro [23]. Macedonia withdrawals are about 380 m^3/capita/yr from surface water accumulations for nonpotable water uses [26], while the estimated value for this type of water withdrawal from surface water accumulations is about 480 m^3/capita/yr for Serbia [24]. Over 95% of nonpotable water in Bosnia and Herzegovina [15], Montenegro [23], and Serbia [24] is withdrawn to support electric power generation. Interestingly, however, one could not help but question the reliability of Montenegrin data when the order-of-magnitude higher water withdrawal values are reported for energy generation than Serbia, for example, yet most of the electricity is imported in Montenegro [18]. Although more detailed classifications about the categories of nonpotable users are provided for Bosnia and Herzegovina [27], Montenegro [23] and Serbia [24], there is no uniform or standardized classification that is employed. For Macedonia, however, the situation is even worse. In the Macedonian reports, available information does not provide a clear description about the different categories of nonpotable water withdrawals. Nonetheless, it can be indicated that the majority of the nonpotable water withdrawals are for energy generation. Industry and agriculture could be considered a small nonpotable water consumer when compared to the energy generation sector, probably because of the industrial decline since 1990s and the dependence on agriculture precipitation as a balanced source of water.

3.3 Water Resources Potential Utilization Rates in Bosnia and Herzegovina, Macedonia, Montenegro, and Serbia

Absence of data to quantify the actual hydrological storage capacity in aquifers and surface accumulations significantly increases uncertainties related to water resource vulnerabilities in the four former Yugoslav republics. Published reports suggest that the existing capacities of surface water accumulations, with ability to cover about 30% of the minimum water demand, are insufficient to strategically address water needs of the public and mitigate water resource vulnerabilities [28–31]. To better understand the vulnerability and risks that stem from potential water resources depletion, water resources potential utilization rates are estimated as a ratio of the annual water withdrawals and available annual water resource potentials.

As it illustrated in Fig. 8.6, Bosnia and Herzegovina utilizes about 29% of the available annual water resource potential [15], while Macedonia and

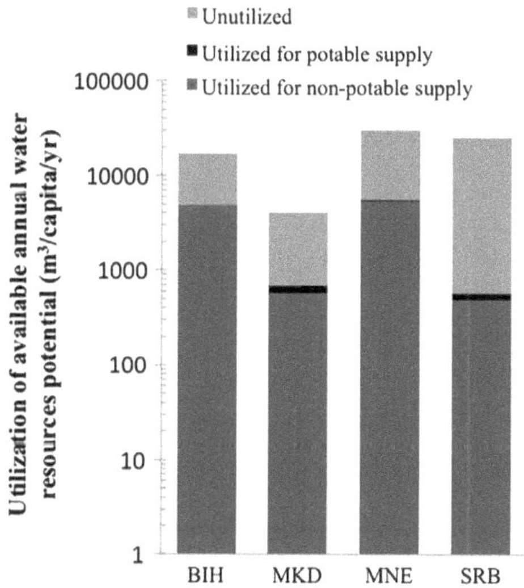

Figure 8.6 Annual utilization of water resources potential in Bosnia and Herzegovina (BIH), Macedonia (MKD), Montenegro (MNE), and Serbia (SRB).

Montenegro utilize about 17% and 19%, respectively [22,23]. In all of these countries, only a small fraction of water is utilized for potable supply. This is particularly emphasized in Bosnia and Herzegovina and Montenegro and consequently reflected through an almost invisible partition, as presented in Fig. 8.6.

The water resources potential utilization rates for Serbia are estimated to be about 3% [24]; however, this small relative value is a direct consequence of the high the surface water influx from rivers originating outside Serbia, such as the Danube River. Because of the high surface water influx from transient rivers, Serbia exhibits the greatest unutilized water resources potential with 24,420 m^3/capita/yr, which is estimated as a difference between available annual water resources potentials and withdrawn water. The same potential is estimated for Montenegro too, while the unutilized water resource potential for Bosnia and Herzegovina is estimated to about 12,100 m^3/capita/yr. With 3320 m^3/capita/yr, this value was significantly lower for Macedonia, which suggests that Macedonia may be the most vulnerable republic.

3.4 Surface Water Quality in Bosnia and Herzegovina, Macedonia, Montenegro, and Serbia

Seasonal variations and geographical features of the terrain have significant impacts on quality of surface waters in former Yugoslav republics. However,

Figure 8.7 Waste dumping in the village stream in western Serbia.

it could be stated as a general trend that isolated upstream segments of mountainous rivers are characterized by excellent water quality parameters, which rapidly deteriorate as the rivers reach populated areas [15,32]. This trend becomes even more pronounced during summer periods when the majority of smaller rivers become dominated by raw sewage discharges and solid waste dumping, as illustrated in Fig. 8.7 [33,34].

In Bosnia and Herzegovina during the 2000–2009 period, typically reported 5-day biochemical oxygen demand (BOD_5) and nitrate values exceeded 10 mg/L O_2 and 5 mg/L $N\text{-}NO_3^-$, respectively [15]. Such high values suggest seriously polluted waters, especially in the downstream river segments. The water quality conditions of many Montenegrin rivers are not much different during the period 2008–2014, with BOD_5 and phosphates frequently reaching 10 mg/L O_2 and 0.85 mg/L $P\text{-}PO_4^{3}$, respectively [32]. The conditions of Macedonian surface waters are probably the worst among the four different republics; however, in the absence of available or reliable data, it could only be stated that the major rivers in Macedonia are categorized as severely polluted, while certain rivers sections are extremely polluted, with values frequently exceeding 15 mg/L (O_2) as BOD_5 [25,26].

A better information pool is available for the quality of Serbian surface waters. However, the reliability and quality of the reported data could be frequently rendered questionable because the published reports do not provide any description of the methodology that was employed in generation of these data sets. Serbian reports provide a water quality index (WQI), which was established by the Serbian Ministry of Environment [35,36]. According to the mean value of this index, an estimated 92% of surface waters exhibited good, very good, or excellent quality during the 2004–2013 period [37], and could potentially be used as sources for potable water [35]. Interestingly, however, if the minimum values of the reported water quality parameters were used to assess the WQI, then about 76% of the waters in Serbia could be classified as exhibiting poor or very poor quality [37]. Detailed examination of the discrete water quality analysis data further confounds the already questionable information in these reports. For example, the water quality data for measurements conducted in August of 2013 on the Velika Morava River shows unreasonable discrepancy with chemical oxygen demand (COD) of 5.4 mg/L when the permanganate method was used and COD of 16 mg/L when the dichromate method was used [38]. To make things worse, the reported total organic carbon (TOC) value for the same sample was 6.3 mg/L, which is unrealistically low to support the unusual and discrepant COD values [38]. Data discrepancies like this example suggest that water quality data collection may not be properly conducted because of absence of adequate and standardized methodologies or trained personnel. To further confound the already questionable methodologies and data sets, it should be noted that the data analysis is often conducted in the context of old regulations, which are an obsolete remnant of the old communist federation.

3.5 Groundwater Quality in Bosnia and Herzegovina, Macedonia, Montenegro, and Serbia

The situation with reported groundwater quality data could be considered worse because groundwater monitoring is frequently overlooked in the absence of national groundwater protection strategies in the four former Yugoslav republics. Macedonia and Bosnia and Herzegovina are probably in the worst position because literature review does not reveal any available groundwater quality data. Limited data is available for Montenegro and Serbia; however, the reported data either create suspicion about the validity of the employed sample collection, analysis methodology, and reported values, or raise serious concerns about the level of groundwater contamination in many regions in these two republics. For example, reported data for groundwater samples collected in the Zeta plains of Montenegro during the period of 2008–2014 consistently exceeded values of 2 mg/L P-PO_4^{3-}, which is surprisingly high [32]. In 2013, samples collected from the aquifers used to supply potable water to the municipalities of Sombor and Kikinda in northern Serbia exceeded TOC values of 16 mg/L and 360 mg/L As [38].

4. OVERVIEW OF WATER SUPPLY AND SANITATION INFRASTRUCTURE OF BOSNIA AND HERZEGOVINA, MACEDONIA, MONTENEGRO, AND SERBIA

4.1 Potable Water Supply, Treatment and Distribution Systems in Bosnia and Herzegovina, Macedonia, Montenegro, and Serbia

With an estimated 82% of the population serviced by public water supply systems in 2013, Serbia is considered a country with a relatively well-developed water distribution system [39]. Based on data from 2000 to 2010, about 70% of the population has access to public water supply in the Macedonia and Montenegro [18,25,40]. Probably because of the 1990s civil war, this number is lower for Bosnia and Herzegovina, which is ranked the worst among the four former Yugoslav republics, with 58% of the population having access to public water supply [15]. In general, it could be stated that the majority of the population that is serviced by public water supply systems in these republics lives in larger urban areas, while the population in the rural areas has limited access to water through public systems [4,15,18].

The poor and deteriorated water infrastructure, which was mainly constructed during the industrialization period of former Yugoslavia and is generally older than 20 years [25], is the primary cause for astounding potable water losses that occur in the water distribution systems. For example, during the 2003–2010 period, the estimated water withdrawals for potable uses in Bosnia and Herzegovina were 87 m^3/capita/yr. Less than half, or 41 m^3/capita/yr, of the withdrawn water actually reached the consumers; the remaining water, or about 53%, was lost during distribution (Fig. 8.8) [15,41,42]. According to the only available information originating from 1994 to 1997, about 47 m^3/capita/yr from the estimated 107 m^3/capita/yr of water for potable use was delivered to the consumers in Macedonia, which represents loss of about 65–70% during distribution [25]. Similarly, during 2014, the public water supply systems in Montenegro delivered estimated 73 m^3/capita/yr of potable water at losses of 59% in distribution system [23]. The situation could be considered to be better in Serbia, where the 2013 estimates suggest only 31% of losses of potable water to distribution, with 63 m^3/capita/yr reaching the Serbian consumers [39]. In contrast, the leading EU countries, like as Germany, Denmark, and Sweden, report significantly lower losses of potable water to transport, which are 3%, 10%, and 17%, respectively [43]. Considering that investments in

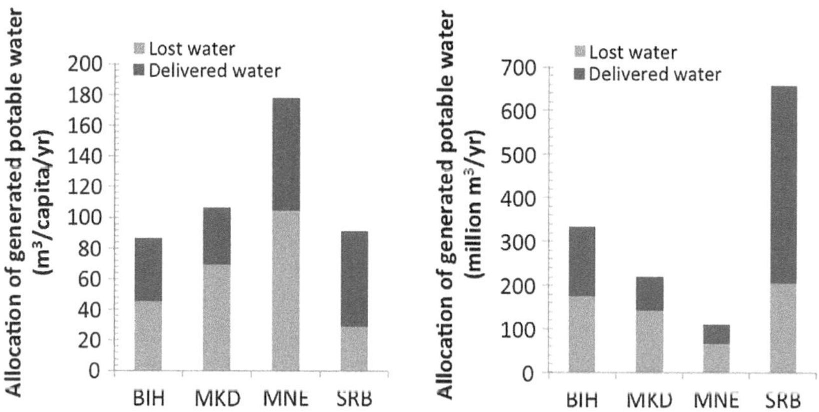

Figure 8.8 Delivery and losses of water generated for public supply in Bosnia and Herzegovina (BIH), Macedonia (MKD), Montenegro (MNE), and Serbia (SRB).

distribution infrastructure have been minimal over the last decade, the situation to distribution water losses is only expected to worsen. High potable water losses could have staggering economic impacts on a small country, and yet these impacts are often neglected and absorbed by the economy. To illustrate these adverse economic effects stemming from potable water losses, a simple back-of-the-envelope calculation could be conducted assuming a realistic water price of $0.3 for each m^3 of potable water [44]. Under the estimated water loss scenarios for the four former Yugoslav countries, Montenegro loses about $19 million/yr in Montenegro and Macedonia loses $43 million/yr, while Bosnia and Herzegovina and Serbia lose $53 million/yr and $62 million/yr, respectively. These large quantities could be easily infused in the local economies or allocated to improve the existing potable water treatment and distribution infrastructure.

The water treatment systems in the four former Yugoslav republics are not cataloged, providing for the very limited pool of reported information that could be used to describe the current situation with water treatment. Generally, however, larger cities or towns may employ trains with standard water treatment technologies like coagulation/flocculation or filtration, which may be old, in disrepair, inefficient, or obsolete, to treat water from polluted water sources or remove emerging contaminants [4]. For example, absence or inadequate water treatment technologies that could remove contaminants like arsenic have left the residents of some parts of northern Serbia without potable water for more than 10 years [45]. The situation may be even direr in smaller communities, as these may lack water treatment beyond chlorination [45–47]. However, absence of trained personnel, monitoring, and standardized chlorination protocols have the potential to create more harm than help because of carcinogens generation of waters with high TOC content.

The insufficient or inadequate water treatment often triggers potable water use restrictions due to health concerns that are issued by the National Institutes of Public Health, which are responsible for ensuring access to safe potable water. According to reports for the 2010–2013 period, more than 75% of collected and analyzed potable water samples from public supply systems in Bosnia and Herzegovina met microbiological and physicochemical water quality standards [42]. Interestingly, however, when surface water samples were excluded from this pool, over 45% of the samples were in violation of the water quality standards, suggesting a decline in groundwater quality during the reported period [42]. In contrast, Macedonia reported that over 90% of the collected potable water samples were in compliance with regulations during the 2010–2014 period, but without providing information about the sample collection and analysis methodology, which leaves room for suspicion and data validity concerns [26]. The average fraction of potable water quality samples reported safe by the Montenegrin public health institutions was 85% in 2014, although the reports show that only 40% of the samples could be rendered safe in certain areas of the country [32]. In Serbia, however, only 59% of the analyzed potable water samples complied with the potable water quality regulations during the first half of 2013 [47].

Similar to the problem with absence of a surface and groundwater monitoring platforms, it could only be postulated that the number of out-of-compliance potable water samples is significantly greater, which further questions the reliability of presented reports, since the sampling and analysis frequency is not standardized [45–47]. Furthermore, existing regulations that mandate water quality are antiquated or remnants from former Yugoslavia and do not provide clear protocols for sampling, transport, and analysis of potable water samples. Consequently, the treatment and water quality monitoring practices could not be considered reliable, which raises the uncertainty levels about the vulnerability of the existing water resources in these former Yugoslav republics.

4.2 Wastewater Collection and Treatment Plants in Bosnia and Herzegovina, Macedonia, Montenegro, and Serbia

Data related to wastewater (i.e., sewage) discharges and treatment from all four former Yugoslav republics not only is scarce, questionable, or obsolete, but also is based on ambiguous estimates. It could be safely assumed, however, that wastewater discharges represent a portion of the potable water withdrawals, and an even smaller fraction of this wastewater is being treated using primary or secondary wastewater treatment before discharging into the surface waters of these republics.

Similar to the disposition of supply and distribution systems, the sewage infrastructure in the four former Yugoslav republics is generally concentrated in urban areas [15,18,39,40]. As reported for the last 15 years, it was estimated that only 33% of the population of Bosnia and Herzegovina was discharging their wastewaters into public sewers [15]. Data from the 2002 census in Macedonia reports that only about 47% of total population, although about 60% of dwellings are connected to sewage collection systems, [26,40]. An even smaller portion of the populous in Montenegro, or 37%, is served by public sewage collection systems [18]. The situation with sewage collection is the most favorable in Serbia, where about 58% of people are connected to some kind of a sewage collection system, according to data from 2013 [39]. All of these values are significantly smaller in comparison to European cities, where about 95% of the population has access to sewage collection [17]. To further exacerbate things, the a large fraction of the sewage collection systems in these republics represent open canal systems or combined storm water/sewage collection systems. Only in the newer and urbanized communities, wastewater collection systems are being constructed to independently collect sewage and storm water. Considering that the new sewer systems are constructed to be generally small and could be considered negligible, the presented estimates could be considered unchanged if they are depicting the actual conditions in the four former Yugoslav republic.

In rural areas, sewage is mainly collected in septic tanks. It is estimated that generation of septic tank discharged wastewater is 15 m^3/capita/yr (58×10^6 m^3/yr) in Bosnia and Herzegovina [41], 12 m^3/capita/yr (24×10^6 m^3/yr) in Macedonia [40], 20 m^3/capita/yr (12×10^6 m^3/yr) in Montenegro [23] and in Serbia, 18 m^3/capita/yr (126×10^6 m^3/yr) [39,48]. Absence of septic tank construction standards and regulations creates a potential environmental and public health hazard because the majority of these septic tanks are simple open-bottom septic tanks that have the ability to contaminate local groundwater, which is frequently used as a source of potable water and threatens public health due to groundwater pollution.

Only a small fraction of the collected wastewaters is treated. As illustrated in Fig. 8.9, this fraction ranges from about 4% ($\sim$1 m^3/capita/yr) for Bosnia

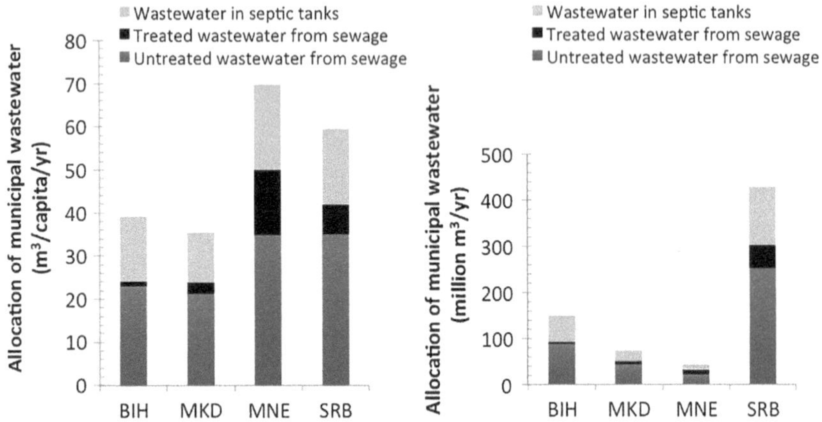

Figure 8.9 Allocation of municipal wastewater in Bosnia and Herzegovina (BIH), Macedonia (MKD), Montenegro (MNE), and Serbia (SRB).

and Herzegovina [41], 11% (~2.5 m^3/capita/yr) for Macedonia [25], and 16% for Serbia (~7 m^3/capita/yr) [39,48] to about 31% (15 m^3/capita/yr) for Montenegro [23]. The majority of the sewage is discharged directly into the rivers and often dominates the flows. About 23 m^3/capita/yr of sewage are directly discharged in Bosnia and Herzegovina's rivers. This value is estimated to about 21 m^3/capita/yr for Macedonia and 35 m^3/capita/yr for both Montenegro and Serbia. These normalized per capita values equate to 88×10^6 m^3/yr for Bosnia and Herzegovina, 44×10^6 m^3/yr for Macedonia, 22×10^6 m^3/yr for Montenegro, and 253×10^6 m^3 for Serbia. For Serbia, such results are not surprising, considering that although the three largest cities in Serbia, Belgrade (central Serbia), Novi Sad (northern Serbia), and Nis (southern Serbia), have well-developed water supply and sewage collection systems, they directly discharge their untreated municipal wastewaters into the rivers flowing through their municipalities [4,39].

When compared to other European countries (Fig. 8.10), Serbia is at the bottom of the list of countries treating their sewage, with only 10% of the collected sewage receiving adequate treatment before discharge, according to data from 2013 [48]. This percentage is even lower in Bosnia and Herzegovina, which is also at the bottom of this list. According to EEA and government reports from 2014, only 2% of the wastewaters are treated in Bosnia and Herzegovina before discharging them into the environment [41]. Probably due to lack of information, secondary treatment of wastewater in the Macedonia and Montenegro is not presented by EEA. Therefore, estimates

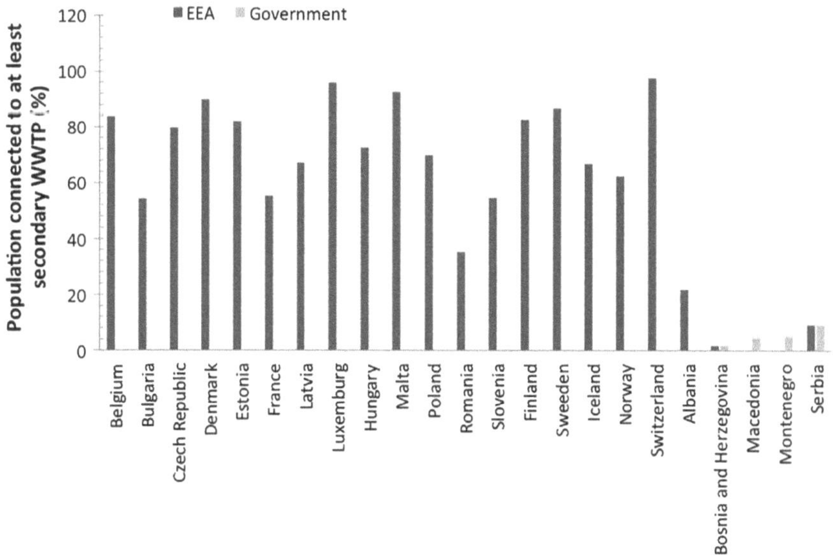

Figure 8.10 Percent of population connected to at least secondary wastewater treatment according to the European Environment Agency (EEA) for the period 2000–2011 and latest national government reports.

of population connected to at least secondary wastewater treatment plants for Macedonia and Montenegro are based on sporadic and obsolete government data as the only available information source [18,23,25]. As illustrated in Fig. 8.10, such an assessment placed Macedonia and Montenegro between Serbia and Bosnia and Herzegovina, but also at the bottom of this list.

This multifaceted problem is far more complicated than is frequently reported in government or international reports. Attempts to address the wastewater collection and treatment by simply constructing sewage collection and treatment systems have yielded much success because of a number of indirect factors that determine long-term functioning of sanitation systems. Absence of proper regulatory foundation, adequate monitoring and compliance platforms, trained personnel, and funding are typically not considered as determinants of success. However, local politicians and government personnel lack knowledge to engage this problem in a systematic manner. Their quick Band-Aid solutions frequently further exacerbate the problem and offer no strategies to mitigate the risks stemming from the climate change.

5. CLIMATE CHARACTERISTICS OF BOSNIA AND HERZEGOVINA, MACEDONIA, MONTENEGRO, AND SERBIA

5.1 General Temperature and Climate Characteristics

Situated in the western/central part of the Balkans peninsula, Bosnia and Herzegovina, Macedonia, Montenegro, and Serbia have four seasons with gradual changes between them. Generally, Serbia and Macedonia have similar climate, characterized as continental and with average annual temperatures typically between 11 and 12 °C [49,50]. A Mediterranean climate prevails in coastal areas of Bosnia and Herzegovina and Montenegro, and transitions into continental climate by moving further from the sea [51,52]. Local climate variations are prevalent in the coastal area of Bosnia and Herzegovina and Montenegro because of differences in elevation, proximity to the Adriatic Sea, and exposure to the winds. The mean annual air temperature in the coastal regions of Bosnia and Herzegovina ranges between 12 and 17°C, and decreases in lowland areas to between 10 and 12°C [51]. In Montenegro, the mean annual air temperature ranges between 4.6°C in the mountainous areas and 15.8°C in the coastal regions [52].

5.2 Temperature and Precipitation Trends in Bosnia and Herzegovina, Macedonia, Montenegro, and Serbia

Many of the available reports that have information about temperature and precipitation in the four former Yugoslav republics provide average temperatures and precipitation, but do not provide information about the existing trends, methodology of data collection and analysis, or actual data. With pressure from the United Nations under the United Nations Framework Convention on Climate Change (UNFCCC)'s National Communication reporting requirements, these four republics were also mandated to provide information about climate change trends [49,50,52,53].

Existing reports for Bosnia and Herzegovina provide a temperature and precipitation comparison between data collected during 1981–2010 and mean climate temperature obtained for the period of 1961–1900 [53]. The reported information shows a temperature increase of up to 1.2°C in some parts of Bosnia and Herzegovina, while significant changes in average precipitation are not observed. Interestingly, however, it was noted that the number of days exhibiting precipitation greater than 1.0 L/m^2/day has declined, while the number of intense precipitation days has increased.

Absence of reported actual data in these reports obscures better understanding of any long-term trends that would help to better assess the climate-induced risks and vulnerabilities to water resources.

Similar temperature and precipitation trends have been also reported for Montenegro. According to government reports, the 2001–2010 decade has been the recorded as the warmest since the beginning temperature tracking in 1961 [52]. With respect to the mean climate temperature obtained for period 1961–1990, the average temperature has increased by 1.4°C in the mountainous region of Montenegro and by 1.3°C in the coastal area. Interestingly, it is also reported that the average precipitation in certain parts of Montenegro has increased by 155 L/m^2, but this increase is primarily because of increase in precipitation intensity rather than overall balanced distribution.

Compared to the mean climate temperature for the period 1961–1990, the average temperature in Serbia for the 1971–2000 period has increased by 0.7°C [50]. The mean annual precipitation in Serbia was lower for the 1971–2000 period than for the 1961–1990 period, with several exceptions in western and northern Serbia, where an increase in precipitation was reported. The Serbian government, however, provides annual reports that contain actual discrete data, but this data is not available in a format that allows easy statistical or trend analyses [54].

Probably the worst situation with access to temperature and precipitation data is in Macedonia. In this republic, changes in annual temperature and precipitation are reported just for two monitoring stations throughout the country, although a number of measuring stations exist. Reports often contain data suggesting temperature and precipitation extremes, but systematic tracking and analyses of trends is not conducted or reported [49]. Recent peer-reviewed publications [33,34], however, report a steady temperature increase in Macedonia, changes in rainfall intensity, and decrease in surface water flows during a 50-year period of meteorological and hydrological data obtained from the Hydrometeorological Service's archives, which are not publically accessible.

Absence, incompleteness, or inaccessibility of empirical meteorological data limit any in-depth comparison of the climate change trends among the four former Yugoslav republics. Spatial and temporal analyses are essential to better understand the risks and vulnerabilities stemming from the climate-induced impacts on regional water resources. Consequently, appropriate mitigative measures on a local level could not be developed to reflect the national specifics.

6. IMPLICATIONS OF CLIMATE CHANGES ON NATIONAL WATER RESOURCES AND SOCIOECONOMIC SITUATIONS OF BOSNIA AND HERZEGOVINA, MACEDONIA, MONTENEGRO, AND SERBIA

The estimated increases in temperature and changes in the annual distribution of rainfall represent some of the key factors that impact occurrence of more frequent and more intense droughts and floods, and consequently affect climate change–induced risks and vulnerabilities. The national reports to UNFCCC frequently address vulnerability and its assessment, climate change impacts, and adaptation measures. However, these assessments and measures primarily rely on model predictions for the Balkans region and do not realistically depict the actual situation on local levels. High-resolution data that can describe the water resource and climate change trends on a local level is needed. Recent studies by Hristovski and collaborators reported the adverse influence of climate changes on quantity of surface water resources in Macedonia [33,34]. Obtained results indicated that the rise of average temperatures was mirrored by a statistically significant decrease of national average surface water resources.

Considering interconnectedness of surface and groundwater hydrology, it can be predicted that climate change will indirectly affect groundwater resources. Due to a shortage of information on aquifer capacity, the sustainability of the groundwater reserves could not be estimated. If anticipated grim scenarios come to fruition, potable water resources will be severely reduced and water restrictions will have to be implemented. The lack of sewage collection and treatment infrastructure would further exacerbate the problem by increasing the potable water treatment requirements and increasing public health expenditures [34]. Intense fluctuations in precipitation intensity, which are already causing problems in these republics, would further add to the cost of the local economies. Fig. 8.11 illustrates flood damages from 2014 that devastated the fragile economy of Bosnia and Herzegovina for $2 billion and Serbia for $1.7 billion [55,56]. Only a year later, in 2015, floods in Macedonia caused contamination of potable water sources and led to a tularemia outbreak [57]. Climate change implications are expected to also adversely affect the energy and agriculture sectors, which represent the pillars of the local economies [58]. The vulnerability of uncertainties in these sectors alone have the potential to undermine the already fragile socioeconomic structure of these four former Yugoslav republics. Looking toward the future, with an increase in projected extreme weather

Figure 8.11 Flood destruction in municipality Krupanj in western Serbia in 2014.

events, building national resilience platforms in each of the four former Yugoslav republics represents an urgent priority, but also a mission impossible due to absence of systematic climate and water resources monitoring approaches.

CONCLUSIONS

Comparative analysis of Western Balkan States indicated that: (1) scarcity of data, (2) questionable and obsolete estimates presented without methodology of data collection, and (3) lack of trends analysis hinder estimation of existing and future state of water resources in Bosnia and Herzegovina, Macedonia, Montenegro, and Serbia. Recently occurred floods, however, clearly demonstrated that current water management systems present in these four republics cannot resist predicted increases of temperatures and fluctuations in the magnitude of precipitation. To further exacerbate things, an anticipated grim climate scenario is going to fuel depletion and deterioration of water resources that will severely burden already fragile socioeconomic situations and increase public health risks in these countries. Similar issues defined in this comparative analysis pointed to common socioeconomic, cultural, demographic, and geopolitical pasts of former Yugoslav republics that adversely affected the existing state of water resources. For building a national sustainable water management system that includes resilience and adaptive responses to inevitable climate change, it is imperative to profoundly transform rooted communist ideology by developing an integrated water management strategy based on actual national data and analysis.

ACKNOWLEDGMENTS

The study was supported in part with funds from Ministry of Education, Science and Technological developments of the Republic of Serbia (Project No. III43009).

REFERENCES

[1] Pahl-Wostl C. Transitions towards adaptive management of water facing climate and global change. Water Resour Manag 2007;21:49–62.
[2] Ludwig R, Roson R, Zografos C, Kallis G. Towards an inter-disciplinary research agenda on climate change, water and security in Southern Europe and neighboring countries. Environ Sci Policy 2011;14:794–803.
[3] Hristovski KD, Hild NR, Young-Hristovski JE. The environmental management system in the Republic of Macedonia—problems, issues and challenges. Communist Post Communist Stud 2010;43:115–24.
[4] Markovski J, Hristovski KD, Rajaković-Ognjanović VN, Marinković AD. Building a sustainable water management system in the Republic of Serbia: challenges and issues. In: Ahuja S, de Andrade JB, Dionysiou D, Hristovski KD, Loganathan BG, editors. Water challenges and solutions on a global scale, ACS symposium Series. Washington, DC: American Chemical Society; 2015. p. 257–83.
[5] Lloyd-Hugnes B, Saunders MA. A drought climatology for Europe. Int J Climatol 2002;22:1571–92.
[6] Olmstead SM. Climate change adaptation and water resource management: a review of the literature. Energy Econ 2004;46:500–9.
[7] Martins G, Brito AG, Nogueira R, Ureña M, Fernández D, Luque FJ, et al. Water resources management in southern Europe: Clues for a research and innovation based regional hypercluster. J Environ Manag 2013;119:76–84.
[8] Encyclopedia Britanica, Balkans, Available from: http://www.britannica.com/place/Balkans.
[9] Eurostat, Population, Available from: http://ec.europa.eu/eurostat/web/population-demography-migration-projections/population-data.
[10] Lukic T. Depopulation in the western Balkan countries. Eur J Geogr 2012;3:6–23.
[11] Pašalić S. The Western Balkans and European integration in the context of modern demographic processes. In: Symposium on science and eurointegration, Sarajevo; May 22–24, 2015. Available from, http://www.porodicnikrug.com/wp-content/uploads/2015/05/Stevo-demograf-Balkana.compressed.pdf.
[12] International Monetary Fund. World economic outlook database. October 2015. Available from: https://www.imf.org/external/pubs/ft/weo/2015/02/weodata/index.aspx.
[13] T. Mills Kelly, GDP in Eastern Europe: 1980–1989, Making the History of 1989, Available from: https://chnm.gmu.edu/1989/items/show/667.
[14] The Liberation, Western Balkans: In Bosnia and Herzegovina and Macedonia highest number of unemployed, Available from: http://www.oslobodjenje.ba/ekonomija/zapadni-balkan-u-bih-i-makedoniji-najvise-nezaposlenih.
[15] Ministry of Foreign Trade and Economic Relations of Bosnia and Herzegovina, The report on the state of environment in Bosnia and Herzegovina 2012, Available from: http://www.fmoit.gov.ba/userfiles/Izvje%C5%A1%C4%87e%20o%20stanju%20okoli%C5%A1a%20u%20BiH.pdf.
[16] Popovska C. Climate change and vulnerability assessment of the water resources. Water Manag 2004;36:219–28.
[17] Eurostat Statistics Explained, Water statistics, Available from: http://ec.europa.eu/eurostat/statistics-explained/index.php/Water_statistics.

[18] Tomovic S. Objectives of water resources engineering in Montenegro. Water Manag 2008;40:231–3.
[19] Institute of Hydrometeorology and Seismology of Montenegro, Climatology, Available from: http://www.hmz.gov.me/misc.php?text=27&sektor=1.
[20] Serbian Environmental Protection Agency. Environment in Serbia: an indicator–based review. 2007. Available from: http://www.sepa.gov.rs/download/Environment_in_Serbia_Full.pdf.
[21] Milovanovic M. Water quality assessment and determination of pollution sources along the Axios/Vardar river, Southeastern Europe. Desalination 2007;213:159–73.
[22] Tančev L. Dams and reservoirs in Macedonia. In: Bengtsson L, Herschy RW, Fairbridge RW, editors. Encyclopedia of lakes and reservoirs geography, geology, hydrology and paleolimnology. US: Springer; 2012. p. 186–91.
[23] Statistical Office of Montenegro. Statistical Yearbook of Montenegro. 2015. Availiable from: http://www.monstat.org/userfiles/file/publikacije/godisnjak%202015/2_klima.pdf.
[24] Statistical Office of Republic of Serbia. Statistical yearbook of the Republic of Serbia. 2014. Available from: http://pod2.stat.gov.rs/ObjavljenePublikacije/G2014/pdf/G20142013.pdf.
[25] Aracadis Ecolas, Institute European Environmental Policy, Task 2- Benefits for the former Yugoslav Republic of Macedonia and other countries of SEE of compliance with the environmental acquis, Final Report—Part II: Country, Available from: http://ec.europa.eu/environment/archives/international_issues/pdf/report_macedonia_feb08.pdf.
[26] State Statistical Office of Republic of Macedona. Environment statistics. 2015. Available from: http://milieukontakt.mk/wp-content/uploads/2016/01/Statistika%20na%20Zivotnata%20Sredina%20vo%20RM%202015.compressed.pdf.
[27] Agency for Statistics of Bosnia and Herzegovina, Utilization and protection of water against pollution in industry. 2014. Available from: http://www.bhas.ba/saopstenja/2015/VOD1_2015_001_01_BA.pdf.
[28] Glass bell, Is Serbia rich in water, Available from: http://staklenozvono.rs/?p=9038.
[29] Aljazeera, Sarajevo threatened by water shortage, Available from: http://balkans.aljazeera.net/vijesti/sarajevu-prijeti-nestasica-vode.
[30] Montenegro media, In Zaljevo water restrictions continue, Available from: http://crna.gora.me/vijesti/opstine/bar/u-zaljevu-i-dalje-restrikcije-vode/.
[31] Aljazeera, Due to water shortages challenged mayor on Facebook, Available from: http://balkans.aljazeera.net/vijesti/zbog-nestasice-vode-izazvao-gradonacelnicu-na-facebooku.
[32] Ministry of Sustainable Development and Tourism of Montenegro, Information about the state of the environment in Montenegro in 2014, Available from: http://www.epa.org.me/images/izvjestaji/Informacija%20o%20stanju%20%C5%BEivotne%20sredine%20u%20Crnoj%20Gori%20u%202014.godini%20(1).pdf.
[33] Mitev T, Custodio T, Hristovski KD, Ulrich JW. The effects of climate change on water resources of small developing countries a case study of the republic of Macedonia. In: Ahuja S, de Andrade JB, Dionysiou D, Hristovski KD, Loganathan BG, editors. Water challenges and solutions on a global Scale, ACS symposium series. Washington, DC: American Chemical Society; 2015. p. 37–52.
[34] Hristovski KD, Pacemska-Atanasova T, Olson LW, Markovski J, Mitev T. Potential health implications of water resources depletion and sewage discharges in the Republic of Macedonia. J Water Health 2016. http://dx.doi.org/10.2166/wh.2016.274.
[35] Agency for Environmental Protection of the Republic of Serbia, Serbian Water quality index, Available from: http://www.sepa.gov.rs/index.php?menu=6&id=8007&akcija=showXlinked&#tabela2.

[36] Republic of Serbia, Regulation on classification of water from Inter-republic Water-courses, Interstate waters and Yugoslav coastal waters, Official Gazette SFRJ #6/78.
[37] Agency for Environmental Protection of the Republic of Serbia, Serbian Water Quality Index. Available from: http://www.sepa.gov.rs/index.php?menu=6&id=8010&akcija=showXlinked.
[38] Ministry of Agriculture and Environmental Protection, Agency for environmental protection of Republic of Serbia, Results of testing the quality of surface and groundwater in 2013, Available from: http://www.sepa.gov.rs/download/KvalitetVoda2013.pdf.
[39] Statistical office of the republic of Serbia, Eco-bulletin. 2013. Available from: http://webrzs.stat.gov.rs/WebSite/repository/documents/00/01/61/00/SB-585-EKO-2013.pdf.
[40] J. Novak, N. Kodre, S. Terpin Savnik, P. Repnik Mah, S. Sisko Novak, D. Dolinar, G. Pipus, N. Pivec, A. Invanc, J. Dodic, P. Erzen, Z. Panovski, National Strategy for Water for republic of Macedonia (2011−2041), Available from: file:///C:/Users/jmarkov2/Downloads/1428578577_water_strategy_for_republic_of_macedonia.pdf.
[41] Agency for statistics of Bosnia and Herzegovina, public sewage system. 2014. Available from: http://www.bhas.ba/saopstenja/2015/VOD2K_2015_001_01_BA.pdf.
[42] Agency for statistics of Bosnia and Herzegovina, Selected environment indicators. 2013. Available from: http://www.bhas.ba/tematskibilteni/OPS_2013_001_01_bh.pdf.
[43] European Environment Agency, Water use efficiency (in cities): leakage, Available from: http://www.eea.europa.eu/data-and-maps/indicators/water-use-efficiency-in-cities-leakage/water-use-efficiency-in-cities-leakage.
[44] PUC Water Supply Sremska Mitrovica, Price of services PUC Water Supply, Available from: http://vodovodsm.rs/korisnici/cene-usluga.
[45] Ministry of Health. Ministry of agriculture and environment of the republic of Serbia, Implementation of the protocol on water and health in the republic of Serbia, situation analysis. 2014. Available from: http://www.sepa.gov.rs/download/Protokolovodiizdravlju_Analiza_stanja.pdf.
[46] Democratic Economic Governance of Water Access, General assessment of the water supply sector and its human development function in Bosnia and Herzegovina, Available from: http://ba.one.un.org/content/dam/unct/bih/PDFs/GLAVNI%20DOKUMENT_UNDP.pdf.
[47] Institute of Public Health of Serbia Dr Milan Jovanovic Batut, Health safety of drinking water from central water supply system in the Republic of Serbia in 2013, Available from: http://www.batut.org.rs/download/izvestaji/higijena/Sestomesecni%20izvestaj%20vode%20za%20pice%202013.pdf.
[48] Statistical office of the Republic of Serbia. Statistical yearbook of the Republic of Serbia. 2014. Environment, Available from: http://webrzs.stat.gov.rs/WebSite/repository/documents/00/01/54/10/11-Zivotna_sredina.pdf.
[49] Ministry of environment and physical planning of Republic of Macedonia, Macedonian's first national communication under the united nations framework convention on climate change. 2013. Available from: http://unfccc.int/resource/docs/natc/macnc1.pdf.
[50] The Ministry of environment and Spatial planning, initial national communication of the republic of Serbia under the united nations framework convention on climate change. 2010. Available from: http://unfccc.int/resource/docs/natc/srbnc1.pdf.
[51] United Nations Framework Convention on Climate Change, Climate characteristics of Bosnia and Herzegovina, Available from: http://www.unfccc.ba/site/pages/karakteristike_klime.php.
[52] Ministry of sustainable development and Tourism of Montenegro, second national report of Montenegro on climate change under the united nations framework convention on climate change. 2014. Available from: file:///C:/Users/jmarkov2/Downloads/Nacrt%20Drugi%20nacionalni%20izvje%C5%A1taj%20Crne%20Gore%20o%20klimatskim%20promjenama%20(1).pdf.

[53] Ministry of Foreign Trade and Economic Relations of Bosnia and Herzegovina, Second National Communication of Bosnia and Herzegovina under the United Nations Framework Convention on Climate Change, Available from: http://www.unfccc.ba/site/upload/PDF_dokumenti/BiH_SNCBiH_BHS-L.pdf.
[54] Republic Hydrometeorological Service of Serbia, Hydrological yearbooks, Available from: http://www.hidmet.gov.rs/ciril/hidrologija/povrsinske_godisnjaci.php .
[55] Aljazeera, Two billion euros of damages caused by floods in Bosnia and Herzegovina, Available from: http://balkans.aljazeera.net/vijesti/dvije-milijarde-eura-steta-od-poplava-u-bih.
[56] Telegraf, The total flood damages of 1.7 billion Euros, Available from: http://www.telegraf.rs/vesti/1150095-ukupna-steta-od-poplava-17-milijardi-evra.
[57] Blic, Macedonia declared an epidemic of tularemia in Lozovo and Karatmanovo, Available from: http://www.blic.rs/vesti/svet/makedonija-proglasena-epidemija-tularemije-u-lozovu-i-karatmanovu/fglhvkx.
[58] Belgrade Chamber of Commerce, Increase share of agriculture in GDP, Available from: http://www.kombeg.org.rs/aktivnosti/zadruzni_savez/Detaljnije.aspx?veza=3354.

CHAPTER NINE

Ion Chromatography Instrumentation for Water Analysis

K. Srinivasan
Thermo Fisher Scientific, Sunnyvale, CA, United States

1. INTRODUCTION

Ion chromatography is the premier and established method for analyzing ions in water samples derived from a variety of sources, such as surface water, groundwater, and well water [1–3]. The concentration of ions present in drinking water is highly dependent on the source of the water. Surface water is relatively free from major ionic contaminants. However, falling leaves and other debris result in accumulation of organic matter in the water and can be transformed into ionic components when reacted with disinfectant products during the water processing step. As opposed to surface waters, groundwater has relatively higher amounts of ions dissolved due to equilibration with the ground matrix and typically has high levels of divalent ions, such as calcium and sulfate. Usually, the presence of organic contaminants is low in groundwater. Drinking water, depending on the availability of suitable water, can be a blend of ground-, surface, and well waters. From an analysis perspective, water poses challenges due to the diversity of the concentration of the common inorganic ions, organic contaminants, and the presence of trace components, such as disinfectant by-products. Analyzing trace components in the presence of high levels of matrix ions can be challenging. The focus of this chapter is on the required instrumentation, sample pretreatment, and setup required for pursuing water analysis.

Environmental Protection Agency (EPA) Method 300.1 Part A is used for determination of inorganic ions, such as fluoride, chloride, bromide, nitrate, nitrite, orthophosphate, and sulfate, and Part B is used for the determination of bromate, bromide, chlorite, and chlorate in drinking water [2,3]. Analysis of components present in mg/L concentrations, such as the ions in Part A listed above, are fairly straightforward, provided adequate resolution is obtained with the separator column used. The injection volume is

Chemistry and Water
ISBN 978-0-12-809330-6
http://dx.doi.org/10.1016/B978-0-12-809330-6.00009-X

usually a full loop injection that is roughly proportional to the column format in use. For example, EPA Method 300.0 specifies a large loop injection volume of 2—3.0 mL with a 4-mm separator column, and EPA Method 300.1 specifies two loop volumes based on the target analytes, namely, 10 μL and 50 μL for Part A and Part B, respectively, with 2-mm internal diameter (ID) format columns. Higher ID columns would require a proportionally higher sample injection loop volume. The challenges in the analysis usually stem from the presence of ions listed in Part A interfering with the analysis of trace components, such as the disinfection by-product ions listed in Part B section of EPA Method 300.1. As regulation drives toward lower concentrations of components such as ionic disinfectant by-products, there is a need for methods that would allow the detection of trace ions in the presence of high levels of matrix ions.

2. ION CHROMATOGRAPHY INSTRUMENTATION

Since its inception [4], the core ion chromatography instrumentation has remained unchanged, with the exception that the type of the suppressor has evolved over the years [5—10]. Fig. 9.1 shows a simple schematic of an ion chromatograph. The ion chromatography system is comprised of a metal-free

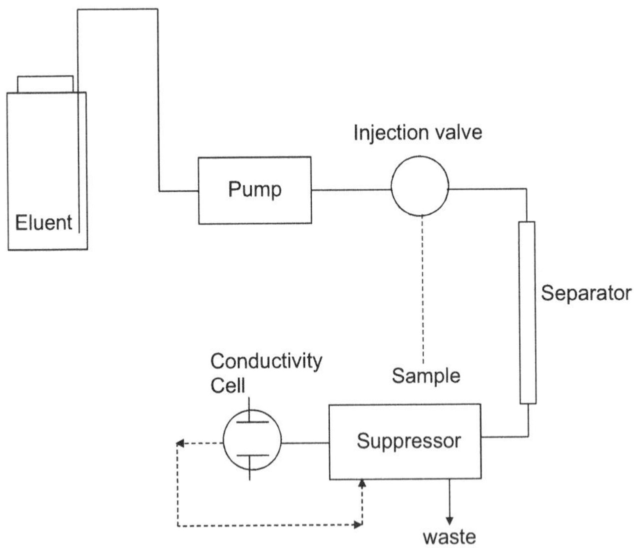

Figure 9.1 The instrumental setup for a standard ion chromatograph is shown above. The system uses manually prepared eluents for operation.

pump that is used to pump the eluent. The eluent is routed through an injection valve to inject the sample and then routed through a separator column. Usually a guard column (not shown) is installed in front of the separator column to protect the separator column from sample matrices and extends the lifetime of the separator column. The pump is operated typically at a flow rate of 1 mL/min for a 4-mm ID- column format. For a 2-mm ID-column, the pump is operated at a flow rate of 0.25 mL/min. The column effluent is routed through a suppressor eluent channel (a continuously regenerated electrolytic suppressor is shown here) for suppressing the eluent, and then the suppressed eluent, along with the analytes, are routed to a detector cell, usually a conductivity cell, for detection. The effluent from the conductivity cell is routed back to the suppressor in the recycle mode to provide the water required for the electrolysis reactions and then routed to waste. In an alternate configuration, an external water source (not shown) is connected to the suppressor for supplying water to the regenerant channel for the electrolysis reaction and for carrying away the suppressor waste. In the chemical mode of operation, a pressurized regenerant reservoir, such as a reservoir filled with sulfuric acid as the regenerant when analyzing for anions, will be in fluidic communication with the regenerant side of the suppressor to supply regenerant ions and for carrying away the suppressor waste [11].

The sample is loaded into an injection loop in the injection valve using a manual syringe or using an autosampler module. The injection valve typically comprises six ports in a two-position switching format. One position is used for loading the sample into the injection loop. In this configuration, the pump outlet is fluidically connected to the separator column, and there is no fluidic contact with the sample loop. In the second position, the valve is used for injecting the contents of the injection loop into the separator column for analysis. In this position, the outlet of the pump is in inline with the injection loop and the contents are diverted to the separator column for analysis.

An expanded drawing of the injection valve is shown in Fig. 9.2. The sample injection loop is installed between ports 1 and 4. The sample loading is accomplished via port 5 and is routed to waste from port 6. The sample loading can be accomplished by manually displacing the sample using a syringe in a manual mode of operation or can be loaded using an autosampler module. Referring back to Fig. 9.2, the pump outlet is connected to port 2, and port 3 is connected to the inlet of the separator column. The sample stream is typically arranged countercurrent to the eluent stream to facilitate partial sample loop loading and injection. In operation in the load position,

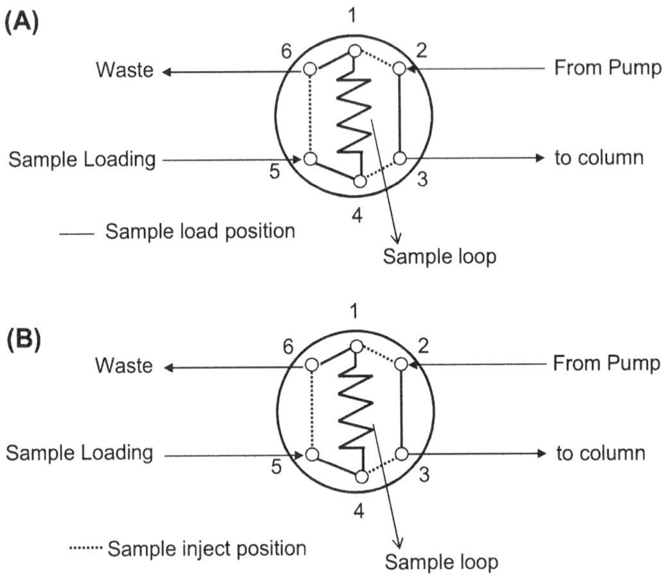

Figure 9.2 A schematic of a six-port valve is shown above with a sample injection loop installed. (A) Shows the configuration when the loop is in the sample load mode. (B) Shows the configuration when the loop is in the inject position.

the sample is loaded into the injection loop, and in the inject position, the content of the injection loop is loaded by the pump flow onto the separator column for analysis. For improved reproducibility of the injections, with full loop injections, it is recommended to keep the injection loop in the inject position for 7–10 sample loop volumes. EPA 300.0 and 300.1 [2,3] specify a large loop injection for the sample loading function.

Fig. 9.3 shows a schematic of a reagent-free ion chromatography (RFIC) system [11]. This system incorporates an eluent generator module that produces eluent from a deionized (DI) water stream. In this approach, the DI water is pumped using the metal-free pump and routed into the inlet of the eluent generator cartridge (EGC) for generating eluents. The output from the EGC is routed to a continuously regenerated trap column (CR-TC) for purifying the eluent and the water stream. The output from the CR-TC is routed to a degasser module to remove any electrolytic gases present in the eluent. To facilitate degassing, an optional restrictor (not shown) may be added to ensure that the system pressure is around 2300 psi. After the degassing process, the eluent is routed to the injection valve. Following this location, the plumbing is similar to Fig. 9.2. The key benefit of the RFIC system is that contaminant-free eluent is produced in an automated manner,

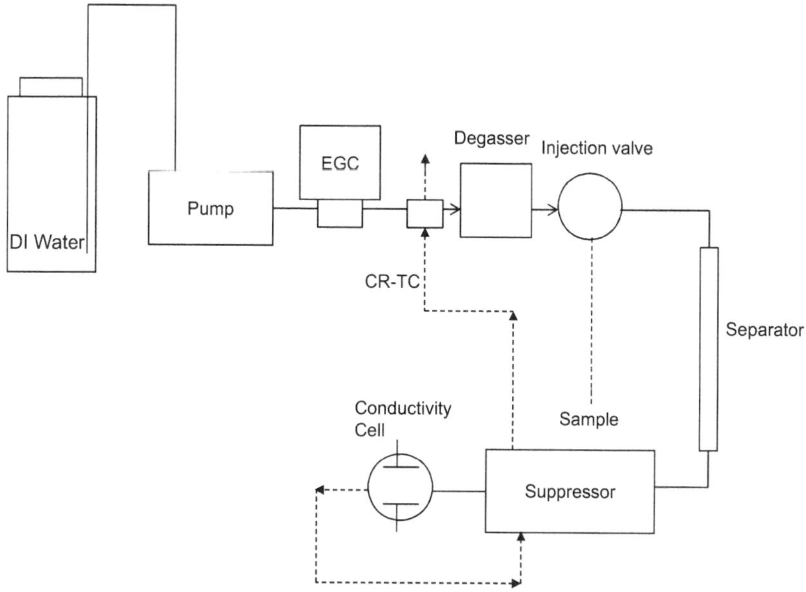

Figure 9.3 The instrumental setup for a reagent-free ion chromatography (RFIC) system is shown above. The system uses deionized (DI) water for operation.

allowing reproducible chromatographic performance from day to day, system to system, and operator to operator [11].

3. CONCENTRATOR COLUMN

A concentrator column can be used in place of the sample injection loop, as shown in Fig. 9.4.

The function of the concentrator column is to concentrate the analytes of interest from the sample stream. This is invaluable when the analytes of interest are present in trace amounts and the sample is present in a large volume. By loading the large sample volume into a concentrator column, the trace components can be focused on the head of the concentrator column. Elution with the eluent in the countercurrent direction allows for a narrow concentrated injection plug for analysis. This approach is typically used in ion chromatography when working with trace components, in order to concentrate the trace components. The concentrator column is loaded in one direction and eluted in the opposite direction to minimize any band dispersion associated with the injection. When working with drinking water

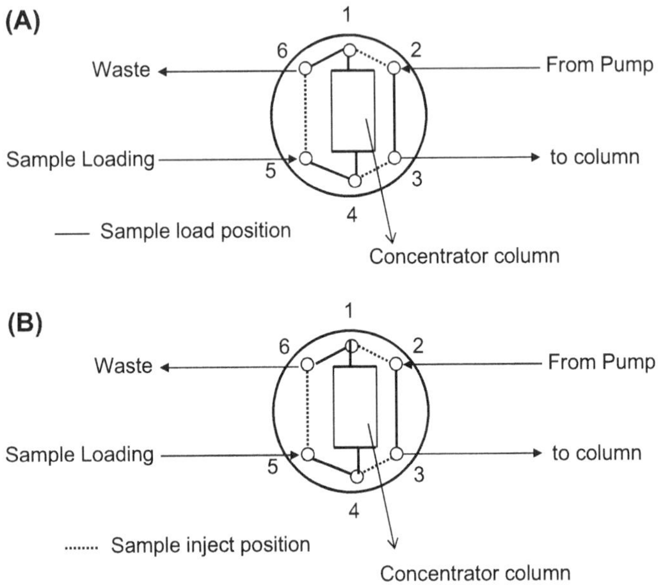

Figure 9.4 A schematic of a six-port valve is shown with a concentrator column installed. (A) Shows the configuration when the concentrator column is in the sample load mode. (B) Shows the configuration when the concentrator is in the inject position.

samples, however, this approach is used only for selective concentration of species of interest. The reason for this stems from the fact that in drinking water, there are matrix ions present. For example, when pursuing anion analysis, relatively high levels of chloride and sulfate are typically present. With the concentrator column approach, the high levels of chloride and sulfate can also be concentrated and may displace the trace components of interest. Also, concentrating large amounts of matrix ion may not be beneficial for the analysis, since they can elute close to the trace component of interest and interfere with the analysis. The concentrator column, however, finds utility in re-concentrating trace components of interest from a suppressed background stream, such as when pursuing matrix elimination ion chromatography (MEIC) work using EPA 314.2 for perchlorate or EPA 302.0 for bromate analysis [12,13].

4. POSTCOLUMN DERIVATIZATION

Postcolumn derivatization techniques have been applied to the detection of certain trace components in drinking water to facilitate detection of

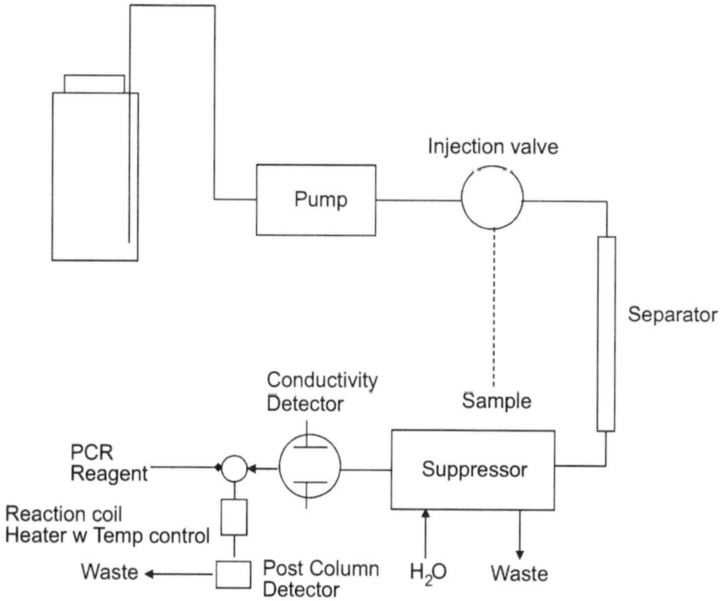

Figure 9.5 A schematic of a postcolumn derivatization setup is shown here. *PCR*, post-column reagent.

trace components, such as bromate, in the presence of high levels of matrix ions, such as chloride and sulfate. The derivatization technique itself is accomplished en route to the detector by selective reaction chemistries employed to transform the component of interest with suitable chromophores. A typical setup is shown in Fig. 9.5. The key difference between Figs. 9.1 and 9.5 is the postdetector plumbing setup. Referring to Fig. 9.5, the line from the conductivity detector is routed to the first leg of a T connector. A postcolumn reagent (PCR) from a pressurized reservoir or pulse-free pump is routed to the T connector at the second leg, and the combined flow is diverted from a third leg of the T connector to a reaction column heater that is set at an appropriate reaction temperature required for the derivatization reaction. The reaction coil is knitted to prevent excessive band dispersion and preserve the peak shape of the reacted components. The reacted species is routed to a suitable detector, such as a UV/visible detector for detecting the derivatized species. When using a pneumatic delivery system, it is critical that the flow rate of the postcolumn reagent is adjusted based on the applied head pressure to achieve the required derivatization. When using a pump, a pulse-free setup is needed to ensure that the baseline noise is low enough to result in the desired signal-to-noise ratio. One example of this application is in bromate detection with EPA Method 317.0, where bromate is reacted with ortho-dianisidine reagent to enhance the visible detection of the bromate ion at 450 nm. A sixfold improvement in the detection limit is feasible with this approach when compared to suppressed conductivity detection [14].

5. SOLID PHASE EXTRACTION

Another strategy for minimizing interference from matrix ions, such as chloride and sulfate, is by pursuing a sample pretreatment using solid phase extraction (SPE) cartridges. For example, high levels of chloride ion can be minimized by treating the sample with a silver-form cation exchange SPE. The precipitation of silver chloride will allow the sample to be free of chloride ions. A cation exchange SPE cartridge in the hydronium form is also used in conjunction with the silver SPE to remove any colloidal silver. A barium-form SPE cartridge can be employed to remove sulfate ions. Thus, trace components can be analyzed without interference from matrix ions. Fig. 9.6A shows the typical format of the SPE cartridge with Luer lock adapters that facilitate easy interface with a syringe, and multiple cartridges can be utilized in series, as shown in Fig. 9.6B, to remove interfering ions, such as sulfate chloride and carbonate, from a drinking water sample prior to determination of bromate. This is a manual setup. Fig. 9.7 shows an automated setup to process multiple drinking water samples. The Dionex OnGuard Sample Prep Station has individual stopcock valves for each 5-cc

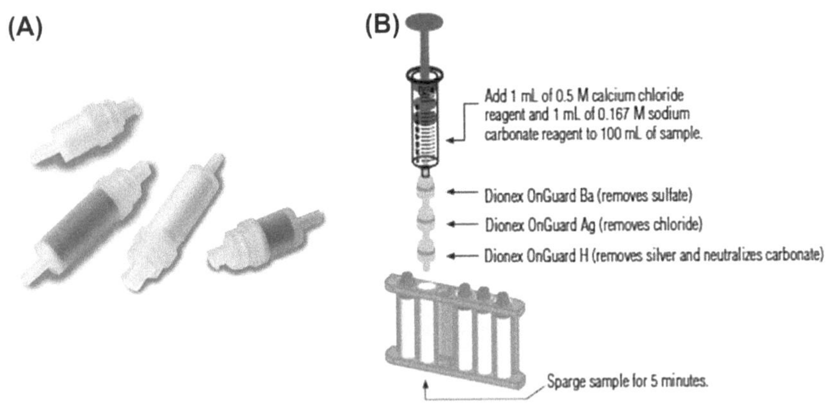

Figure 9.6 A schematic of a Dionex OnGuard sample pretreatment cartridge is shown here in (A). (B) Shows the setup for loading the sample with multiple sample pretreatment cartridges.

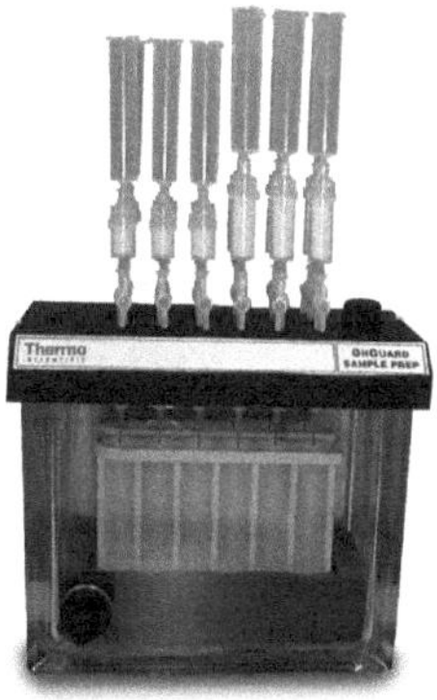

Figure 9.7 The setup for a vacuum-driven accessory for simultaneous pretreatment of multiple samples with the Dionex OnGuard sample pretreatment cartridges.

sample reservoir to allow control of individual flow rates. It should be noted that the SPE approach requires that the cartridges are cleaned prior to use and may retain species such as thiosulfate, which can impact sulfate analysis. Consequently, this approach can be used for specific ions that are not retained by the SPE matrix. For example, trace level analysis of perchlorate ion is feasible with this approach of matrix removal. [15,16].

6. INLINE POLISHER

Inline sample pretreatment has also been used to eliminate matrix effects when pursuing ion chromatography. For example, when pursuing analysis of well water, which tends to have high levels of divalent cations such as calcium and magnesium, the presence of these ions has a deleterious effect on the peak shape of anions, such as phosphate. Inline sample pretreatment with a sodium form cation polisher column results in improved peak shape for the phosphate ion. An IonPac CP1 Na + form cation polisher column (6 × 16 mm) from Thermo Scientific was used for this application and installed between the autosampler and the injection valve. The sample was routed through the CP1 column, and the divalent ions in the sample were replaced with the sodium ions. Fig. 9.8 shows a comparison of anion analysis with a Dionex IonPac 4 × 250 mm AS18 column and 32-mM potassium hydroxide eluent at a flow rate of 1 mL/min and using a 1-mL injection loop. A Dionex 4-mm AERS 500 suppressor was used in this work at a setting of 88 mA. The sample comprised a five-anion mixture spiked with high levels of calcium and magnesium chloride. Fig. 9.8 shows the analysis for a simulated sample before (A) and after the polishing step (B). It is clear

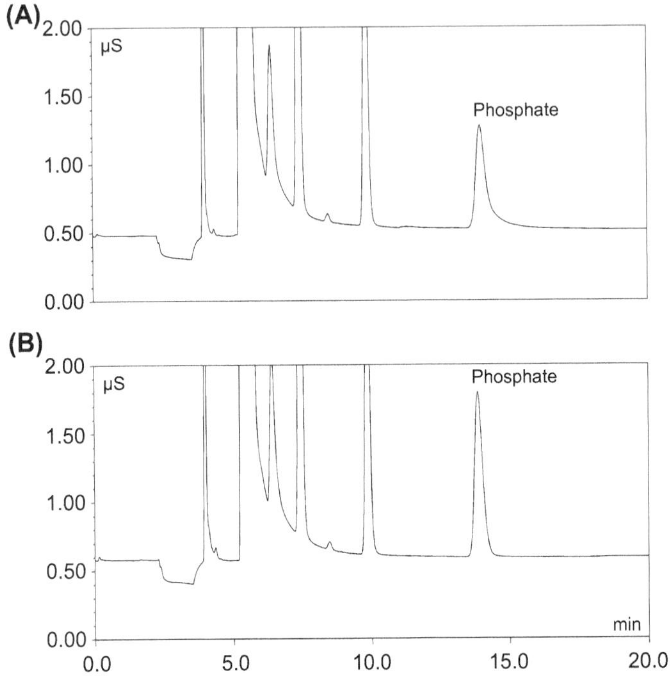

Figure 9.8 Chromatogram A shows the analysis of a five-anion standard with added calcium and magnesium. Poor peak shape for phosphate can be inferred from above. When the sample stream was polished using a Dionex CP1 (6 × 16 mm) polisher column, the recovery for phosphate improved and excellent peak shapes were achieved, as shown in chromatogram B.

that by eliminating the matrix ions, the chromatography is improved [17]. It is clear from the improved peak shape that better quantitation of phosphate is feasible by this approach.

7. MATRIX IONS AND HEART CUT METHODS

The impact of matrix ions on the chromatography of a highly retained ion, such as perchlorate, can be seen in Fig. 9.9. This set of chromatograms illustrates the performance of Method 314.0 with a Dionex IonPac AS16 column and using a 65-mM potassium hydroxide eluent at 1.2 mL/min and an injection volume of 1 mL for determining 1 µg/L of perchlorate in increasing concentrations of matrix anions on the ICS-2000 system. A Dionex 4-mm AERS 500 was used in this work at a setting of 193 mA. The peak shape of perchlorate broadens and suffers as the matrix

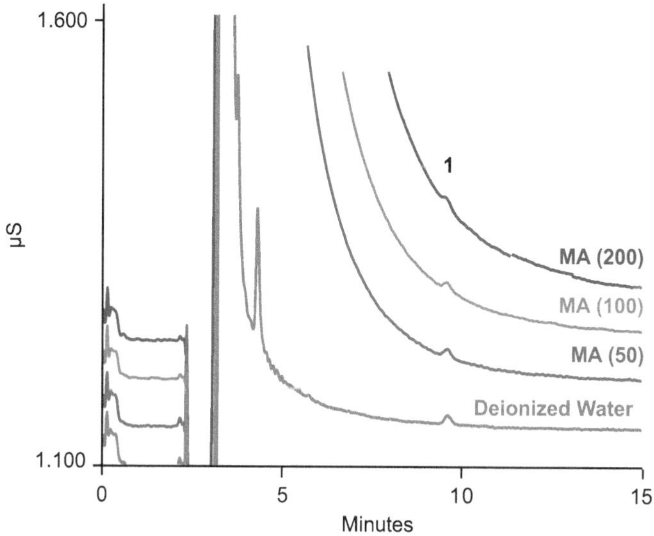

Figure 9.9 This set of chromatograms illustrates the performance of Method 314.0 for analyzing perchlorate in the presence of matrix ions. Notice that at 200 ppm each of chloride, carbonate, and sulfate, there is almost a complete loss of signal from perchlorate. In this case, the perchlorate is essential, eluting with the matrix components, and its presence is masked.

composition is increased, as shown, from 50 to 100–200 mg/L of chloride, sulfate, and carbonate ions. The retention time is shifted to an earlier time, the peak becomes smeared, and recovery suffers due to the sloping baselines. As discussed previously, one setup to address this issue would be to pursue a sample cleanup using SPE cartridges, as shown in Figs. 9.6 and 9.7. The SPE approach, however, requires multiple steps, and can be labor-intensive and cumbersome. This was a primary consideration for developing an improved, more robust method for sub-ppb determinations of perchlorate.

Another approach in addressing this issue is by heart cutting the peak of interest, perchlorate in this case, pursuing reanalysis. It should be noted that this strategy adds to the analysis time [18–20]. Fig. 9.10 shows this setup. In this case, a concentrator column is installed into a second valve and plumbed to facilitate collection of the heart cut and subsequent analysis in the same separator column. In operation, the sample is loaded into the injection loop and is loaded on the separator column for analysis. When the peak of interest is observed in the conductivity cell, valve 2 is switched to capture the peak of interest on the concentrator column. The analysis is pursued by eluting the concentrated species in the concentrator column in valve 2. This

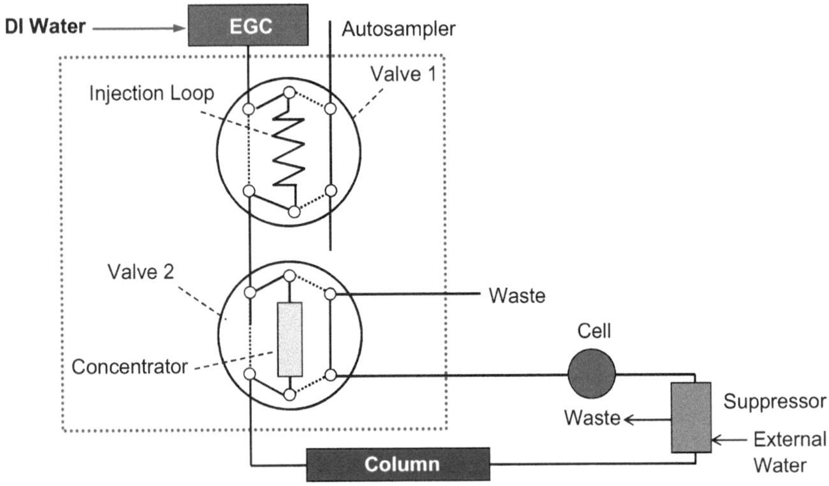

Figure 9.10 A schematic of a valve-switching approach for pursuing single and multiple pass of the sample is shown here. The concentrator column can be switched for analysis or for postconductivity concentration of the target sample. *DI*, deionized.

is the first-pass collection and analysis. Fig. 9.11A shows the analysis of a sample containing 1 μg/L of perchlorate in a low level of matrix ions in the 100-mg/L regime of chloride, sulfate, and bicarbonate anions. The chromatographic conditions of operation were similar to Fig. 9.9. The heart cut window of 7–10 min, where the perchlorate peak eluted, was routed to the concentrator column for preconcentration. The concentrated sample was analyzed in the next pass designated here as first-pass capture and analysis, and showed a recovery of about 93%. Essentially the sample was run twice, thus doubling the run time. Fig. 9.11B shows the same analysis, however, with a higher level of matrix-laden sample of 1000 mg/L of chloride, sulfate, and bicarbonate spiked into a drinking water sample. The recovery in this case suffered, and after the first-pass heart cut collection and analysis, the recovery was in the 60% regime for a 1-μg/L sample of perchlorate. The same analysis protocol, when repeated for a second-pass capture and analysis, is seen in Fig. 9.11C and shows a recovery of only 80% for the 1-μg/L sample of perchlorate. In this case, the sample was run three times with matrix elimination pursued on two analysis cycles, thus tripling the analysis time. These results indicated that the method was not robust for high concentrations of matrix ions. The previously mentioned method, although automated, has limited utility, but can be used for samples containing low levels of matrix ions at the expense of analysis time [18–20].

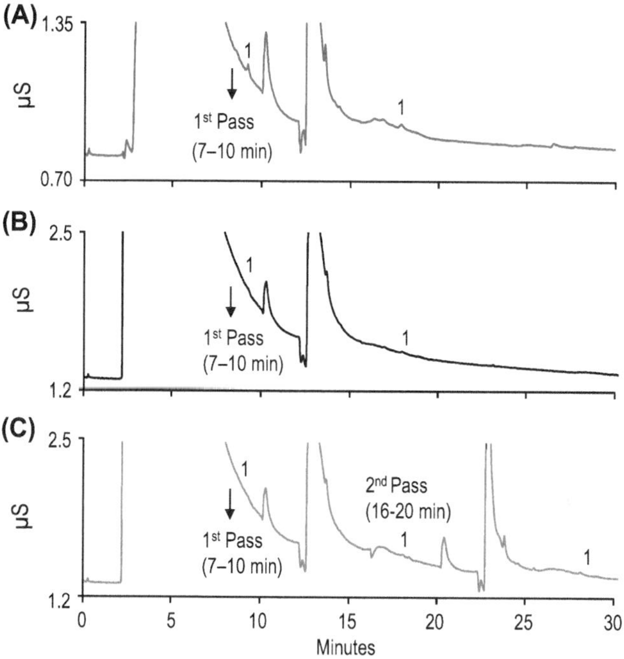

Figure 9.11 The analysis of 1 μg/L of perchlorate was pursued similarly to the conditions shown in Fig. 9.9. (A) The sample contained a relatively low level of matrix ions at 100 mg/L of chloride, sulfate, and bicarbonate ions. The results showed a recovery of 93% for perchlorate after the first-pass capture and analysis. (B) The sample contained a high level of matrix ions at 1000 mg/L of chloride, sulfate, and bicarbonate, and was spiked to a tap water sample. The results from the first-pass capture and analysis showed a recovery of about 60% for perchlorate. (C) The sample contained a high level of matrix ions at 1000 mg/L of chloride, sulfate, and bicarbonate, and was spiked to a tap water sample. The results from the second-pass capture and analysis showed a recovery of about 80% for perchlorate.

Fig. 9.12 shows the setup for MEIC [21,22]. This is a 2-D setup where the first dimension is used for matrix elimination and the second dimension is used for pursuing analysis of a selected heart cut from the first dimension. The heart cut is refocused on the concentrator column and analyzed with a smaller cross-sectional column that is operated at a smaller flow rate, hence aiding in less dilution of the sample during the chromatographic process. The net impact is that the sensitivity with a concentration-sensitive detector, such as conductivity, is increased proportionally to the ratio of the cross-sectional area of the first-dimension column to the second-dimension column. A further benefit of this setup is that the selectivity of the two

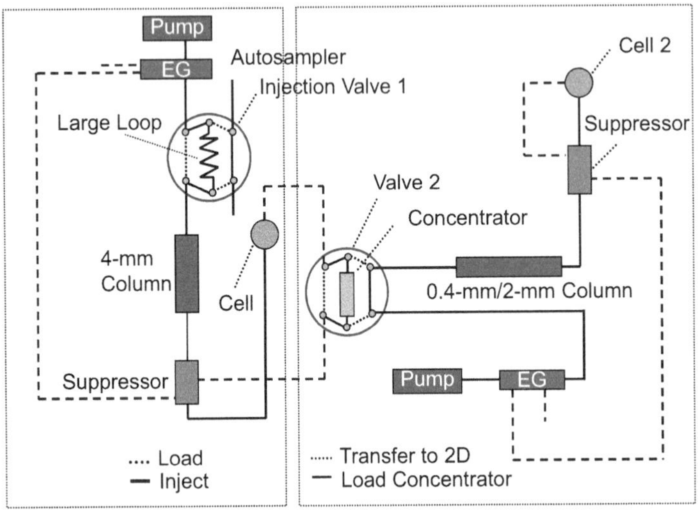

Figure 9.12 A schematic of the matrix elimination ion chromatography method is shown here. In the first dimension, the sample is analyzed and the matrix ions are diverted to waste. The peak of interest from the first dimension is focused on a concentrator column and analyzed in the second dimension using a column with a smaller internal diameter.

dimensions can be optimized to achieve a unique selectivity for the separation. By running the two dimensions in parallel, the average analysis time remains close to that of a single dimension, analysis method. Lastly, the setup is fully automated with RFIC and does not require any user intervention. Referring to Fig. 9.12, a pump is used to pump DI water into the EG module (CR-TC/degasser not shown) for eluent generation and is routed to injection valve 1. The sample is loaded into the large loop (2–4 mL) in the injection valve 1. The analysis proceeds on the 4-mm column (IonPac AS20 4 × 250 mm), and a predetermined cut window focuses the sample window of interest onto the concentrator column. When the concentrator column is offline, the matrix ions are diverted to waste. The focused sample of interest is analyzed in the second dimension, using a smaller cross-sectional column (IonPac AS16 2 × 250 mm). Fig. 9.13 shows an example of the analysis of a 0.5-μg/L spike of perchlorate ion in the highest sample matrix of 1000 mg/L of chloride, bicarbonate, and sulfate. Chromatogram labeled A was generated in the first dimension, using a Dionex IonPac AS20 (4 × 250 mm) column operated at 1 mL/min and using an eluent comprising of 35 mM potassium hydroxide (EGC-generated). A Dionex 4-mm AERS 500 suppressor was used in this work and operated at

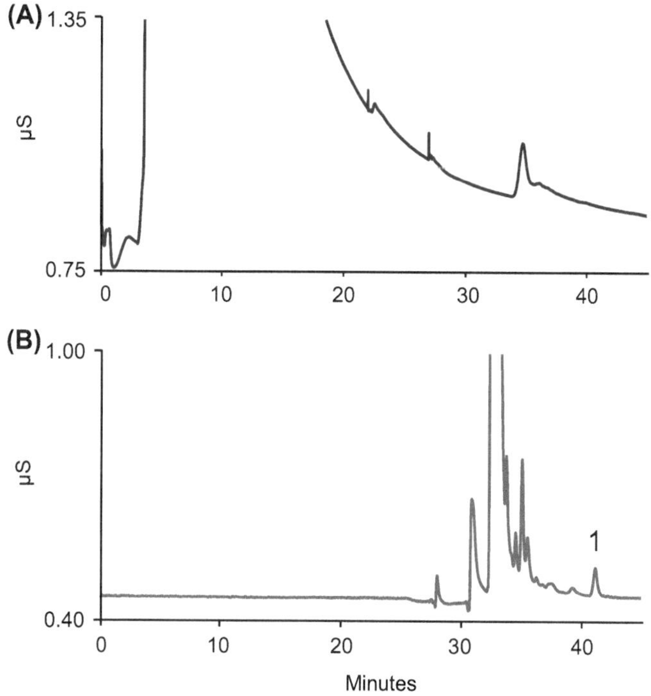

Figure 9.13 The results obtained from the matrix elimination ion chromatography method are shown here for 0.5 μg/L of perchlorate spiked to a high ionic strength matrix containing 1000 mg/L of chloride, sulfate, and bicarbonate ions. Chromatogram labeled A was generated in the first dimension, and no peak for perchlorate was observed. The chromatogram labeled B was obtained in the second dimension and showed excellent recovery and peak shape for perchlorate in this high ionic strength matrix.

150 mA. The injection volume was 4 mL in this work. Because of the high concentration of matrix ions, detection of perchlorate, the peak of interest, was not possible in this method. The spikes in the chromatogram show the valve-switching times. Chromatogram labeled B was obtained in the second dimension using a Dionex IonPac AS16 (2 × 250 mm) column and operated at 0.25 mL/min using a 65 mM potassium hydroxide eluent (EGC-generated). A Dionex 2-mm AERS 500 was used in this work as a suppressor and operated at 41 mA. The second-dimension analysis, however, showed an excellent peak shape and recovery for perchlorate, as shown in Fig. 9.13B. Table 9.1 summarizes the impact of matrix ions on the MEIC method, and it is clear that the matrix composition has very little impact on the peak shape and recovery, and excellent performance is inferred. It

Table 9.1 The Effect of Matrix Concentration on Perchlorate Peak Area and Recovery Is Shown as Follows. The Matrix Ions Comprised of Chloride, Sulfate, and Bicarbonate and the mg/L Values Listed in the Following Are for Each Individual Matrix Ion

Matrix Concentration* (mg/L)	Perchlorate Peak Area Counts	Recovery (%)
0	0.3522	100
50	0.3560	101.1
100	0.3567	101.3
200	0.3509	99.6
500	0.3505	99.5
800	0.3468	98.5
1000	0.3438	97.6

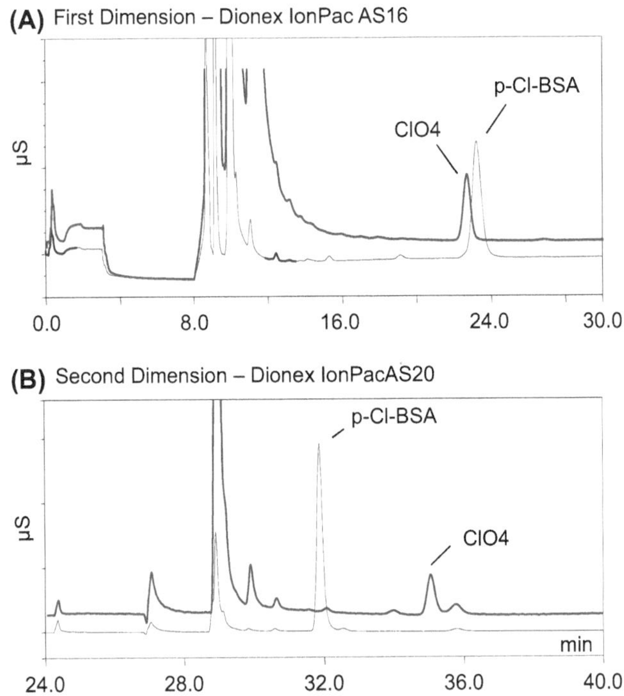

Figure 9.14 The results obtained with the matrix elimination ion chromatography method with Dionex IonPac AS16 chemistry for the first dimension and Dionex IonPac AS20 chemistry for the second dimension. *Para*-chlorobenzenesulfonate elutes close to perchlorate in the first dimension.

is clear that the MEIC method offers superior performance to the standard method for perchlorate, as shown in EPA Method 314.2 [12].

It should be noted that selectivity optimization was pursued in the separator column choices for the first and second dimension. Fig. 9.14A shows the analysis of perchlorate and *para*-chlorobenzene sulfonic acid, and because of the close proximity of the retention times, a single broad peak is expected if these compounds are present in a water sample. Fig. 9.14B shows the second-dimension separation, which resolves the two peaks substantially. By choosing the right selectivity in the two dimensions, one can minimize the probability of coelution when pursuing analysis by the MEIC methodology. For example, Fig. 9.15A shows the separation of the two peaks of interest in the first dimension. Pursuing the heart cut and analysis in the second dimension shows that there is no chance of coelution with perchlorate, as

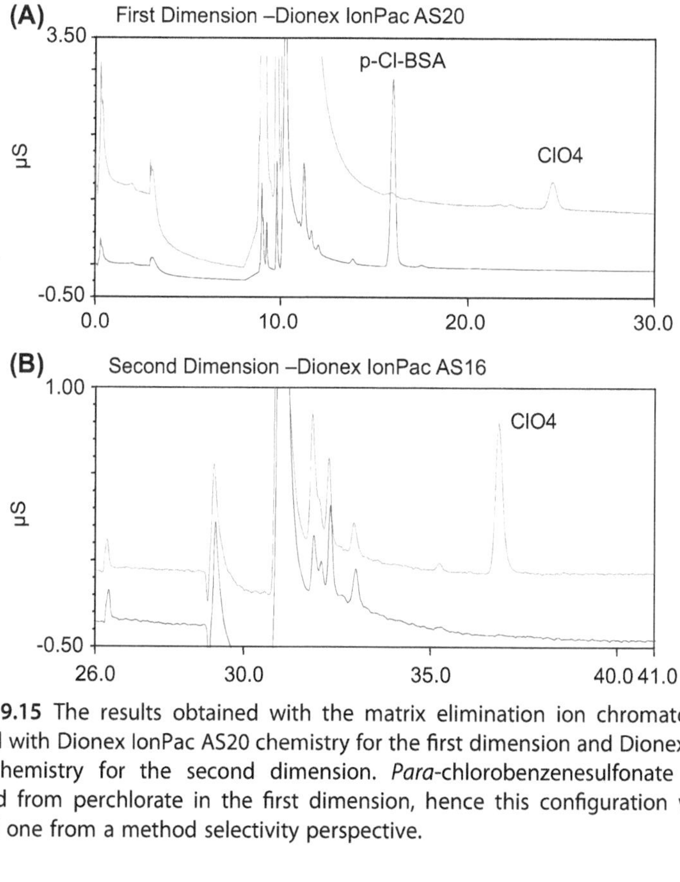

Figure 9.15 The results obtained with the matrix elimination ion chromatography method with Dionex IonPac AS20 chemistry for the first dimension and Dionex IonPac AS16 chemistry for the second dimension. *Para*-chlorobenzenesulfonate is well resolved from perchlorate in the first dimension, hence this configuration was the optimal one from a method selectivity perspective.

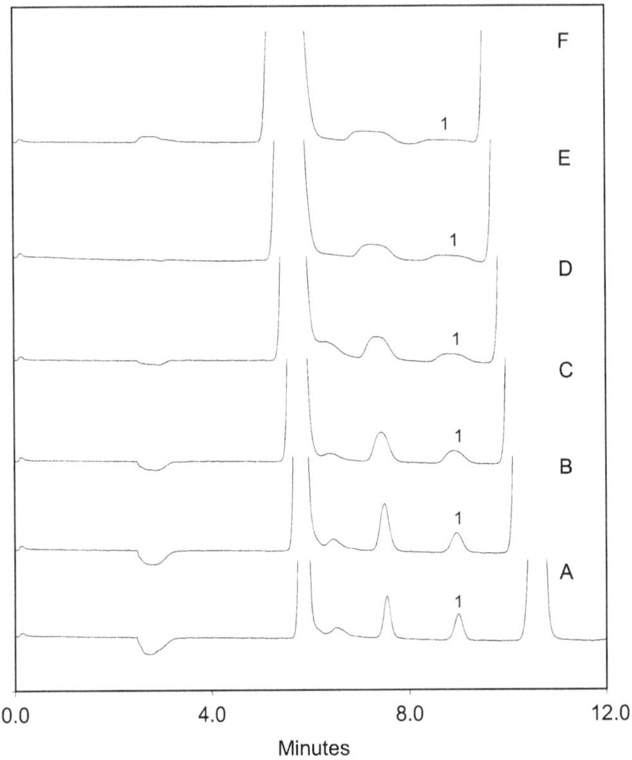

Figure 9.16 This set of chromatograms illustrates the impact of matrix on the quantitation of 5-μg/L bromate ion. Notice that at 250 ppm each of chloride and sulfate, there is almost a complete loss of signal from bromate.

shown in Fig. 9.15B. Thus EPA 314.2 specifies this order for the column chemistry when pursuing the MEIC analysis of perchlorate [12].

Figs. 9.16(A–F) shows the impact of matrix ions on the peak shape and recovery performance of 5 μg/L of bromate ion in increasing concentrations of matrix anions. A Dionex IonPac AS19 column was used with a 10–45 mM potassium hydroxide eluent gradient at a flow rate of 1.0 mL/min and an injection volume of 0.5 mL. A Dionex 4-mm AERS 500 was used in this work at a setting of 113 mA. Notice that at 250 ppm each of chloride and sulfate, there is almost a complete loss of signal from bromate. In this case, the presence of bromate is masked, and a broadened baseline smear is observed. It would be difficult to detect and quantify bromate using this method.

Fig. 9.17A shows the first-dimension analysis of bromate at 5-μg/L concentration in a high ionic strength matrix. Fig. 9.17B shows the same

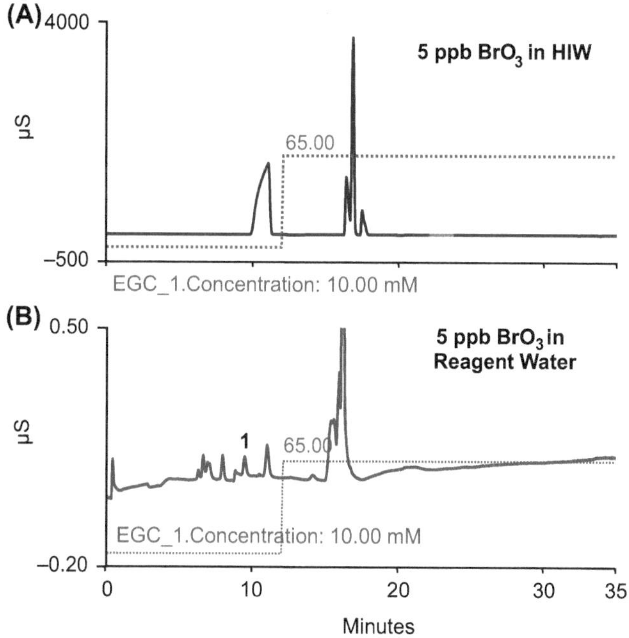

Figure 9.17 The results obtained by the matrix elimination ion chromatography method is shown here for the first dimension for a reagent water sample and for a sample containing high levels of matrix ions referred to as high ionic strength water (HIW). Detection of bromate in the HIW was not possible.

analysis, but without the matrix present. The matrix concentrations were 100 mg/L of chloride, carbonate, and sulfate, and 10 mg/L of N as nitrate and P as phosphate. A Dionex IonPac AS19 (4 × 250 mm) column was used with a gradient from 10 mM to 65 mM potassium hydroxide concentration. A Dionex 4-mm AERS 500 suppressor was used at a current setting of 161 mA. The sample injection volume was 1 mL. It is clear that the bromate peak cannot be detected in the high ionic strength sample in the first-dimension method.

Figs. 9.18A and B compares the second dimension results for the same analysis. A Dionex IonPac 2 × 250 mm column was used at 0.25 mL/min and with a potassium hydroxide gradient of 10–65 mM. A Dionex 2-mm AERS 500 suppressor was used at a current setting of 41 mA. Excellent peak shape and recovery can be inferred for the 5-μg/L bromate sample. Clearly the MEIC setup, as depicted in EPA Method 302.0, eliminates the impact of the matrix as seen previously in Fig. 9.16 [13].

A more sophisticated setup for matrix elimination is needed when pursuing concentration of multiple heart cuts, such as when pursuing analysis of

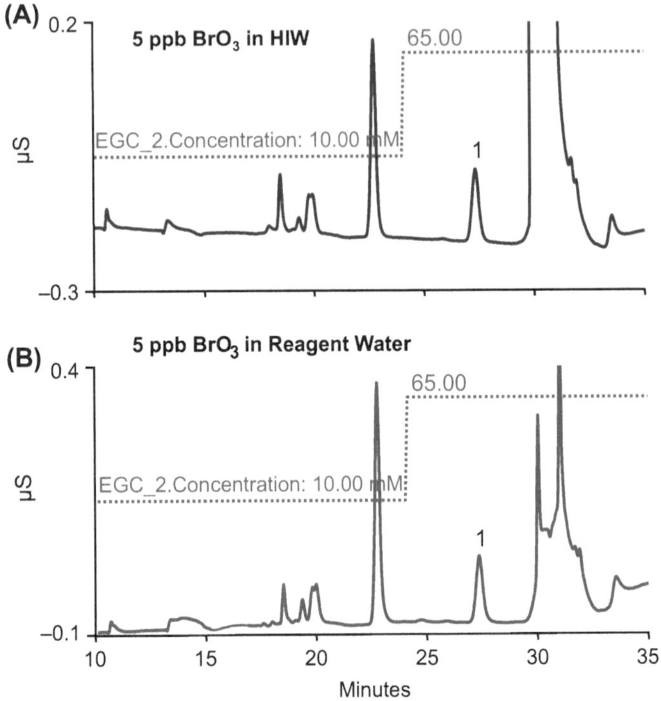

Figure 9.18 The results obtained by the matrix elimination ion chromatography method is shown here for the second dimension for a reagent water sample and for a sample containing high levels of matrix ions referred to as high ionic strength water (HIW). Excellent recovery of bromate was obtained as evident from the peak shape and peak area of the peak labeled "1" in the two chromatograms.

nine haloacetic acid samples. Fig. 9.19 shows the setup for MEIC implementation. In the simplest setup, an additional diverter valve is added between the first and second dimension to facilitate diversion of various heart cuts into the concentrator column. In order to facilitate improved sensitivity in the second dimension, the second-dimension separator is a capillary format so that a 100 x improvement in sensitivity can be realized with a concentrator sensitive detection, such as conductivity detection. Referring back to Fig. 9.19, multiple heart cuts are routed from the first dimension to the concentrator column while diverting the matrix ions to waste. Three heart cut portions representing the mono-, di- and trihaloacetic acids were routed to the concentrator column. The results from the MEIC method second dimension is shown in Figs. 9.20A and B. The first dimension was operated with a Dionex IonPac AS24 4 ×250 mm column operated at 1 mL/min with a potassium hydroxide gradient from 7 to 65 mM. A Dionex 4-mm

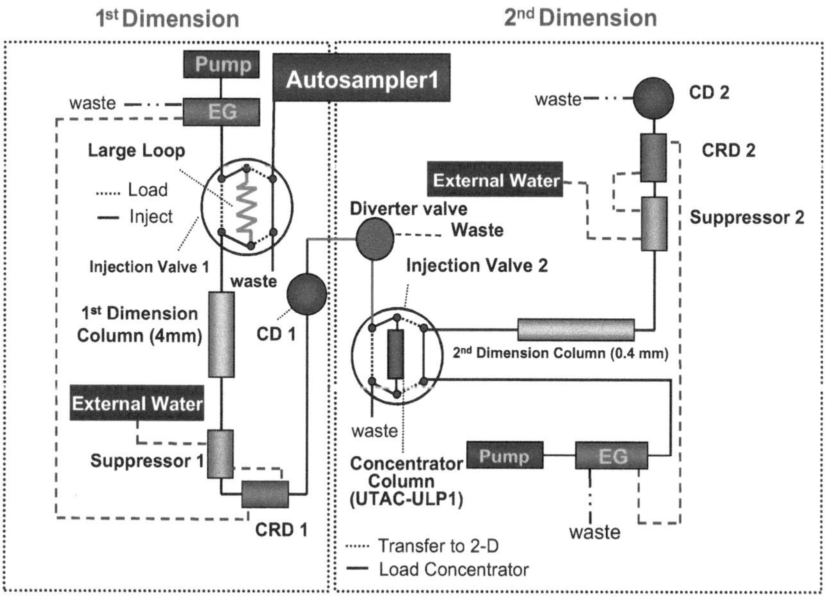

Figure 9.19 A schematic of the matrix elimination ion chromatography method is shown here. In the first dimension, the sample is analyzed and the matrix ions are diverted to waste. Three zones of interest are routed from the first dimension and are focused onto a concentrator column and analyzed in the second dimension using a column with a smaller internal diameter. A diverter valve is used for the routing the three zones from the first dimension to the second dimension.

AERS 500 suppressor was used at a current setting of 161 mA. The sample loop size was 0.5 mL. Sample A comprised nine haloacetic acid (HAA) standards at a final concentration of 10 μg/L spiked to an HIW matrix containing 250 mg/L of chloride and sulfate and 150 mg/L of bicarbonate and 10 mg/L of N as nitrate and P as phosphate. One hundred mg/L of ammonium chloride was added as a preservative to the water sample. Sample B comprised nine HAA standards at a final concentration of 10 μg/L spiked to 100 mg/L of ammonium chloride prepared in DI water. A Dionex MAC200 concentrator column was used in this work. The second dimension is operated with a Dionex IonPac AS26 0.4 × 250 mm capillary column operated at 0.012 mL/min. A potassium hydroxide gradient from 6 to 160 mM was used. A Dionex ACES suppressor was used in this work at a current setting of 25 mA. Figs. 9.20A and B show the comparison of the recovery obtained with samples containing matrix and no matrix. Comparable results are seen for the nine HAA standards from the two chromatograms illustrating excellent performance of the MEIC method in the presence of matrix ions [23].

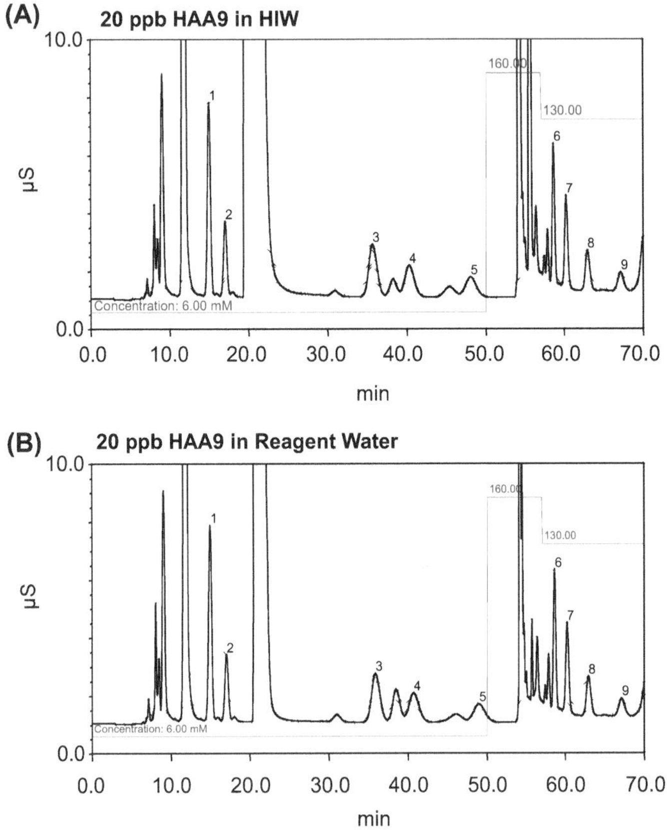

Figure 9.20 The results from the matrix elimination ion chromatography method second dimension is shown here comparing the performance of a haloacetic acid sample that had no matrix ions versus a sample containing high ionic strength water (HIW) matrix.

REFERENCES

[1] Understanding the safe water drinking acthttps, www.epa.gov/sites/production/files/2015-04/documents/epa816f04030.pdf.

[2] Pfaff JD. Determination of inorganic ions in drinking water by ion chromatography. 1993. U.S. EPA Method 300.0.

[3] Hautman DP, Munch DJ. Determination of inorganic ions in drinking water by ion chromatography. 1997. U.S. EPA Method 300.1.

[4] Small H, Stevens TS, Bauman WC. Novel ion exchange chromatographic method using conductimetric detection. Anal Chem 1975;47:1801–9.

[5] Dasgupta PK. Linear and helical flow in a perfluorosulfonate membrane of annular geometry as a continuous cation exchanger. Anal Chem 1984;56:96–103.

[6] Dasgupta PK. Annular helical suppressor for ion chromatography. Anal Chem 1984; 56:103–5.
[7] Dasgupta PK, Bligh RQ, Mercurio MA. Dual membrane annular helical suppressors in ion chromatography. Anal Chem 1985;57:484–9.
[8] Stillian JR. An improved suppressor for ion chromatography. LC Liq Chromatogr 1985;3:802–6.
[9] Srinivasan K, Barreto VB, Pohl CA, Thayer JR, Avdalovic N. Inventors; Dionex Corporation, assignee. Current-efficient suppressors and method of use. US patent 6,077,434. June 20, 2000.
[10] Srinivasan K, Barreto VB, Pohl CA, Thayer JR, Avdalovic N. inventors; Dionex Corporation, assignee. Current-efficient suppressors. US patent 6,328,885. December 11, 2001.
[11] Liu Y, Srinivasan K, Pohl C, Avdalovic N. Recent developments in electrolytic devices for ion chromatography. J Biochem Biophys Methods 2004;60:205–32.
[12] Wagner HP, Pepich BV, Later D, Pohl C, Srinivasan K, Borba BD, Lin R, Munch DJ. Determination of perchlorate in drinking water using two- dimensional ion chromatography with suppressed conductivity detection. May 2008. U.S. EPA Method 314.2.
[13] Wagner HP, Pepich BV, Later D, Pohl C, Srinivasan K, Borba BD, Lin R, Munch DJ. Determination of bromate in drinking water using two- dimensional ion chromatography with suppressed conductivity detection. September 2009. U.S. EPA Method 302.
[14] Wagner HP, Pepich BV, Hautman DP, Munch DJ. Determination of inorganic oxyhalide disinfection by-products in drinking water using ion chromatography with the addition of a postcolumn reagent for trace bromate analysis. May 2000. U.S. EPA Method 317.
[15] Wagner HP, Suarez FX, Pepich BV, Hautman DP, Munch DJ. Challenges encountered in extending the sensitivity of US Environmental Protection Agency Method 314.0 for perchlorate in drinking water. J Chromatogr A 2004;1039(1–2):97–104.
[16] Lin R, et al. Technical Programme, 1960–1965. Pittcon; 2010.
[17] Srinivasan K, Bhardwaj S, Lin R, Pohl, C. Weak Cation-exchange resin based concentrator and trap column for ion chromatography, Published at: http://www.dionex.com/en-us/webdocs/52824-LPN%201924_Cation-exchange.pdf.
[18] Peldszus S, Huck PM, Andrews SA. Quantitative determination of oxalate and other organic acids in drinking water at low microgram/1 concentrations. J Chromatogr A 1998;793(1):198–203.
[19] Huang Y, Mou S, Yan Y. Determination of bromate in drinking water at the low μg/L level by column switching ion chromatography. J Liq Chrom Rel Technol 1999; 22(14):2235–45.
[20] Srinivasan K, Lin R, Pohl CP. IICS 2008 Scientific Programme. 2008. P32.
[21] Lin R, De Borba B, Srinivasan K, Woodruff A, Pohl C. Matrix diversion methods for improved analysis of perchlorate by suppressed ion and conductivity detection. Anal Chim Acta May 10, 2006;567(1):135–42.
[22] Srinivasan K, Lin R, Pohl CP. Application of Multidimensional matrix elimination ion chromatography for bromate analysis. 2011. LC-GC Europe June 2–7.
[23] Srinivasan K, Lin R, Wagner HP. Multi-dimensional ion chromatography for trace ion analysis. Atlanta, Georgia, USA: Pittcon; March 7, 2016. p. 360–2.

CHAPTER TEN

Use of Ion Chromatography for Monitoring Ionic Contaminants in Water

C. Pohl
Thermo Fisher Scientific, Sunnyvale, CA, United States

1. INTRODUCTION

From its earliest days, ion chromatography (IC) has been used for the analysis of ionic components in a wide variety of sample types. IC has been used in analytical applications ranging from the analysis of biological fluids to air monitoring. Even the first paper published [1] by the inventor of IC in 1975 described its use with a number of different applications. This paper pioneered the use of an ion exchange separation in conjunction with a "suppressor column" and conductivity detection as a general approach to the analysis of ionic species (technical properties of the suppressor column will be discussed in more detail in later sections of this chapter). In spite of the broad analytical application range of IC, use of the technique for the analysis of ionic contaminants in drinking water and wastewater represents the largest single application of IC. IC provides the combined benefits of high sensitivity, broad dynamic range, and general suitability to the analysis of ionic compounds, making it ideally suited for this area of analytical chemistry.

2. INSTRUMENTATION AND COLUMN TECHNOLOGY FOR ION CHROMATOGRAPHY

At first glance, IC instrumentation appears to be just another form of high pressure liquid chromatography (HPLC), needlessly differentiated from HPLC in name only. However, direct comparison between the two techniques reveals a number of substantial differences that warrant differentiating the nomenclature for the two techniques. For example, while both techniques make use of high-pressure pumping technology to deliver the mobile phase to the chromatographic column, IC instrumentation invariably makes exclusive use of nonmetallic fluidic components in the flow path,

Chemistry and Water
ISBN: 978-0-12-809330-6
http://dx.doi.org/10.1016/B978-0-12-809330-6.00010-6

while HPLC generally makes use of metallic fluidic components for most of the flow path. The differences mainly arise from the types of mobile phases commonly used in the two techniques. In IC, optimum performance of the suppressor function dictate the use of highly aqueous mobile phases based on alkali metal hydroxides, such as potassium hydroxide, or strong acids, such as methanesulfonic acid. Nonmetallic flow path components are the optimal choice for eluents of this sort, as fluidic components based on engineering polymers, such as polyether ether ketone, can be exposed to such mobile phases indefinitely without damage. On the other hand, metallic flow path components are subject to corrosion upon exposure to these mobile phases, and corrosion frequently results in contamination of the chromatographic column upon extended exposure of the metallic flow path components to IC mobile phases. In contrast, HPLC typically makes use of water—organic solvent mixtures. The presence of organic solvents in the mobile phase dictate the use of metallic fluidic components, as such components are optimally compatible with water—organic solvent mixtures. Use of polymeric components in HPLC is generally avoided, at least in high-pressure region of the flow path, due to the fact that organic solvents reduce the pressure rating of nonmetallic fluid components.

Another major difference between IC and HPLC is the chemical composition of the stationary phase. Most stationary phases in HPLC are based on a silica substrate to which a variety of different ligands have been covalently attached. Silica represents an ideal substrate for modes of chromatography where organic solvent is a common component, since silica does not undergo swelling upon exposure to organic solvent. As evidence of this, while there are a few notable exceptions, such as in the case of gel permeation chromatography, very few commercially available HPLC columns are based on polymeric substrates. In contrast, polymeric substrates generally are to be avoided when using mobile phases containing organic solvents, as polymeric substrates generally swell in organic solvent, especially if the solubility parameters of the solvent is close to those of the polymeric substrate. Most stationary phases in IC are based on polymeric substrates. Because most IC mobile phases are aqueous based, swelling of the substrate in organic solvents is irrelevant. At the same time, polymeric substrates based on styrenic monomers are extremely stable with respect to exposure to extremes of pH. As previously mentioned, mobile phases in IC make extensive of highly alkaline or highly acidic mobile phases, so it is critical that substrates used in IC are compatible to such pH extremes without measurable degradation. In contrast, silica-based substrates are notoriously unstable under alkaline conditions, while covalently bonded ligands typically used in HPLC are readily hydrolyzed under the low pH conditions commonly utilized in IC. Taken together, these two attributes render silica unsuitable for use in IC, and as such, there are very few commercially available IC columns based on silica.

A third distinction between IC and HPLC is related to differences in the design of the stationary phase typical of columns used for each technique. By far, the most common retention mode in HPLC is reversed phase chromatography. Reversed phase retention is predominantly associated with van der Waals interactions between relatively hydrophobic analytes and hydrophobic ligands attached to the surface of the stationary phase. Since van der Waals forces are relatively weak, stationary phases designed for reverse phase require relatively high surface areas in order to provide sufficient retention. As such, HPLC chromatographic substrates tend to make use of relatively high surface area materials. Since 2005 the use of core shell particles in HPLC with lower surface area has become popular, the use of such materials did not become prevalent until technologies for making these materials with relatively high surface area became available. In contrast, the most common retention mode in IC is ion exchange. In the case of ion exchange, the dominant retention factor is electrostatic interaction between the fixed ion exchange site and the oppositely charged exchangeable ion. Because electrostatic forces are comparatively strong forces, the number of ion exchange sites required to achieve acceptable chromatographic performance is substantially lower than required for reversed phase retention. At the same time, the greater intensity of the electrostatic attraction associated with retention in ion exchange results in slow exchange kinetics if the stationary phase film thickness is too great. As such, commonly used materials for IC tend to be low surface area materials, which have been highly optimized to maximize ion exchange kinetics.

A fourth distinction between IC and HPLC is the choice of detector typically used for the two techniques. While both techniques make use of a wide variety of different detectors, the vast majority of all IC is accomplished with a conductivity detector. In the case of HPLC, the vast majority of applications utilize a UV detector. Because the magnitude of the conductivity signal is highly temperature dependent while UV signal is relatively free from temperature sensitivity, this dictates that the design of instrumentation for IC is typically highly optimized with respect to temperature control in order to provide the best detection limits. In HPLC, temperature control is common, but the stringency of such control is typically inferior to what is employed in

IC as a consequence of the differences in temperature sensitivity for the most commonly used detectors associated with each technique.

As a result of these four distinguishing characteristics, instruments optimized for IC are substantially different than instruments optimized for HPLC. While nothing prevents one from practicing IC on an HPLC instrument or HPLC on an IC instrument, the best overall performance with respect to ruggedness and reliability as well as sensitivity, precision, and accuracy is achieved when using an instrument optimized specifically for IC or optimized specifically for HPLC.

3. CHROMATOGRAPHIC COLUMNS IN ION CHROMATOGRAPHY

As mentioned previously, the design of chromatographic materials for IC is substantially different than the design of chromatographic materials in HPLC. The optimal mobile phase for IC is either a strong acid, such as methanesulfonic acid when separating cations, or a strong base, such as potassium hydroxide when separating anions. Preference for mobile phases of this type is based on the fact that the product of suppression for such mobile phases is water, which minimizes suppressed mobile phase conductivity and thus detector noise. Because IC stationary phases are typically exposed to such mobile phases during the entirety of the column life, it is imperative that stationary phases used in IC are stable with respect to both acid and base hydrolysis. This means that neither ester nor amide linkage chemistries can be utilized in such phases. In addition, virtually all inorganic substrates commonly used in HPLC, such as silica, alumina, or zirconia, are incompatible with such mobile phases, especially under alkaline conditions. For that reason, as previously mentioned, most IC stationary phases are based on hydrolytically stable polymeric materials, which can withstand such extremes in pH indefinitely. Polymeric materials most often used for this application are usually styrene—divinylbenzene (DVB) copolymers or, more commonly in recent stationary phases, ethylvinylbenzene—DVB copolymers. The latter are usually prepared from technical grade 55% DVB, which produces a particle that is extremely hard and durable, with mechanical properties similar to that of a billiard ball.

IC stationary phases tend to be relatively low capacity compared to conventional ion exchange resins, which have ion exchange sites distributed homogeneously throughout the particle. Low capacity is needed in order to minimize the concentration of mobile phase necessary to elute analytes of interest so that suppressor life is maximized. This was particularly true in the early days of IC when packed bed suppressors were in wide use. Newer continuous regeneration suppressor technology has a greatly expanded operating concentration range, and as such, many of the newer columns in use today for IC applications have significantly higher capacity while still maintaining high chromatographic performance. The earliest stationary phase architecture used in IC, made use of pellicular architecture, as described in the original IC paper by Hamish Small. The cation-exchange phase was based on polystyrene beads cross-linked with 2% DVB, which was lightly surface-sulfonated to introduce cation-exchange sites on the exterior surface of the bead. The first commercial columns made use of relatively large particle size media by today's standards: 25—35 μm particles. Anion exchange columns were prepared by coating the surface-sulfonated cation-exchange resin with a monolayer of colloidal anion exchange particles, attached to the surface of the particle via electrostatic interaction. At first glance, such an attachment mechanism might appear to be impermanent, with the anion exchange particles easily removed by contacting the ion exchange composite with a suitably high electrolyte concentration. In fact, the attraction between the colloidal particles coating and the surface-sulfonated substrate is so strong that no eluent is capable of removing the coating. Furthermore, given the fact that the surface of the substrate has the same charge as analytes of interest, the negatively charged surface serves the secondary function of repelling analytes from the substrate surface, affectively focusing retention in an extremely thin film on the outer surface of the particle and preventing any unwanted secondary interactions with the substrate surface.

4. SUPPRESSORS IN ION CHROMATOGRAPHY

The suppressor serves two parallel roles in this analytical technique. First, the suppressor reduces the conductivity of the mobile phase via an exchange reaction, which replaces the eluent counter ion with hydronium in the case of anion analysis or hydroxide in the case of cation analysis. In either case, the replacement ion combines with the eluent species to produce a low conductivity, minimizing the conductivity of the mobile phase and greatly reducing the detector noise associated with the eluent conductivity. Second, the suppressor converts the analyte species into a more conductive form, which increases sensitivity. For example, in the case of anion analysis, the detected species shifts from the potassium salt of the anion in question to the hydronium form of the anion. Since hydronium is much more conductive than any other ion, the signal associated with the hydronium form of the anion is significantly enhanced. Taken together, the reduced baseline noise and the enhanced analyte detection combine to provide substantial enhancement of detection sensitivity and substantially reduced detection limits. Typical detection limits are 10–50 times lower than analogous detection limits obtained in the absence of a suppressor column [2].

Initially, the primary appeal of IC was in the analysis of ions such as chloride and sulfate that lack a suitable chromophore for UV/visible detection. Likewise, although elemental spectroscopy methods are suitable for the quantitation of alkali metals and alkaline earth metals, IC tends to be used for applications involving the quantitation of cations, such as ammonia, that are incompatible with the various forms of elemental spectroscopy. Soon after the introduction of IC, the use of the analytical technique expanded to include a wide variety of anions and cations, regardless of whether such ions contain a chromophore or are amenable to elemental spectroscopy. For example, early work focused on methodologies for the quantitation of anions commonly found in drinking water, such as fluoride, chloride, nitrite, bromide, nitrate, sulfate, and phosphate [3] in spite of the fact that nitrite, bromide, and nitrate can be readily measured using UV detection. Suppression can also be of direct benefit in UV detection in cases where the eluent is rendered weakly absorbing via the suppression process. Hydroxide anion absorbs at low UV wavelengths, so passing such an eluent through a neutralizing cation exchanger will substantially reduce the detector background, reducing noise and improving detection limits.

5. COLUMN DIMENSIONS IN ION CHROMATOGRAPHY

IC column dimensions tend to be somewhat larger than is common in modern HPLC. The majority of IC columns are 4-mm internal diameter (ID), with most columns being either 250 mm or 150 mm in length. This is despite the fact that 2-mm ID columns have been commercially available for more than 20 years, and capillary (0.4 mm ID) columns have been commercially available for more than 6 years. One reason that IC practitioners have been somewhat slow to convert to smaller ID columns is that, as previously mentioned, most IC mobile phases are 100% aqueous. The cost of purchasing organic solvents and the subsequent cost of solvent disposal has been a significant driving force in HPLC toward the adoption of smaller ID columns. Because IC mobile phases tend to be exclusively aqueous based, the extent of cost-driven column miniaturization has been comparatively slow. In addition, sample size limitations, another major driving force for the miniaturization of chromatographic columns, is rarely relevant to IC applications since sub-1 mL sample volumes tend to be rather uncommon. Generally, both drinking water and wastewater sample volumes tend to be more than adequate for use with 4-mm ID columns, providing relatively little motivation to use smaller ID columns for samples of this type.

5.1 Other Ion Chromatography Specific Components

In addition to the suppressor device, IC utilizes a number of other components that are unique, including:

1. eluent generators
2. continuously regenerated trap columns
3. carbon dioxide removal devices

Eluent generator technology makes use of electrolysis to produce high-purity eluents on demand. Because eluent generation is considerably more precise than mechanically prepared mobile phases, eluent generation provides not only high purity eluents, but also better precision and accuracy with regard to eluent concentration. While eluent generation is not a requirement, it is widely used, with more than half of all instruments configured to enable this operating mode.

Continuously regenerated trap columns serve the purpose of further purifying mobile phases to remove contaminating impurities. Hydroxide is particularly prone to contamination with carbon dioxide arising from contact with air. In order to achieve the highest stability with minimal baseline drift when operating in gradient mode, it is critically important to minimize the amount of carbon dioxide dissolved in the mobile phase. For that reason, most practitioners of IC utilize not only the electrolytic eluent generator technology, but also continuously regenerated trap columns.

Carbon dioxide removal devices are also relatively common when using IC for anion analysis. After the suppressor device, any carbonate contamination in the mobile phase is converted to carbonic acid. A variety of membrane-based devices are commercially available for removing these trace carbon dioxide contaminants. When using eluent generator in conjunction with the continuously regenerated trap column to produce high-purity hydroxide-based mobile phases, the removal of carbon dioxide from the mobile phase is of minimal importance. In this case, the major factor driving the use of carbon dioxide removal devices is the prevalence of carbon dioxide—contaminated samples. In addition, carbon dioxide removal devices can also be used in conjunction with carbonate-based mobile phases. In such cases, the magnitude of the carbonate that must be removed is significantly higher than in the case of hydroxide eluents. For that reason, the removal of carbonate with such devices is not as quantitative as in the case of hydroxide eluents. As a consequence, carbon dioxide removal devices are generally only used with hydroxide eluents.

6. DRINKING WATER STANDARDS

Globally, there are a wide variety of drinking water standards regulating the concentration of a variety of ions commonly found in drinking water. In some regions of the world, such regulations are minimal. In other regions of the world, such regulations are in place, but lack consistent enforcement. In the developed world, there tends to be a fairly similar range of ions that are regulated. The Environmental Protection Agency (EPA) in the US has promulgated a series of regulations related to drinking water, and in many cases throughout the world, similar regulations are in place modeled after the EPA drinking water standards. For that reason, before delving into the various IC columns and methods for specific ionic components commonly found in drinking water, it is useful to consider the specific regulatory framework associated with such analytes. Table 10.1 shows a list of the ions commonly found in drinking water, either as a naturally occurring component or as a by-product of disinfection. Note that the majority of the anions commonly found in drinking water are subject to either maximum contaminant level regulations or secondary drinking water guidance. Notable exceptions include bromide, chlorate, and phosphate. In the case of cations, very few of the alkali metals or alkaline earth metals have any maximum contaminant level regulations or secondary drinking water guidance. The notable exception here is barium. It might seem surprising that there is no concentration guidance or contaminant limits for ions such as lithium or iodide, given that excesses of these two species will undoubtedly have health consequences at sufficiently high concentrations. However, EPA contaminant limits and drinking water secondary standards are guided by not only health effects and secondary water quality characteristics, but also the probability of detecting set species at a concentration that might be of concern. For example, elevated natural lithium levels in drinking water have been associated with lower suicide rates [4], presumably because lithium is a commonly utilized medication for treatment of bipolar disorder, whereas concentrations as low as 3 mmol/L in blood may be indicative of acute overdose conditions [5]. However, in spite of these well-known health effects, prevalence of lithium at concentrations that might be a health concern has been deemed by the EPA to be so low as to obviate the need for such a regulation.

Table 10.1 EPA Regulatory Status for Ions Found in Drinking Water Commonly Analyzed by Ion Chromatography

Ions Found in Drinking Water	Regulated Contaminant	Maximum Contaminant Limit (mg/L)	Secondary Standard	Recommended Concentration Limit (mg/L)	Unregulated Component
Fluoride	X	4	X	2	
Chlorite	X	1			
Cyanide (free)	X	0.2			
Bromate	X	0.01			
Chloride			X	250	
Nitrite (as N)	X	1			
Bromide					X
Nitrate (as N)	X	10			
Chlorate					X
Haloacetic acids	X	0.06			
Arsenic	X	0.01			
Phosphate					X
Chromium (total)	X	0.1			
Selenium	X	0.05			
Iodide					X
Sulfate			X	250	
Lithium					X
Sodium					X
Ammonium					X
Potassium					X
Rubidium					X
Cesium					X
Magnesium					X
Calcium					X
Strontium					X
Barium	X	2			

7. ANALYSIS OF IONS SUBJECT TO SECONDARY DRINKING WATER STANDARDS

Initial application of IC was to the analysis of the anions commonly found in drinking water: fluoride, chloride, nitrite, bromide, nitrate, phosphate, and sulfate as well as the cations commonly found in drinking water: lithium, sodium, potassium, ammonium, magnesium, and calcium. Of these ions, the concentrations of chloride and sulfate are generally regulated via secondary drinking water standards, [6] given that concentrations of these ions rarely exceed levels that will present a health hazard. Such water quality standards mainly serve the purpose of assuring that drinking water has acceptable taste. In contrast to regulated drinking water contaminants, which are associated with identified health effects and are subject to specific maximum contaminant levels, adherence to secondary drinking water standards is on a voluntary basis, given the fact that secondary drinking water standards are based upon aesthetic, cosmetic, or technical considerations rather than health considerations. EPA 300.0, Part A covers the use of IC for the analysis of fluoride, chloride, nitrite, bromide, nitrate, phosphate, and sulfate using an Dionex@ IonPac@ AS4A column. This ion exchange column, introduced in 1984, was designed specifically for this application. It makes use of a colloidal coating of 1% DVB cross-linked anion exchange latex functionalized with diethylethanolamine on a surface-sulfonated 15-μm nonporous substrate particle—based polystyrene cross-linked with 4% DVB. The combination of low cross-linking with a hydrophobic amination reagent results in significantly improved peak shape for bromide and nitrate compared to the earliest anion exchange phases used in IC when it was first introduced as an analytical technique. Although there have been many improvements in stationary phases for the analysis of anions in IC, the AS4A column is still widely used today. In 1992, the AS4A was further improved by switching to a smaller (13 μm) particle utilizing surface-sulfonated 55% DVB cross-linked particle. This version of the column, known as the IonPac AS4A-SC column (shown in Fig. 10.1), is also still widely used in drinking water applications. The smaller particle size provides improved resolution of target analytes, while the 55% cross-link of the substrate particle renders the column compatible with all common HPLC solvents. The latter feature is beneficial when attempting removal of polymeric anionic contaminants, such as humic substances, from the column. Such polymers may be present in certain water samples and will result in loss of effective column capacity and shifting retention time behavior, compromising analytical performance

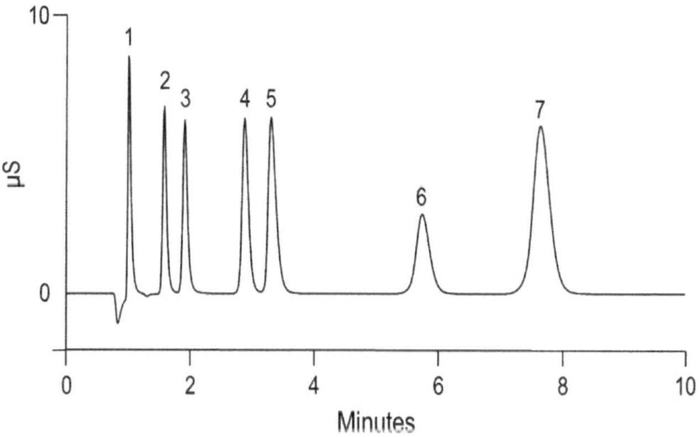

Figure 10.1 Columns: 4 mm × 250 mm IonPac AS4A-SC and 4 mm × 50 mm IonPac AG4A-SC, eluent: 1.7 mM sodium bicarbonate and 1.8 mM sodium carbonate, flow rate: 2 mL/min, injection volume: 50 μL, suppressor: 4 mm ASRS ULTRA, recycle mode, analytes: 1 – fluoride (2 mg/L); 2 – chloride (3 mg/L); 3 – nitrite (5 mg/L); 4 – bromide (10 mg/L); 5 – nitrate (10 mg/L); 6 – phosphate (15 mg/L); 7 – sulfate (15 mg/L).

of the column. Removal of such polymers from the column requires the use of cleaning protocols making use of 0.1 molar HCl in either 80% acetonitrile or 40% tetrahydrofuran THF. Such cleaning reagents result in excessive swelling of the AS4A 4% cross-link substrate particles, irreversibly damaging the column, while the highly cross-linked AS4A-SC substrate particles do not swell in such solvents and as such are compatible with these advanced cleaning protocols.

8. APPLICATION OF ION CHROMATOGRAPHY TO FLUORIDE ANALYSIS

Fluoride is naturally occurring in drinking water in many parts of the world. At high concentrations, it can result in a condition known as fluorosis. This condition is mainly associated with streaks of discoloration on teeth. Health effects are known to occur at higher levels, but cosmetic effects are at concentrations as low as one part per million. It is for that reason that EPA regulations with regard to fluoride include a maximum concentration limit of 4 ppm, but a secondary drinking water standard of 2 ppm due to cosmetic effects associated with higher concentrations. In fact, in April 2015, the recommended maximum concentration of fluoride in drinking water in cases where fluoride is intentionally introduced into the drinking water was changed to 0.7 ppm from the prior recommendation of 0.7–1.2 ppm. Before the advent of IC, the most common technique for analysis of fluoride in drinking water involved the use of ion selective electrodes. Fluoride ion selective electrodes are somewhat unique in that they are both highly sensitive to fluoride and subject relatively few interferences compared to other ion-selective electrodes [7]. The primary advantage of IC with respect to the analysis of fluoride stems from the ability to measure multiple anionic species of interest in a single analysis, whereas ion selective electrodes are generally restricted to the analysis of a single analyte.

Early anion exchange columns developed for use in IC provide relatively poor retention of fluoride due to the fact that fluoride is among the most highly hydrated anions with the possible exception of hydroxide anion. Highly hydrated ions are generally weakly retained in ion exchange due to the restricted access of such ions to the anion exchange matrix. The limited retention of fluoride presents a challenge to quantitation at sub part per million levels due the proximity of the fluoride peak to the system void, especially when using a carbonate eluent system, since such systems result in a relatively large negative baseline disturbance at the system void due to the carbonic acid background associated with the suppression reaction (see, for example, Fig. 10.1). In addition, a number of weakly retained carboxylic acids are known to elute in the proximity of fluoride, sometimes compromising the quantitation of fluoride. While carboxylic acids are relatively uncommon in drinking water, the presence of such acids in wastewater analysis frequently results in erroneous fluoride measurements in such samples with first-generation IC columns.

In the late 1980s, Doug Gjerde developed a new type of ion exchange column for use in IC based on that absorbed coating: the Sarasep AN1. Because absorbed coatings present a virtually unrestricted environment for ion exchange site hydration, columns prepared using this technology provided substantially improved resolution of fluoride from the system void, enhancing the reliability of fluoride measurements. This development stimulated a stream of new columns designed to provide similar or improved resolution of fluoride from the system void while at the same time avoiding the limited solvent compatibility associated with absorbed coating–based phases. The first Dionex column developed with the goal of achieving enhanced fluoride retention was the IonPac AS12A column. This column, still available today, makes use of a porous surface-sulfonated particle with average pore size of approximately 2000 Å. This substrate is coated with a monolayer of electrostatically attached colloidal anion exchange particles. The anion exchange particles were optimized so as to increase the water content of the ion exchange phase by making use of an extremely low cross-link polymer matrix. Because this ion exchange phase is electrostatically attached to a highly cross-linked substrate, the column is compatible with virtually all common HPLC solvents. A secondary benefit of this new column technology was improved resolution of fluoride from acetate, which sometimes compromised the quantitation of fluoride in earlier columns.

Following the development of the IonPac AS12A column, there was continued development of new stationary phases providing even higher water content and superior fluoride retention. The second column in the series was the IonPac AS14 column. This column utilized a highly cross-linked macroporous substrate so as to provide the means of attaching a thin film of highly hydrated ion exchange material covering the entire surface. The combination of high surface area (450 m^2 per gram) and small pore size (80−100 Å) provides an ideal platform for producing a high water content phase with reasonably high capacity, something that is hard to achieve using earlier ion exchange materials based on colloidal coatings. Instead, the stationary phase is covalently attached to the surface using radical grafting of the ion exchange monomer to the surface. The resulting stationary phase consists of a high density of ion exchange strands projecting from the surface, providing ample room for hydration of the ion exchange polymer. The polymer backbone of the stationary phase is based on a methacrylate monomers system, which, due to its relatively high polarity, further enhances the hydration of the stationary phase. Fig. 10.2 illustrates the use of the IonPac AS14 column for the separation of common anions, including fluoride. As can be seen from the figure, this column provides excellent resolution of fluoride from the system void while at the same time providing improved acetate−fluoride resolution.

While the IonPac AS14 demonstrates improvement over the IonPac AS12A, the use of a methacrylate-based polymer backbone for the stationary phase brings with it limitations in terms of the hydrolytic stability of the stationary phase. Because the concentration of ion in the stationary phase can be orders of magnitude higher than the concentration of the mobile phase components, the pH of the stationary phase can easily exceed the pH limit of the polymer matrix under alkaline conditions. This is especially relevant when the pH of the mobile phase exceeds 11 or greater. For that reason, there was interest in producing an improved version of the IonPac AS14

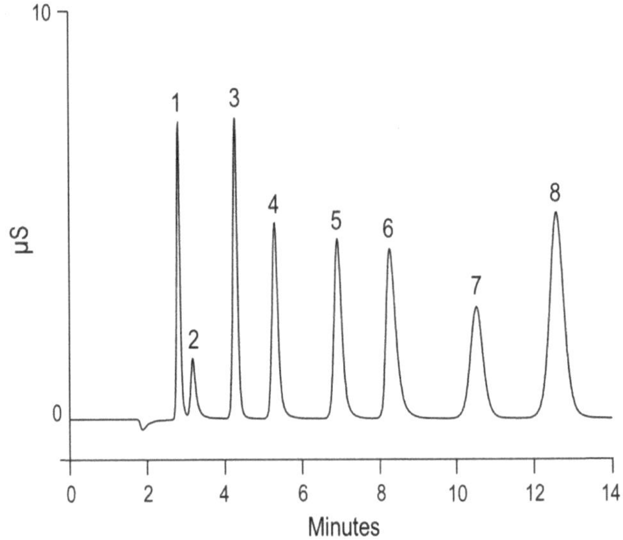

Figure 10.2 Columns: 4 mm × 250 mm IonPac AS14 and 4 mm × 50 mm IonPac AG14, eluent: 1 mM sodium bicarbonate and 3.5 mM sodium carbonate, flow rate: 1.2 mL/min, injection volume: 10 μL, suppressor: 4 mm ASRS ULTRA, recycle mode, analytes: 1 – fluoride (5 mg/L); 2 – acetate (20 mg/L); 3 – chloride (10 mg/L), 4 – nitrite (15 mg/L); 5 – bromide (25 mg/L); 6 – nitrate (25 mg/L); 7 – phosphate (40 mg/L); 8 – sulfate (30 mg/L).

using a styrenic monomer system in order to avoid issues with stationary phase hydrolysis when operating under elevated pH conditions. One of the problems with grafting ionic monomers directly to a hydrophobic substrate is that there is a tendency for some of the monomer to graft subsurface in the dense polymer matrix—producing ion exchange sites, which are highly restricted with regard to hydration. Such ion exchange sites have totally different ion exchange properties than the polymer strands projecting away from the surface, which are in a highly hydrated environment. As a consequence, grafting directly to the surface tends to result in a relatively heterogeneous ion exchange phase. Achieving high efficiency and narrow, Gaussian peaks in a chromatographic stationary phase requires a high degree of homogeneity in terms of the stationary phase. In the case of the IonPac AS14, a methacrylate polymer backbone was chosen partially because methacrylate monomers, especially ion exchange monomers, are relatively insoluble in the highly cross-linked styrenic substrate, minimizing the tendency of the grafting process to produce ion exchange sites subsurface. However, in this follow-up development, making use of styrenic monomers while avoiding propensity of such monomers to dissolve in the substrate was desirable to avoid introducing a subsurface ion exchange site with undesirable ion exchange properties. So, in order to achieve the combination of good hydrolytic stability and good peak shape, it was necessary to develop a new technique for preparing the stationary phase. This was accomplished by introduction of a hydrophilic boundary layer on the substrate surface prior to the attachment of the ion exchange polymer strands to the surface. The realization of this new synthetic approach was the IonPac AS14A. The IonPac AS14A makes use of the same type of substrate particles used in the preparation of the IonPac AS14, but instead of attaching the styrenic ion exchange monomers directly to the substrate surface, the surface is first modified to introduce polar groups, and in a subsequent reaction, styrenic monomer units attached directly to the surface of the substrate via a hydrogen abstraction reaction with polar substituents on the surface. The resulting polymer has chromatographic performance similar to the IonPac AS14A with the added benefit that the stationary phase is compatible with mobile phases spanning pH 0 to pH 14. Fig. 10.3 illustrates the chromatography of to the IonPac AS14A.

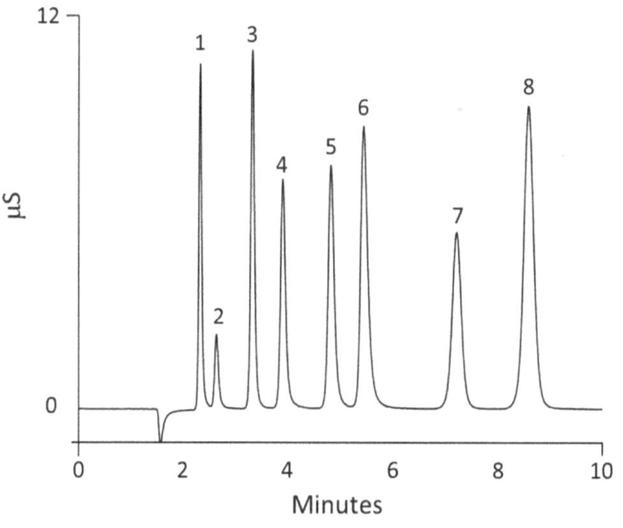

Figure 10.3 Columns: 3 mm × 150 mm IonPac AS14A-5μm and 3 mm × 30 mm IonPac AG14A-5 μm, eluent: 1 mM sodium bicarbonate and 8 mM sodium carbonate, flow rate: 0.5 mL/min, injection volume: 5 μL, suppressor: 2 mm ASRS ULTRA, recycle mode, analytes: one- fluoride (5 mg/L); two- acetate (20 mg/L); three- chloride (10 mg/L); four- nitrite (15 mg/L); five- bromide (25 mg/L); six- nitrate (25 mg/L); seven-phosphate (40 mg/L); eight- sulfate (30 mg/L).

All of the previously mentioned columns provide approved fluoride resolution primarily through the development of stationary phases with relatively unlimited hydration of the stationary phase. Such conditions provide improved resolution for fluoride primarily because they provide greater accessibility for highly hydrated ions, such as fluoride, into the stationary phase. This approach is particularly useful in the case of fluoride analysis when the mobile phase is based on bicarbonate/carbonate mobile phases. However, another entirely different approach has proven to be even more versatile: the development of stationary phase optimized for the use of hydroxide eluent systems. Early IC columns made use of carbonate mobile phases, not because carbonate provided better chromatography for better detection limits, but because conventional ion exchange sites resulted in stationary phases with relatively limited affinity for hydroxide anion. As mentioned previously, hydroxide is of the few ions more highly hydrated than fluoride, and just like fluoride, its elution power tends to be limited by the restricted access of highly hydrated hydroxide. In the mid-1980s, work began on the development of new ion exchange materials optimized for use with hydroxide eluents. The first column in this series was the IonPac AS5 column. Although it was designed for use with hydroxide eluent, the fiber suppressors in commercial use at the time did not provide adequate suppression capacity to allow the use of hydroxide eluents with this column. However, 1985 saw the introduction of the first commercial membrane suppressor, the AMMS-1. This suppressor was capable of continuously suppressing up to 200 mM hydroxide, and as such, it brought the suppression capacity in line with the hydroxide selectivity technology under development at that time. Thus, in 1986, the field of IC saw the first column specifically developed for the use with hydroxide eluents: the IonPac AS5A. The architecture of this new column was similar to that of the IonPac AS4A except that the particle size was reduced to 5 μm and the colloidal anion exchange phase is attached to the surface-sulfonated substrate and was specifically designed for use with hydroxide eluents. Fig. 10.4 shows a separation of a wide variety of organic acids and common inorganic anions with a more recently developed column designed for use with hydroxide eluents, the AS11-HC column, making use of a hydroxide gradient.

Gradient elution conditions are generally beneficial in chromatography when samples contain both weakly retained ions and highly retained ions. In particular, gradient elution provide significantly better resolution of fluoride from weakly retained monocarboxylic acids that might be found in a wastewater sample. In contrast to carbonate/bicarbonate eluent

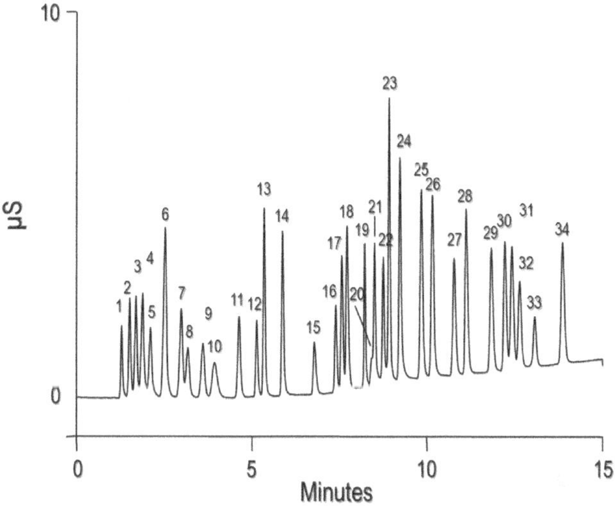

Figure 10.4 Column: 4 mm × 250 mm IonPac AS11, eluent: 0.5 mM NaOH 0–2 min, 0.5–5 mM NaOH 2.5–5 min, 5–38.25 mM NaOH 6–15 min, flow rate: 2 mL/min, injection volume: 10 μL, suppressor: 4 mm ASRS ULTRA, recycle mode, analytes: ions: 1 – isopropyl methylphosphonate (5 mg/L); 2 – quinate (5 mg/L); 3 – fluoride (1 mg/L); 4 – acetate (5 mg/L); 5 – propionate (5 mg/L); 6 – formate (5 mg/L); 7 – methylsulfonate (5 mg/L); 8 – pyruvate (5 mg/L); 9 – chlorite (5 mg/L); 10 – valerate (5 mg/L); 11 – monochloroacetate (5 mg/L); 12 – bromate (5 mg/L); 1 – chloride (2 mg/L); 14 – nitrite (5 mg/L); 15 – trifluoroacetate (5 mg/L); 16 – bromide (3 mg/L); 17 – nitrate (3 mg/L); 18 –chlorate (3 mg/L); 19 – selenite (5 mg/L); 20 – carbonate (5 mg/L); 21 – malonate (5 mg/L); 22 – maleate (5 mg/L); 23 – sulfate (5 mg/L); 24 – oxalate (5 mg/L); 25 – ketomalonate (10 mg/L); 26 – tungstate (10 mg/L); 27 – phthalate (10 mg/L); 28 – phosphate (10 mg/L); 29 – chromate (10 mg/L); 30 – citrate (10 mg/L); 31 – tricarballylate (10 mg/L); 32 – isocitrate (10 mg/L); 33 – *cis*-aconitate; and 34 – *trans*-aconitate (10 mg/L combined).

systems, hydroxide is substantially superior to carbonate for gradient applications due to the low background conductivity associated with hydroxide. Furthermore, salt form changes associated with a bicarbonate/carbonate gradient are relatively incompatible with true gradient conditions due to the fact that the affinity of the stationary phase for bicarbonate is much lower than the affinity of the stationary phase for carbonate. Even if one executes the gradient starting with, for example, 5 mM bicarbonate and increases the concentration of carbonate to 5 mM while simultaneously reducing the concentration of bicarbonate to zero, none of the carbonate entering the column will exit the column until virtually all of the bicarbonate in the stationary phase has been converted to the carbonate form. As a result, attempting to accomplish a gradient using carbonate eluent has because the transition of the column from the bicarbonate form to the carbonate form in a discontinuous process rather than a continuous process as in the case of a hydroxide concentration gradient. As a result, a composition gradient starting with bicarbonate and ending with carbonate tends to produce two quasiisocratic zones, the first zone where bicarbonate elution strength is dictating the elution process and a second zone where essentially all of the bicarbonate has been displaced from the stationary phase and carbonate is dictating the elution process. Because of the displacement process associated with conversion of the stationary phase on the bicarbonate to the carbonate form, there is a limited focusing effect associated with a bicarbonate/carbonate composition gradient. In the case of a hydroxide concentration gradient, however, elution power is directly proportional to the concentration of hydroxide, and as such, analyte peaks can be highly focused during the gradient process. This focusing process results in peaks which elute with substantially reduced peak width along with a corresponding increase in peak height in comparison to isocratic conditions.

9. APPLICATION OF ION CHROMATOGRAPHY TO NITRATE AND NITRITE ANALYSIS

Nitrite and nitrate are contaminants of concern in drinking water and, to a lesser extent, in wastewater. Both nitrite and nitrate can enter water supplies from the air as well as via direct contamination of groundwater as well as from rainwater sources in areas heavily contaminated with atmospheric nitrate. In most of the developed world, the major source of nitrate and nitrite in the water supply is agricultural runoff as well as direct groundwater contamination in agricultural areas. Nitrate-based fertilizer is commonly the source of nitrogen in industrialized agriculture. Considering the high water solubility of nitrate and its minimal affinity for soils and minerals present in soils, nitrate is highly mobile in groundwater. Reports of high levels of groundwater contamination in agricultural areas are very common, and for many living in such areas, bottled water has replaced groundwater as a source of drinking water in order to avoid exposure to excessive levels of nitrate and nitrite. Nitrate is not as particularly toxic as nitrite, although accidental exposure to toxic levels of nitrate have been reported. Toxicity of nitrate is primarily associated with conversion in vivo to nitrite, which is significantly more toxic than nitrate. High levels of nitrate or nitrite can result in a condition known as methemoglobinemia. Infants are especially susceptible to high levels of nitrate or nitrite. Methemoglobinemia in infants is more commonly known as blue baby syndrome.

Given the aforementioned toxicity issues associated with nitrite and nitrate and the prevalence of nitrate and nitrite contamination of both groundwater and surface water, it's not surprising that restrictions in the concentration of these ions in drinking water is a common feature of the regulatory environment for drinking water. Spectrophotometric and ion selective electrode analytical methods are widely used for the determination of both nitrate and nitrite in drinking water. IC has become a generally preferred alternative to these methods for water analysis laboratories equipped with a full range of analytical instrumentation, given the fact that IC allows for the determination of not just nitrite and nitrite, but a whole range of ionic contaminants commonly found in drinking water.

Given the relatively high concentrations of nitrite and nitrate commonly found in surface and groundwater, analysis is typically straightforward. Generally, any commercially available anion exchange column designed for IC is suitable for the analysis of nitrite and nitrate. Nonetheless, there are special aspects to the chromatography of both nitrite and nitrate that warrant special attention. Nitrate, for example, is a relatively poorly hydrated, polarizable anion that interacts with anion exchange materials both via electrostatic interaction and through interaction with the polymer backbone of the anion exchange material. As mentioned previously, the high pH associated with mobile phases commonly used in IC necessitates the use of a polymer backbone in order to avoid hydrolytic damage to the stationary phase. Most of the early anion exchange materials used in IC made use of styrene monomers cross-linked with divinylbenzene, subsequently functionalized with a tertiary amine to form quaternary ion exchange sites. Polarizable ions, such as nitrate, interact with the π electrons as the aromatic backbone. The presence of a secondary retention mechanism of this sort in and of itself doesn't represent an issue with the chromatography. However, one artifact associated with DVB cross-linking is that cross-links tend to be somewhat nonuniform due to the fact that DVB is much more reactive than other styrenic monomers. This gives rise to areas in the anion exchange polymer that are relatively poorly hydrated due to the presence of relatively high cross-link density compared to the overall average cross-link density. Since polarizable anions such as nitrate are highly sensitive to the hydration environment associated with the ion exchange site and because of the fact that hydration is significantly restricted in regions of high cross-link density, polarizable ions tend to exhibit much higher affinity for ion exchange sites in such environments. Because such sites are high affinity but relatively low in abundance, they result in relatively poor peak shape for polarizable ions. At low concentrations, the local concentration of nitrate in the ion exchange polymer will be predominantly in these areas of low hydration since the affinity is highest for polarizable ions at poorly hydrated ion exchange sites. The retention time associated with these high retention sites will be higher than the average ion exchange site with a higher degree of hydration. Once the concentration of nitrate in the injected sample is high enough to overload these relatively few high retention sites, nitrate is forced to occupy ion exchange sites with higher degrees of hydration, which are less retentive than the relatively few poorly hydrated sites. As a result, as the concentration of polarizable ions increases in the sample, retention time shifts earlier while the peak tends to exhibit a "tail" ending at the retention time associated with the poorly hydrated sites, whereas highly hydrated ions such as fluoride or sulfate will tend to exhibit normal Gaussian peak shape under the same conditions due to the fact that highly hydrated ions have low affinity for poorly hydrated ion exchange sites and high affinity for the predominant better hydrated sites in the anion exchange polymer.

All of the preceding results in peak shape for polarizable ions such as nitrate or perchlorate that tend to suffer when using styrenic monomers to form the ion exchange polymer, especially when DVB cross-linking is used. Non-Gaussian peak shape adversely affects quantitation. For Gaussian peaks, peak height and peak area are equally amenable as the basis for quantitative analysis. Peak areas are generally preferred because the measurement of peak area is generally more precise than the measurement of peak height. However, as concentrations approach the detection limit where peak response factors fall within a factor of three of the baseline noise, peak height measurements generally provide better precision. Because non-Gaussian peaks exhibit shifting retention time for the peak apex as concentration increases, peak heights exhibit a nonlinear relationship with analyte concentration. Peak area is, of course, a linear function regardless of peak shape; however, a high degree of peak asymmetry generally compromises the accuracy of peak area measurements. Consequently, when using chromatography for analyte concentration measurement, it is highly preferable to utilize chromatographic media that provide Gaussian peak shape. The earliest ion exchange phases had relatively poor peak shape for nitrate. The IonPac AS4A column shown in Fig. 10.1 is an example of one of the earliest attempts to improve the peak shape for polarizable anions. This was largely accomplished by utilizing a low cross-link ion exchange polymer matrix; in this case, only 1% DVB is used to form the ion exchange polymer. Generally, retention for all ions decreases as the cross-link decreases due to the fact that there is relatively unrestricted hydration in low cross-link materials, and hydration restriction is a key factor in driving ion exchange selectivity. In order to offset this limited retention, it was necessary to form the quaternary ion exchange site using a more hydrophobic amine; in this case, diethylethanolamine was used. Earlier materials made use of dimethylethanolamine, which is more suitable for higher cross-link materials.

Another option that provides improved peak shape for nitrate is to prepare ion exchange materials based on a methacrylate polymer backbone. Methacrylate base polymers have several advantages over anion exchange materials based on styrenic monomers. For example, the reactivity of methacrylate cross-linkers such as ethylene glycol dimethacrylate are very close to the reactivity of other methacrylate monomers such as glycidyl methacrylate, commonly used as the functionalizable monomer in such polymer systems. This tends to reduce the cross-link heterogeneity associated with styrenic monomers cross-linked with DVB. Second, while there are π electrons associated with the ester group in methacrylate polymers, polarizable ions

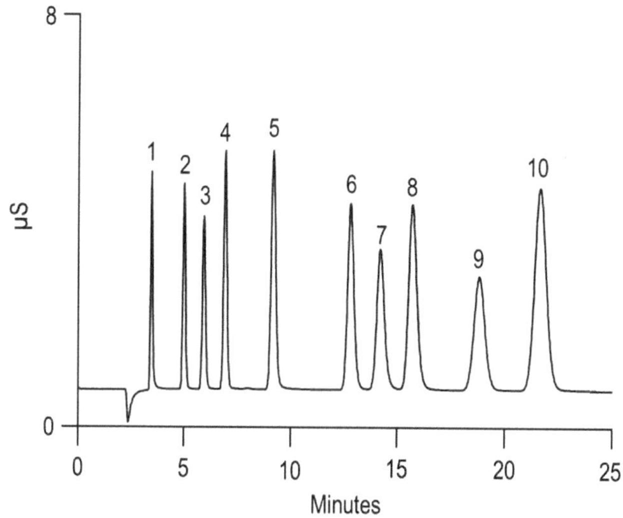

Figure 10.5 Columns: 4 mm × 250 mm IonPac AS9-HC and 4 mm × 50 mm IonPac AS9-HC, eluent: 9 mM sodium carbonate, flow rate: 1 mL/min, injection volume: 25 µL, suppressor: 4 mm ASRS II, recycle mode, analytes: 1 – fluoride (3 mg/L); 2 – chlorite (10 mg/L); 3 – bromate (20 mg/L); 4 – chloride (6 mg/L); 5 – nitrite (15 mg/L); 6 – bromide (25 mg/L); 7 – chlorate (25 mg/L); 8 – nitrate (25 mg/L); 9 – phosphate (40 mg/L); 10 – sulfate (30 mg/L).

interact much more weakly with the polymer matrix than in the case of styrenic ion exchange polymers. Consequently, peak shape of polarizable ions such as nitrate is generally significantly superior. An example of this chemistry is the IonPac AS9-HC column, shown in Fig. 10.5. Note that shape for nitrate is much better than is observed for anion exchange materials based on styrenic monomers with comparable cross-link.

A third option that results in even better peak shape for polarizable anions such as nitrate over an even wider concentration range is to make use of even more polar polymer backbones for the preparation of anion exchange materials. An example of this is the IC SI-50 4E column from Shodex (also available as an OEM material sold by Metrohm under the name Metrosep A Supp 5). This anion exchange material makes use of a hydrophilic cross-linked polyvinyl alcohol polymer backbone with ion exchange sites introduced via reaction with epichlorohydrin and subsequent reaction with a tertiary amine. Columns based on this polymer architecture exhibit excellent nitrate peak shape. The one disadvantage of this polymer architecture is that the polymer matrix is significantly softer than polymers based on methacrylate or styrenic monomers. As a consequence, the recommended flow rate for a 4-mm ID column (0.7 mL/min) is very close to the maximum recommended flow rate (0.8 mL/min). As a result, columns based on this chemistry platform are not as robust as more conventional IC phases. For example, the column bed is relatively easily disturbed by fluctuations in flow rate associated with normal operating conditions, such as when the pump loses prime or during initial instrument startup.

A fourth option, which also provides considerably improved chromatography for polarizable anions, utilizes an entirely new type of chemical synthesis based on condensation polymers electrostatically attached to surface-sulfonated resin. Synthesis of the stationary phase [8,9] is accomplished by passing a solution containing an equal molar mixture of methylamine and butanediol diglycidyl ether through a column packed with 2000-Å pore size surface-sulfonated resin. These two reagents react to produce a highly hydrophilic cationic linear polymer, which binds electrostatically to the surface of the resin. Since the condensation polymer contains a tertiary amine group at regular intervals along the backbone of the polymer, these tertiary amine groups can be used as the starting point for repairing ion exchange sites projecting from the resin surface. Synthesis of these projecting groups is accomplished by first rinsing the column to remove any excess linear polymer remaining from the first step in the reaction. After this, a solution of butanediol diglycidyl ether is passed through the column. This step quaternizes all of the tertiary amines on the polymer backbone, intensifying the electrostatic bond to the resin surface. At the same time, this step produces a pendant epoxy group projecting from the surface of the resin directly above the quaternary ion exchange site. Following this, the remaining diepoxide solution is rinsed from the column and replaced with a methyl amine solution. This step results in reaction of the primary amine with the pendant epoxy group, resulting in the formation of a secondary amine group directly above each quaternary ion exchange site in the polymer backbone. Following this reaction, the excess primary amine solution is rinsed from the column and replaced with the diepoxide solution. Because the product of the previous reaction is a secondary amine, each pendant secondary amine will react consecutively with two diepoxide molecules, producing a quaternary anion exchange site with two separate epoxy groups attached. The remainder of the diepoxide solution is rinsed from the column, and then a fresh solution of methylamine is passed over the column to produce a fresh crop of secondary amine groups projecting from the surface. Because each secondary amine group will react with two diepoxide molecules, this

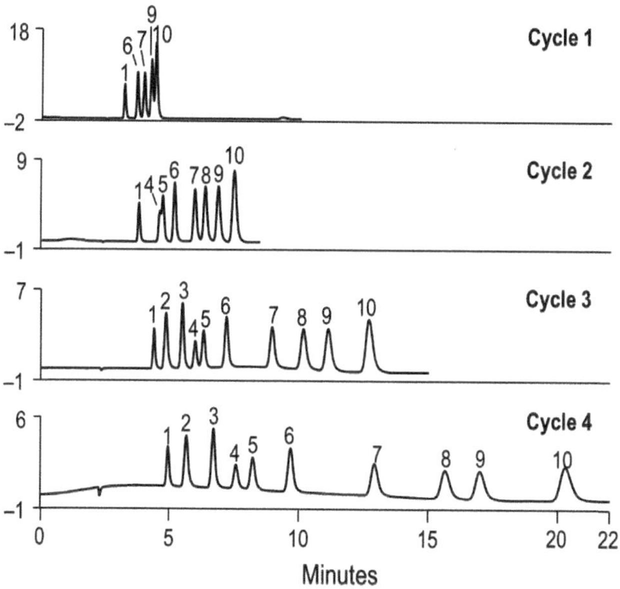

Figure 10.6 Column: 4 mm × 250 mm prototype, eluent: 5 mM potassium hydroxide (KOH), flow rate: 1 mL/min, injection volume: 25 μL, suppressor: 4 mm ASRS ULTRA, recycle mode, analytes: 1 – fluoride (1 mg/L); 2 – acetate (10 mg/L); 3 – formate (5 mg/L); 4 – chlorite (5 mg/L); 5 – bromate (10 mg/L); 6 – chloride (3 mg/L); 7 – nitrite (5 mg/L); 8 – bromide (10 mg/L); 9 – chlorate (10 mg/L); 10 – nitrate (10 mg/L).

introduces a branch point in each polymer strand projecting from the resin surface. Further repetition of the exact same reaction sequence will introduce progressively more branch points, doubling the number of ion exchange sites with each reaction sequence. Fig. 10.6 illustrates the progressive shifts in capacity and changes in selectivity associated with each reaction cycle, where reaction first with a diepoxide monomer followed by reaction with methylamine comprises a reaction cycle.

Because the functional components of this hyperbranched anion exchange polymer contains only relatively polar substituents, contains no hydrolyzable linkages, and has no π electron-containing substituents, this polymer architecture exhibits excellent peak shape for nitrate. Since the underlying substrate is based on the styrenic polymer, such materials are physically much more robust than comparable materials based on polyvinyl alcohol. Fig. 10.7 illustrates the selectivity of such materials utilizing carbonate eluent.

With regard to nitrite quantitation, factors surrounding the optimum column chemistry are considerably different. Unlike nitrate, nitrite is

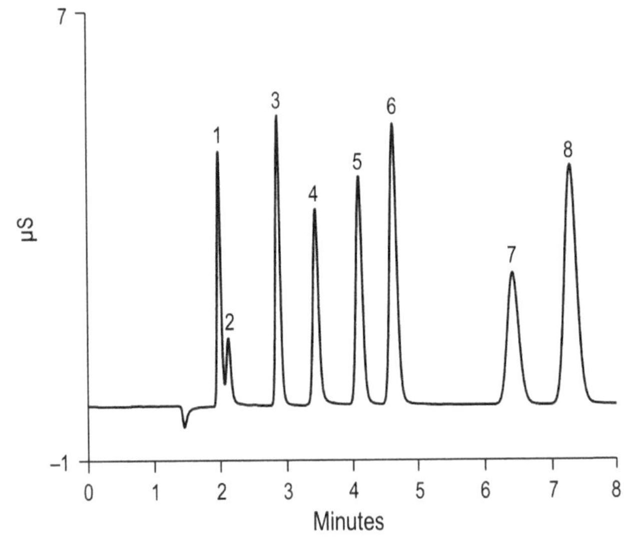

Figure 10.7 Columns: 2 mm × 250 mm IonPac AS22 and 2 mm × 50 mm IonPac AG22, eluent: 1.4 mM sodium bicarbonate and 4.5 mM sodium carbonate, flow rate: 0.25 mL/min, temperature: 30°C, injection volume: 2.5 μL, suppressor: 2 mm ASRS ULTRA II, recycle mode, analytes: 1 – fluoride (5 mg/L); 2 – acetate (20 mg/L); 3 – chloride (10 mg/L); 4 – nitrite (15 mg/L); 5 – bromide (25 mg/L); 6 – nitrate (25 mg/L); 7 – phosphate (40 mg/L); 8 – sulfate (30 mg/L).

much better hydrated and much less polarizable than nitrate. As a consequence, stationary phase design factors described previously are not particularly relevant to the challenge of measuring nitrite in drinking water samples. The principal difficulty in column selection with regard to nitrite quantitation stems from two factors: the relatively low concentration of nitrite in drinking water samples and the general tendency of nitrite to elute just after chloride. The combination of these two features can make it difficult to quantitate trace levels of nitrite, given the fact that chloride is often 1000–10,000 times more concentrated than nitrite in drinking water samples. As a consequence, the optimum column choice for quantitation of nitrite is a column that maximizes the separation from chloride while at the same time providing good peak shape for chloride under moderately overloaded conditions. Generally, columns such as the IonPac AS22, shown in Fig. 10.7, or columns based on polyvinyl alcohol substrates provide the best combination of features with regard to this analytical challenge. As such, the more modern phases are ideally suited to the quantitation of both nitrite and nitrate.

An additional consideration with regard to quantitation of nitrite and nitrate in drinking water is the choice of detector. While conductivity detection is certainly compatible with detection of nitrate and nitrite, 5- to 10-fold lower detection limits can be achieved by using UV detection rather than conductivity detection. In certain parts of the world, standard IC methods specify UV detection for nitrite and nitrate. One such example is International Organization for Standardization (ISO) 10304-1:2007 Part 1, which targets the determination of bromide, chloride, fluoride, nitrate, nitrite, phosphate, and sulfate in drinking water. Both nitrite and nitrate absorb in the UV, so one would expect that detection of these two anions should be straightforward using UV detection. One not so obvious nuance is a requirement to detect these anions after the suppression reaction. Several literature references suggest that carbonate eluent absorbs in the UV and that this absorbance introduces excessive baseline noise and interferes with quantitation unless a suppressor is used to convert carbonate to carbonic. In fact, sodium carbonate and sodium bicarbonate are only very weakly absorbing in the ultraviolet. The actual underlying cause of the elevated background and increased baseline noise is due to the fact that the hydroxide anion absorbs at the low UV wavelengths and hydroxide is always present at appreciable levels when carbonate is present. The suppression reaction removes this hydroxide and thus improves the detection limit when using UV detection with basic eluents.

10. APPLICATION OF ION CHROMATOGRAPHY TO DISINFECTANT BY-PRODUCTS ANALYSIS

Chlorine has been widely used as a disinfectant for drinking water supplies since the early 20th century. Chlorine was first used in the US as a major disinfectant in Jersey City, New Jersey in 1908. Chlorine has a number of distinct advantages, including persistence throughout the water distribution system and a high degree of efficacy against a wide variety of biohazards. At the same time, chlorine is known to produce a number of suspect carcinogens, such as methylene chloride and chloroform, when organic carbon–containing components are present in drinking water, as is often the case in many parts of the world. Another issue with chlorine is that it is a highly toxic gas and a potential terrorist target. Recognition of the presence of such carcinogens and consideration of hazards associated with handling chlorine has led to a push to replace chlorine with other disinfectants. Disinfectants used today include chlorine, chloramine (produced by reacting chlorine with ammonia), hypochlorite (widely considered to be a safer alternative to chlorine for water treatment in spite of the fact that the actual active ingredient is identical to disinfection with chlorine), chlorine dioxide, and ozone. The latter three disinfectants have the advantage that they do not produce chlorinated organic disinfectant by-products, but they do produce oxyhalide disinfectant by-products, although hypochlorite solutions may contain bromate and perchlorate, depending upon production and storage conditions. As a consequence, there are a number of columns that have been developed specifically for disinfectant by-products using IC instrumentation. Initially, the primary interest in this application area focused on chlorite and chlorate, which are disinfectant by-products associated with chlorine dioxide and hypochlorite. The IonPac AS9-SC was developed specifically for this application. Chlorate is a polarizable anion with properties similar to nitrate, and the two ions tend to be poorly resolved on styrenic anion exchange materials due to this similarity. Since chlorate is more prone to interacting with π electrons associated with styrenic polymers, one solution to the selectivity issue associated with chlorate and nitrate is to make use of methacrylate-based polymers. As previously mentioned, methacrylate polymers are much less prone to adverse polymer interactions between polarizable anions and the anion exchange phase polymer backbone. As a consequence, chlorate elutes significantly earlier than nitrate on a methacrylate-based anion exchange polymer. Furthermore, bromide almost always elutes just before nitrate, so a stationary phase suitable for chlorate analysis must have adequate selectivity to allow for baseline resolution of bromide, chlorate, and nitrate. The IonPac AS9-SC column achieves the required selectivity by making use of the fact that selectivity can be fine-tuned through the use of the appropriate tertiary amine. Experimental work with various tertiary amines resulted in selectivity where chlorate was partially resolved from bromide in one case and partially resolved from nitrate in another case. By blending together latex formulations based on each of these two amines, it is possible to fine-tune the selectivity so that chlorate elutes precisely in between bromide and nitrate with baseline resolution of all three analytes. Chlorite invariably elutes well before chloride, so the optimum selectivity for bromide, chlorate, and nitrate proved to be optimal for chlorite as well. EPA 300.0, EPA 300.1, and ISO 10304:1997 all cover use of suppressed conductivity detection for the determination of both chlorite and chlorate in drinking water.

Later the EPA asked for a column that was optimized for the separation of bromate. The reason for this interest was connected to the transition to the use of ozone for disinfection instead of chlorine, hypochlorite, or chlorine dioxide. In theory, ozone should be free from the risk of forming any chlorine-based disinfection by-products and at the same time is much safer to handle than chlorine or chlorine dioxide, albeit it is more capital intensive to implement. The one disinfection by-product identified with ozone disinfection is bromate. Bromate is only formed in conjunction with ozone disinfection in drinking water that contains bromide. All drinking water contains detectable levels of bromide, but generally, the concentration of bromide is too low to produce significant concentrations of bromate upon treatment of the water with ozone. A notable exception to this rule is areas with significant halide concentrations in the drinking water, such as is often the case for coastal regions due to seawater intrusion into groundwater. Higher infrastructure costs combined with reduced operating costs associated with ozone treatment of drinking water results in the most favorable economics for relatively large water treatment facilities, which tend to be located in large population centers. As a consequence, the use of ozone disinfection tends to be associated with large population centers. However, since the majority of large population centers, at least in the US, tend to be in coastal regions, bromate disinfection by-products are often a concern in ozone treatment facilities. Bromate was of particular concern to the EPA due to the fact that bromate had recently been declared a suspect carcinogen [10]. Thus, the EPA was interested in better methods for determination of bromate, and virtually all commercially available columns at the time suffered from poor resolution of bromate from chloride. Fortunately, it was discovered that bromate was baseline resolved on the IonPac AS9-SC, eluting relatively equidistant from chlorite and chloride.

Because bromate is a suspect carcinogen, it has a relatively low maximum contaminant limit of 10 ppb, more than 100-fold lower than chlorite. This maximum contaminant limit presents a significant challenge to quantitation of bromate in drinking water, given the proximity of bromate to chloride and the fact that drinking water invariably contains significant concentrations of chloride. Concentration ratios of chloride to bromate in drinking water are often 10,000:1 or higher, presenting significant difficulties when it comes to quantitate in such low levels of bromate. For that reason, there are a substantial number of different IC methods for bromate. Two of the earliest methods for bromate utilized postcolumn derivatization following eluent suppression. EPA 317.0 rev. 2 specifies the use of *o*-Dianisidine as the postcolumn reagent, whereas EPA 326.0 specifies the use of hydroiodic acid as the postcolumn reagent (prepared by passage of the KI postcolumn reagent through a second, separate suppressor). Both methods provide for highly selective detection of bromate with minimal interferences. However, both methods have the disadvantage of requiring the use of a dedicated instrument, since switching back and forth between conventional suppressed IC and such postcolumn methods is a nontrivial task. In addition to the two EPA methods, there are several other postcolumn methods that are variants on EPA 326.0, such as ISO 11206:2011, which makes use of an acidic eluent and a molybdate-based catalyst system.

In addition to the previous postcolumn reaction detection methods, there are a number of methods for detecting bromate directly via conductivity detection, including EPA 300.0 and EPA 300.1. Because the IonPac AS9-SC had somewhat limited capacity, sample prep to remove chloride was necessary in cases where the chloride level in the sample was high enough to make bromate quantitation difficult. In an effort to minimize sample prep requirements when utilizing standard IC instrumentation and conductivity detection, a high-capacity version of the IonPac AS9-SC column was developed: the IonPac AS9-HC. Both EPA 317.0 rev. 2 and EPA 326.0 specify use of the IonPac AS9-HC. As mentioned earlier, hydroxide eluent systems provide lower detection limits than carbonate-based eluent systems. Since both the IonPac AS9-SC column and the IonPac AS9-HC column are based on methacrylate-based stationary phases, these columns are incompatible with hydroxide eluents due to the fact that methacrylate-based stationary phases are prone to hydrolysis under the alkaline conditions associated with hydroxide eluents. As a consequence, a significant effort was undertaken in order to develop stationary phases compatible with hydroxide mobile phases with useful selectivity for bromate, chlorite, and chlorate. Using the same architecture described previously for the IonPac AS22, it was possible to develop an entirely new stationary phase specifically optimized for separation and quantitation of bromate using hydroxide eluent systems: the IonPac AS19 column. Fig. 10.8 shows an example chromatogram from this column. Sensitivity for bromate with hydroxide eluent using this column is roughly fivefold better than previous carbonate eluent based systems when using suppressed conductivity detection. As a consequence, for most samples, detection limits below one part per billion can be achieved without any sample prep when using this eluent/column combination.

In addition to postcolumn reaction methods and direct conductivity methods, there are also a number of methods utilizing spectroscopic detectors, including EPA 321.8, which uses IC in conjunction with inductively coupled plasma mass spectrometry (ICP-MS) and EPA 557, which uses IC with electrospray ionization tandem MS. Both of these methods have

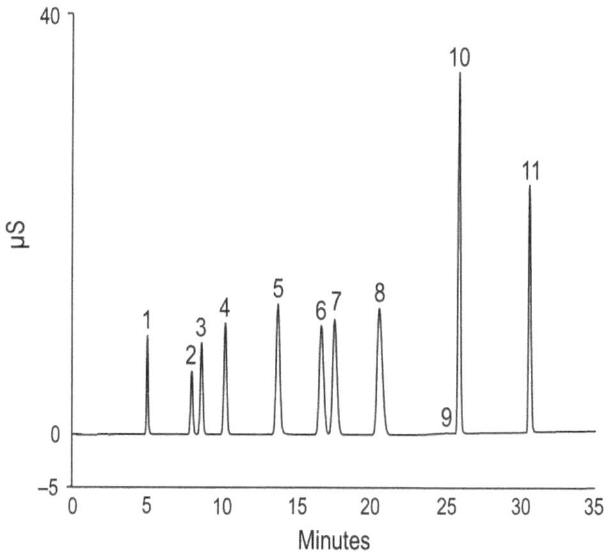

Figure 10.8 Column: 4 mm × 250 mm IonPac AS19, eluent: 10 mM potassium hydroxide (KOH) from 0 to 10 min and 10–55 mM KOH from 10 to 25 min, flow rate: 1 mL/min, temperature: 30°C, injection volume: 25 μL, suppressor: 4 mm ASRS ULTRA, external water mode, analytes: 1 – fluoride (1 mg/L); 2 – chlorite (10 mg/L); 3 – bromate (20 mg/L); 4 – chloride (6 mg/L); 5 – nitrite (15 mg/L); 6 – bromide (25 mg/L); 7 – chlorate (25 mg/L); 8 – nitrate (25 mg/L); 9 – sulfate (25 mg/L); 10 – phosphate (30 mg/L).

the advantage of being significantly more sensitive than any of the previously mentioned methods. While all of the methods mentioned so far demonstrate adequate sensitivity for determination of bromate in drinking water samples EPA 321.8 and EPA 557 use their higher sensitivity to avoid the necessity of any sample prep. In fact, both methods provide significantly better detection limits than the previously mentioned methods, even with the most difficult samples without sample prep. Problems with column overloading are avoided by injecting smaller samples.

Finally, there is EPA 302.0, which makes use of heart cutting in conjunction with 2-D chromatography. This method specifies the use of the IonPac AS19 column in the first dimension and the IonPac AS24 column in the second dimension. By utilizing a 4-mm ID column in the first dimension, using a heart cut to transfer bromate to a 2-mm ID second-dimension column, matrix affects are eliminated, while at the same time, the change in column ID between the first- and second-dimension column results in a fourfold enhancement in detection limit. Furthermore, even greater improvement in the detection limit can be achieved by using a capillary column for the second dimension. As a consequence, this method provides detection limits comparable to EPA 321.8 and EPA 557 with much less expensive instrumentation.

Another class of disinfectant byproducts applicable to IC are the haloacetic acids. This class of disinfectant by-products includes monochloroacetic acid, dichloroacetatic acid, trichloroacetic acid, monobromoacetic acid, dibromoacetic acid, tribromoacetic acid, bromochloroacetic acid, dichlorobromoacetic acid, and chlorodibromoacetic acid. These toxic and potentially carcinogenic acids are by-products of disinfection using chlorine, chloramine, or ozone. While chlorine is the primary causal element in the formation of haloacetic acids, both chloramine and ozone can also induce the formation of haloacetic acids indirectly through the presence of chlorine during the formation process of chloramine or due to the presence of chlorine formation through the reaction of ozone with chloride. Nonetheless, use of chloramine or ozone significantly reduces the concentrations of by-products formed during the disinfection process, presumably due to the lower concentrations of free chlorine in the latter disinfection processes. The specific haloacetic acids regulated varies from country to country. In the US, five of the nine possible haloacetic acids (commonly referred to as the HAA_5) are regulated. Specifically, the HAA_5 include monochloroacetic acid, dichloroacetate acid, trichloroacetic acid, monobromoacetic acid, and dibromoacetic acid. The maximum concentration limit for the HAA_5 is 60 ppb. The premise for selecting these five haloacetic acids is that these five are the predominant forms formed during the disinfection process. There has been some discussion of expanding the regulated haloacetic acids to include the entire suite of nine haloacetic acids, but no timetable has been given for this transition, and there is no evidence that such work is currently underway.

The earliest methods for quantitation of haloacetic acids in drinking water make use of liquid–liquid extraction followed by formation of haloacetic acid methyl esters and gas chromatography with electron capture detector (GC-ECD). The latest version of this method is incorporated in EPA 552.3 rev. 1. The GC method provides excellent sensitivity and chromatographic performance, but suffers from relatively long derivatization times and relatively poor reproducibility. In contrast, EPA 557 uses anion IC to separate haloacetic acids directly without the need for derivatization. Detection sensitivity and analysis speed are comparable to the GC method. In addition, work is currently underway on the development of a new method for haloacetic acid determination utilizing 2-D IC via a multiple heart cut method. In this case, a 4-mm ID IonPac AS24 A column is used in the first dimension, while a 0.4 mm ID IonPac AS26 column is used in the second dimension. The combination of a 4-mm ID first-dimension column with a 0.4-mm s dimension column results in a 100 times increase in concentration sensitivity as a consequence of the 100:1 ratio of the cross-sectional areas of the two columns.

11. APPLICATION OF ION CHROMATOGRAPHY TO PERCHLORATE ANALYSIS

In the past 20 years, perchlorate has become of increasing analytical interest. Perchlorate has been included in the EPA contaminant candidate list since 1998. Interest in perchlorate stands initially from the discovery of perchlorate in groundwater in several areas surrounding a former flare manufacturing site in northern California. Initially, it was assumed that perchlorate contamination arises solely from environmental contamination due to man-made sources, such as perchlorate-containing fuel used as a rocket propellant and in flares. Perchlorate contamination has also been detected in Lake Mead, and most of the perchlorate contamination in southern Nevada has been attributed to contamination of Lake Mead. While the exact formation process has not yet been identified, there are also natural processes that produce perchlorate. For example, there are a number of well-documented examples of perchlorate contamination in the southwestern portion of the US that cannot be tied to any specific man-made perchlorate source. It is speculated that perchlorate might arise from lightning discharges in areas with high concentrations of aerosol sodium chloride particulate in the atmosphere.

Health risks associated with perchlorate are primarily associated with interference with proper thyroid function. Iodine is transported to the thyroid via the thyroid sodium–iodide symporter, and perchlorate binds competitively to this iodide transport protein. There is a related bioaccumulation pathway associated with mammalian milk production. As a consequence, perchlorate contamination of milk throughout the US is widespread with concentrations typically significantly higher than perchlorate levels found in drinking water. For those that have an adequate intake of iodide in their diet, the levels of perchlorate in the environment probably do not represent a risk, but for many people who do not have an adequate iodide supply, there is an increased risk of thyroid problems. Thus, so far, US government regulatory agencies have been resistant to reducing the regulatory limits for perchlorate from the current 11 ppb screening recommendation, although both California and Massachusetts have introduced lower concentration limits (6 ppb and 2 ppb respectively).

For many years, quantitation of perchlorate using IC was problematic. In comparison to nitrate, perchlorate is much more polarizable, and as such, chromatographic performance as manifested by chromatographic peak shape of perchlorate was poor. Early columns typically necessitated the use of organic solvent to at least partially mitigate the relatively poor peak shape associated with perchlorate with styrenic stationary phases. In addition, *p*-cyanophenol was often used in the eluent to improve peak shape for perchlorate. Even with both of these measures taken together, efficiency and asymmetry for perchlorate with older stationary phases were typically much worse than for other inorganic anions commonly analyzed by IC. However, with increasing interest in quantitation of perchlorate taking place in the late 1990s, efforts were made to improve the chromatographic performance of perchlorate. The IonPac AS16 column, shown in Fig. 10.9, was the first column designed specifically for analysis of perchlorate in environmental samples. It makes use of three features that improve peak shape compared to earlier columns: a more hydrophilic styrenic monomer (vinylbenzyl glycidyl ether), low cross-link (0.5% cross-link), and a hydrophilic quaternary ion exchange site. The combination of these three features resulted in significant improvement in peak shape for perchlorate with a simple hydroxide eluent system. The IonPac AS16 column is specified as part of EPA 314.0. Later, the IonPac AS20, also shown in Fig. 10.9, which uses the same previously described hyperbranched condensation polymer

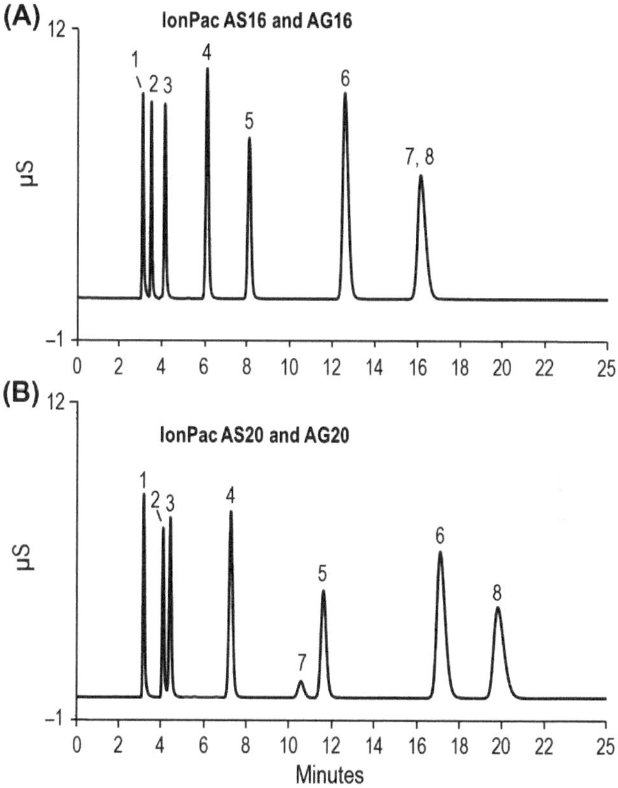

Figure 10.9 Column dimensions: 2 × 250, and 2 × 50 mm, eluent: 35 mM NaOH, flow rate: 0.25 mL/min, temperature: 30°C, injection volume: 2.5 μL, suppressor: 4 mm ASRS ULTRA II, recycle mode, analytes: 1 — fluoride (2 mg/L); 2 — chloride (3 mg/L); 3 — sulfate (5 mg/L); 4 — thiosulfate (10 mg/L); 5 — iodide (20 mg/L); 6 — thiocyanate (20 mg/L); 7 — *p*-chlorobenzenesulfonate (5 mg/L); 8 — perchlorate (30 mg/L).

chemistry used in the IonPac AS19 and the IonPac AS22, was developed as a secondary column to complement the IonPac AS16. The IonPac AS20 provides even better chromatographic performance for perchlorate and has significantly different overall selectivity. The combination of these two columns are specified in EPA 314.1, with the IonPac AS16 to be used as the primary analysis column, while the alternative selectivity IonPac AS20 column is to be used as the confirmation column in cases where perchlorate is detected in water samples.

In addition to these methods based on suppressed conductivity detection, there are two additional EPA methods for analysis of perchlorate making use of the combination of ion exchange with spectroscopic detection. EPA 332.0 specifies the combination of suppressed conductivity detection and electrospray ionization MS. This method specifies the use of either the IonPac AS16 column or the Metrosep A Supp 5 column for the analysis. EPA 331.0 is an liquid chromatography-mass spectrometry (LC-MS) method utilizing anion exchange in conjunction with electrospray ionization MS. This method specifies the IonPac AS21 column, shown in Fig. 10.10. This column was also developed specifically for perchlorate analysis, but in this case, capacity and selectivity was optimized for use with a volatile mobile phase: methylamine. Thus, this column can be used without a suppressor, the column effluent sent directly into the mass spectrometer.

In addition to the methods previously mentioned utilizing mass spectrometric detection, 2-D IC utilizing a heart cut transfer of perchlorate from the first dimension to the second dimension has also been developed as an alternative to MS detection methods. One such method utilizes a 4-mm ID IonPac AS20 as the first-dimension column and a 2-mm ID IonPac AS16 column in the second dimension [11]. The combination of these two columns in a single method provides freedom from interferences, which might arise in a 1-D method while at the same time providing a fourfold sensitivity enhancement due to the differences in the cross-sectional area of the two columns. EPA 314.2 utilizes a similar methodology.

12. APPLICATION OF ION CHROMATOGRAPHY TO ARSENIC SPECIATION

Arsenic is found in groundwater in many parts of the world. While there are cases where the groundwater has been contaminated due to pollution from human activities, the vast majority of groundwater arsenic contamination is due to the natural sources [12]. A 2007 study estimated that more than 137 million people in 70 countries are subject to arsenic poisoning due to high levels of arsenic in their drinking water [13]. As a consequence of this widespread contamination of drinking water sources with arsenic, arsenic speciation is an area of analytical interest in IC. Speciation of arsenic is of particular interest due to the fact that certain forms of arsenic, such as As(III) and As(V), are considerably more toxic than organometallic forms of this element. A number of methods have been published using IC for the analysis of arsenic species in drinking water [14]. Typically, such methods utilize anion exchange for the separation of the arsenic species coupled to either ICP or ICP-MS. The most commonly columns used for this separation are the PRP-X100 column or the IonPac AS7 column.

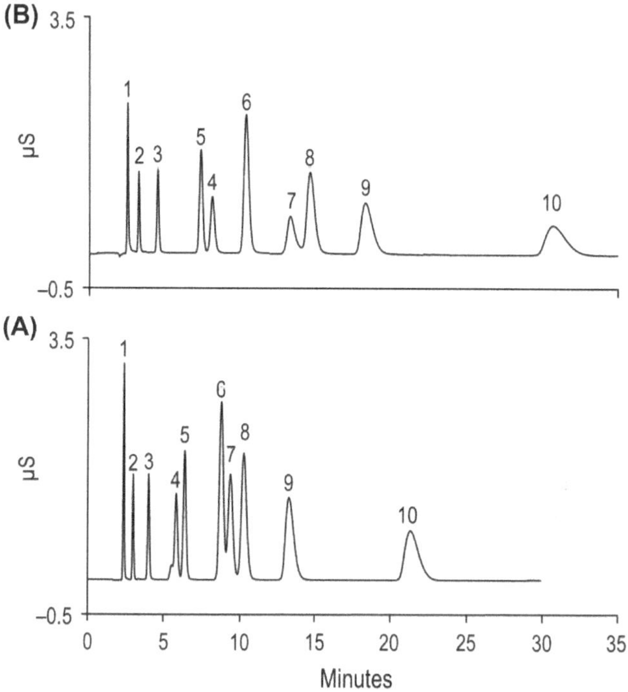

Figure 10.10 Columns: 2 mm × 250 mm IonPac AS21 and 2 mm × 50 mm IonPac AG21, eluent A: 12.5 mM potassium hydroxide (KOH), eluent B: 200 mM methyl amine, flow rate: 0.35 mL/min, temperature: 30°C, injection volume: 2.5 µL, suppressor: 2 mm ASRS ULTRA, recycle mode, analytes: 1 – fluoride (2 mg/L); 2 – chloride (2 mg/L); 3 – nitrate (5 mg/L); 4 – sulfate (5 mg/L); 5 – iodide (20 mg/L); 6 – perchlorate (30 mg/L); 7 – chromate (30 mg/L); 8 – tungstate (30 mg/L); 9 – phosphate (30 mg/L), 10 – arsenate (30 mg/L).

13. APPLICATION OF ION CHROMATOGRAPHY TO SELENIUM SPECIATION

Selenium contamination of drinking water is far less common than arsenic contamination. Contamination of drinking water sources can occur in cases where agricultural runoff passes over normally dry areas. Common selenium species, including selenate and selenite, can be detected by suppressed conductivity detection. However, as in the case of arsenic speciation, the more common practice is to couple IC to either ICP or ICP-MS, since these two techniques provide superior detection limits [15]. A number of IC columns have been applied to this application. The most common columns used for this separation are the PRP-X100 column or the IonPac AS7 column.

14. APPLICATION OF ION CHROMATOGRAPHY TO CHROMIUM SPECIATION

In the case of chromium speciation, the primary species of interest is Cr(VI), as this is the most toxic form of chromium. Contamination of drinking water sources nearly always arise from man-made sources, such as waste products from plating or leather production. Assessment of chromium contamination of drinking water has long been in place, but generally, regulations only stipulate the total chromium maximum contaminant level (100 ppb) rather than regulating chromium species. However, recently, California has set a public health goal of 0.02 ppb for Cr(VI) with a maximum contaminant level of 10 ppb for Cr(VI). There have been a number of IC methods in place for many years [14]. As in the case of arsenic and selenium speciation, direct detection using suppressed conductivity detection, while possible, is not widely used. In some cases, chromium speciation is accomplished by making use of IC coupled to either ICP or ICP-MS. However, a more common practice is to use a postcolumn reaction method for measuring Cr(VI) in drinking water and wastewater. EPA 218.6 and EPA 1636 both use an IonPac AS7 column for the analytical separation in conjunction with the postcolumn reagent: diphenylcarbohydrazide. The reaction product is detected at 530 nm with a UV-visible detector.

15. APPLICATION OF ION CHROMATOGRAPHY TO AMMONIA AND BARIUM ANALYSIS

As mentioned previously, for the most part, spectroscopic methods are used for the analysis of cationic impurities in drinking water. ICP or ICP-MS are the most commonly utilized methods for measurement of barium in drinking water. Neither ICP nor ICP-MS are suitable for the determination of ammonia or other amines in drinking water. Colorimetric methods are commonly employed for the determination of ammonia in drinking water. IC presents an opportunity to quantitate common inorganic cations as well as ammonia in a single analysis, which is advantageous for labs that can't justify the cost of purchasing much more expensive ICP or ICP-MS instrumentation.

Although in the case of IC anion analysis, suppressors are nearly universally used, in the case of cation analysis, the nonsuppressed mode is sometimes used. This is due primarily to the widely promoted misconception that nonsuppressed IC provides better detection limits than its suppressed counterpart when it comes to cations. However, even Metrohm, the chief proponent of the nonsuppressed mode of operation, recently introduced a cation suppressor. In a recently published brochure, [16] they state: "Trace analysis of cations, amines, and transition metals is possible with or without suppression. However, some applications require particularly high sensitivity of the analysis. This can only be achieved by sequential suppression, as suppression considerably lowers the detection limits for the analytes."

While there are a number of IC columns designed specifically for the analysis of alkali metals, alkaline earth metals, and ammonia using suppressed IC, the most widely used column for this application is the IonPac CS12A column (see Fig. 10.11). Another column that is widely used in applications requiring detection of ammonia in the presence of large sodium concentration excesses is the IonPac CS16 column (see Fig. 10.12). It allows for

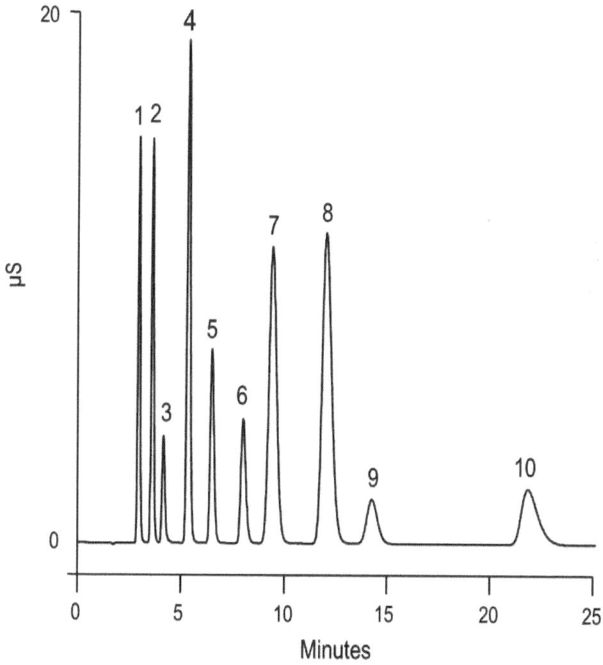

Figure 10.11 Column: 4 mm × 250 mm IonPac CS12A, eluent: 18 mM methanesulfonic acid, flow rate: 1 mL/min, injection volume: 25 µL, suppressor: 4 mm CSRS, recycle mode, analytes: 1 – lithium (1 mg/L); 2 – sodium (4 mg/L); 3 – ammonium (5 mg/L); 4 – potassium (10 mg/L); 5 – rubidium (10 mg/L); 6 – cesium (10 mg/L); 7 – magnesium (5 mg/L); 8 – calcium (10 mg/L); 9 – strontium (10 mg/L); 10 – barium (10 mg/L).

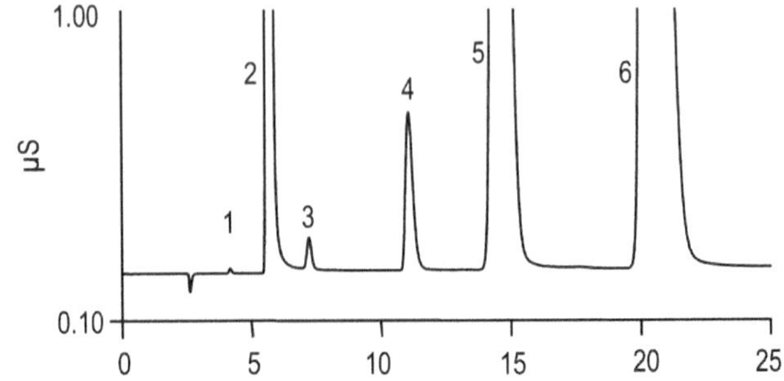

Figure 10.12 Column: 5 mm × 250 mm IonPac CS16, eluent: 30 mM methanesulfonic acid, flow rate: 1 mL/min, temperature: 40°C, injection volume: 25 µL, suppressor: 4 mm CSRS ULTRA, recycle mode, drinking water analytes detected: 1 – lithium (0.002 mg/L); 2 – sodium (18.7 mg/L); 3 – ammonium (0.065 mg/L); 4 – potassium (0.99 mg/L); 7 – magnesium (7.2 mg/L); 8 – calcium (18.5 mg/L).

quantitation of ammonia in samples where the sodium-to-ammonium concentration ratio is as high as 10,000:1 without any sample preparation.

REFERENCES

[1] Small H, Stevens TS, Bauman WC. Novel ion exchange chromatographic method using conductometric detection. Anal Chem 1975;47:1801–9.

[2] Rey M, Bordunov A, De Borba B, Slingsby R, Pohl C. Cation exchange chromatography with conductivity detection: suppressed or nonsuppressed poster presented at IC Symposium. 2003. San Diego (CA).

[3] Edgell KW, Longbottom JE, Pfaff JD. Determination of inorganic anions in water by ion chromatography: a collaborative study. J AOAC Int 1994;77(5):1253–63.

[4] Ohgami H, Terao T, Shiotsuki I, Ishii N, Iwata N. Lithium levels in drinking water and risk of suicide. Br J Psychiatry J Ment Sci 2009;194(5):464–5.

[5] Amdisen A. Clinical and serum level monitoring in lithium therapy and lithium intoxication. J Anal Toxicol 1978;2:193–202.

[6] https://www.epa.gov/dwstandardsregulations/secondary-drinking-water-standards-guidance-nuisance-chemicals.

[7] Frant MS, Ross Jr JW. Electrode for sensing fluoride ion activity in solution. Science 1966;154(3756):1553–5.

[8] Pohl CA, Saini C. Coated ion exchanged substrate and method of forming. November 6, 2007. U. S. Patent 7,291,395.

[9] Pohl C, Saini C. New developments in the preparation of anion exchange media based on hyperbranched condensation polymers. J Chromatogr A 2008;1213(1):37–44.

[10] Kurokawa Y, Maekawa A, Takahashi M, Hayashi Y. Toxicity and carcinogenicity of potassium bromate—a new renal carcinogen. Environ Health Perspect 1990;87:309–35.

[11] Lin R, De Borba B, Srinivasan K, Woodruff A, Pohl C. Matrix diversion methods for improved analysis of perchlorate by suppressed ion and conductivity detection. Anal Chim Acta 2006;567(1):135–42.

[12] Smedley PL, Kinniburgh DG. A review of the source, behaviour and distribution of arsenic in natural waters. Appl Geochem 2002;17(5):517–68.

[13] Arsenic in drinking water seen as threat. USATodayCom August 30, 2007. http://usatoday30.usatoday.com/news/world/2007-08-30-553404631_x.htm.

[14] Pantsar-Kallio M, Manninen PKG. Simultaneous determination of toxic arsenic and chromium species in water samples by ion chromatography-inductively coupled plasma mass spectrometry. J Chromatogr A 1997;779(1–2):139–46.
[15] Martínez-Bravo Y, Roig-Navarro AF, López FJ, Hernández F. Multielemental determination of arsenic, selenium and chromium(VI) species in water by high-performance liquid chromatography-inductively coupled plasma mass spectrometry. J Chromatogr A 2001;926(2):265–74.
[16] Cation suppression in ion chromatography, Metrohm brochure (8.000.5163EN – 2015-10).

CHAPTER ELEVEN

Can Incongruent Studies Effectively Characterize Long-Term Water Quality?

J.R. Peller[1], R.L. Whitman[2], J.K. Schoer[1]
[1]Valparaiso University, Valparaiso, IN, United States
[2]Great Lakes Science Center, Porter, IN, United States

1. INTRODUCTION

The role of science in the assurance of worldwide public and environmental health and safety is more relevant now than ever before. This is due to the tremendous environmental changes imposed by increasing populations, rapid industrial and technological developments, unprecedented use of resources, massive transportation of goods, people, and organisms, and many other modern pressures. In the mix, social and economic influences often overshadow scientific evidence for appropriate management, decision-making, and progress [1]. For the protection of public and environmental health, science-centered strategies create the best pathways to understanding problems and finding solutions.

In environmental science and services, programs are often not built around rigorous science standards, resulting in disconnected information with limited utility for current and future decisions. Applied science, which is ultimately intended to impact society, is a principal purpose of the US Environmental Protection Agency (USEPA), other federal agencies such as the Center for Disease Control (CDC) and the Department of Agriculture (USDA), and corresponding state agencies. These agencies are tasked with protecting public health and the environment. They rely on sound science for informing and advising the public and for important decision-making. Since applied environmental science is utilized to guide decision-making on behalf of the environment and the public, it is crucial that the work is directed by the scientific community [2]. Examples of problems or failures in the function of science for public health protection are abundant for a variety of reasons, most notably when scientific investigations and data are not given appropriate consideration [3–7]. In a 2005 report entitled *Decision*

Chemistry and Water
ISBN 978-0-12-809330-6
http://dx.doi.org/10.1016/B978-0-12-809330-6.00011-8

Making for the Environment, the National Research Council (NRC) expressed "when science is gathered to inform environmental decisions, it is often not the right science." [8] The scientific community must be key players in all levels of decisions. Questions such as what must be measured, studied, analyzed, modeled, and how? and where and how often must it be measured? require input and guidance from scientists in cooperation with managers, community members, and other invested people. The outcomes of scientific measurements and investigations should provide logical conclusions and directions for future work and thus provide long-term value [9].

Fresh water is a primary focus area of many federal and state agencies and an active area of research for many scientists. Healthy water is dependent not only on modern water-treatment systems, but also on ecological services provided by natural water ecosystems that function to pretreat water resources. Science provides the required understanding of the natural and technological services, and science-based work delivers both warnings and solutions when natural services are disrupted, polluted, or destroyed. To address threats to living organisms and public health in surface waters, properly planned (and often long-termed) scientific projects are required for: (1) understanding these systems, (2) identifying and modeling changing systems, and (3) identifying appropriate remedial actions. Program structures that utilize scientists, as opposed to program managers, are especially important when the scientific information is used for decision-making to avoid serious mishaps, as recently experienced in the Flint, Michigan drinking water crisis [10]. When the planning, scrutiny, and data assessments of freshwater systems are done with appropriate input by the scientific community, the outcome is more profitable for environmental and public health.

For the protection of freshwater resources, a useful water quality monitoring program is likely to be described differently by research scientists than by management officials. Individual work perspectives and abundant professional experiences lead to different views on requirements and procedures for programs that require scientific measurements. As an example, watershed-management plans (WMPs), developed on the local level following US EPA and state guidelines, constitute one of the major current efforts to protect freshwater resources from nonpoint source pollution. Watershed-management plans are programs designed to serve communities, ensure healthy waterways, and promote environmental stewardship in a number of ways. However, WMPs often incorporate a water quality monitoring protocol lacking scientific rigor and guided mostly by management practices rather than scientific principles. The data outcome from the management plan approach is typically not suited for scientific publication.

Water quality monitoring needs to be designed and supervised by qualified scientists. This level of scientific expertise requires an organized funding mechanism with long-term goals. From an environmental and public health protection perspective, the science—assessment and monitoring of the water quality—should be the guiding factor, not just the assessment mechanism in the protocol. The US EPA lists nine required elements for funded watershed projects in their publication for developing WMPs; these elements begin with a requirement for a thorough description of the watershed, critical information for understanding potential pollution sources [11]. The *last* of the nine elements is the water quality monitoring component, used to determine the effectiveness of the management efforts. However, it is not possible to assess changes and improvements if initial conditions are not measured. Water quality data are necessary for the initial planning and for determining the success of implemented actions [12].

This chapter will highlight necessary elements of applied environmental programs that rely on science, but effectively and beneficially utilize managers/citizens/community to assist in solving environmental and public health challenges. The role of science in the complex management and planning of these projects is discussed in context with a Lake Michigan watershed and its history of water quality monitoring. The importance of coordination, the value of past and future work, and the significance of community input and cooperation are addressed.

2. DISCUSSION

2.1 Watershed-Management Plans: Roles for Science, Management, and Citizens

Science-based studies and measurements are required to serve and protect freshwater supplies, systems, and habitats, and to prevent the tainting or destruction of freshwater bodies and supplies. In the United States, the 1972 Clean Water Act (CWA) effectively initiated the requirement of regular monitoring of surface waters to ensure ecological and public health. The regulations developed as the CWA was enforced based on peer-reviewed scientific studies and well-established sampling and analytical approaches. Regulations have been modified as scientific advancements have increased understanding of aquatic systems. The responsibilities detailed in the federal regulations are relayed to state government agencies that create and enforce effective programs to ensure adequate water quality and ecosystem health. The public relies on these science-based programs to guarantee safe drinking water and to maintain the health of surface water. The agencies are responsible for responding to the multitude of new challenges associated with human progress, increasing/shifting populations, and technology advancements. Therefore programs that generate scientifically sound datasets are necessary to determine changes in water quality over time [13]. Data of questionable quality or of inconsistent representation is rarely useful and can lead to improper decisions and management. The best resources for planning and implementation of water quality programs for state government environmental agencies, in accordance with the US EPA, include scientists familiar and experienced with established protocol and current concerns.

Watershed management plans are broad efforts by government agencies to protect fresh surface water and are management-oriented. They represent an economically favored, yet scientifically weakened, approach to understanding, identifying, and addressing freshwater challenges [14]. Many parts of the WMPs are useful for identifying and understanding the sources of pollution and for developing strategies for community stewardship; these are, indeed, critical components of environmental protection and a means for building social capital for the environment. The process for WMP development emphasizes alterations in the causes or sources of nonpoint source pollution, which can potentially and dramatically improve water quality. However, proper adaptive management can only be achieved through effective monitoring [15]. Appropriate management, in turn, requires a planned, systematic approach for the continual and consistent collection and use of appropriate, quality data. In many cases, there is no enforcement element or required remediation, just voluntary modifications toward healthier waterways.

Research-based efforts in many areas of environmental science and research, such as water quality monitoring, are challenged by economic and societal barriers. As a consequence, there has been an increasing reliance on trained workers from nonprofits or trained volunteers for the collection of water quality measurements [2,16–18]. Citizens can and often do play significant roles in environmental and public health investigations and/or monitoring. Community members who live in strategic areas are more readily available for making everyday observations than scientists or other professionals who visit target sites as part of investigative studies. For example, if a pollution source creates observable data, citizens who live near the source or frequent the site are able to collect a large dataset for a better understanding of the extent of pollution. When scientists can organize citizens to collect data as part of studies, citizens become a collaborative and useful part of the work. Moreover, information collected by citizen volunteers lessens the financial load of the work. When these collaborative efforts are in place, funding agencies can more appropriately design and support programs with the necessary professionals providing leadership and oversight.

For a scientifically sound WMP, experienced scientists should provide guidance for the most suitable monitoring measurements and protocol. The nine necessary elements of WMPs required for EPA funding, from Section 319 Nonpoint Source Grants to States and Territories, are listed in Table 11.1. Baseline measurements must be used to determine the types

Table 11.1 US EPA's Nine Minimum Elements for a Watershed Plan for Impaired Waters Funded Using Incremental Section 319 Funds and the authors' Perspectives on Personnel Roles

Nine Elements	Professional Requirement	Citizen Participation?
1. Pollutant sources identification	Managers, scientists	√
2. Estimated pollution reduction from *management* measures	Managers, scientists	
3. Nonpoint source *management* measures to achieve #2	Managers, scientists	
4. Estimates of technical and financial needs	Managers, scientists	
5. Community education and interaction	Educators, scientists, managers	√
6. Nonpoint source *management* schedule	Managers, scientists	
7. Interim measures to determine the effectiveness of the *management* measures	Managers, scientists	
8. Criteria to determine load reductions	Scientists	
9. Monitoring component to determine progress	Scientists	√

of pollution-reduction processes required for an area of the watershed. From a science perspective, the ninth element (the monitoring component) of the EPA's nine total elements should be at the top of the list and guide the structure of the watershed pollution management, when appropriate improvements are desired.

2.2 Science Components for Public and Environmental Protection: Focus on Salt Creek Watershed, Porter County (USA)

When society provides the scientific community with opportunity and support, scientists have historically responded well to challenges. Science for wartime efforts tops the list of examples, and science to address a shrinking ozone layer is another compelling case. However, forces and pressure both within and outside the world of science for public benefit can weaken the technical effectiveness of science for public good [19]. The requirements for valuable science undertakings, which benefit the environment and public health, can be broken into five fundamental categories, as shown in Table 11.2. Each will be addressed and put into context with the difficulties encountered by the authors in attempting to assess the changes in water quality in the Salt Creek watershed in Northwest, Indiana (USA) over the past four decades. For other failed science-based scenarios around the United States, including the recent example of lead contamination in drinking water in Flint, Michigan and other cities faced with contaminated drinking water, similar problems in one or more of these factors have been identified [10].

Table 11.2 Five Requirements for Effective Science for Public Benefit

1. Quality Assurance (QA)/Quality Control (QC) in methodology, equipment/calibrations [20–22]
2. Consistency and long-term planning—monitoring sites, parameters [22]
3. Training and oversight—requires qualified scientists
4. Scientific motivation—guiding factors are promotion, recognition [23]
5. Funding—fragmented, limited long-term support

2.2.1 *Case Study: Salt Creek Watershed, Porter County, IN*

For decades, water quality concerns of southern Lake Michigan tributaries in the state of Indiana have led to numerous studies and projects focused on the health of the streams, creeks, lakes, and surrounding watersheds [24–27]. An area of interest is the Salt Creek watershed in Porter County, Northwest Indiana consisting of 77.5 square miles of land of multiple uses and 24 miles of water flow, ultimately discharging into southern Lake Michigan (Fig. 11.1). Over the past three decades, land use in the Salt Creek watershed has significantly changed, with recent additions of a county hospital, a paper recycling plant, and numerous housing developments. According to the Indiana Department of Natural Resources, approximately 54% of the Salt Creek watershed was devoted to agricultural activity in the early 1970s. The 2006 data from the National Land Cover Database measured agricultural land use at 29% of the watershed. More current data has not been located, but in the last decade further reduction in agricultural activity has taken place in the Salt Creek watershed, as residential and industrial development has continued to develop.

This change in land use is not surprising considering the fact that the population of Porter County increased by 45% from 1980 to 2000, with the number of households within the unincorporated portion of the county increasing by approximately 18,000. The unincorporated population growth is projected to increase to 91,000 by 2020, from the 2010 value of 79,288 [28]. The total population in Porter County increased from 120,000 to 167,000 from 1980 to 2013.

Since a thorough water quality study was done in strategic portions of the Salt Creek watershed in 1980–1981, comparisons of surface water

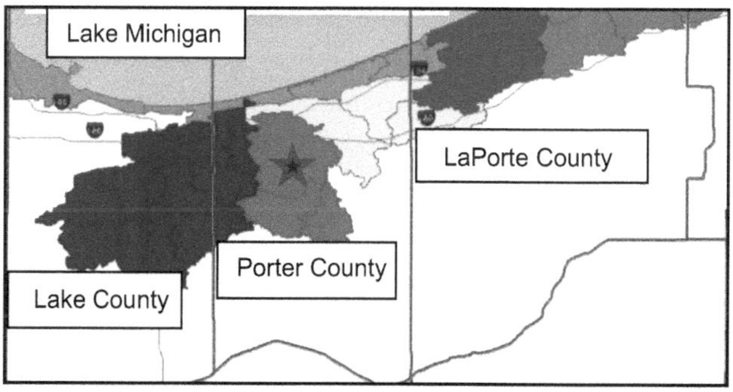

Figure 11.1 Lake Michigan watersheds in Northwest Indiana are highlighted in different colors or shades. The Salt Creek watershed is located in Porter County, and is notated with a star.

quality changes over nearly four decades were expected to be informative. As mentioned above, significant land-use changes have taken place in the Salt Creek watershed, and the state has directed water quality studies over the past two decades. A systematic comparison of water quality parameters can be a means for the assessment of infrastructure upgrades and other improvements, in addition to the effects of land use and other changes. It is important to distinguish the two broad categories of water pollution, which have different report/oversight requirements: point-source pollution and nonpoint source pollution; the generators of point-source pollution are required to monitor and report certain parameters of their discharged water. Point sources of water pollutants are regulated under the US federal program called the National Pollutant Discharge Elimination System (NPDES). This data is to be made available to the public and is defined in terms of location, frequency of sampling, and methodology. Common point sources include wastewater treatment facilities, industry, and storm sewers. In the Salt Creek watershed, eight facilities have NPDES permits dictating acceptable levels of components in their discharge, and seven facilities have Municipal Separate Storm Sewer System permits [29].

Nonpoint sources of pollution are much more complex and origins are often diffuse. The multitude of stressors that contribute to nonpoint-source pollution can include pollution from urbanized areas, changes in land use or structure, agriculture, severe weather events, climate change and invasive species, among others. The monitoring, assessment, and management of water affected predominantly by nonpoint-source pollution requires a different approach than point-source monitoring. While the monitoring of fresh, surface water has been part of the required protective provisions for lakes, rivers, and streams since the establishment of the Clean Water Act's pollution control measures, the recorded Water Quality Standards (WQS) are determined by states and based on the use of the body of water. The water quality standards are provisions of state, territorial, authorized tribal, or federal law, but are ultimately approved by the US EPA. They require a designation of the desired use of a body of water, such as *full-body contact* or *well-balanced aquatic communities*. This category establishes the level of protection and remediation required, which is accomplished through appropriate and accurate monitoring and measurements. In the state of Indiana, for bodies of water that do not meet the WQS, the Office of Water Quality, which is part of the Indiana Department of Environmental Management (IDEM), is mandated to develop Total Maximum Daily Loads (TMDL), according to section 303(d) of the EPA's Clean Water Act. The

EPA defines the process of establishing TMDLs, in part, as "a planning tool and potential starting point for restoration or protection activities with the ultimate goal of attaining or maintaining water quality standards." The TMDLs for the state of Indiana are limits on the measured values for *Escherichia coli* (*E. coli*) bacteria, total nitrogen (TN), total phosphorus (TP), and total suspended solids (TSS). These parameters were deemed by the agency to be the most useful water quality measures as a result of failures of Indiana's waterways to meet the minimal biotic communities and *E. coli* standards. For the Salt Creek watershed, the TMDL targets are shown in Table 11.3, according to the Salt Creek Watershed TMDL Program, Office of Water Quality Report [30]. For areas in the watershed that are classified as impaired biotic communities, the relevant TMDLs are TSS and total phosphorus.

The water quality parameters measured over the past three decades in the Salt Creek Watershed, found in state documents and other studies, include the TMDL parameters of TSS, TP, TN, and *E. coli* along with additional data. Some of the collected, stored, and published data and other important aspects of the water quality monitoring in the Salt Creek watershed will be described within the context of each of the following sections.

2.2.1.1 Quality Assurance and Quality Control

The collection of water quality data relies on scientifically sound and consistent methods [20,21,31]. For surface-water monitoring, established and consistent procedures are required for each phase of the work: onsite measurements, appropriate containers, sample collections, sample preservation/transportation/storage, laboratory measurements, and sample

Table 11.3 TMLDs for Salt Creek, Porter County, Indiana, USA

Parameter	Units	TMDL Targets	Designated Use
E. coli	#/100 mL	235 (single sample maximum)	Total body contact recreation and aquatic life
Total suspended solids (TSS)	mg/L	30	Well-balanced aquatic communities
Total phosphorus (TP)	mg/L	0.3	Well-balanced aquatic communities
Total nitrogen (TN)	mg/L	10	Well-balanced aquatic communities

analyses. All analytical instruments, including field-measuring equipment, need to be maintained and calibrated using reliable standards as part of the quality assurance and control protocols (QAQC). For federally funded water-monitoring programs, an EPA-approved Quality Assurance Project Plan (QAPP) is required to guarantee scientifically sound procedures and data collection. The validity of these science-based activities is crucial when considered as part of a larger water quality monitoring program. A thorough description of a water quality monitoring protocol was broken down into six activities by Sanders et al.—network design, sample collection, laboratory analysis, data handling, data analysis, and information utilization [32]. If the design, sampling, and laboratory work are not done according to scientific standards, data handling, analysis, and utilization are of limited value.

Since sound methodology is critically important in the collection of water quality data, we revisited the methods used in the Whitman study for comparison to both state of Indiana historical data and more current data collected by other groups for sites in the Salt Creek watershed [25]. Many of the methods utilized by Whitman and his research team are described in detail as part of their project report and most are summarized in Table 11.4. In a more recent Salt Creek water quality study by Peller et al., the documented measurement methods were comparable to the Whitman project for stream velocity and TSS determinations [24]. Conductivity, temperature, dissolved oxygen, and pH were measured using a Hach Quanta DataSonde, which was calibrated prior to each sampling and then used onsite, similar to the Whitman team. Anion concentrations (Cl^-, NO_3^-, NO_2^-, SO_4^{2-}) were determined using ion chromatography (IC), and total organic (TC), and inorganic carbon concentrations (TIC) were measured using a total organic carbon (TOC) analyzer. In both of these studies, the procedures for water sampling and the on-site and laboratory measurements are clearly described in the manuscripts, as expected for scientific publications. In unpublished studies recently completed by Schoer et al., the measurement protocols were similar to those of Peller and Whitman, except for the use of YSI sondes for onsite collection of ammonium and nitrate ion concentrations. Detailed experimental information enables other investigators to adopt the same or similar methodology, which can confidently be used for comparative purposes.

The state of Indiana's Water Quality Assessment Branch lists monitoring protocol for the different state programs [33]. While the methods used for the major funded programs (IDEM Office of Water Quality Assessment Branch Monitoring Programs) are EPA-approved, the state-sponsored

Table 11.4 Selected Water Quality Parameters and Methods Used by Whitman, IDEM, and Indiana Volunteer Water Quality Monitoring Programs

	Methods		
Parameter	**Whitman [25]**	**IDEM [33]—Current**	**Volunteer Programs (State of Indiana)**
Stream area and discharge	Conventional morphometric	ISCO automatic sampler with flowmeter	Unclear
Stream velocity	Digital flowmeter	Digital flowmeter	Unclear
Temperature	Portable digital Temperature/pH meter		
pH	Hach portable digital Temperature/pH meter	DataSonde: EPA 150.2 or 150.1 (standard methods)	pH test strip
Specific conductance	Lectro MHO meter (Wheatstone bridge)	Hydrolab sonde	NA
Dissolved oxygen	Winkler technique and digital titrator	Variety of standard methods: Hydrolab or YSI datasonde method: Standard methods 4500-OG	Test strip (CHEMetrics dissolved oxygen)
Total ammonia nitrogen	Kjeldahl analyzer and Nessler's reagent	Variety of standard methods: Selective electrode, ion chromatography	NA

(*Continued*)

Table 11.4 Selected Water Quality Parameters and Methods Used by Whitman, IDEM, and Indiana Volunteer Water Quality Monitoring Programs—cont'd

	Methods		
Parameter	**Whitman [25]**	**IDEM [33]—Current**	**Volunteer Programs (State of Indiana)**
Nitrate nitrogen	Orion ion specific electrode and Leeds & Northrup specific ion/mv meter	EPA standard laboratory methods: EPA 353.1, 353.2, 353.3	Test strip
Phosphorus as total P or phosphate	Spectrophotometric analysis (ammonium molybdate and stannous chloride)	EPA standard laboratory methods: EPA 300., 365.1, 365.2	Orthophosphate test strip
Chlorophyll a	Modified artificial glass slide periphyton apparatus; 90% buffered acetone extraction; spectrophotometric determinations at 630, 645, and 663 nm	EPA 445.0 EPA standard methods 445.0 10200H	NA
Total suspended solids (TSS)	Gravimetric	EPA method 160.2	NA

volunteer programs, which provide test kits and training to citizens, do not meet science standards (Table 11.4). The IDEM-funded water chemistry work performed in the creation of the Salt Creek WSMP (2006–2007) references *Standard Methods for the Examination of Water and Wastewater*, published by the American Public Health Association (APHA), 1998, as the methodology. No specific descriptions on the methods utilized in measuring water quality parameters are described in the plan. However, the management plan does state that a QAPP for the water quality monitoring work was approved by the IDEM.

2.2.1.2 Consistency and Long-Term Planning

Crucial factors in the planning, implementation, and oversight of water quality monitoring are the selection of monitoring locations and the consistency of testing at these sites. Data comparison cannot be done if sample locations and times of the day/year are not the same. A stream's flow varies considerably with location, season, and weather, and is an important factor in water quality monitoring. One study might sample downstream of a bridge in a riffle, another upstream of that same bridge in a pool, while still another along the bank in shallow easily accessible backwater. These locations may have different physical parameters, such as flow rate, depth, sun exposure, sediment loading, bottom type, temperature, plant life, etc., that significantly affect water quality measurements. Furthermore, the season and the time of day often have considerable effects on certain measurements, such as dissolved oxygen. Over time, the physical characteristics of selected monitoring locations may change due to natural occurrences; examples include formation/disruption of debris jams, bank erosion, and sediment deposition. Consequently, these types of observations are also important components of regular watershed monitoring.

One must make a distinction between monitoring and characterization of a sampling area or stream. Consistent sampling of a reference site does not characterize the stream or even that stretch of water. It provides some idea of the particular reference site, and if replicates are used, some of the variation within that site. It also identifies trends over time and interannual variation. Statistical analyses of these sites are hampered by a statistical problem termed autocorrelation. Here, the sites are presumably related usually by time, space, or some event. Care should be exercised therefore to consult with a statistician to use the proper statistical inferences (repeated measures vs. ANOVA). If, on the other hand, it is the creek that one wishes to characterize as in a TMDL, then locations, times, and replicates should be randomized. Again, specific statistics are appropriate when randomized designs are employed and a trained statistician should be consulted.

Consistency is critical in both the sampling locations and sampling frequency for the meaningful comparison of data over the long term. For scientifically based datasets designed to protect fresh surface water, programs must (1) monitor at specific locations, (2) measure a regular set of parameters to assess water quality and rationalize changes to these parameters, (3) ensure consistent or comparable scientific methodology in both the collection and measurement procedures, and (4) sample and measure with adequate frequency at the appropriately determined monitoring sites. If frequent sampling does not take place, the data can only be considered a snapshot of the actual water quality and it becomes difficult or impossible to assess long-term trends. A watershed is a complex, dynamic ecosystem, and occasional monitoring cannot adequately assess its health. Sampling frequency must also take into consideration seasonal changes, weather factors, and natural geological processes.

In an effort to compare data from the past few decades in the Salt Creek watershed using the 1980 Whitman study, the water sampling sites monitored by the state and other groups needed to be identified. Since the Whitman study was supported by a planning grant from the US EPA, Section 208 of the Federal Water Pollution Control Act Amendments of 1972, there was an objective for the selected study sites. In the IDEM database for the Salt Creek watershed, water quality data can be found dating back to the 1970s, but the locations are different from those studied by Whitman. The sampling sites selected for the collection of data compiled for the 2008 Salt Creek Watershed Management Plan (SCWMP) by the nonprofit organization Save the Dunes "were selected based upon land use, site accessibility and historic data, including information from IDEM on past and current water sampling sites." The map in Fig. 11.2 shows the sampling sites for the EPA-funded 1980 study and the IDEM funded 2006–2008 work. Furthermore, the SCWMP lists historical water quality data collections in Appendix I, but does not list the Whitman study or data collected prior to 1990. This signals another limitation concerning the scientific integrity in the development of WMPs. Published scientific studies and pertinent historical data compiled by the state environmental agency should be evaluated prior to the watershed-management planning.

Of the eight sampling sites targeted in the 1980-1 Whitman project, none were used for the 2006 SCWMP project, which selected 16 sampling sites for the 24 miles of waterways in the Salt Creek watershed. For

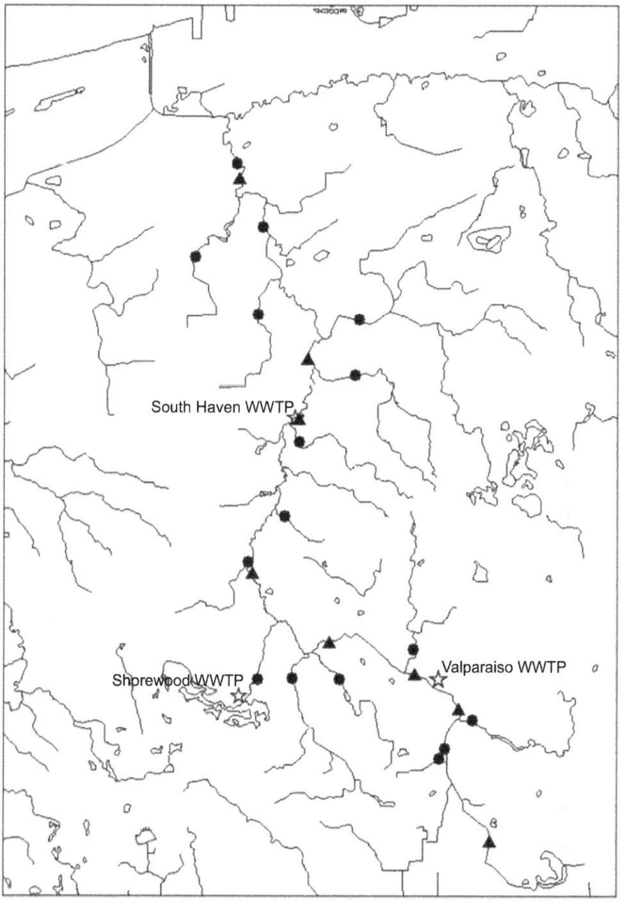

Figure 11.2 The eight sites monitored in the 1980–1981 Whitman study (▲), and the 16 sites monitored in the preparation of the watershed management plan (●). Also noted on the map are the locations of the wastewater treatment plants that discharge into Salt Creek.

Whitman's project, four of the eight sampling sites were above and below the two wastewater-treatment facilities for the local communities of Valparaiso and South Haven within the watershed. Station II in the report is described as "1.3 km upstream of the Valparaiso Combined Sewage Outfall." The description of Station III is, in part, "500 m downstream of the Valparaiso CSO." Stations VI and VII were similarly upstream and downstream of the South Haven wastewater-treatment plant, and were, in part, an assessment of the wastewater treatment plant's influence on the watershed. The remaining sites were strategically located as controls and used to track downstream trend information [34]. For the SCWMP, the sampling sites were determined by the IDEM to be appropriate for creation of the WMP and several were the same sites used by the agency for many years, as per the state database. Six of these 16 sampling locations were used for the study by Peller et al. The university-led study focused on the major watershed-pollution challenges and conducted several thorough samplings for the six sites over the course of a year. Similarly, two of the sites utilized in the study by Schoer et al. were selected in conjunction with the SCWMP and the Whitman study.

The inconsistencies in the sampling locations for water quality monitoring between the EPA-supported study (1980–1981) and the IDEM-supported Salt Creek Watershed Management Plan project (2006–2008) and other IDEM monitoring meant that the Whitman datasets could not be compared in any meaningful manner to the state's data. The data compiled by the state's water quality program was not taken at any of the locations monitored in the Whitman study. This fact may signal an organizational disconnect concerning funded watershed studies in the state, or may be a matter of changing priorities. At the time of the 1980 study, combined sewage outfalls (CSOs) were common and considered to be a major source of watershed contamination. When heavy rains add to the volume of sewage water, untreated wastewater potentially discharges into outfall creeks and rivers (CSO); this has been a problem in the Salt Creek watershed.

Outfall creeks and rivers are often significant contributors to water quality impairments in many areas around the Great Lakes, and are predicted to increase as a result of more violent storms caused by climate change [35,36]. A 2007 EPA report to Congress on CSOs in the Lake Michigan basin stated that Indiana had 18 CSO communities, and 17 of the 18 communities required the development of a long-term control plan to reduce the CSO events [37]. Maps showing the extent of CSOs have been created by the Great Lakes Environmental Mapping and Assessment Project (GLEAM), and can be found on their website.

In Porter County, Indiana, the cities of Valparaiso (pop. 32,261) and South Haven (pop. 4,378) discharge their treated sewage into Salt Creek. As required by law for point-source pollution, monitoring is done by the wastewater-treatment plants, both of which have flows greater than 0.1 million gallons per day (mgd). Until recently, these facilities had numerous CSOs (as noted in the 2007 EPA report), which were reported to be the main contributor to the reduction in water quality in Salt Creek in 1981, according to Whitman's study [25]. In 2012, the Valparaiso wastewater-treatment plant completed construction of a combined sewer overflow disinfection facility, aimed at eliminating the discharge of untreated sanitary sewage into Salt Creek. The Whitman study can serve as a very useful dataset to accurately assess the city's wastewater-treatment plant improvements on the Salt Creek water quality, especially downstream from the discharge areas. Thorough, scientifically sound projects that measure important water quality parameters are beneficial for establishing the water quality benefits of major infrastructure upgrades, when the measurements are performed pre- and postimprovements. Lack of sampling and location consistency translated into poor options for comparison. Thus there is no current manner to learn from the remediation or to evaluate efficacy of expensive wastewater improvements.

This disconnect in watershed work is due in part to the different objectives of the two major Salt Creek dataset collections. The EPA-funded study was intended to evaluate the impact of sewage-treatment plants and the problematic CSOs on Salt Creek, and Whitman was hired to perform the work. The IDEM contracted Save the Dunes in an effort to characterize land use-based components of the watershed for remediation planning and to support the routine monitoring done by the state. Consequently, the current follow-up work to the watershed plan is contracted to a different organization.

While the selection of sites for future water quality monitoring needs to be more consistent and rationalized in the Salt Creek watershed, the sampling frequency is an additional challenge for monitoring and understanding the effects of changes in the watershed. Appropriate frequency in monitoring requires an investment by communities or state and federal agencies. For the implementation of science-based water monitoring programs, professional resources are vital and required. In the 2012 document "Monitoring Water in Indiana: Choices for Nonpoint Source and Other Watershed Projects," the authors calculate the estimated cost for the "most economical" option for weekly sampling performed at one site for a year at $14,500 [38]. The state of Indiana's monitoring protocol (IDEM's core parameters) includes nitrate, total phosphorus, *E. coli*, turbidity, dissolved oxygen, pH, and temperature. Since weekly monitoring is clearly cost-prohibitive by these calculations, the state of Indiana's environmental agency, the IDEM, has broken the state into nine major water-management basins. Each year, one of the water basins is monitored and the data analysis is performed the following year. Essentially, the state ensures that 11% of the watersheds are monitored each year, and then relies on other groups (universities, nonprofits, local government entities, and volunteers) to assist or fill in the gaps. However, no true coordination for these efforts currently exists.

For long-term and scientifically sound planning of water quality monitoring, the number of sites and the frequency of the monitoring are critical for understanding the effects of pollution and determining and implementing the most appropriate remediation efforts. The state of Indiana has 63,000 miles of streams, creeks, and rivers. It is reasonable to monitor more sites in watersheds with greater pollution challenges than watersheds with fewer pollution sources. In the work for the Salt Creek Watershed Management plan project, 16 sites were selected for approximately 24 miles of primary waterway. For sustainable programs, the selection and focus on a limited number of sites are necessary for both adequate consistency and frequency (#2 and #5, Table 11.2). Furthermore, the importance of careful planning for long-term programs is clear in the EPA's description of nonpoint-source monitoring. Specifically, the EPA's Clean Water Act section 319 Nonpoint Source National Water Monitoring Program's Document of Successes and Recommendations suggests two years of data are needed to establish baseline conditions, followed by implementation of improved practices and further monitoring for 3–6 years [39]. For these types of meaningful programs, a practical number of monitoring sites is a requirement to ensure sampling frequency and replications for useful, ongoing studies of water quality.

2.2.1.3 Training and Oversight

Experienced scientists who plan, train, and oversee programmatic work are a necessary component for programs intended to ensure watershed and surface water health. The most useful indicators of either threats to water quality or improvements in water quality are measurements of the chemical, biological, and/or physical conditions of the water. Since these parameters are mostly quantitative in terms of validity and reliability, measurements are best conducted or supervised by professionals experienced in the protocol. Training and oversight should also be carried out by those who have the required knowledge and background. The World Health Organization's 1996 publication, "Water Quality Monitoring—A Practical Guide to the Design and Implementation of Freshwater Quality Studies and Monitoring Programmes," describes measurements and methodologies required for useful datasets. "Complete assessment of the quality of the aquatic environment requires that water quality, biological life, particulate matter and the physical characteristics of the water body be investigated and evaluated" [13].

These assessments require experienced scientists for meaningful planning, data collection, and data evaluation.

Prudent practices must be in place for valid data collection, with the utilization of expertise for the recognition of uncertain or unexpected circumstances. The monitoring of natural systems is subject to a multitude of natural variations. Sampling, analytical, and/or interpretation error can compromise data and its findings. The accumulation of data requires analysis in a timely manner for long-term understanding of freshwater quality in the midst of ever-changing environments. If analyses are performed several months or even years after the data collections, efforts to recognize or remediate pollution sources may be critically delayed. Timely analyses can also signal problems in the process or equipment, which can be remedied prior to further work.

Data compiled using questionable protocols cannot be used for scientific evaluations, and certainly should not be used for decision-making. Programs that rely heavily on nonscientist supervision and/or volunteer monitoring are often conducted using inferior water quality monitoring practices. These are not designed to respond systematically to variables and lack the required quality assurance and control (#1, Table 11.2). Also, the recognition of invalid measurements, which may lead to false conclusions, is unlikely without scientific supervision. Thorough training coupled with knowledge of the watershed is absolutely required for the valid collection of data, but the process must also include adequate oversight. Citizen programs are extremely effective for promoting environmental stewardship. However, when scientific measurements are needed for purposeful water quality work, scientists should provide training and supervision. Water quality monitoring programs should not be established under the pretense that a basic colorimetric testing kit will provide adequate data, which is often the case for volunteer water quality monitoring programs. Test strips may have severe limitations that untrained water monitors may not be aware nor feel qualified to correct. Many of these tests are designed for approximate measurements in drinking or swimming pool water, and are inadequate or invalid for natural or polluted waters. These types of measurements do not to take into effect the complexity of real-world fresh, surface-water systems, which require experienced scientists (#3, Table 11.2).

The published studies on the Salt Creek Watershed by Whitman and Peller et al. and the unpublished work by Schoer et al. were designed and implemented by research scientists, who worked with university students from the sample collection steps through the laboratory analyses. Other benefits of utilizing academic facilities and expertise include the students' learning of scientific methodology (training) and the cost effectiveness and collaborative nature of the programs. Training and oversight are strong aspects of most university-led water quality monitoring projects. Interaction with scientific peers or a multidisciplinary research team is a significant advantage. When planned, performed, and analyzed in a timely manner and according to scientific standards (e.g., EPA-certified methods), the studies can be published in peer-reviewed venues, and are then readily available for the scientific community, regulatory agencies, and the public. If legal corrective action is necessary, a more successful case against a polluter is plausible. However, many of these types of funded projects are short-term and then subject to failure in the category of consistency and long-term planning (#2, Table 11.2).

2.2.1.4 Scientific Motivation

The incentive and inspiration for scientific studies carried out for public good may be due to a number of factors. Nonetheless, similar to other professionals, scientists who work on studies carried out for public good pursue accomplishments that lead to promotion, tenure, and advancement. Scientists are motivated to provide solid information for the purpose of effective decision-making when appropriate programs are in place, those that involve scientific methodology. The science community that takes part in science for public benefit works with the same intent as other scientists: to advance knowledge and understanding. This is the main reason people choose to be educated and trained to work in areas of science. These types of pursuits can lead to publications and/or funding, and corresponding achievements can lead to promotions in the work environment.

Projects intended to generate science to benefit the public are often not rigorously designed or carried out for peer review journal publications. Also, they may not contribute substantially to the subject discipline or may not in actuality be testing scientific questions. This does not suggest that the efforts are necessarily below scientific standards, only that they serve an appropriately gray literature report for a selective goal rather than a contribution to the discipline. The directives for many supported environmental or public health projects (such as WMPs) regularly consist of gathering information in an omnibus fashion. While these project type efforts use scientific techniques, they often do not employ the scientific method. Nonetheless, the data may still be used to answer questions of status and trends and may be used to test important scientific questions if acceptable and consistent practices are employed. Routine data compiling that is expected or typical, without a rationale, defendable hypothesis, is not only inefficient, it dampens the motivation of the scientific community, discouraging professional researchers and encouraging more monitoring by nonscientists or volunteers. In other words, if the scientific method, which requires the formation of a hypothesis to guide an investigation, is not the influencing parameter for a project, scientists tend not to be eager to participate (#4, Table 11.2). This type of work typically does not lead to publishable, purposeful outcomes; in such cases, there is little motivation to pursue science for public good. In well-funded academic disciplines, paid paraprofessionals, who are well-trained, gather data and monitor experiments and outcomes while researchers interpret data, test the initial hypothesis, and publish the results. Typically, only the best-funded programs that are focused on environmental and/or public health-related science have similar opportunities for their faculty, postdocs, and graduate students.

When science for public good is not viewed as academically or professionally rigorous or lauded, scientists may not seek to perform the work if it does not lead to professional goals. The scientific community focuses on distinctions such as the number of publications, grants, publication citations, and professional recognitions. Still, the pursuit of knowledge pertaining to surface freshwater quality is an active area of study. When a typical literature search is performed on surface-water quality, tens of thousands of publications are found. One way to evaluate the extent of scientific investigations of surface-water quality in context with public good on a research level is to focus this literature search. For example, when the literature search results are refined with the term "public health," the number of published studies decreases to several hundred. This seems to indicate that either few researchers directly connect their research with public needs or the work done in this domain is not published or pursued for publication.

From a broader scope, the scientific community is actively involved in work associated with public and environmental health. Many scientists are hired to perform work investigating natural systems affected by challenges to ecological, environmental, and public health. They are professionally and personally motivated, and the occupations are geared toward understanding and solving problems for public benefit. Many government labs are charged with studying and addressing many aspects of natural systems as they relate to public health and well-being. However, the outcomes of the work are frequently removed from the public eye, or not well communicated with the lay public. This creates a divide between the lay public and the scientific community, and a perception that scientists struggle to relate their work on the public realm. There are plenty of examples where researchers in the scientific community conduct investigations on a local level and interact with interested citizens. This type of organization and communication is difficult to formalize, but more effort should be made to lessen the communication gap. Another reason for the communication gap may be related to overall science literacy or science phobia among/within the public. Furthermore, when community members request the expertise of local scientists from either academic institutions or research labs, these undertakings are often considered by scientists as low priority, due, in part, to the reduced professional output or reward.

Several scientific investigations have been, and continue to be, performed in the Salt Creek watershed in Porter County, Indiana based on a variety of motivating factors. The main logistic barrier for Salt Creek watershed projects is funding, but the incentives include educational experiences, scientific curiosity, and environmental concern. The authors have unsuccessfully sought funding over the years to reevaluate the watershed to determine whether remediation efforts have been successful. The political and bureaucratic reality is that new initiatives are favored rather than revisiting and evaluating the success of past efforts, or long-term goals. In some cases, particularly with politically sensitive issues, the motivation is that "no news is better than potential bad news." The result is that every few years or decades, brand new initiatives are started under the partial rational that a new water quality baseline is needed for decision-making and planning. Thus adaptive management through monitoring or follow-up studies is defeated until once again an issue or specific funding initiative arises.

The Salt Creek Watershed management plan lists many of the groups that have been involved in watershed monitoring over the past two decades. Projects, in general, were short-term to satisfy short-term goals or compile data. None of the work resulting from the creation of the WMP has been published in scientific publications. Furthermore, only a few of the projects associated with water quality in the Salt Creek watershed over the past 3–4 decades, have been driven by hypotheses with long-term intentions. While there is a great deal of interest in tracking changes associated with rapid growth of local communities, there is no science-based comprehensive long-term plan in place for the watershed (#2, Table 11.2). The supported monitoring has been and continues to be associated with TMDLs, which deal only with the parameters listed in Table 11.3, to satisfy the state's requirements.

2.2.1.5 Funding

Most of the funding for applied science work (e.g., water and watershed-monitoring programs) originates from federal departments, such as the EPA, USGS, or NOAA, and often is then distributed through state agencies. When the allocation of support (technical, financial) is in the hands of nonscientists, the project is generally not science-based, and will likely not be guided by the scientific community. Often, this is because funding is insufficient to bring in trained scientists or to pay for professional analyses. Even though a community or organization may recognize specific needs and obvious pollution problems in the watershed, securing funding for sound science-based investigations and solutions for specific areas is extremely difficult. Most programs associated with science for public benefit are defined by government agencies and other nongovernmental funding sources (i.e., businesses, foundations). In many ways, funding unfortunately dictates the scientific agenda. In most cases where a need for scientific studies is identified, and there is undoubtedly a great quantity, there is only a small likelihood that monetary support is available.

Another major problem in the funding for science intended to benefit society is the timeline of the financial support. Projects that require long-term support are often not funded in a manner that creates a sustained, worthwhile program. The majority of funded water quality monitoring-related projects receive short-term support and, consequently, the funding process is inefficient and ineffective [40]. If long-term studies (several years) are required, and projects are funded for 2 or 3 years with little or no possibility for continuation, the overall goals of the work cannot be achieved. This seems to indicate that agencies don't recognize or rationalize the value of the long-term needs of scientific studies for public benefit or are unable to commit to long-term support. More often, it is due to budgetary cycles, constraints, priorities, and program longevity.

In efforts to address environmental problems in a scientifically sound manner, researchers often integrate the work as part of another funded project. Many academic researchers have secured partial funds to perform beneficial surface-water research within their educational grants. This approach, again, is often subject to fragmented, unsustainable plans, leading to unfinished work. Priorities for science programs for public benefit need to be reevaluated to more effectively utilize resources. Programs should be directed by science, with rational objectives, not established to conduct basic surveillance work to acquire numbers for the compilation of databases. Useful water quality monitoring programs necessitate investigative approaches that can be supported and sustained long-term, guided by the scientific community. These types of programs and projects reflect the most current science and are best suited to provide solutions to benefit public and environmental health.

2.3 Models of Successful, Integrated Science-Based Projects

A number of Watershed Districts consist of well-organized groups of professionals who effectively meet the five requirements for effective science listed in Table 11.2. Community participation often compliments science-based projects, integrating interested citizens in manners that affect their lives. Other organizations, such as nonprofit, governmental, and outreach groups, offer valuable assistance in compiling additional information from their different areas of expertise, such as policy, planning, or historical research. There are numerous examples of science-based watershed districts or programs based on this type of collaborative model, which integrate different entities for a more complete and sustainable public benefit [41,42].

Watershed and water quality monitoring can be designed to acquire and analyze quality data and also benefit from the input of appropriately trained and supervised volunteers. When projects require frequent observations, citizens can record important data, such as odors or noted discolorations of air or water, on a more routine basis than scientists or paid workers. These aspects of volunteer work are highly valuable. Several examples showing the value of citizen science have been, and continue to be, part of the Hudson River monitoring program. Scientists from Columbia University's Lamont–Doherty Earth Observatory and Queens College, CUNY work with citizens in monitoring water quality. In other projects, volunteers assist in tallying numbers of alewife fish or glass eels or relocate salamanders, toads, and frogs during migration season. In the Great Lakes region, Adopt-a-Beach programs, organized by the Alliance for the Great Lakes, rely on volunteers to scour shorelines and collect information on pollution. Other shoreline observations that are well tallied by volunteers are bird and fish kills and algal growth. A successful program that collects water quality data in the polluted tributary of Maryland's South River called Broad Creek is science-based and utilizes volunteers who are supervised by scientists. The program has effectively collected data for the past 8 years, which is valuable to the state and others involved in water quality work [43].

In the Salt Creek, IN watershed, a current EPA Environmental Education-funded project is an integrated watershed-monitoring program that educates middle-school students and is supervised by university faculty.

The project is directed by three university scientists, trains and utilizes undergraduate researchers and science education students, and incorporates the strengths of nonprofit groups and other interested community members. Over 175 middle-school students and their science teachers monitor the physical, biological, and geographical parameters at waterways near their schools. The university faculty communicate regularly with the middle-school science teachers and direct and oversee the data (observations and measurements) taken by the young students. Additionally, undergraduate research students take part in more extensive research investigations and measurements. Nonprofit environmental groups and city departments compliment the project in a number of ways, with substantial efforts in place to increase community environmental stewardship. While the short-term successes of the program are impressive—increased interest and awareness of the watershed by educators, students, and community and constructive research training and output—these types of projects can be vastly more effective and sustainable when supported for the longer term. This current program is funded for 2 years, only long enough to initiate a comprehensive program, but with the prospect for renewal. The project is also representative of surface-water research funding made possible through a larger educational grant. The lack of funding opportunities specifically for long-term, local water quality monitoring and research continues to create disconnected datasets and mostly just short-term benefits.

CONCLUSIONS

Interested citizens, environmental advocates, volunteers, planners, managers, political officials, scientists, and other professionals all have a role in the planning, remediation, engineering, management, monitoring, and research of a challenged or sensitive watershed. Each entity offers an important perspective on the problem, and collectively can achieve the desired outcomes. In most cases, when the experiences and training are integrated with one another, the outcome is a more balanced strategy toward the ultimate solutions. An interdisciplinary team creates a well-balanced, effectual program for public benefit with efficient use of resources. We argue that environmental groups or volunteers may engage in scientific work, but not without adequate training or supervision by the scientific community. The lack of professional direction often leads to inadequacies in planning, site selection, analyses, interpretation and ultimately the suggestion of management alternatives. Likewise, researchers are poorly equipped to make management, policy, financial, or politically sensitive decisions. We suggest that projects should include a long-term monitoring component so temporal variation can be accounted for and trends established. Often, officials want solutions resolved during the duration of their tenure or when the pressure of the issue subsides. We reviewed the motivation, funding, and approach to local environmental projects and found them inadequate in many cases. Finally, we suggest that local environmental managers should establish a new paradigm that incorporates continuous monitoring and correction when problems arise. These are areas where scientists are best suited to plan and advise. Adaptive management, a process based on continual learning and adjustments, may be the most reasonable way of approaching long-term monitoring. It incorporates necessary corrective actions before they become issues that are difficult or impossible to correct [15,44].

REFERENCES

[1] Lubchenco J. Entering the century of the environment: a new social contract for science. Science 1998;279(5350):491–7.

[2] Weichselgartner J, Kasperson R. Barriers in the science-policy-practice interface: toward a knowledge-action-system in global environmental change research. Glob Environ Change 2010;20(2):266–77.

[3] Goldstein BD, Kriesky J, Pavliakova B. Missing from the table: role of the environmental public health community in governmental advisory Commissions related to Marcellus shale Drilling. Environ Health Perspect 2012;120(4):483–6.

[4] Jenner S, Lamadrid AJ. Shale gas vs. coal: policy implications from environmental impact comparisons of shale gas, conventional gas, and coal on air, water, and land in the United States. Energy Policy 2013;53:442–53.

[5] Atlas RM, Hazen TC. Oil biodegradation and bioremediation: a tale of the two Worst Spills in US history. Environ Sci Technol 2011;45(16):6709–15.

[6] Oliaei F, et al. PFOS and PFC releases and associated pollution from a PFC production plant in Minnesota (USA). Environ Sci Pollut Res 2013;20(4):1977–92.

[7] Sarewitz D, Pielke Jr RA. The neglected heart of science policy: reconciling supply of and demand for science. Environ Sci Policy 2007;10(1):5–16.

[8] Decision making for the environment: social and behavioral science research priorities. Washington, DC, USA: National Academies Press; 2005.

[9] Lindenmayer DB, Likens GE. Adaptive monitoring: a new paradigm for long-term research and monitoring. Trends Ecol Evol 2009;24(9):482–6.

[10] Bellinger DC. Lead contamination in Flint — an Abject failure to protect public health. N. Engl J Med 2016;374(0):1101–3.

[11] Water Oo, editor. Handbook for developing watershed plans to Restore and protect Our waters; 2008. Washington, DC.

[12] Likens GE. Some perspectives on long-term biogeochemical research from the hubbard brook ecosystem study. Ecology 2004;85(9):2355–62.

[13] UNEP/WHO. Water quality monitoring—a practical guide to the design and implementation of freshwater quality studies and monitoring programmes. In: Bartram J, Ballance R, editors. United Nations environmental Programme and world health organization; 1996. p. 348.

[14] Chaffin BC, et al. The role of agency partnerships in collaborative watershed groups: lessons from the Pacific Northwest experience. Environ Manag 2015;55(1):56–68.

[15] Holling CS. Adaptive environmental assessment and management. Chichester (UK): John Wiley; 1978.

[16] Water EOo, editor. Volunteer stream monitoring: a methods manual; 1997.

[17] Penrose D, Call SM. Volunteer monitoring of benthic macroinvertebrates: regulatory biologists' perspective. J N Am Benthol Soc 1995;14(1):203 9.

[18] de França Doria M. Factors influencing public perception of drinking water quality. Water Policy 2010;12(1):1–19.

[19] McNie EC. Reconciling the supply of scientific information with user demands: an analysis of the problem and review of the literature. Environ Sci Policy 2007;10(1): 17–38.

[20] Facchi A, Gandolfi C, Whelan MJ. A comparison of river water quality sampling methodologies under highly variable load conditions. Chemosphere 2007;66(4): 746–56.

[21] Rubio S, Perez-Bendito D. Recent advances in environmental analysis. Anal Chem 2009;81(12):4601–22.

[22] Richardson SD, Ternes TA. Water analysis: emerging contaminants and current issues. Anal Chem 2011;83(12):4614–48.

[23] Sarewitz D, Pielke RA. The neglected heart of science policy: reconciling supply of and demand for science. Environ Sci Policy 2007;10(1):5–16.

[24] Peller JR, et al. Analytical measurements to improve nonpoint pollution assessments in Indiana's lake Michigan watershed. In: Ahuja S, editor. Monitoring water quality: pollution assessment, analysis, and remediation. Elsevier; 2013. p. 171–86.

[25] Whitman R. Environmental quality assessment of Salt creek, porter county, in. Northwest Indiana Regional Planning Commission; 1982.

[26] Whitman RL, Nevers MB, Byappanahalli MN. Examination of the watershed-wide distribution of Escherichia coli along southern Lake Michigan: an integrated approach. Appl Environ Microbiol 2006;72(11):7301–10.

[27] Haack SK, Duris JW. Dynamics of fecal indicator bacteria, bacterial pathogen genes, and organic wastewater contaminants in the little calumet river-portage burns waterway, Indiana. J Gt Lakes Res 2013;39(2):317–26.

[28] Porter county land Use Thoroughfare plan. HNTB Corporation; 2001.

[29] IDEM. Decision document for the Salt creek watershed TMDL, Indiana. Indiana Department of Environmental Management; 2012. p. 1–33.

[30] Quality OoW. Total Maximum Daily load for Esherichia coli (E. coli) and impaired biotic communities (IBC) in the Salt creek watershed in porter county. Indianapolis, IN: Indiana Department of Environmental Management; 2012.

[31] Strobl RO, Robillard PD. Network design for water quality monitoring of surface freshwaters: a review. J Environ Manag 2008;87(4):639–48.

[32] Sanders TG, et al. In: Design of networks for monitoring water quality. Littleton Colorado, USA: Water Resources Publications; 1983.

[33] IDEM. Monitoring protocols. Available from: http://monitoringprotocols.pbworks.com/w/page/21630005/IDEM%20Office%20of%20Water%20Quality%2C%20Assessment%20Branch%20Monitoring%20Programs.

[34] Whitman R. The influence of a combined sewage outfall on the chemical and biological quality of Salt creek, porter county, Indiana. Proc Indiana Acad Sci 1982;92:275–82.

[35] Bi EG, et al. Quantitative and qualitative assessment of the impact of climate change on a combined sewer overflow and its receiving water body. Environ Sci Pollut Res 2015; 22(15):11905–21.

[36] Nilsen V, et al. Analysing urban floods and combined sewer overflows in a changing climate. J Water Clim Change 2011;2(4):260–71.

[37] Combined sewer overflows to the lake Michigan basin. Washington, DC: United States Environmental Protection Agency; 2007.
[38] Frankenberger J, Esman L. Monitoring water in Indiana: choices for nonpoint source and other watershed projects. Purdue University and Indiana Department of Environmental Management; 2012. p. 1–134.
[39] Lombardo LA, et al. Section 319 nonpoint source National monitoring program successes and recommendations. Raleigh, North Carolina: NCSU Water Quality Group, Biological and Agricultural Engineering Department, NC State University; 2000.
[40] Nichols JD, Williams BK. Monitoring for conservation. Trends Ecol Evol 2006; 21(12):668–73.
[41] Capitol Region Watershed District. Available from: http://www.capitolregionwd.org/.
[42] Ramsey Washington Metro Watershed District. Available from: http://www.rwmwd.org/.
[43] Gibs SK. Volunteers against pollution. In: Chemical and engineering news. American Chemical Society; 2015. p. 12–7.
[44] Walter CJ. Adaptive management of renewable resources. USA: MacMillan; 1986.

CHAPTER TWELVE

Historical Perspectives on Water Purification

L. Smith
Filters for Families, Wheat Ridge, CO, United States

1. INTRODUCTION

The pursuit of water has left an indelible record in every culture since the Upper Paleolithic period. Prehistoric cultures searched for "sweet" water to sustain their settlements and for travel routes. This basic need for water resources and clean water encouraged the development of charcoal, clay-fired vessels, filter cloths, carved antlers, and animal skins for water storage, hand dug, and wood-reinforced wells and botanicals to add to water for purification. Artistic designs flourished alongside primitive yet highly skilled crafts/technologies such as wooden and stone corkscrews to secure the tops of animal skin water pouches, dove-tailed fitted slates for wooden wells, and elaborate hooks and patterns on clay-fired pots used to carry and hang water vessels. The Japanese skillfully created chrysanthemum-shaped charcoal for Samurai tea ceremonies and purifying water.

As early as the Upper Paleolithic period (50,000−10,000 BC), water was associated with sacred rituals. Even today most Indigenous cultures hold a reverence for the cleansing and healing properties of water through their offerings, celebrations, and sacred vessels. The oldest functioning shrine or temple in Sumer was the Temple of Enki, whose name is associated with the God of Water and male fertility. First built about 6500 years ago the temple was operational for almost 2000 years. Temple worship provided a profitable business for entrepreneurs dealing with sacred vessels, incantations, linens, clean water, and holy water for worshipers. Heron of Alexandria (AD 10−AD 70), a Greek mathematician and inventor, designed the first vending machine operated by depositing a coin to dispense holy water. Early medical practitioners often held surgeries and clinics within a temple complex. Some of the first recorded practices of water purification were implemented by these professionals. Water was boiled for sterilization, hot copper daggers were plunged into water, alum was added to water, and other methods were used to provide clean water for patient care and good health.

Chemistry and Water
ISBN 978-0-12-809330-6
http://dx.doi.org/10.1016/B978-0-12-809330-6.00012-X

The medicinal and water purification benefits of materials used in ancient cultures have been studied and are found to be so effective that they are important components in modern filters. Research has shown that the disinfection character of bone char, charcoal, clay, and metals relates to the adsorption properties and germicide effect of the materials. This is discussed in more detail in Part II.

The demand for household water filters is found in developing, emergent, and high-income economy countries. The market is awash with a plethora of technologies that range from pathogen elimination to removal of organic and inorganic compounds, metals, and pesticides. Shifting through these products can be a daunting task. Part III of this chapter will examine the more common household filters and discuss the science behind the technology, the pros and cons of the technology if the information is available, and comments regarding the maintenance of the units. Although the list of filter technologies provided here is not exhaustive, understanding the technology will provide an understanding for evaluation of most technologies.

PART I

2. HISTORICAL ACCOUNTS OF WATER PURIFICATION

Little if any information is available regarding Upper Paleolithic period water purification, although archeological discoveries from Upper Paleolithic sites in Old Europe (35,000–10,000 BC) where sophisticated stone Venus figures, ivory corkscrews used to cover skin water bags, weaving tools, fine cloth, netting, clay shards, and a plethora of other artifacts suggest a skilled artisan community (Figs. 12.1 and 12.2). Materials for water storage such as animal skins and unfired clay decompose unless preserved in a bog, so little if any evidence remains. Based on the beautifully carved ivory toppers found in three European sites, Hitchcock [1] suggests goatskins were most likely used to carry water. The water skin was sealed on top by the corkscrew topper. Other local materials were also available: reeds lined with mud, moss-lined wood sticks, hollow bones, or animal horns. There is no indication that these water skins or mud-covered reeds purified water. However, filtering water could been done through the exquisite "narrow gauge linen weavings found imprinted on Moravian ceramic fragments" [2,3].

FIG. 1. *Upper Paleolithic Europe, with location of sites discussed in text.* 1, *Brassempouy;* 2, *Lespugue;* 3, *Laussel;* 4, *Grimaldi Caves;* 5, *Dolní Věstonice;* 6, *Pavlov I;* 7, *Willendorf;* 8, *Avdeevo;* 9, *Kostenki I;* 10, *Gagarino.*

Figure 12.1 Location of European Paleolithic sites. *After Soffer O, Adovasio JM, Hyland DC. The "venus" figurines: textiles, basketry, gender, and status in the upper paleolithic. Curr Anthropol August–October 2000;41(4). The Wenner-Gren Foundation for Anthropological Research. 0011–3204/2000/4104-0002. http://www.unl.edu/rhames/courses/current/venus1.pdf.*

Not until the Neolithic Age (neo = new; lithic = stone) more than 10,000 years later were clay-fired vessels abundant among the Chinese, Japanese, and Indus Valley cultures and Egyptian, Cyprus, Eastern Mediterranean, Balkan, and Greek people. Neolithic sites in Japan (14,000–8000 BC) display examples of well-established clay-fired pottery. The clay vessels from the Japanese Neolithic Age are named the Jomon (cord pattern) Pottery Culture. Clay was coiled and smoothed and then simple decorations with bean imprints and fingernails were added to the outside of pots [4]. The elemental contents of mica, iron, and aluminum have antimicrobial properties that would provide some purification to water stored in the vessels. The development of Chuko type pottery is significant since they were formed with spouts and may have been the start of boiling water for tea, again providing some protection against bacteria. The Jomon clay artifacts have holes in the upper section of the pot to thread twine or rope. The twine may have hung the vessel on a pole balanced across the shoulders for stable transport of water or grain. This method is similar to wooden water jugs hung by ropes on bamboo poles. Subsequent clay vessels (Figs. 12.3 and 12.4B–D) are more intricate with large open holds for ropes or sticks.

(A) (B) (C)

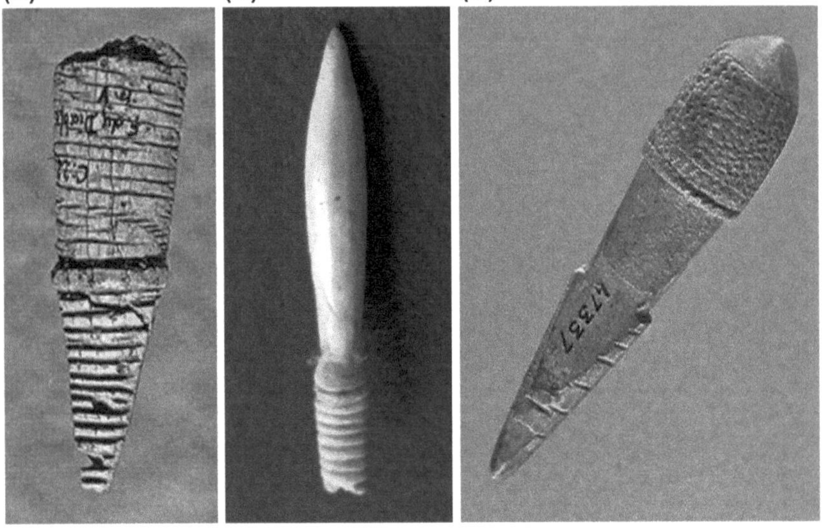

Figure 12.2 Ivory corks for goatskin bags were discovered from Perigordian IV period in the European sites (30,000–28,000 BP). (A) From Roc de Combe-Capelle. (B) From Fourneau du Diable. They are both in the Dordogne area, about 90 km apart. (C) From Laugerie Haute Estate: notice the rounded helix [3,80]. Reproduced from http://donsmaps.com/changes.html, http://paleobox.forumactif.com/t2049p40-sortie-paleobox-le-29-septembre-au-man. *Display at Musée d'Archéologie nationale, Saint-Germain-en-Layeand http://donsmaps.com/images30/stopper.jpg.*

(A) (B) (C) (D) (E) (F)

Figure 12.3 The pottery is divided into five categories: (1) "fukabachi"—deep bowls or jars; (2) "hachi"—bowls of medium depth; (3) "asabachi"; (4) "tsubo"—containers with narrow mouths and long necks; and (5) "chuko"—vessels with spouts. Six sites: Tohoku, Kushu, Kanto Kyushu, Kanto, and Shikoku Island have dates for pottery ranging from 14,540 BC to 10,000 BC. (A) Incipient Jōmon Pottery (14,540–8000 BC): cord-patterned deep bowl with rounded bottom. An example of the earliest art of its type in Japan. Tokyo National Museum, Japan. (B) Middle Jōmon (3000–2000 BC). (C) Middle Jomon (3000–2000 BC). (D) Final Jōmon (1000–300 BC): jar with spirals. Final Jomon, Kamegaoka style. (E) Final Jomon (1100–400 BC) Chuko: This spouted vessel and an example of black-burnished pottery. http://www.tufts.edu/programs/mma/fah188/ospina/jomon/. (F) Final Jomon (900 BC–400 BC) Tsubo style with a long neck made from clay and burnished black.

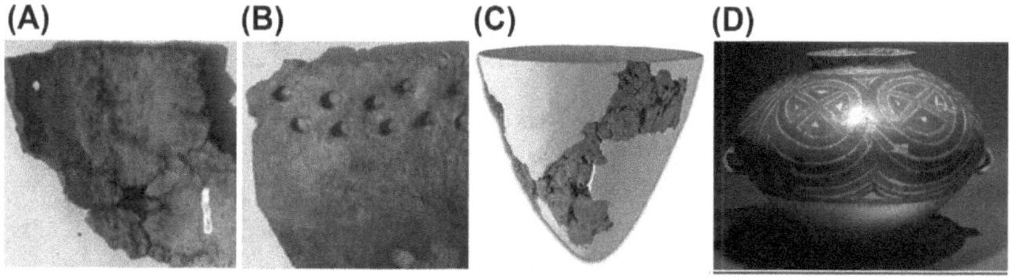

Figure 12.4 (A) and (B) Xianrendong Cave, Jiangxi Province. Two artifacts of clay-fired pottery dated 18,000 BC. Reproduced from http://www.visual-arts-cork.com/east-asian-art/xianrendong-pottery.htm. Science magazine (June 2012) (C) Yuchanyan Cave, Hunan province, China. A reconstructed clay-fired pot,16,000 BC. Reproduced from http://www.visual-arts-cork.com/east-asian-art/yuchanyan-cave.htm. (D). Yangshao pained terracotta vessel, c. 3000 BC to 1500 BC, 12.75 Inches. The geometric patterns are thought to represent some of the earliest evidence of calligraphy in China. The flat bottom also made it easier to store water; the rings on the outside were used for rope or twine to carry the pot or maybe they fit into a stand. Reproduced from http://www.artfromancientlands.com/NeolithicYangshaoTerracottaVesselH778.html.

This would allow the carrier to bring water from a greater distance. The Tsubo style is characterized by a long neck at the top of the vessel, which prevented water spillage as the water was carried. The long neck may have also enhanced boiling and steeping herbs.

The Neolithic period in China (18,000–2000 BC) reflects the development of clay pottery around 18,000 BC. In 2012 pottery found in Xianrendong Cave, Jiangxi Province was dated at 18,000 BC, making it the oldest clay-fired pottery in the world (Fig. 12.5; [4,5]). Two significant

Figure 12.5 Six water wells discovered in Cyrus in 2009 dated 8500 to 7000 BC. The shafts are 16 ft deep. This suggests a high level of environmental understanding. The settlers might have migrated from Levant and brought the well technology with them. http://news.bbc.co.uk/2/hi/europe/8118318.

consequences of clay-fired pottery is that soup could be cooked in these containers, thus providing more nutrients from starchy plants and meat, and water could be heated and boiled to provide a warm drink with minimal purification [4,6]. This was the height of the last Glacial Maximum of the Ice Age; warm brewed teas may have proved a healthier choice than cold river water. Perhaps this was the beginning of China's enormous tea industry, with over 800 different types of tea, totaling a consumption of 1,100,000,000 kg of tea in 2010 [7].

Neolithic peoples were also skilled in well construction. In 2009, archeologists from Edinburgh University discovered six water wells on western Cyprus that date 8500–7000 BC (Fig. 12.5). They suggest migrants from Levant settled Cyprus several thousand years before the Biblical patriots fought over wells in the lands of the Old Testament. Did they bring the knowledge of well technology or was it developed on this island environment? If yes, then well technology was transferred from Levant. The oldest (5206–5098 BC) timber-constructed water wells were discovered in the Leipzig region in Germany (Altscherbitz, Brodau, and Eythra; Fig. 12.6; [8]). According to Tegel et al. [9] "these early carpenters were very skilled, making sophisticated joints with stone tools and were probably were part of a migration from the Balkan Peninsula and the Carpathian Basin ~7500 years ago leaving a uniform archaeological record of settlement structures with longhouses, pottery and stone tools, called the Linear Pottery Culture (LBK; Linearbandkeramik) after the typically decorated ceramics. LBK settlements rapidly spread across the continent's fertile loess regions." Similar technology is observed in the Leipzig sites, including numerous clay vessels (some mended with birch tar), medicinal herbs, fruit, peas, and several grains. The clay vessels have numerous clay rings on the outer surface that could guide a rope as the pot was lowered to retrieve water from the well. These sites were inhabited for at least four generations.

The late Neolithic Period is distinguished by the rise of three great delta cultures who made a giant leap forward in water storage and purification: Mesopotamia (5000–1700 BC; climax 3500–2000 BC), Egypt (2575–323 BC), and the Indus Valley cultures (3300–1300 BCE; climax 2600–1900 BCE). Although occupation of the Fertile Crescent started around 8000 BC, the development of formal city states wasn't until 4000 BC. Massive fortifications constructed for 12 individual city states founded near one of the rivers, each city was ruled by a Priest King. The main ziggurat or temple was at the center, but smaller shrines to other gods dotted the landscape. It wasn't until 3500 BC that the Sumerian culture began to

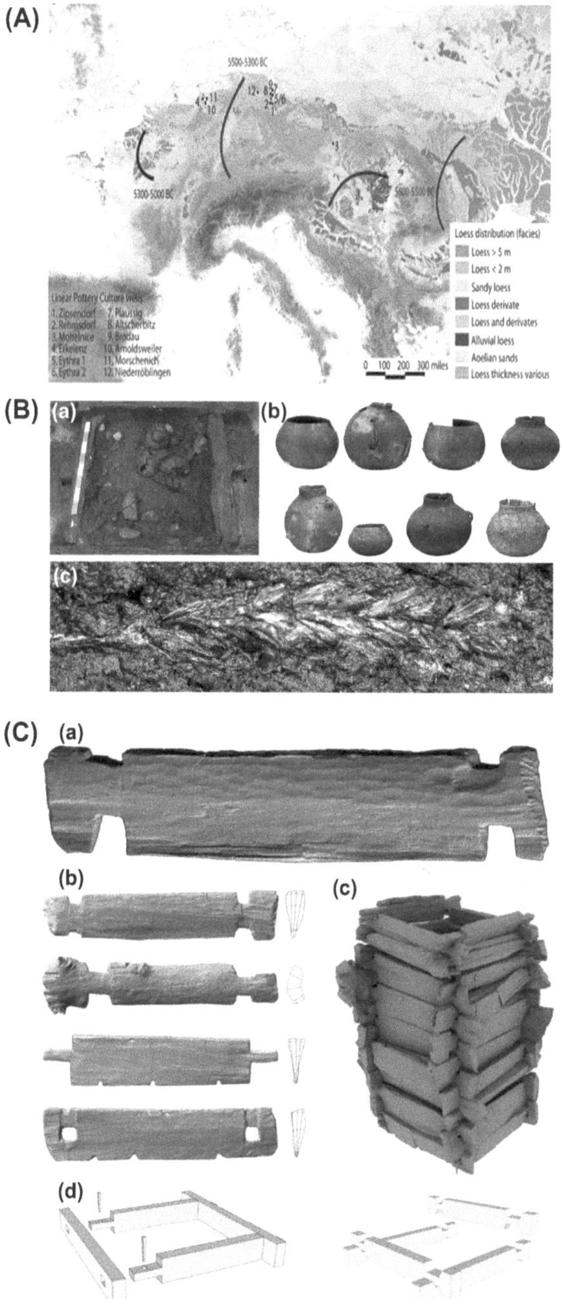

Figure 12.6 (A) Geographic locations of Early Neolithic cultures in Northern Europe. (B) Ancient wooden constructed water wells and early Neolithic artifacts found in wooden wells from the Leipzig Region Germany (5206–5098 BC) [9]. Pottery characteristics show high necks and clay rings that could support ropes to carry jugs or use in collecting water from the well. Preserved Einkorn ear and wheat berries in the mud. http://www.natureworldnews.com/articles/428/20121223/worlds-oldest-wooden-water-wells-discovered.htm. (C) 3D laser rendering of (a) timber bearing tool marks on the surface, (b) various timbers with cogging joints, (c) 3D model of the well lining, and (d) joints. http://journals.plos.org/plosone/article?id=10.1371/journal.pone.0051374.

flourish [10–12]. As the population grew, the demand for water increased. To meet the water concerns of this burgeoning population, Gudea (2144–2124 BC) designed large tunnels made from crushed stone to transport clean water from the hills of northern Mesopotamia to Sameria. Romans later used this technology to provide both water supply and clean water to their cities.

Around this time water regulations, irrigation canals, and sacred vessels associated with water became the foremost commercial venture in Sumerian culture. They were the first to harness the flood waters of the Tigris and Euphrates rivers by developing large irrigation systems and canals that diverted the water through artificial levees and regulators or gate-like structures for agriculture and drinking water (Figs. 12.7–12.9) [11]. To move water from one side of the canal to the field, a shaduf was invented. It's a large pole balanced on a crossbeam, with a rope and bucket on one end and a heavy counter weight at the other. By pulling the rope it lowered the bucket into a canal or river. The operator would then raise the full bucket of water by pushing down on the counter weight. The pole could be swung around and the bucket emptied in a field or different canal. This development enabled irrigation when a river wasn't flooded, which meant higher ground could be used for farming. To deal with the logistics

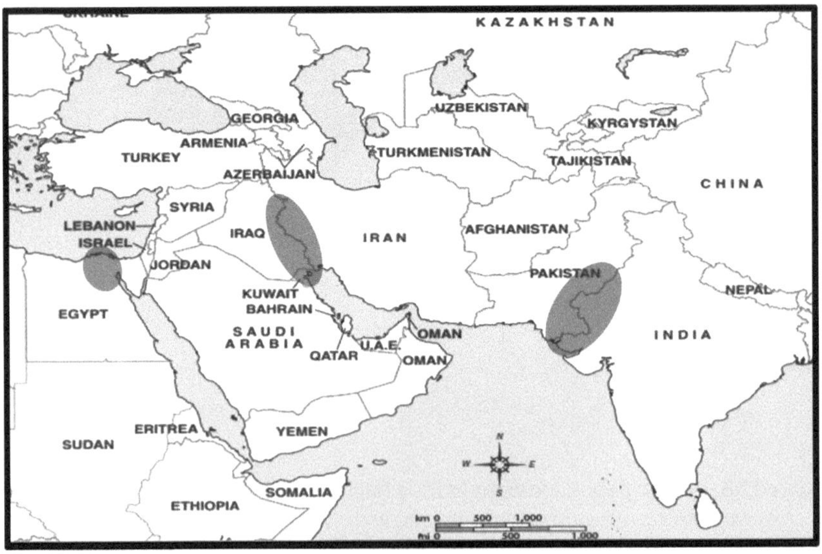

Figure 12.7 Map. *Modified from https://www.eduplace.com/ss/maps/pdf/cent_swasia.pdf.*

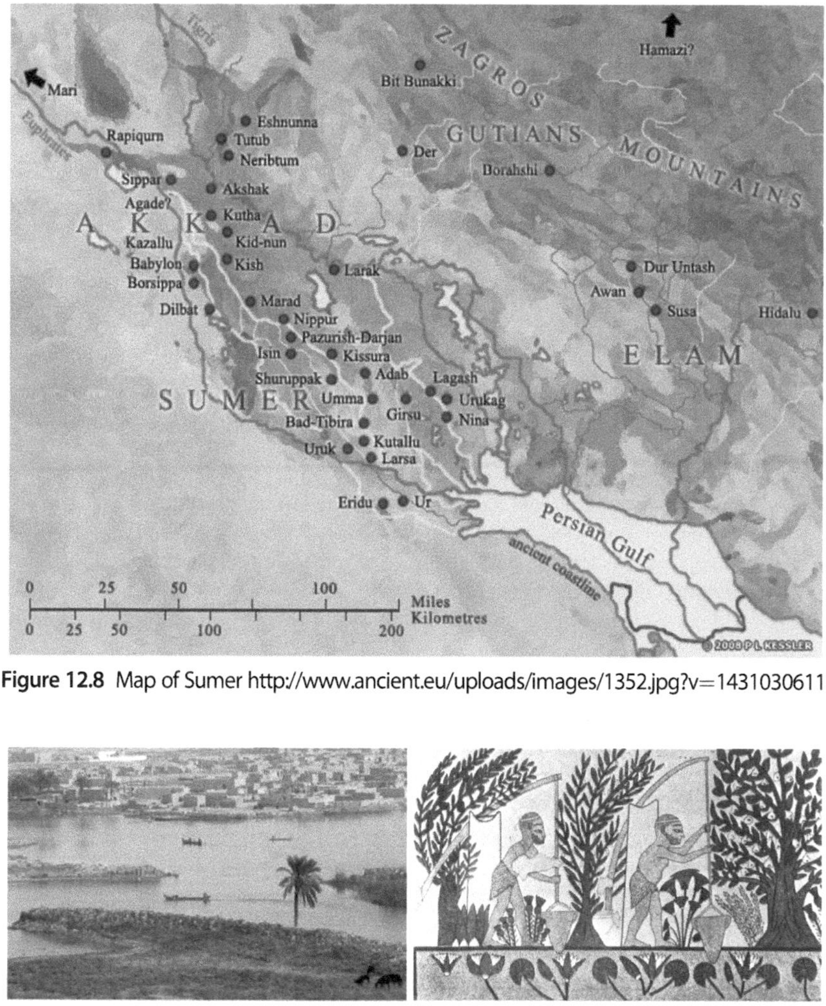

Figure 12.8 Map of Sumer http://www.ancient.eu/uploads/images/1352.jpg?v=1431030611.

Figure 12.9 Left: Al Qurnah in today's Iraq marks the northern tip of the ancient Sumerian canal and levee system. Right: In many places where fields were too high to receive water from the canals, water was drawn from the canals or the Nile directly by a swape or a shaduf. These consisted of a bucket on the end of a cord that hung from the long end of a pivoted boom, counterweighted. The frieze is dated around 2000 BC. In 1258 the Mongols conquered Mesopotamia and destroyed the irrigation systems [9].

of such a vast system, regulations and laws were established to limit the amount of water a farmer could withdraw from the canal. The Babylonian King Hammurabi (reigned 1792–1750 BC) was the first to institute water regulations within his kingdom. The Hammurabi Code ruled that: (1) the

Figure 12.10 The Temple of Enki, variously called the E-Abzu or the Egur, was the first known to have been built in Southern Iraq. The first shrine from the Ubaid period is over 6500 years old. Over the following 4500 years, the temple was expanded 18 times, until it was abandoned during the Persian period.

distribution of water was proportionally based on the acres farmed; (2) farmers were responsible for maintaining canals on their property; and (3) all users were responsible for the collective administration of canals. Babylon was the largest city in the world from c.1770–1670 BC and again between c.612 and 320 BC. It was perhaps the first city to reach a population above 200,000.

One of the first shrines was built in 4500 BC, in Uruk. It was dedicated to Enki, the god of freshwater, male fertility, and additional attributes (Fig. 12.10). His prestige is emphasized by the temple's expansion over 18 times in 2500 years [10]. The religious system provided homage to a pantheon of gods where devotees were required to bathe in clean/sacred water before they received a healing ritual, which was performed by the physician (asu) or an exorcism by the "sorcerer" (asipu); these rituals also made full use of consecrated water [13–16].

One manuscript notes that the "consecrated water " entrepreneurs had a thriving business [16]. The business of the "scared" provided a demand for hundreds of clay bowls (small ones for offerings and large ones for water storage), an assortment of herbs, and writers of incantations and hymns. This mass production of unpainted pottery is the dominate marker for the beginning of the Uruk period around 3900 BC [17]. Although the potter's wheel wasn't invented in Sumeria until 3500 BC the mass production of these small offering bowls indicates the invention was likely 400 years earlier. At this time there was a shift from women potters to a male-dominated business. One of advantages of the potter's wheel included the design of larger vessels to harvest rainwater and to store water at temples, shrines, homes, and other buildings. The wheel allowed the potter to design pots with a flat base, making it easier to stand alone. Some pottery was found to be sealed with tar; these vessels were used to store grain. Large unglazed pottery was preferable for water storage since the clay was porous and allowed for evaporation and cooling of the water, a nice touch in hot, dry climates.

3. EARLY MEDICAL PRACTICES FOR WATER PURIFICATION

Health workers shaved their heads to signify their profession, which limited the number of women physicians. The physician (asu) would tend to the care for symptoms of the patient and conduct surgeries. The exorcists (asipu) performed incantations and rituals to remove the spirit from the diseased part of the body [16]. The earliest written record of medical practices appears around 2100 BC, describing the treatment of wounds [16]. The asu (physician) generally relied on three fundamental techniques: washing, bandaging, and making plasters. In surgical practices it was noted to use boiled water. The Bronze Age (2500 BC) exploded in Mesopotamia, Levant, Cyprus, and Turkey, metallurgy and allowed the use of copper, silver, and bronze (alloys tin and copper) to be used for medicine, purification, and beautiful art creations. Translations of recently discovered clay tablets noted that silver healed wounds and the use of hot copper daggers placed in water for "purification." A preserved copy of the *Treatise of Medical Diagnosis and Prognoses* was dated 1600 BC, which is a collection of practices from both oral and written traditions. Fig. 12.11 shows some physician's tools such as bronze knives, cups, and needles.

Sumerian Hymn that expresses the importance of skilled well technology [18],
Verse 15.
Lord, son of the prince of the sea who understands the digging of water sources.

3.1 Indus Valley Civilization

The ancient Indus Valley civilization was centered around two cities, Harappa and Mohenjo-daro, which were established on high ground within

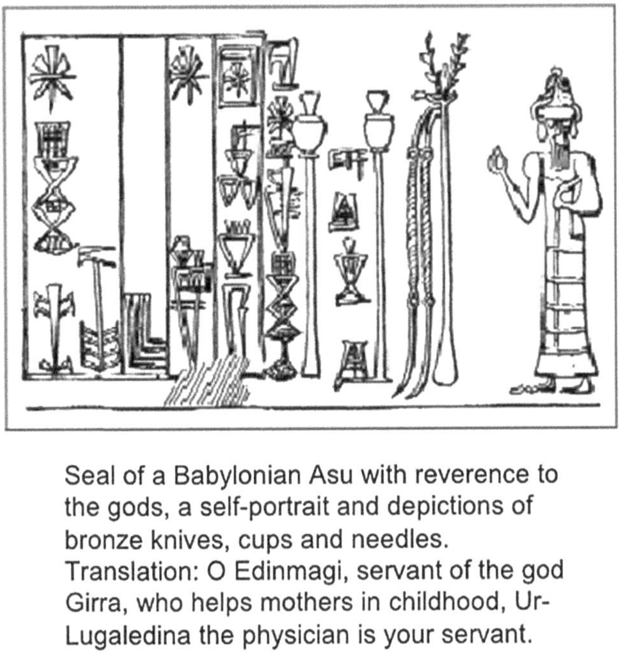

Figure 12.11 Seal of Babylonian Asu with reverence to the gods, a self-portrait and depictions of bronze knives, cups, and needles used in medical practice. Translation: O Edinmagi, servant of god Girra, who helps mothers in childhood, Ur-Lugaledina the physician is your servant.

the Indus River valley (Pakistan and the Indian states of Gujarat, Rajasthan, Haryana, and Punjab) (Figs. 12.12 and 12.13). Religious artifacts unearthed from these sites indicate isolated settlements were established from 5500 BC, agricultural settlements began around 4000 BC, and urbanization may have begun around 3000 BC. By 2600 BC, dozens of towns and cities had been established, and between 2500 and 2000 BC the Indus Valley Civilization was at its peak. There appears to be a rapid decline from 1900 to 1300 BC, although no single source is the trigger. There is some evidence of change in monsoon precipitation and Indus volcanism in Rajasthan 2100–1500 BC, a mega earthquake, and Aryan immigration into the area [18–25].

The infrastructure and sophistication of the Indus cities around 3500 BC indicate previous knowledge of flood mitigation, water purification, well construction, and hygiene [24]. Dwellings near the river were built on giant platforms and laid out in a rectangular pattern (Figs. 12.14 and 12.15). There is also evidence that running water and indoor plumbing were connected to

Figure 12.12 Left: A healing ritual taking place in a reed hut. The healing goddess's dog on the roof. The sick person is stretched out on a bed, and two priests carry out the rite. Right: Mesopotamians believed that the rivers had the power to expel evil substance. A small hut was set up for the patient near the river. The family would care for the patient.

a city-wide sewer system. Commoner houses had baths and drains that emptied into underground "soakage" jars. Sophisticated drainage systems throughout the city carried dirty water and sewage outside of living spaces. Elevation drops at regular intervals acted like automatic cleaning devices. A wooden screen at the end of the drains held back solid wastes. Liquids entered a cesspool made of radial bricks where tunnels carried liquid waste to the main channel that connected the dockyard with the estuary. This system might be the first example of "wetland" waste mitigation.

In tandem with good sanitation practices was deep concern over both the medicinal properties of water and the bodily harm from contaminated water. Knowledge found in the ancient "Ayurveda" practice (*ayur* = life and *veda* = science or knowledge) was passed down by word of mouth and incorporated in the text centuries later. Three ancient books known as the Great Trilogy were written in Sanskrit more than 2000 years ago and are considered the main texts on Ayurvedic medicine (traditional medicine)—*Caraka Samhita*, *Sushruta Samhita*, and *Astanga Hridaya*

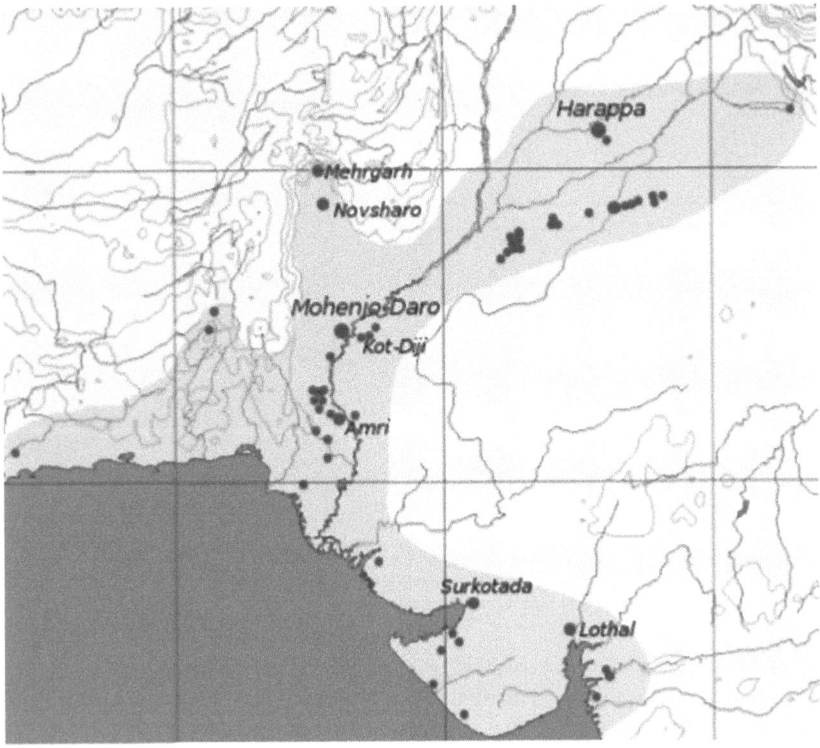

Figure 12.13 Location of Indus valley civilization.

[23,26,27]. The text notes the medicinal properties of the "waters of the principal rivers, lakes, water-falls and mineral springs." Specifically, the "characteristics and the purifications of different types of poisoned water" are discussed. To purify water, it was boiled and made alkaline. Herbs were boiled in water and used for patients' baths. If a person came into contact with a chemical poison or polluted water or foul and stagnant water the body was to be dusted with the powders of different substances. Warnings were issued not to drink water from a broken vessel or with blended palm leaves. Many rules of drinking water were listed related to drinking hot or cold water in specific seasons: "Water may be sparingly taken during the first four months of the rainy season...Water boiled and subsequently cooled should be drunk in summer... Similarly well water (scented) with Us'ira should be used for the purpose of bathing (Sndna). Poisoned water can be purified by the cold ashes, of Dhava, Aŝvakarnâ Asana, Pdribhadrâ Pdtalđ Siddhakâ Mokshaka, Rdja-druma and Somavalka (plants) burnt

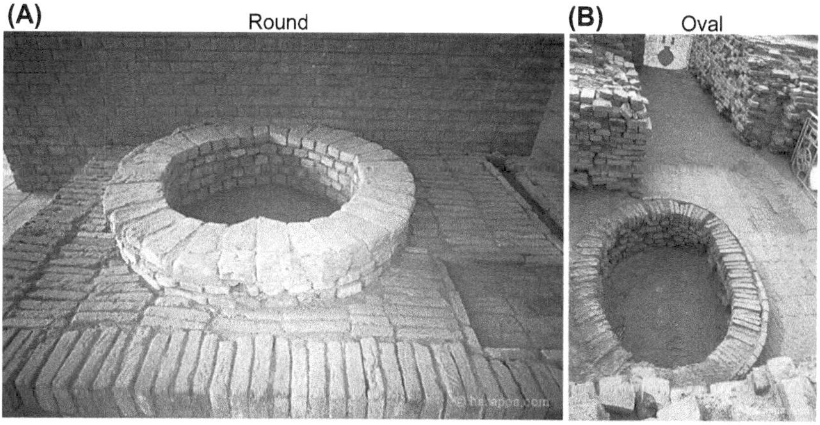

Figure 12.14 This well was excavated from Mohenjo-daro where 700 similar wells have been discovered. According to Dr. Michael Jansen, the well locations throughout the city were planned from the beginning. "Wells are lined for two reasons. First the bricks support the earth and guard against cave-ins and erosion, especially of the well opening. Brick-lined wells also deliver clean, sweet water with a minimum of silt and other large particle-contaminants. Properly dug and maintained brick-lined wells can deliver perfectly clear water, even on a river floodplain." He continues: "The area served by each well and the mean distance between wells is remarkably consistent in the Lower Town; another reflection of the planning that went into Mohenjo-daro from the very beginning. A few wells were abandoned during the life of the city. Excavator Ernest Mackay suggests that this took place because someone jumped into them." *Possehl G. The Indus civilization: a contemporary perspective. p. 101. Reproduced from https://www.harappa.com/blog/mohenjo-daro-city-wells-ii.*

Figure 12.15 (A) Harappa bathing platform. (B) The most unique aspect of planning during the Indus Valley civilization was the system of underground drainage systems.

together, then cast into the poisoned pool or tank, whereby its water would be purified. Just as the atmospheric water which is of imperceptible and undeveloped taste acquires the specific taste of the ground or soil it falls on, so it is the very nature of the (tasteless) Visha that it partakes of the specific taste."

This is an important concept that highlights the connection between water and nearby surroundings, even leaching of materials into water. Other texts stress the use of boiled water, with gems, medical drugs, and mantra to heal a person bitten by a rabid animal. Descriptions of water purification by storing water in copper vessels, exposing water to sunlight, and by filtering through charcoal are also found in the texts [23]. The practice of adding alum to clean water is practiced today in Ayurvedic medicine, but the origins of this method of cleaning water date back to the Egyptian and Indus civilizations [27]. Alum in the form of powdered potassium aluminum sulfate or $KAl(SO_4)_2$ will purify muddy water; it is done by adding about 1 g of powdered alum for every liter of muddy water [28,29]. The mixture will react with the bicarbonate alkalinities in water and form a gelatinous precipitate (flocculent). The flocculent binds with more fine particles and suspended material in raw water, and settles down at the bottom of the container [29]. It is very interesting to note that Varahamihira in AD 550 presented a simple method for obtaining potable water from a contaminated source of water. Various plant materials along with solar heating, aeration, and quenching of water with fire heated stones, gold, silver, iron, or sand were suggested for this purpose. The Vedas also discussed the changes in water quality during the year generally due to monsoons and dry periods, and the benefits and problems of water from various sources was examined in detail, i.e., groundwater, river, spring, and the locations of these sources is examined [30]. These texts should be extensively studied for naturally drinking water practices.

3.2 Egyptian Water Practices

Ancient Egyptians practiced water purification using boiling, botanicals, chemicals, copper, and special well construction. The use of shadufs like those that operated along the Tigris and Euphrates rivers were also used along the Nile. In addition, dikes and sluices transferred river water for agricultural and household purposes [30].

A large siphon mechanism was discovered in frescos on walls within the tomb of Amenophis II at Thebes, dated at 1450 BC (Fig. 12.16; [31]). Originally the mechanism was thought to be a type of water purification, but now we know it is a fermenting device for alcohol.

One of the more popular botanicals used in Egypt to purify water were the Moringa oleifera tree seeds [32a,b,33]. Until recently little was known about the processes that removed bacterium. Researchers found that proteins from the seeds bind the bacterium cytoplasm into clumps and form a

FIG. 1. EGYPTIANS SIPHONING OFF WATER OR WINE CLARIFIED BY SEDIMENTATION
Pictured on wall of tomb of Amenophis II at Thebes, 1450 B.C.
(From Wilkinson's *Manners and Customs of the Ancient Egyptians*, 1879)

Figure 12.16 Egyptian siphon for beer or to clean water is drawn on one of the tomb walls in Egypt. Painting on wall of tomb of Amenophis II at Thebes, 1450 BC [34].

flocculent that kills the organisms [33]. Chemical agents used to clean water were the same as those found in Sumer and the Indus Valley; alum was used in Egypt to removed fine particles from water.

Villagers living in desert climates near large flood plains struggle with too much water in monsoon season and too little in the dry season as well as with dirty water. In 1940 a well was discovered in the modern village of Samana, about 140 km NE of Cairo on the delta, which dated to the time of Ramesses II (1279–1213 BC) [34]. Not until 2005 when the well was excavated to the original lowest level was the ingenious design revealed, a design that alleviated all three water problems. The lowest level was filled with two layers of broken pottery. The shards provided a "filter" that allowed sediment particles to collect at the bottom of well. Thus the water had very low turbidity. In large delta areas, such as the Nile and Ganges, very few if any gravel sources are available, clay bricks or pottery shards are an excellent substitute. The well was built on a geologic feature called a gezira or "turtleback," which had a slightly higher elevation than the floodplain (Fig. 12.17) [34]. Sands were deposited above the Holocene alluvium and thus the clay formed a seal below the sand. The well was dug to the alluvium seal. During flood stage water would flow above the clay and diffuse into the

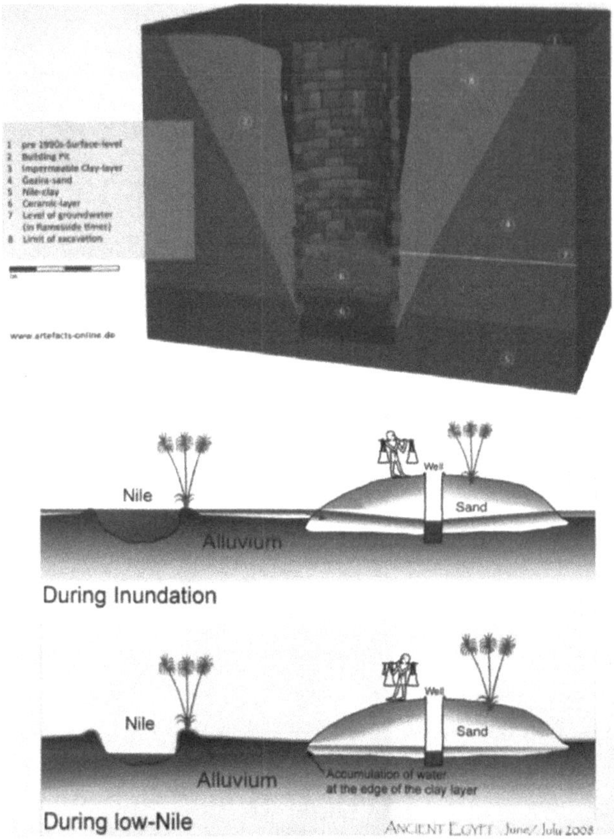

Figure 12.17 Well design during the time of Ramesses II, which allowed water to remain in the well and surrounding areas in dry seasons and sealed it from flood waters in monsoons. Two layers of broken pottery at the bottom of the well provided a mechanism to trap sediment in the bottom of the well, thus providing clean well water [34].

sand. In the low stage, the water table would stay above the clay, thus flowing into the well.

3.3 Greek Inventors

In 460–370 BC Hippocrates, the Father of Medicine, acknowledged that the water flowing through the Greek aqueducts was unclean and required treatment. Along with his theory of "four humors" he also conducted experiments in water purification. He designed the theory of the "four humors," or essential fluids of the body that related directly to the four temperatures of the seasons. He suggested that to maintain good health, the four humors should be in balance. An important tenet of this theory

Figure 12.18 Mural painting of Galenus/Galen (AD 129–AD 200) and Hippocrates (460–370 BC), from the 12th century. Hippocratic sleeve filter is shown on the right side of the painting.

was the recognition of the healing power of water. For example, feverish patients were told to bathe in cool water because the bath would realign the temperature and harmony of the four humors. Knowing the water supply was contaminated, he designed a water filter to "purify" the water he used for his patients. Later known as the "Hippocratic sleeve," this filter was a cloth bag hung upright on a metal stand. Boiling water was poured into the cloth to filter large particles from the water (Fig. 12.18) [35]. The cloth would trap any sentiments in the water that were causing bad taste or smell; he believed good taste meant cleanliness and purity of that water. Today we know that many pollutants such as arsenic are invisible, tasteless, and do not have a smell.

Various materials were added to water to clean the smell and taste of water. Diophanes of Nicaea of the first century BC advised putting macerated laurel into rainwater. Paxamus proposed that immersing bruised coral or pounded barley in a bag would clear the taste of bad water [30].

As with all the great civilizations, clean water and ritual were woven together. Holy or scared water was offered in prayer bowls and vessels, and water was also used to purify oneself from sin. Heron, a Greek inventor who lived in Alexandria from AD 10 to 70, designed a coin-operated Holy Water dispenser (Fig. 12.19) [35]. This may be the first vending machine for dispensing holy water.

Aristotle (384–322 BC) performed experiments that involved immersing hollow wax balls into saltwater for a full day and night. The wax ball filtered

Figure 12.19 Coin-operated holy water dispensing Machine designed by the Greek inventor Heron. This coin-operated holy water dispenser was used in Egyptian temples to dispense water for ritual washings. Worshipers would place a coin into the machine and receive holy water to bathe themselves with before entering the temple. At the end of the day, the slot machine would be emptied of its coins and refilled with holy water for the next day's worshipers. Dropping a coin into the slot machine initiated a chain reaction: the weight of the coin depressed a metal pan, which in turn resulted in the opening of a valve, which in turn allowed the water to flow out for the worshiper. http://www.smith.edu/hsc/museum/ancient_inventions/hsc18b.htm and https://explorable.com/heron-inventions.

the water leaving a salt residue on the wax. Aristotle claimed that the "water would be potable as the salty and earthy residues would be sieved off." Aristotle also discussed distillation by "boiling seawater then its 'sweet vapors' were entirely drinkable."

This section on historical accounts of water purification only discusses some of the more common technologies invented by early civilizations in Europe, Mesopotamia, Indus Valley, Egypt, and Greece. Many other cultures including Roman, Arabian, African, North and South American, and indigenous peoples invented their own technologies for water purification. Further investigation into these resources would be valuable journey worth sharing, but it is out of the scope of this chapter.

It wasn't until AD 1595 that the groundwork for a compound microscope was laid by Zacharia and Hans Jansen. Anton van Leeuwenhoek perfected glass lenses that magnified 270×, and later in AD 1665 Robert Hook's publication *Micrographia* laid the foundation for the study of microbiology [36]. One of the first leaps in epidemiology came when Dr. John Snow in 1854 created a map showing that the clusters of cholera cases in Broad Street in the Soho area of London were related to water drawn from 13 public wells. He identified an "unknown" bacteria from his microscope that he suspected as the culprit [37]. The cholera outbreak from 1848 to 1849 killed approximately 54,000–62,000 in London, and the outbreak from 1853 to 1854 killed an estimated 31,000 in London [37].

PART II

4. PURIFICATION MATERIALS

4.1 Bone Char

Bone char is a black, porous, natural granular carbon material that looks similar to activated carbon. The char is produced by charring animal bones, usually cow bones. To produce bone char the bones are heated in a sealed vessel up to 700°C (1292° F) with low oxygen concentration to avoid ashing. The organic particles in the bone emit volatile compounds through evaporation. The residual organic carbon mineralize to graphite where it stays in the porous apatite structure [38].

Manufacturers exclude the spine and skull bones to avoid the spread of Creutzfeldt-Jakob disease. The composition is generally 57–80% tricalcium phosphate (or hydroxylapatite), 6–10% calcium carbonate (CaCO3), and 7–10% activated carbon.

Bone char may be the oldest and most cost-effective material to remove fluoride. Jacosen and Dahi [38] designed a low-tech and low-cost charcoal furnace from steel drums to make bone char for use as defluoroidation in rural Tanzania. Bone char can be reused by washing with hot water to remove impurities and heating to 500°C (932°F).

The high surface area and high adsorptive property of heavy metals from group 12 (copper, zinc, and cadmium) [39], arsenic [40], and lead [41] make it an excellent inexpensive filter. The best metal-removal results are found in high pH waters (see Table 12.1). Organo-chlorine pesticides are also removed by bone char. Bone char is used as a color-sorbing agent in refining sugar and as a de-ashing agent, but use has decreased due to concerns by

Table 12.1a Metal Removal of Bone Char [34]

Aluminum	Very effective in removing soluble aluminum. Adsorption influenced by pH and retention time. Best results at pH 7 or above.
Arsenic	Largely dependent on form of arsenic (i.e., AsIII or AsV). pH has considerable influence. At low pH almost no adsorption occurs. Affected by protonation.
Cadmium	Bone char is better than conventional precipitation and/or coagulation processes. Superior results in soft water. Retention time on par with aluminum.
Chromium	Much dependent on form present, i.e., cation or anion, valency (chrome III or chrome VI). Chrome III adsorbs well.
Copper	Adsorption dependent on pH and retention time.
Iron	Dependent on form of iron. Better results with organic iron than with inorganic. Inorganic results dependent on valency. Short retention times suffice.
Lead	Very effective with short retention time (i.e., 1 min or less).
Manganese	Very effective. Pilot results good. Competition from other methods of oxidation.
Mercury	Best with organic mercury. Inorganic mercury is dependent on valency (mercury I and II). Although similar, the higher charged monatomic mercury II has a greater affinity for bone char than mercury I.
Nickel	Adsorption best at higher pH. Requires high retention time. At 5 pH almost no adsorption.
Silver	Quite rapidly adsorbed. Better at low concentrations in solute. High deposits of silver on bone char inhibit adsorption. Adsorption better at pH 5.
Zinc	Efficiency dictated by contact time and pH. Both pilot plant and full-scale plant performance good.

Table 12.1b Bone Char Adsorption of Nonmetals [34]

Bromine	Bone char is effective in removing free bromine. Bromine form is also important. In the presence of potassium and sodium adsorption decreases.
Chlorine chloramines	Bone char is minimally effectively in reducing chlorine or chloramine.
Color	Retention time required is usually about 6–15 min, depending on chemical makeup of color. Can be used in slow sand, rapid gravity, or pressure filters.
Fluoride	Adsorption onto bone char is good, but influenced by pH—better at 6.5 or lower. Concentrations of fluoride present also influence performance. Activated carbons do not effectively adsorb fluoride.
Pesticides	Good at removing organo-chlorine pesticides (i.e., Dieldrin, septachlor, etc.); not very good at atrazine, etc.

vegetarians and vegans, religious concerns, and increased cost. Substitutions are activated carbon and other ion-exchange resins.

4.1.1 Charcoal

Archeological evidence of charcoal's presence was found in cave paintings from the Chauvet-Point-d'Arc Cave in southern France. These paintings, which date back to about 32,000−30,000 years ago, were drawn with charcoal and red/yellow ochre [42]. Sumerians (3000 BC), Egyptians (2000 BC), Indus Valley cultures (1500 BC), Israelites (~1550 BC), and Greeks (400−300 BC) record the use of charcoal for both medicinal purposes and water purification. Charcoal was responsible for the onset of both the Bronze (3000 BC) and Iron ages, due to its high fire temperatures, which can exceed 1,000°C, allowing separation of a metal from ore. In fact, charcoal was used in the iron industry until the 1700s in Britain and 1800s in America. It was an English iron worker, Abraham Darby, who found a way to make cast iron using coke produced from bituminous coal (Harris). Until then, colliers made a living producing charcoal in the spring and summer months for the thriving iron factories in Pennsylvania, all the while devastating local forests. One of the first sustainable charcoal traditions was developed in Japan. As early as the 13th century, charcoal making had a sustainable and long tradition in Ishikawa Prefecture, located on the west coast of Japan. The region is still known for sustainable charcoal production (http://ourworld.unu.edu/en/japans-charcoal-making-traditions-still-alive). It was here that the Samurai, who honored the tea ceremony, respected the role charcoal played in making tea. This appreciation encouraged "charcoal" artisans to produce a chrysanthemum-like pattern called "tea charcoal" (*ochazumi*), which is still in use today (Figs. 12.20−12.24);

Figure 12.20 Covered food container from 6th century BC.

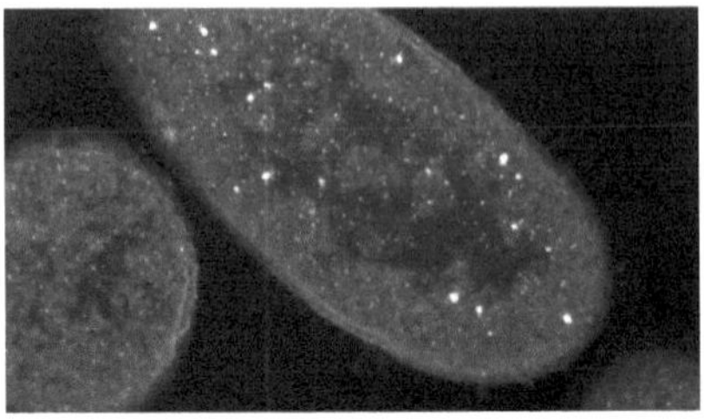

Figure 12.21 Bacteria killed by silver store it in their cells, making them deadly to other bacteria. Silver nanoparticles are visible inside the bacteria as white spots. *Reproduced from Wakshlak RBK, Pedahzur R, Avnir D. Anitbacterial activity of silver-killed bacteria: the "zombies" effect. Scientific Reports 5, No 9555. Nature Publishing Groups (open), http://www.thesilveredge.com/pdf/Antibacterial%20activity%20of%20silver-killed%20bacteria.pdf; April 2015.*

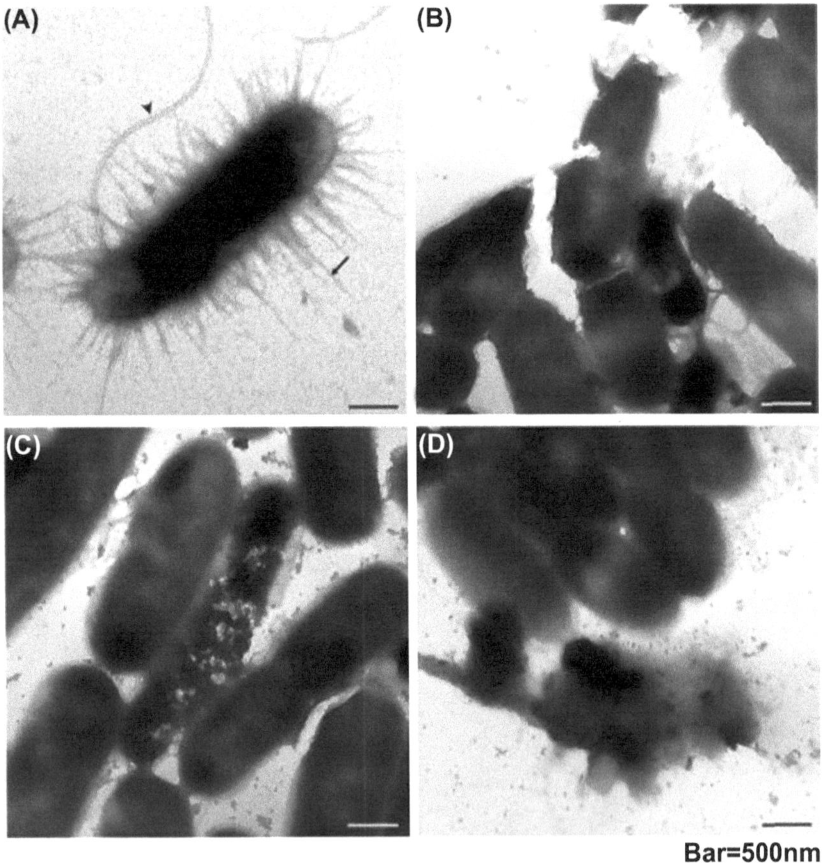

Figure 12.22 Nanoparticles are effective at delivering silver ions (shown here destroying *E. coli*), but their nano nature does not appear to imbue them with additional antimicrobial properties [34]; *Jung WK, et al. Antibacterial activity and mechanism of action of the silver ion in Staphylococcus aureus and Escherichia coli. Appl Environ Microbiol 2008;74(7):2171-2178. http://aem.asm.org/content/74/7/2171.*

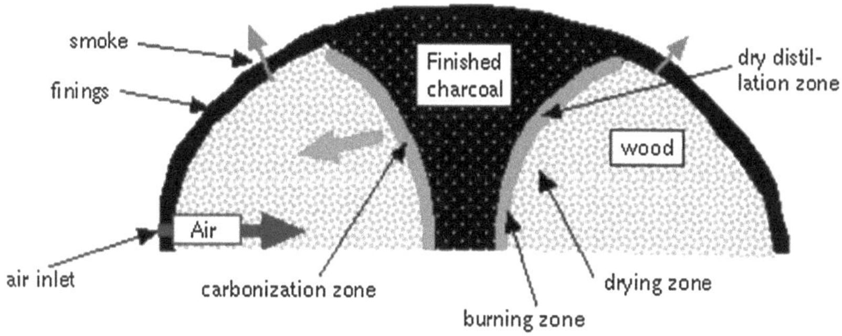

Figure 12.23 Cross-section of wood to charcoal. Reproduced from *Harris P. A version of this article appeared in "Interdiscip Sci Rev" 1999;24(4):301–6. http://www.dailymail.co.uk/news/article-3046274/Charcoal-makes-water-healthier-Adding-stick-improve-taste-remove-impurities-chlorine.html#ixzz46H5KkFdc.*

Figure 12.24 Japanese chrysanthemum charcoal for tea ceremonies.

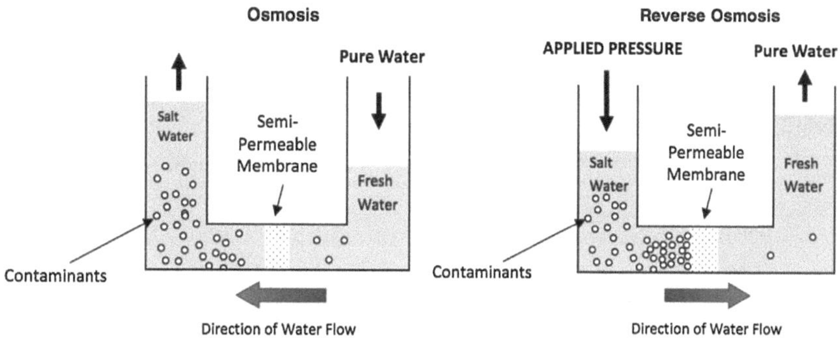

Figure 12.25 Osmosis on the left and reverse osmosis on the right. *From www.lenntech.com/systems/reverse-osmosis.*

[43]. In the 17th century "white" charcoal cultivated from Ubame oak was made in Kyoto Prefecture. The charcoal was used for cooking, smoking food, and water purification. Unlike most softer charcoals found in the market, the white charcoal is very hard and will ring like slate when tapped. To purify water, charred branches 1 to 2″ in diameter and about 6″ long are placed in a vessel. After a few hours the water is cleaned. The adsorption properties of charcoal draw cations to the charcoal surface and remove toxins and impurities. At the same time, alkali ions are released in the water, increasing the pH 8–8.5 and creating naturally made mineral water. Binchotan sticks are still produced today under various names and are used mainly to clean tap water. They can be reused by rinsing and boiling the stick for 10 min. After several uses, the stick can be mixed with soil for fertilizer or put in the refrigerator to remove odors.

4.1.2 Activated Charcoal

Charcoal is simple to make. It can be made in a pit covered with straw and soil to form a mound. A wood structure is build around the center of the mound and covered with weeds or straw and finally capped with soil. Charcoal can also be made in a large steel drum. Inside the large drum is a smaller drum that holds the wood to be charred. The small drum is placed open end down and paper and wood are stacked around the small drum. The large drum can be covered in fireproof insulation. A pipe can be inserted into the drum to let out wood gas or the drum can be left open. Large kilns, retorts, and furnaces are used to produce large quantities of coal. In Pennsylvania, coal was a big business in the iron industry. Once anthracite replaced charcoal the demand for charcoal dramatically decreased (1830s). The best wood for charcoal is hardwood, but bamboo and coconut husks are also good sources.

In the 1920s, Henry Ford, Thomas Edison, and EB Kingsford started making briquettes from sawdust and wood scraps from Ford's Detroit auto plants, and Kingsford became the household name for BBQ charcoal.

Little was known about the chemistry of charcoal until 1780, when the Swedish scientist Carl Wilhelm Scheele described how the "vapors adsorbed by charcoal could be expelled by heating, and taken up again during cooling." During World War I, the German use of sulfur mustard gas prompted research on the effective absorption properties of charcoal for use in gas masks. American research found that charcoal made from coconut shells created a macroporous structure, which allowed rapid air flow. Pioneering research by Rosalind Franklin in the 1950s [44,45] showed that charcoal was 65% carbon in a highly structured lattice within individual graphite layers (1.6 nm in diameter) with the remaining carbon in a disordered array. Using an early induction furnace at the French Laboratoire de Haute she also found that cokes could be graphitized by heat above 2200°C, but charcoal was not graphitized by temperatures around 3000°C. Instead, they formed a porous, isotropic material which only contained tiny domains of graphite-like structure. Franklin's seminal paper on graphitizing and nongraphitizing carbons published in the *Proceedings of the Royal Society* in 1951 [45] is one of the classic pieces of literature on carbon. In this paper she coined the terms *graphitizing carbons* and *nongraphitizing carbons* to describe the two classes of material she had identified, and proposed models for their microstructures. More recent research by Kroto et al. and Shibuya et al. has shown that C_{60} can be removed from charcoal and that "charcoal has a structure made up of fragments of randomly curved carbon sheets, containing pentagonal and a heptagonal rings dispersed throughout a hexagonal network" [46,47]."

Like clay, carbon has a negative charge, attracting cations from metals, organic contaminates (such as benzene, chlordane, etc.), dissolved gases (including hydrogen sulfide), and other toxins. These adsorption properties are enhanced by activating charcoal and by increasing the surface area by steam activation. By heating charcoal to high temperature (1700–1800°F) all volatile compounds and some carbon are driven off, thus enlarging the pores. About three pounds of raw charcoal will produce 1 pound of activated charcoal [48]. After the activated charcoal cools, a water wash or acid wash is used to remove ash content. A second water wash removes acid. Chemical activation produces the same end result—increased internal surface area—but uses a chemical solution to dissolve the internal structure. Typically wood-based activated charcoals are commercially made using heat (450–900°C) and phosphoric acid. This "activation" process results in the creation of an enormous surface area—on the order of 600–1200 square meters per gram (m^2/g), depending on the raw material used and the process. Once again the industrial model is a large steel rotisserie. The wood may be first pyrolized, then the char is saturated with phosphoric acid, followed by controlled reheating to enhance the chemical erosion of carbon atoms, and then an elaborate washing cycle to remove the acid. This process is sometimes preferred because it requires less heat and time. Homemade activated charcoal is possible, but it's generally not as effective as the industrial products. There are many YouTube videos on how to make it at home. Commercially produced activated charcoal has such a high surface area and will adsorb 200 times its weight in toxins, or put another way, one standard 50 g dose of activated charcoal has the surface area of 10 football fields [49].

A large granular activated charcoal filter can remove solid radioactive radon, trichloromethane (chloroform), and odors released by glues used to assemble plastic water-supply pipes. A tightly compressed form of charcoal known as microfine carbon or carbon block is effective in removing protozoan cysts (Cryptosporidium parvum oocysts) and mechanically filtering lead and asbestos [50]. The block filters prevent water bypass and increase the resident time water is in contact with the charcoal. Bower discusses several synthetic enhancement manufacturers use in charcoal filters such as imbedded polypropylene to produce a positive electrostatic charge produced by the movement of the water stream to attract oppositely charged pollutant.

Charcoal filter units can be purchased as whole-house units or point-of-use devices that fit under the sink, or for use in showerheads. Plastic cartridges or carbon-impregnated ceramic units are also common. Some units have stainless-steel housings and can sit on a counter and don't use plastic cartridges. Some manufacturers add a bacteriostatic additive that inhibits growth of bacteria, such a silver nitrate. Some research indicates that silver coatings are not effective after 3 months, since they are removed by filtering water. Periodic back washing can also help remove bacteria buildup in the media, but contaminants are also washed into the sewer or ground. Eventually the charcoal is saturated and the media has to be replaced. Some manufacturers suggest replacing the cartridge every 3 months, which can become expensive over time.

4.1.3 Special Chemical Properties of Clay

Clay was one of the first materials used to craft vessels that carry and store water. Paleolithic communities knew nothing of clay's remarkable ability to kill waterborne pathogens, but its medicinal properties were recorded in clay tablets from Nippur (south of Bagdad) dated around 3000 BC and in Egyptian writings called the Eber's Papyrus from 1550 BC [50]. During the last several hundred years over 200 indigenous cultures have used clay for medicinal therapies [51]. Several modern-day filters use ceramic candles to filter bacteria from water, or clay pots impregnated with nanosilver particles to kill bacteria, even small disks with clay and sawdust can kill bacteria for several months, but clay's chemical and physical processes responsible for these antibacterial activities were not understood until the last decade.

Clay is a weathering product of volcanic ash, feldspars, and other minerals and rocks. What makes this phyllosilicate mineral an excellent pathogen killer? Clay's small size (0.002 mm/<2 μm) and enormous surface area (100's m2/g) with dominantly negative charged sites within the crystal lattice. This valence promotes the free exchange of positively charged (cations) metals, toxins, and nutrients. Cations bonded to the clay can be released into the surrounding medium, thus releasing toxins such as arsenic, cadmium, lead, and mercury to the water. Therefore chemical analysis of clay used for medicinal or pathogen removal should be considered before use. On the other hand, some metal ions are responsible for the antibacterial activity of clays. Otto et al. [50] notes that the *microbicidal* activities of Cu^{+2}, Co^{+2}, Ni^{+2}, and Zn^{+2} are effective against *Escherichia coli*, while Cu^{+2}, Co^{+2}, and Zn^{+2} are effective against methicillin-resistant *Staphylococcus aureus* (MRSA) bacterium responsible for several antibiotic resistant infections in humans. Toxicity to pathogens is critically dependent on pH, ion solubility, osmotic strength, and temperature [50]. Similar studies were conducted on Kisameet clays (KC) used as medicinal treatments for several centuries by the Hieltsuk First Nations people from coastal British Columbia. The sedimentary deposit is composed of about 24% clay, the mineral biotite $[K(Mg,Fe)_{2\text{-}3}Al_{1\text{-}2}Si_{2\text{-}3}O_{10}(OH,F)_2]$, and the naturally occurring antibiotic, Actinobacteria [52]. KC also has a potent broad-spectrum antibacterial effect on multidrug-resistant bacteria called ESKAPE (*Enterococcus faecium, S. aureus, Klebsiella pneumoniae, Acinetobacter baumannii, Pseu strains*) [52].

Natural antibacterial clays such as smectite have reduced iron phases that buffer the aqueous pH and oxidation states to conditions that promote Fe^{2+} solubility. Williams [53] suggests " that extracellular processes do not cause cell death, rather Fe^{2+} overwhelms outer membrane regulatory proteins and is oxidized when it enters the cell, precipitating Fe^{3+} and producing lethal hydroxyl radicals." Hydrated Oregon clay and leachates completely eliminate *E. coli, S. aureus, Pseudomonas aeruginosa, Salmonella typhimurium*, and antibiotic-resistant extended-spectrum beta lactamase (ESBL) *E. coli* and MRSA within 24 h. Williams [53] identified high Al concentrations in the leachates, along with widespread in pH (<4 or >10).

Iron-rich clays from West Africa also showed a strong antibacterial effect on patients with Buruli ulcers [54]. The critical antibacterial components are soluble Fe^{2+} and Al^{3+} that attack multiple cellular systems in pathogens. Al^{3+} will misfold cell membrane proteins, while Fe^{2+} evokes membrane oxidation and enters the cytoplasm, inflicting hydroxyl radical attack on intracellular proteins and DNA. The lethal reaction precipitates Fe^{3+} oxides as biomolecular damage proceeds. This is especially important due to the emergence of antibiotic-resistant human pathogens including MRSA. Some researchers have found that the difference in the clay's ability to kill bacteria is related to the amount of iron, copper, nickel, cobalt, zinc, and pH. Not all clays are the same and should be tested before used as poultice or used to make clay pots that will kill bacteria. Metals attached to the clays can enter the water within the pot and contaminate it beyond EPA-safe values. Caution should be taken when claiming that clay vessels are able to kill bacteria if the proper analysis hasn't been conducted on the clay and leachate.

4.1.4 Copper, Silver

Archeologists unearthed copper water containers from Persian, Arabian, Jewish, and Turkish civilizations that flourished around 2500–2000 BC (Copper Age). Pure copper is soft and thus it's usually alloyed with other metals to produce more strength. Copper ores are sometimes associated with other minerals such as arsenic, bismuth, and antimony. These accessory minerals can leach from the vessel into the drinking water.

Copper and silver have a long history of use in water purification. The following sections will discuss how these metals inactivate bacterium in water and make it potable.

4.1.4.1 Copper

The dawn of the Bronze Age (3000 BC) introduced new materials for water purification. Copper and bronze vessels were designed for water-purification rituals as offering bowls and vases. As the price of copper decreased, larger vessels were molded for use in the home and given as traditional wedding gifts, not unlike today's Newari copper gagris in Nepal. Early Sumerian and Indus Valley records show that copper was used to clean water by those in the medical and Ayurvedic professions. Sumerian tablets recommended using Copper spikes inserted in water vessels to clean the water. Until recently, little was known about how copper kills bacteria. Although the majority of research is devoted to dry copper surfaces, extrapolation to copper water vessels does apply. Espirito [55] notes that possible bacterial resistance to copper can occur in wet environments.

Over the last decade there has been remarkable research explaining copper's unique bacterial killing property. It is a two-step process: First, copper causes massive membrane damage to the cell, and second, through holes punched in the outer membrane, the cell loses vital nutrients and water, causing a general weakening of the cell [55,56]. According to the CuVerra website, "every cell's outer membrane, including that of a single cell organism like a bacterium, is characterized by a stable electrical microcurrent, called 'transmembrane potential,' which is literally a voltage difference between the inside and the outside of a cell." It is strongly suspected that when a bacterium comes in contact with a copper surface, short circuiting of the current in the cell membrane occurs, which weakens the membrane and creates holes. Another way to make a hole in a membrane is by localized oxidation. This happens when a single copper molecule, or copper ion, is released from the copper surface and hits a building block of the cell membrane (either a protein or a fatty acid). If the "hit" occurs in the presence of oxygen, we call it oxidative damage. Once the cell's main defense (its outer membrane) is breached, an unopposed stream of copper ions enters the cell. This puts several vital processes within the cell in danger. Copper literally overwhelms the inside of the cell and obstructs cell metabolism (i.e., the biochemical reactions needed for life). These reactions are accomplished and catalyzed by enzymes. When excess copper binds to these enzymes, their activity grinds to a halt. After membrane perforation, copper can inhibit enzymes and stop the cell from transporting or digesting nutrients, from repairing its damaged membrane, and from breathing or multiplying. It is also thought that this is why such a wide range of bacteria[1] are susceptible to contact action by copper. Bacteria that was tested showed effective antibacterial activity against: MRSA, vancomycin-resistant *Enterococcus faecalis* (VRE), *S. aureus*, *Enterobacter aerogenes*, *P. aeruginosa*, and *E. coli* [56].

While traveling in remote areas, it may be wise to carry a copper penny in our pocket to purify water.

4.1.4.2 Silver

Recorded use of silver goes back to the Egyptians (1500 BC) and Hippocrates (3rd century BC) who used it for the treatment of ulcers and open wounds. Today silver's antimicrobial properties are of keen interest in the medical field's pressing problem of antibiotic-resistant microbes.

In the late 1800s the use of colloidal silver started a new field of silver research and application [57]. Colloidal silver is generally formed by electrolysis when an electric current is passed through a series circuit of a silver electrode and deionized (DI) water. The current flow causes Ag^0 (metal) and Ag^+ (ions) to migrate from the electrode into the DI water as H ions replace

Ag^+ on the cathode. This process produces colloidal silver, which comes in two distinctly different forms, metallic silver particles and ions. The total amount of silver that is reported as the silver concentration (in parts per million) is the sum total of the silver contained in the particles and the silver contained in the silver ions. A detailed explanation can be found in [57]. Some people may remember the small electrolysis systems used by travelers over 30 years ago to purify water. The market for colloidal silver has exploded over the last 15 years. To understand more about the effectiveness of a specific brand and its total silver concentration visit http://silver-colloids.com/Reports/reports.html.

Why is silver so effective at killing bacteria? The answers are found in publications by several innovative research teams. Silver launches a multi-pronged attack on microbes. The first pathway is bonding to sulfurs in proteins. These sulfur groups normally bond to each other and sustain the protein's folded shape. If silver ions interact with the sulfur then the protein cannot fold correctly, weakening it. Second, silver interferes with bacteria's metabolism. Iron is stabilized by binding to sulfur and thus silver changes the shape of the microbe. Finally, silver causes bacteria to produce extremely toxic substances called reactive oxygen species [58]. These go on to cause damage inside the cell, harming the DNA, proteins, and even the membrane that surrounds cells. The net result of this silver onslaught is bacteria with severely damaged defenses. Most importantly, the membranes and walls that surround it are leakier after the silver treatment. Once weakened, they are much more susceptible to conventional antibiotics. Collins and his team showed that when boosted with a small amount of silver these drugs could kill between 10 and 1000 times as many bacteria. The increased membrane permeability also allows more antibiotics to enter the bacterial cells, which may overwhelm the resistance mechanisms that rely on shuttling the drug back out. That disruption to the cell membrane also increased the effectiveness of vancomycin, a large-molecule antibiotic, on Gram-negative bacteria, which have a protective outer coating. Gram-negative bacterial cells can often be impenetrable to antibiotics made of larger molecules. Collins and his colleagues saw good results in mice using nontoxic amounts of silver, but notes that there are ways to reduce the risk even further. "We're also encouraging people to look at what features of silver caused the helpful effects, so they can look for nontoxic versions." Alvarez noted that concentrations of silver ions lower than 15 μg/L appeared to boost bacterial growth instead of killing it [59,60]. According to Alvarez, this stimulatory response at low concentrations resembles suboptimal treatment with antibiotics, which can create resistant microbes. This observation suggests that "we may have to be careful not to undertreat with silver nanoparticles" [59].

There still seems to be discussion regarding the use of colloidal silver-coated clay pots like those used in Cambodia and South Africa. Some studies show that the antibacterial effect decreases dramatically after 3 months, due to the loss of the colloidal silver. Discussions regarding the safe limits of colloidal silver are heated among some research teams, so caution should be taken when using it for personal or developmental use. At least until the verdict is clear.

PART III

5. HOUSEHOLD TECHNOLOGIES FOR WATER PURIFICATION

Anyone looking for a household water-purification system knows the plethora of units on the market. This chapter is not an evaluation of these units but a summary of the various methods to purify water, to help readers make an informed decision. The first step is to know what bacteria and elements should be removed from a drinking source. If you have access to a water lab or field kit, test the water. If you don't have access, then find water-testing data from your local water department or neighbors. Armed with these results, the elimination process can begin, i.e., if the water is contaminated with only bacteria, metal-removing media isn't required. Second, do you need a point of use (POU) filter for the entire house, or only drinking water? Determine the necessary volume of clean water per day or gallons (liters) per minute. This will identify the minimum flow rate. Units will vary as to their delivery path: gravity flow, source of electricity, or pressure. The next consideration is to assess electrical charge: intermittent, full load, and source (solar, wind, etc.). The water resource is also important. Some systems use large volumes of water to flush the system before purification even gets started. After the 2015 earthquake in Nepal, several groups installed chlorination systems, which run water for 30 min before purification, wasting precious water. These units and the reverse osmosis units ran on electricity. In Nepal there was only 4–6 h of electricity a day (usually in the early hours of the day when people are sleeping) before the earthquake, and after the earthquake electricity was even more scare. The next step is price consideration. The user must carefully consider the total cost of a unit. Many household water filters require cartridge change out as frequent as every 3 months. The cartridge, paper filters, and lifetime cost of removal media should be considered. Using a 5-year cost analysis will help determine the total cost of a filter system. When researching a filter, ask two questions: is the filter easy to maintain? and is a technician required to test the water? Check out the manufacturer's website and Youtube.com for an analysis of specific brand-name filters.

The following section will discuss the basic technology for eight purification methods: adsorption, boiling, chemical, distillation, electrolysis, microntube separation, reverse osmosis, and ultraviolet. Remember, there is a range of filter units among each type, so do your homework and read the fine print.

5.1 Adsorption/Absorption

Adsorption is a surface reaction in which atoms, ions, or molecules bind to the top layer of a material by electrostatic attraction, chemisorption, or physisorption. This differs from absorption, which involves the entire volume not just the surface [61]. Three major classes of industrial adsorbents are carbon-based compounds (graphite, activated charcoal), oxygen-based compounds (zeolites, silica), and polymer-base compounds. Typically, adsorbents have small pore diameters with a high surface area to facilitate adsorption. The pore size usually ranges between 0.25 and 5 mm. In water treatment, adsorption is an efficient removal process for pathogens, metals, and pesticides from a solution, whereby the pollutant molecules in the fluid are trapped inside the pore structure of the carbon or other substrate [62]. Both phosphate and silica in groundwater can compete for the same sites, thus reducing adsorption.

The World Health Organization (WHO) has called arsenic in groundwater the world's greatest natural disaster, affecting more than 250 million people worldwide. Understandably there is a great need for affordable water filters in these areas. Adsorption technologies have focused on adsorbents such as alumina, composite iron matrix [63,64] (Table 12.2), zero valent iron [65], and a host of experimental media.

Boiling water will kill thermally sensitive microbes, but will not remove metals and other toxins. In fact, boiling arsenic contaminated water will increase arsenic content in food. This simple method of purifying water is effective if the water temperature remains above 160°F (70°C) for 30 min, above 185°F (85°C) for 3 min, and at a rolling boil 212°F (100°C) for

Table 12.2 Phases of Reaction on CIM for Arsenic Removal [34,79]

Corrosion of FeO-composite continuously generates hydrated iron oxides with high specific surface area.
Mn (ca. 0.2%) in the FeO-composite acts as a catalyst for rapid conversion As(III) to As(V).
Arsenate is removed by surface complexation with HFO (possibly magnetite and maghemite (γ-Fe_2O_3).
Removal process is independent of the concentration of As(III) + As(V) input concentration.
Excess Ca^{2+}, Fe(II)/Fe(III) and other divalent cations enhance adsorption and complex formation through double-layer charging.
Further, cementation reactions produce a porous high surface area insoluble spent material very similar to natural HFO with a high capacity for arsenic removal.

1 min. At high altitudes water will boil at lower temperatures due to lower air pressure, thus the Wilderness Medical Society suggests boiling water at higher altitudes for 3 min (i.e., water boils around 185.9°F (85.5°C) at 14,000 ft (4267 m) (Chart 1). In the mid-1800s the British government employed Indian surveyors disguised as religious pundits to explore Tibet. These courageous men, measured altitude by measuring the boiling temperature of water as they trekked through the Himalayas [66].

In recent times, traditional clay ovens (*chullis*) in Bangladesh were fitted with a hollow aluminum coil to purify surface water. Water flows into the coil from a bucket filled with surface water and water is heated as it moves through the coil during normal cooking (Fig. 12.26). The heated coil

(A) (B)

Figure 13.26 (A) Hollow coil inserted into clay oven. (B) All components of the Chulli water-treatment system [66a,b].

maintains a constant water temperature around 70°C by adjusting the flow rate. The *chulli* system produces 30 L of treated water during a typical cooking cycle of 1.5–2 h, providing up to 90 L per day of treated water, without any additional time or fuel requirement [66a,b]. If the water is stored in clay vessels, additional purification occurs. One downside of boiling water as a daily routine is the use of extensive use of diminishing supplies of wood.

5.2 Chemical

The effectiveness of all chemical treatments in water is dependent on temperature, pH, and turbidity of the raw water. Generally, the colder the water, the less effective the chemical is as a purifying agent. Research has shown that water temperatures $\leq$40°F (4°C) require at least 60 min to kill bacteria. At 50°F (10°C) only 90% of *Giardia* cysts were inactivated after 30 min of exposure. The most effective water temperatures are at least 60°F (16°C) before treating. Chlorine loses effectiveness as pH rises above 7.8; optimum pH is 7.2–7.8. High turbidity will decrease chemical treatments; it's best to filter before disinfecting. Raw water can be poured through a cloth or bucket of sand, which will easily remove particles and clear the water.

5.3 Chlorine Treatment

Chlorination can kill microbes by disrupting the cell-wall barrier, releasing vital cellular constituents from the cell that terminate membrane-associated functions and cellular functions within the cell. The microorganisms die and cannot transfer disease or grow. Scientists have studied how chlorine affects the cell walls of certain gram-negative bacteria that cause typhoid fever, dysentery, cholera, and Legionnaires' disease. They found that chlorination increased the permeability of the outer membrane, leaving the bacteria vulnerable to destruction. Scientists do not understand much about spore-forming bacteria or gram-positive bacteria, which have no outer membrane. Although these bacterial types are, in general, more chlorine tolerant than gram-negative bacteria, most waterborne species do not normally pose a health threat. Certain waterborne viruses, such as enteric viruses and hepatitis A, may be even more tolerant to chlorine disinfection than some bacterial species. Calomiris [66] noted that the means by which chlorine inactivates viruses is not well understood. In the same study, Dorworth [66], noted that "the concern with chlorinated water is its tendency to form trihalomethanes (THMs), carcinogenic by-products of the disinfection process. In 1979, the EPA adopted the THM regulation, limiting their allowable level in drinking water supplies. In 1992, the EPA established federally enforceable standards that now cover 83 contaminants, including THMs, that may be found in drinking water."

Chlorine doses that inactivate specific bacteria have been calculated by the Center for Disease Control and Prevention (CDC and Environmental Protection Agency (EPA)) for drinking water and swimming pools. Chlorine is available as compressed elemental gas, sodium hypochlorite solution (NaOCl,) or solid calcium hypochlorite ($Ca(OCl)_2$). According to the CDC all forms of chlorine, when applied to water, form hypochlorous acid (HOCl). Gaseous chlorine acidifies the water and reduces the alkalinity, whereas the liquid and solid forms of chlorine increase the pH and the alkalinity at the application point. The pH of the water will affect the dominating chlorine species such that HOCl dominates at lower pH, while the hypochlorite ion (OCl^-) dominates at higher pH. Of the two species, HOCl is the stronger oxidant. Therefore chlorine is more effective as an oxidant and a disinfectant at lower pH. Both forms, HOCl and OCl−, are referred to as free chlorine.

After the April 25, 2016 Nepal earthquake, there were three complaints by Nepalis who were given chlorine tablets or chlorine systems for water treatment. The first was the strong chemical taste, the second was the high water waste during the 30 min flushing period, and the third was the need for electricity and Nepal's lack of electricity.

The use of a small filter with activated carbon/charcoal could remove residual chlorine and chloramines. Load shedding in Nepal is extremely high in late fall and winter in the best of times, and it was even worse after the earthquake. Chlorine treatment systems that depend on electricity laid unused for the majority of a day, using a third of their potential.

5.4 Iodine Treatment

Most likely it was the ancient Greeks that first used iodine as a medical agent. They found that wounds wrapped in seaweed (rich in iodine) kept them clean and helped them heal. Iodine can kill all classes of pathogens including MRSA: gram-positive and gram-negative bacteria, mycobacteria, fungi, yeasts, viruses, and protozoa. Most bacteria are killed within 15−30 s of contact. Skelton suggests iodine kills single-celled organisms by combining with the amino acids tyrosine or histidine when they are exposed to the extracellular environment [67]. All single cells showing tyrosine on their outer cell membranes are killed instantly by a simple chemical reaction with iodine that denatures proteins [68].

Iodine is light sensitive and must always be stored in a dark bottle. It works best if the water is over 68°F (21°C). Iodine has been shown to be more effective than chlorine-based treatments in inactivating *Giardia* cysts (Tables 12.2 and 12.3). Caution should be exercised for people who are allergic to iodine or shellfish, those with thyroid problems, lithium users, women over 50, and pregnant women.

5.5 Flocculation Coagulation

A powdered chemical mix is added to water that will cause a flocculent to remove metals and a disinfecting chemical that will kill bacteria and clarify the water. The powder is usually ferric sulfate (flocculent) and calcium hypochlorite (disinfectant). After adding the packet, the water is stirred for about 5 min to let the flocculent settle to the bottom of a bucket, then the water is strained through a cotton cloth into a second bucket and let stand 20 min for the disinfectant to work. The majority of bacteria, viruses, protozoa, and heavy metals and pesticides are removed from water [69]. Each packet or sachet treats 10 L of water. In Pakistan the small packets are sold for 3.5 cents, making it rather expensive if used on a monthly basis, but it's a quick solution in emergency situations. One disadvantage of this technique is that flocculent should be disposed of at a hazard waste site.

5.6 Distillation

Distillation for water purification involves boiling untreated water, collecting the steam, and capturing the liquid as it cools. Water has a lower boiling temperature than most contaminates and thus will separate from inorganic contaminates and bacteria. As the steam cools it flows into a third vessel

Table 12.3 Common Microorganisms and the Filter Size Needed

Organism	Examples	General Size (μm)	Filter Type	Particle Size Rating (μm)
Protozoa	*Giardia, Cryptosporidium*	5 or larger	Water filter	1.0–4.0
Bacteria	Cholera, *E. coli, Salmonella*	0.2–0.5	Microfilter	0.2–1.0
Viruses	Hepatitis A, *Rotavirus, norwalk virus*	0.004	Water purifier	To 0.004

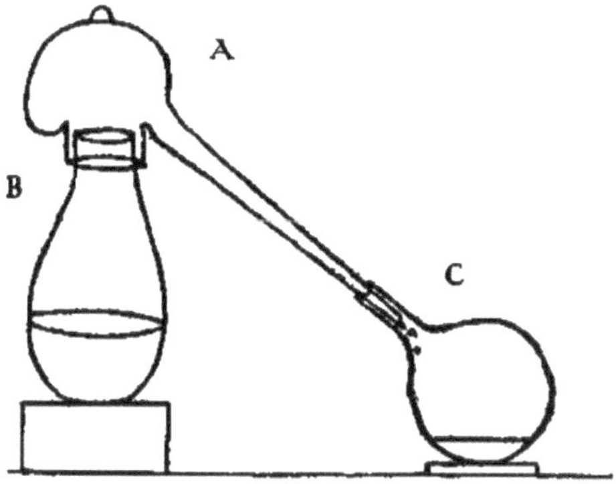

FIG. 1.—Traditional form of still.

Figure 12.27 Traditional form of a still [34,78].

called the condenser (Fig. 12.27). The process can be repeated to ensure complete purification [70].

This technology is not a new. In fact, Aristotle describes distillation in *Meteorologica* in 350 BC. "Salt water when it turns into vapor becomes sweet, and the vapor does not form salt water when it condenses again."

An inexpensive solar distillation can be built by digging a hole, putting an empty pan in the bottom, setting a bucket full of impure water into the middle of the pan, and then setting a peaked clear plastic sheet over top. Pure evaporated water will collect and condense it inside the plastic, and let it drip down into the empty pan. The problem with this method is that it is very slow and produces relatively little water. Large solar distillation arrays are used in many arid climates where saline groundwater is a problem.

5.7 Filtration

5.7.1 Biosand

Household sand gravity flow filtration systems have existed for over 100 years. The basic principle is the use of sand's natural filtering ability to remove particles, clean turbidity, reduce iron and manganese, and develop a biofilm on the surface of the sand that kills pathogens. There is no pretreatment required and thus no chemicals are involved in removing pathogens and cysts. In the last 30 years these systems became popular when Dr. Manz from the University of Calgary developed a thin, plastic, inexpensive unit that was used to remove bacteria in developing countries. Since then a multitude of different types have been developed for household or community use. The effectiveness of biosand filters is discussed in several publications [71]. In 2003, a group from MIT who observed Dr. Munir's research in Bangladesh decided to substitute small iron nails for the effective composite iron matrix. Unfortunately, this led to another filter disaster in SE Asia. The arsenic biosand filter failed after 6 months to remove arsenic >100 ppb. Problems were recognized with the 10-day generation time of the biofilm which left the filter user vulnerable to pathogens 1/3 of a month if the filter was cleaned monthly. Five papers discuss problems with both pathogen and arsenic biosand filters used in Nepal and Cambodia [72–74]. As noted, pathogen removal was only 45%, length of time to reestablish the biofilm after cleaning was 10 days, taps inside cement filters commonly broke, and cement leaked water.

Field tests by Singh [73] showed only 20% of arsenic was removed by nails in raw water with over >50 ppb. These results were confirmed by the Bangladesh Environment Testing Verification program in which the filter failed both in pathogen removal and arsenic removal (http://www.verification-unit.org/not_certified.htm).

5.7.2 Ceramic Filters

The first ceramic filters were made in 1827 using clay by John Doulton, a manufacturer of China in Lambeth, England. By 1862, Henry Doulton started manufacturing the "Manganous" carbon water filter, which led to development of the porous ceramic candle style ceramic drip filter that has become a popular worldwide household water filter. In Asia, diatomaceous earth was used for the porcelain and pottery of shoguns and kings; this silica-rich microfossil algae also proved to be the perfect material to make porous ceramic filters. Water flows through the candle's maze of 0.2 μm size pores removing 99.999% of waterborne bacteria and suspended solids. The ceramic candle is placed inside a housing and water flows by gravity. The candles have become more advanced, using silver impregnation and carbon centers (Fig. 12.28). The filter can be cleaned by boiling and scrubbing the outside with a soft brush. This will remove some of the material, so eventually after about 100 scrubs, the candle will need to be replaced. It's recommended that only water be used to clean it. The candles can develop cracks, which allow water to bypass the labyrinth of pores, reducing or eliminating the effectiveness of bacterium removal. The ceramic candle can remove aluminum, iron, lead, pesticides, herbicides, phenols, MTBE, perchlorate, chlorine and chloramines, aromatic hydrocarbons, trihalomethanes, particles

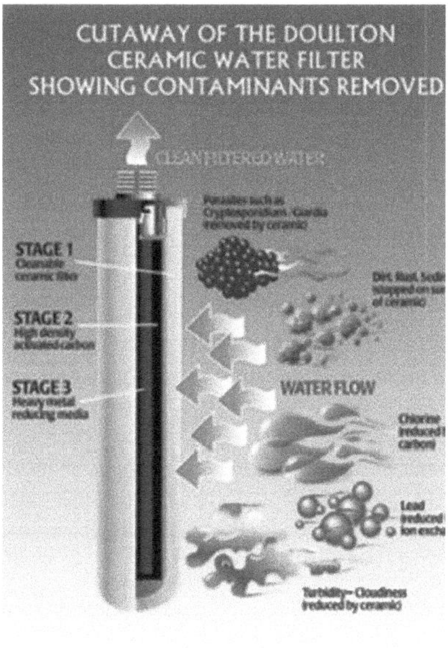

Figure 12.28 Ceramic filter with three stages: outside ceramic shell, high-density activated carbon, and heavy metal-reducing media (Doulton) http://blog.waterfilters.net/blog/what-are-ceramic-filters/.

larger than 0.5 to 0.2 μm, *Giarda lamblia, Cyclospora, Cryptosporidium, E. coli, Vibrio cholerae, Salmonella typhi, Shigella dysenteria, Klebsiella terrigena*. If the raw water contains high silt or other particles, it can be prefiltered through paper or cloth before pouring into the filter housing, which will extend the life of the filter [75].

5.8 Hollow-Fiber Membrane Filters

These water filter units use a lightweight hollow-fiber membrane tube to separate microorganisms from water by size (Tables 12.4–12.6; http://www.water-research.net/index.php/water-treatment/water-disinfection/chlorine-disinfection; https://www.epa.ie/pubs/advice/drinkingwater/Disinfection2_web.pdf). Depending on the micron rating of the tubes, some smaller organisms (like viruses) can pass through the system so it's best to know the size of the organisms that need to be removed and buy the appropriate micron filter. A 0.1 μm tube will remove 99.9% of bacteria from water. A gravity-flow system has a flow rate of 40 L/h. The cost of a filter unit will increase as the tube size decreases. These filters are easy to assemble, require a simple back flushing depending on the turbidity of the water, and have a long lifetime over 10 years.

Table 12.4 Chlorine Disinfection Timetable (CDC)

Agent	Disinfectant Times for Fecal Contaminants in Chlorinated Water
E. coli 0157:H7 (bacterium)	Less than 1 min
Hepatitis A (virus)	Approximately 16 min
Giardia (parasite)	Approximately 45 min
Cryptosporidium (parasite)	Approximately 15,300 min (10.6 days)

Notes:
- 1 mg/L (1 ppm) free chlorine at pH 7.5 and 25°C (77°F)
- These disinfectant times are only for pools that do not use chlorine stabilizers such as cyanuric acid. Disinfection times would be expected to be longer in the presence of a chlorine stabilizer.

Free chlorine kills most bacteria such as *E. coli* 0157:H7 in less than a minute if it is maintained at proper disinfection levels [1–3 parts per million (ppm) free chlorine] and under ideal water conditions [pH at 7.5 or less; temperature at 77°F (25°C) or more] throughout the pool. However, a few germs are moderately (*Giardia*, Hepatitis A) to highly (*Cryptosporidium*) chlorine-tolerant. Table 12.5 shows the approximate disinfection times for these germs in chlorinated water.

Table 12.5 Disinfectant Doses for Drinking Water

Available Chlorine Bleach[a] (%)	Drops Per Quart of Clear Water[b]	Granular CaOHCl Calcium Hypochlorite	Water
1	10	1 heaping teaspoon	Per 2 gallons water
4–6	2	1 pint	12.5 gallons water
7–10	1		

Treated water should be mixed thoroughly and let to stand for 30 min. If there is no odor, repeat the dosage and allow water to stand for an additional 15 min. If treated water is too strong, allow water to stand exposed to air for a few hours or pour it from one clean container to another several times (EPA).
[a]If strength is unknown, add 10 drops per quart of water.
[b]Double amount for cloudy or colored water.

Table 12.6 Common Iodine Dose to Disinfect Clear Water

Tincture of Iodine	Clear Water	Cloudy Water
5 drops of 2%	1 quart	
10 drops of 2%		1 quart

Let the water stand for 30 min (EPA).

5.9 Reverse Osmosis

A reverse osmosis (RO) membrane is a semipermeable membrane that allows the passage of water molecules but not the majority of dissolved salts, organics, bacteria, and pyrogens. A high-pressure pump increases the pressure on the salt side of an RO and forces the water across the semipermeable membrane. The amount of pressure required depends on the salt concentration of the raw water. The more concentrated the raw water, the more pressure is required to overcome the osmotic pressure. This high pressure is required to desalinate (demineralize or deionize) water by allowing pure water through while holding back the majority of contaminants (Fig. 12.25) [76]. The remaining residue is composed of 95–99% dissolved salts, 99% particles, colloids, organics, bacteria, and pyrogens from the raw water. The membrane blocks contaminants with molecular weights >200 and those that are highly charged. Gases such as CO_2 are not removed very well because they are not highly ionized and have a very low molecular weight. CO_2, which does not pass through, will hydrolyze to carbonic acid, which lowers the pH of the rejected water.

5.10 Ultraviolet

Ultraviolet disinfection or UV germicidal irradiation (UVGI) can be used for water, food, and air purification. The short-wavelength ultraviolet light kills pathogens by destroying nucleic acids, disrupting their DNA and preventing them from reproducing [77]. The raw water should have low turbidity, which sometimes requires prefiltering. This method of water purification requires batteries for the ultraviolet light.

The UV light used for disinfection is generally a mercury-vapor lamp that generates a wavelength of 254 nm close to the optimal wavelength, which is 260 nm. LED UV light allows the user to select specific wavelengths for specific pathogens. These units are much smaller and take only 10 ns to warm up and have a lifecycle of 10,000 h. They also require less power to operate (6–30 volts) and can use direct current (DC) and solar power [52]. Lamps are easily transported to remote locations and operated with batteries and solar power. However, replacing the batteries may not be easy in remote areas.

CONCLUSIONS

The quest for water is a universal requirement for community survival as is the need for safe drinking water. Contrary to popular belief, even the earliest human cultures had to treat their water to make it safe. Although bacteria were not identified until the 17th century with the invention of the compound microscope, water sources were classified as "sweet" (good) or "sour" (bad) water. If someone drank water and became sick or died the water was "sour." Unfortunately, this involved a bit of trial and error. Sweet water sources were so important that maps were carved into stone (Jordan) or stone cairns marked wells with good water. It's important to mention the strong relationship between water and the sacred. Water is a purifying element, a symbol of life and fertility, a pathway to the Divine, and a host of other beautifully woven connections with worship. Connection to the sacred has long been associated with clean water. Shrines and temples were placed near the confluence of streams and rivers, wells, and clean water flowed to them through manmade canals. This demand for clean water thrust the sacred into the realm of commercial enterprise, ingenious technology, and artistic design. Throughout history large clay-fired urns stored water and killed bacteria, unbeknownst to worshipers, and copper offering bowls provided a natural germicide.

The use of charcoal to purify water started in Sumeria but became an art form in 17th century Japan when "white" charcoal was cultivated from Ubame oak in Kyoto Prefecture. One small 6″ stick added to water will clean it in only a few hours. Charcoal is still used today for water purification and for cooking.

Henry Doulton's invention of the ceramic candle in 1827 was the watershed moment for modern-day filter technology. The ceramic candle filter became the first standardized filter to remove bacteria and particles. This technology has exploded in recent years, with the addition of carbon and other materials to remove organics and metals. Many simple versions of these systems are used today in developing countries.

Since there are a plethora of filter systems available on the market, users need to research the systems using online websites, youtube.com, etc. There are a few things to consider when purchasing a unit for personal use, developmental work, or emergency use. Research on the type of removal material is key to choosing the right filter. If the area has inconsistent electricity, this will be a problem for units that require electricity. While small solar systems are available they cost more. Gravity-flow systems are sometimes the best choice. Calculate the daily liters or gallons required and make sure the unit will filter enough clean water in an hour (i.e., 30 L/h vs. 5 L/h). If the water only has bacteria, a system that removes metals or other inorganic contaminates is not needed. These systems cost more and require more maintenance. Read the fine print! If the filter requires an expert to change a cartridge or maintain the system, there will be a significant cost increase. Companies make money by providing free or low-cost housing units but charge for monthly or quarterly cartridge replacement. Replacing a cartridge every few months will add considerable cost to your filter. Calculate the cost over 5 years to get a good estimate of total cost. If using a chemical addition to water, check to see if the flocculent can be disposed of in regular waste or hazard waste. There is neither hazard waste nor waste disposal site in most remote areas. Check the leachate produced from a unit, since the values need to conform to EPA and WHO guidelines. Lastly, don't use a filter longer than the recommended lifetime of the media because some contaminates may leach into your drinking water.

REFERENCES

[1] Soffer O, Adovasio JM, Hyland DC. The "venus" figurines: textiles, basketry, gender, and status in the upper paleolithic. Curr Anthropol August–October 2000;41(4). The Wenner-Gren Foundation for Anthropological Research. 0011–3204/2000/4104-0002, http://www.unl.edu/rhames/courses/current/venus1.pdf.
[2] Hitchcock D. Venus figures from Russia, the Ukraine and sites East of the Donau mouth. 2014. http://www.donsmaps.com/ukrainevenus.html#yelisee.
[3] Bordes F. Bordeaux. Gallia préhistoire 1959;2:156–67; Bordes F. Le Protomagdalénien de Laugerie-Haute-Est (fouilles F. Bordes). Bulletin de la Société préhistorique française 1978;75(11–12):501–21.https://en.wikipedia.org/wiki/San_people.
[4] http://www.visual-arts-cork.com/east-asian-art/xianrendong-pottery.htm.
[5] http://emerald.tufts.edu/programs/mma/fah188/ospina/jomon/.
[6] http://www.artfromancientlands.com/NeolithicYangshaoTerracottaVesselH778.html.
[7] http://www.uschina.usc.edu/.
[8] http://news.bbc.co.uk/2/hi/europe/8118318.stm.
[9] Tegel W, Elburg R, Hakelberg D, Stauble H, Buntgen U. Early Neolithic water wells reveal the World's oldest wood architecture. PLoS One 2012;7(12):e51374. http://dx.doi.org/10.1371/journal.pone.0051374.
[10] http://www.amnh.org/exhibitions/water-h2o-life/regeneration/mesopotamia.
[11] Kielmas, M., Ancient sumerian levees & canals, Demand Media http://classroom.synonym.com/ancient-sumerian-levees-canals-16874.html.
[12] Jacobsen T. Treasures of darkness: a history of Mesopotamian religion. London, New Heaven (CT): Yale University Press; 1976.
[13] Guisepi RA, Roy Willis F. The history of ancient sumeria (sumer) including its cities, kings, religions culture and contributions or civilization of the University of California. 1980/2003. http://history-world.org/sumeria.htm.
[14] http://www.wildturmeric.net/2014/06/10-amazing-uses-benefits-of-alum.html.
[15] http://www.waterencyclopedia.com/Hy-La/Irrigation-Systems-Ancient.html.
[16] Emelianov, V. The ritual of water-consecration in sumerian texts. Analysis of its materials and structure. https://www.academia.edu/238629/The_Ritual_of_Water-Consecration_in_Sumerian_Texts?s=tvia@academia.
[17] Collon D. The sumerians of mesopotamia: one of the World's earliest and most influential civilizations. Anc Hist Encycl January 18, 2012. http://www.ancient.eu/article/37/.
[18] Langdon S. Babylonian liturgies: sumerian tests from the early period and from the library of Ashurbanipal, for the most part transliterated and translated, with introduction and index. 1913. Paris.
[19] http://www.indiana.edu/~ancmed/meso.HTM and J des Médecines Cunéiformes http://www.oriental.cam.ac.uk/jmc.
[20] http://www.slideshare.net/AkramJaffar/mesopotamian-medicine.

[21] http://www.slideshare.net/079393/medicine-in-mesopotamia?next_slideshow=1.
[22] Bryson RA. On integrating climatic change and culture change studies. Hum Ecol 1994;22(1):115–28.
[23] Bryson RA. Proxy indications of Holocene winter rains in Southwest Asia compared with simulated rainfall. In: Dalfes HN, Kukla G, Weiss H, editors. Third millennium BC climate change and old World collapse (proceedings of the NA to advanced research workshop held at Kemer, Turkey, September 19–24, 1994, NATO ASI, vol. 49(1). Berlin, London: Springer; 1997. p. 465–73.
[24] Wright R, Bryson RA, Schuldenrein J. Water supply and history: harappa an the Beas regional survey. Antiquity February 2008;82(315):37–48.
[25] Staubwasser MF, Sirocko PM, Grootes M, Segl M. Climate change at the 4.2 ka BP termination of the Indus Valley civilization and Holocene South Asian monsoon variability. Geophys Res Lett 2003;30:1425–9.
[26] Narayanaswamy V. Origin and development of Ayurveda: a brief history. Anc Sci Life July 1981;1(1):1–7. http://www.ncbi.nlm.nih.gov/pmc/articles/PMC3336651/pdf/ASL-1–1.pdf.
[27] Violatti, C. "The Vedas," Anc Hist Encycl. .http://www.ancient.eu/The_Vedas/.
[28] www.instructables.com/id/How-to-Purify-Muddy-Water/.
[29] http://chemistry.stackexchange.com/questions/5681/how-does-alum-improve-the-settling-of-particles-to-produce-clean-water.
[30] http://www.nih.ernet.in/rbis/vedic.htm.
[31] http://www.freedrinkingwater.com/resource-history-of-clean-drinking-water.htm.
[32] a Wilkerson JG. Manners and customs of ancient Egyptians, 1837, vol I, II, III. Cambridge Library Collection; 2013. 556 p.
b http://www.sciencealert.com/this-ancient-egyptian-practice-can-cheaply-purify-dirty-water.
[33] Hellsing MS, Habauka M, Kwaambwa FM, Nermark BM, Nkoane AJ, Jackson MJ, et al. Structure of flocs of latex particles formed by addition of protein from Moringa seeds, Colloids and Surfaces A: physicochemical and engineering aspects. Sci Direct December 1, 2013 ISSN: 0927-7757. http://dx.doi.org/10.1016/j.colsurfa.2013.11.038.
[34] Franzmeier H. Sherds, Clay and Clean Water: using the latest information from excavations, Henning Franzmeier looks at the technology of well-making in ancient Egypt. Anc Egypt June/July 2008:45–9. https://www.academia.edu/543256/Sherds_clay_and_clean_Water_Ancient_Egyptia n_Well-technology.
[35] www.swiss-cleanwater-group.com/en/history-of-water.html (Hippocratic Sleeve) https://explorable.com/heron-inventions and http://www.smith.edu/hsc/museum/ancient_inventions/hsc18b.htm (Heron).
[36] http://www.history-of-the-microscope.org/history-of-the-microscope-who-invented-the-microscope.php.
[37] Johnson S. The Ghost map: the story of London's most terrifying epidemic, and how it changed science, cities and the modern world. (NY): Riverhead Books; 2006. 300 p. https://www1.udel.edu/johnmack/frec682/cholera/.
[38] Jacobsen P, Dahi E. Charcoal Packed Furnace for low-tech charring of bone. In: Proceedings of the 2nd International workshop on fluorosis prevention and defluoridation of water, Nazreth, Ethiopia; November 19–25, 1997. p. 151–5. Dunedin, (New Zealand).
[39] Ko DCK, Porter JF, McKay G. Optimised correlations for the fixed-bed adsorption of metal ions on bone char. Chem Eng Sci December 2000;55(23):5819–29.
[40] Chen Y-N, Chai Li-Y, Shu Yu-D. Study of arsenic(V) adsorption on bone char from aqueous solution. J Hazard Mater December 2008;160(1):168–72.

[41] Deydier E, Guilet R, Sharrock P. Beneficial use of meat and bone meal combustion residue: "an efficient low cost material to remove lead from aqueous effluent". J Hazard Mater July 2003;101(1):55–64.
[42] Chauvet J-M, Deschamps EB, Hillaire C. The dawn of art: the Chauvet cave. New York: 1996; Clottes J, Courtain J. The cave beneath the sea: palaeolithic images at Cosquer, New York. New York: Review of Books; November 14, 1996. p. 8–12.
[43] http://www.casterntea.com/ceremony/tea_art.htm.
[44] Franklin RE. The interpretation of diffuse X-ray diagrams of carbon. Acta Cryst 1950; 3:107–21. https://en.wikipedia.org/wiki/Rosalind_Franklin.
[45] Franklin RE. On the structure of carbon (PDF). J de Chimie Physique de Physico-Chimie Biol 1950;47(5,6):573–5. Retrieved 14 January 2011 Per National Library of Medicine above. Citation count 16. Note: this journal ceased publication in 1999.
[46] Kroto HW, Heath JR, O'Brien SC, Curl RF, Smalley RE. C_{60}: buckminsterfullerene. Nature 1985;318:162–3.
[47] Shibuya M, Kato M, Ozawa M, Fang PH, Osawa E. Detection of buckminsterfullerene in usual soots and commercial charcoals. Fullerene Sci Technol 1999;7: 181–93.
[48] http://www.buyactivatedcharcoal.com/how_to_make_activated_charcoal.
[49] Bower LM. Clays Clay Min August 2008;56(4):437–52.
[50] Otto CC, Haydel SE. Exchangeable ions are responsible for the in vitro antibacterial properties of natural clay mixtures. PLoS One 2013;8(5):e64068. http://dx.doi.org/10.1371/journal.pone.0064068.
[51] Knishinsky R. The clay cure. Healing Arts Press; 1998.
[52] Behroozian S, Svensson SL, Davies J. Kisameet clay exhibits potent antibacterial activity against the ESKAPE pathogens. Open Access 2016. http://dx.doi.org/10.1128/mBio.01842-15 26.
[53] Williams LB, Haydel SE, Rossman, Giese Jr F, Eberl DD. Chemical and mineralogical characteristics of French green clays used for healing. 2011.
[54] Haydel et al 2008.
[55] Santo EC, Lam EW, Elowsky CG, Quaranta D, Domaille DW, Chang CJ, et al. Bacterial killing by dry metallic copper surfaces. Appl Environ Microbiol 2011;77: 794–802.
[56] CuVerra 2016 Website: http://www.cuverro.com/about-cuverro/FAQs/how-does-copper-kill-bacteria#sthash.KjEa8iPX.dpuf.
[57] Gibbs RJ. Silver colloids: do they work? Paperback. Published by R. Gibbs; 1999, ISBN 0967699207. 40 p.
[58] Collins: silver kills. http://www.bullionstreet.com/news/how-silver-kills-bacteria-finally-revealed/5042.
[59] http://ehp.niehs.nih.gov/120-a386/.
[60] Xiu Z-M, Zhang Q-B, Puppala HL, Colvin VL, Alvarez PJ. Negligible particle-specific antibacterial activity of silver nanoparticles. Nano Lett July 5, 2012;12(8): 4271–5. http://dx.doi.org/10.1021/nl301934w. Publication Date (Web).
[61] Helmenstine AM, 2015, http://chemistry.about.com/od/chemistryglossary/g/adsorption.htm.
[62] Worch E. Adsorption technology in water treatment: fundamentals, processes, and modeling. Berline/Boston: Walter de Gruyter GmbH & Co KG; 2012. p. 345.
[63] Hussam A, Munir AKM. A simple and effective arsenic filter based on composite iron matrix: development and deployment studies for groundwater of Bangladesh. J Environmetal Sci Helath Part A 2007;42:1869–78.
[64] Tchieda V, D'Amato E, Chiavola A, Parisi M, Chianese A, Amamra M, et al. Removal of arsenic by alumina: effects of material size, additives, and water contaminants. Weinheim: Wiley-VCH Verlag GmbH & Co. KG; January 22, 2016. http://dx.doi.org/10.1002/clen.201400599. online.

[65] Kanel SR, Manning B, Charlet L, Choi H. Removal of arsenic(III) from groundwater by nanoscale zero-valent iron. Environ Sci Technol 2005;39(5):1291–8. http://dx.doi.org/10.1021/es048991u. American Chemical Society.

[66] Calomiris JJ, Christman KA, Dorworth LD. How does chlorine added to drinking water kill bacteria and other harmful organisms? Why doesn't it harm us? Scientific American. 1998.
[a] Johnston MD, Brown MH. An investigation into the changed physiological state of *Vibrio* bacteria in response to cold temperatures and studies on their sensitivity to heating and freezing. J Appl Microbiol 2002;92:1066–77.
[b] http://www.dwi.gov.uk/stakeholders/guidance-and-codes-of-practice/Boiling-water01-15.pdf.

[67] Skelton, S., 2013 http://iodinen13.imascientist.org.uk/2013/11/16/what-in-iodine-makes-it-a-good-antiseptic-and-how/.

[68] Java C. et al; http://drsircus.com/medicine/cancer/the-end-of-antibiotics.

[69] Crump JA, Otieno PO, Slutsker L, Keswick BH, Rosen DH, Hoekstra RM, et al. Household based treatment of drinking water with flocculant-disinfectant for preventing diarrhoea in areas with turbid source water in rural western Kenya: cluster randomised controlled trial. BMJ 2005; Sep 3;331(7515):478.

[70] Holland FA, Siqueiros J, Santoyo S, Heard CL, Santoyo ER. Water purification using heat pumps. New York: Routledge; 1999.

[71] Slow sand filtration, http://www.cdc.gov/safewater/sand-filtration.html, Patent US 6123858 A.

[72] Smith L. Living in the shadow of death: arsenic contaminated drinking water, article taken from the mirror magazine. 2007. p. 20–1.

[73] Singh A, Smith LS, Shrestha S, Maden N. Efficacy of arsenic filtration by Kanchan arsenic filter in Nepal. J Water Health IWA 2014. 2014.

[74] Chiew H, Sampson ML, Sokhan ML, Sreymon H, Bostick DB. Effect of groundwater, iron and phosphate on the efficacy of arsenic removal by iron – Amended BioSand filters. Environ Sci Technol 2009;43:6295–300.

[75] http://www.doulton.ca/ceramicwaterfilter.html and http://blog.waterfilters.net/blog/what-are-ceramic-filters/#sthash.504badOK.dpuf).

[76] www.lenntech.com/systems/reverse-osmosis.

[77] https://en.wikipedia.org/wiki/Ultraviolet_germicidal_irradiation.

[78] Taylor SF. The evolution of the still, Annals of Science 1945;5:3:185–202. http://dx.doi.org/10.1080/00033794500201451. http://www.tandfonline.com/doi/pdf/10.1080/00033794500201451.

[79] Hussam A, Munir KMA. Development of a simple arsenic filter for groundwater of Bangladesh based on composite iron matrix in arsenic contamination of groundwater: Ahija s, editor. Mechanism, Analysis, and Remediation. Wiley Interscience: 2008; pp. 278e289.

[80] Delluc B, Delluc G. Prehistoric Hunters (Signposts S) Board book e Import, May 1982.

CHAPTER THIRTEEN

Evaluation of Animal Manure Composition for Protection of Sensitive Water Supplies Through Nutrient Recovery Processes

U. Shashvatt, H. Aris, L. Blaney
University of Maryland Baltimore County, Baltimore, MD, United States

1. INTRODUCTION

The chronic development of hypoxic zones around the world due to increased nutrient flux is of growing importance. The US Environmental Protection Agency (EPA) estimates negative impacts on the following US water systems: more than 100,000 miles of rivers; 2,500,000 acres of lakes, reservoirs, and ponds; and 512,000 acres of bays and estuaries [1]. Following the 2014 algal bloom in Lake Erie, which shut down the drinking water supply of Toledo, Ohio for three days, Mayor Collins stated, "...we had an environmental attack. I look at this as a wake-up call" [2].

The wake-up call advocated by Mayor Collins requires a sustainable, feasible, and economical response. For this reason, we advocate nutrient recovery from municipal and agricultural waste streams to prevent both ongoing and future hypoxic events in the US and around the world. As municipal wastewater treatment plants (WWTPs) begin shifting to the new paradigm of water resource recovery facilities [3], the implementation of nutrient recovery processes represents a natural evolution. Indeed, such facilities have recently begun full-scale operation [4–6].

While few obstacles exist for incorporation of nutrient recovery into municipal WWTPs, enacting nutrient recovery for agricultural waste is more challenging since these streams do not currently undergo centralized treatment. Rather, most agricultural operations employ de facto treatment during manure storage, such as stockpiling of poultry litter or anaerobic lagooning of cattle/swine manure [7], and offer partial attenuation of

Chemistry and Water
ISBN 978-0-12-809330-6
http://dx.doi.org/10.1016/B978-0-12-809330-6.00013-1

organic carbon and nitrogen, as well as waste minimization. Several reviews have been written on the fate of carbon, nitrogen, and phosphorus in these systems [8–11]. Depending on configuration, these processes facilitate conversion of organic carbon to carbon dioxide ($CO_2(g)$) or methane ($CH_4(g)$), and tend to volatilize ammonia ($NH_3(g)$). Furthermore, these manure storage techniques do not provide any removal of phosphorus present in agricultural waste. For these reasons, traditional manure management does not sufficiently eliminate environmental concerns regarding nitrogen and phosphorus loads in agricultural waste. The loss of other important resources, namely organic carbon and ammonia, highlights the need for sustainable resource recovery processes.

For millennia, humans have been recovering nitrogen and phosphorus in the form of animal manure for use as a soil additive or fertilizer. One of the major limitations to this approach is that the ratio of nitrogen to phosphorus (and to a lesser extent, potassium) is fixed in these products. However, different soils and crops require specific nutrient ratios to ensure optimal growth conditions [12]. Isolating these essential nutrients from agricultural waste allows for on-site formulation of fertilizer solutions for particular crop and soil types.

The main objective of this chapter is to comprehensively examine the aqueous chemistry of animal waste streams to optimize nutrient recovery systems. As such, this chapter:

1. describes the need for nutrient recovery from agricultural waste;
2. evaluates the composition of various animal manures (i.e., poultry, swine, and dairy cow) for potential implementation of nutrient recovery processes;
3. reports existing nutrient recovery technologies, their inputs, and the resulting products; and
4. discusses properties of recovered solids, as well as their value and potential use as agrochemicals.

2. BACKGROUND ON NUTRIENT ISSUES

Misuse of agricultural fertilizers generates three interrelated problems: eutrophication of sensitive water bodies, compromised drinking water supplies, and depletion of phosphate rock ores (along with excessive energy demands for production of ammonia-based fertilizers). These issues are inherently tied to the US EPA's P3 concept [13]: planet—eutrophication; people—drinking water quality; and prosperity—volatility of the global phosphorus market. Each of these problems involves a suite of environmental, political, agricultural, public health, social, and economic facets; however, in all three cases, specific priority concerns are clear. Eutrophication of inland and coastal water bodies negatively affects water quality and biodiversity [14], among other environmental health metrics. Compromised drinking water supplies represent a public health concern [15], especially in low-income or developing communities without access to alternative water supplies. With a majority of the world's phosphate rock reserves located in just a few countries, concerns abound over food, economic, and political security in the coming decades [16]. These three priority issues are further described in the following subsections to highlight the need for nutrient recovery in the agricultural arena.

2.1 Growing Importance of Algal Blooms

When high nutrient loads are discharged into slow-moving water bodies, excessive algal growth occurs. This process is termed eutrophication [14,17,18]. As those algae die, they consume dissolved oxygen and can result in the formation of dead zones (see Section 2.1.1). Hypoxia, which typically occurs during eutrophication, is defined as dissolved oxygen concentrations less than 2 mg/L. In extreme cases, anoxia (i.e., dissolved oxygen levels approaching 0 mg/L) is observed. The major nutrients that cause algal blooms are nitrogen and phosphorus. These elements are critical for growth of agricultural crops, as well as development of aquatic/terrestrial organisms. For example, nitrogen and phosphorus are important components of DNA, proteins, and enzymes; furthermore, phosphorus is critical for energy production (e.g., adenosine triphosphate) in cells.

In the mid-20th century, investigation of the causes of eutrophication commenced. Algal abundance has been correlated with nitrogen and phosphorus, among other dissolves species [19]. The ratio of nitrogen to phosphorus in the environment, compared with that in algal cells, suggested nitrogen limitations during algal blooms [20]. However, recent studies have proposed that phosphorus loading is the primary cause of eutrophication events. For example, lake experiments have shown that eutrophication was not impacted by decreased nitrogen inputs, thereby indicating that phytoplankton growth was limited by phosphorus [21]. Michalak et al. [22] correlated phosphorus loading to Lake Erie with the 2011 record-setting algal bloom, which covered 20% of the 9910 square mile lake. The size of the hypoxic zone in Lake Erie has increased since the 1990s due to increased dissolved reactive phosphorus (DRP) loads stemming from agricultural activity [23]. As further evidence of the sustained impact of agriculture on lake health, the National Oceanic and Atmospheric Administration Lake Erie Severity Index, which is calculated from the amount of lake biomass, was higher in 2015 (10.5) compared to 2011 (10) [24].

In marine environments, early oceanographers hypothesized that nitrogen limitations could be alleviated by planktonic nitrogen fixation and, therefore, believed phosphorus to be responsible for the primary production of algae in oceans [25,26]. However, planktonic activity is also limited by high salinity [27–29], and specific regional characteristics can affect nutrient availability. For example, nutrient limitations display seasonal swings in estuaries like the Chesapeake Bay [30]. In addition, phosphorus limitation has been reported in estuaries with high nitrogen loading and stringent phosphorus controls [31]. Conley [32] reported that reducing phosphorus inputs into heavily nutrient-loaded systems helps to control oxygen depletion in deep estuaries. Clearly, nutrient limitations to development of algal blooms are complex and vary by water body (e.g., inland lake, marine, etc.), region, season, and extent of human intervention.

Currently, the general consensus is that eutrophication in lakes is generally phosphorus limited, whereas nitrogen limitations typically control eutrophication in oceans and estuaries. For this reason, integrated nutrient management systems that allow control of both nitrogen (coastal ecosystems) and phosphorus (inland water bodies and select estuaries) flux are required. The EU's Water Framework Directive [33] and similar efforts in the US [34,35] have emphasized the need to control both nitrogen and phosphorus inputs to coastal water bodies to prevent the growth of existing, and development of, new, dead zones. We posit that a paradigm shift from nutrient treatment to nutrient recovery will reduce both nitrogen and phosphorus loading into sensitive inland and marine water bodies.

2.1.1 Formation of Dead Zones

Dead zones are regions of inland or coastal water bodies, which have become hypoxic or anoxic due to algal decomposition processes; note that hypoxic conditions in dead zones are a symptom of eutrophication. In coastal ocean environments, dead zones have spread exponentially since the 1960s and have detrimental ecological consequences. Dead zone formation has been accelerated by urbanization and industrial agriculture, resulting in increased nutrient loading (e.g., wastewater effluent, agricultural runoff) to specific receiving bodies of water. Climate change has also been implicated in the acceleration of dead zone formation and size [36–38]. For instance, Cloern et al. [36] suggested that the 2004 dinoflagellate bloom in the San Francisco Bay was driven by large-scale atmospheric processes operating at event and seasonal timescales. In 2008, dead zones were reported for 400 water systems around the world, affecting a total area of more than 245,000 square kilometers [39]. That number now exceeds 550 systems [40], indicating that nutrients are increasingly important stressors on marine ecosystems. These conditions are driven by increased biogeochemical flux of nitrogen and phosphorus from anthropogenic activity. One prime example of this interplay between human activity (i.e., agriculture) and ecological response (i.e., dead zone formation) is the Gulf of Mexico.

Hypoxia in the Gulf of Mexico was initially documented in 1972 [41]. The first concerted efforts to monitor eutrophication came in 1985 [41] and have continued until now. In 2015, the Gulf of Mexico dead zone covered 6474 square miles [40]. The primary cause of hypoxia stems from nutrient runoff in the Mississippi River watershed, which drains more than 1.2 million square miles between the Rocky and Appalachian mountain ranges. This land is predominantly used for agricultural purposes (i.e., the North American breadbasket) and runoff contains high levels of nutrients from fertilizer application. Ultimately, these nutrients are transported into the Mississippi River and deposited into the Gulf of Mexico. Data from the US Geological Survey have shown that the 5-year average of nitrogen inputs to the Gulf of Mexico were flat for 2004 to 2007, increased for 2007 to 2011, and decreased from 2011 to 2015; however, the 5-year average from 2015 (~130,000 metric tons) is still much higher than the goal (~70,000 metric tons) [42]. Unsurprisingly, the size of the Gulf of Mexico dead zone over the past decade has not shown any appreciable decrease, as indicated by Fig. 13.1. The total US phosphorus consumption (as fertilizer) is also plotted in Fig. 13.1. Trends in dead zone size [40] and US phosphorus consumption [43] align reasonably well; furthermore, similar correlations are apparent for nitrogen-based fertilizer consumption (not shown). The alignment of fertilizer use and dead zone size corroborate the general understanding of nutrient causes of eutrophication.

Apart from the Gulf of Mexico, many other North American water resources have been impacted by nutrient loading and hypoxia. The degree of impact varies by region. For example, estuaries in the South Atlantic and Pacific regions generally show a low extent of hypoxia (<20%) [44]. Estuaries in the North Atlantic states have demonstrated variable levels of eutrophication over the past 20 years [44,45]; however, the development of toxic algal blooms in these regions may be caused by nutrient transport

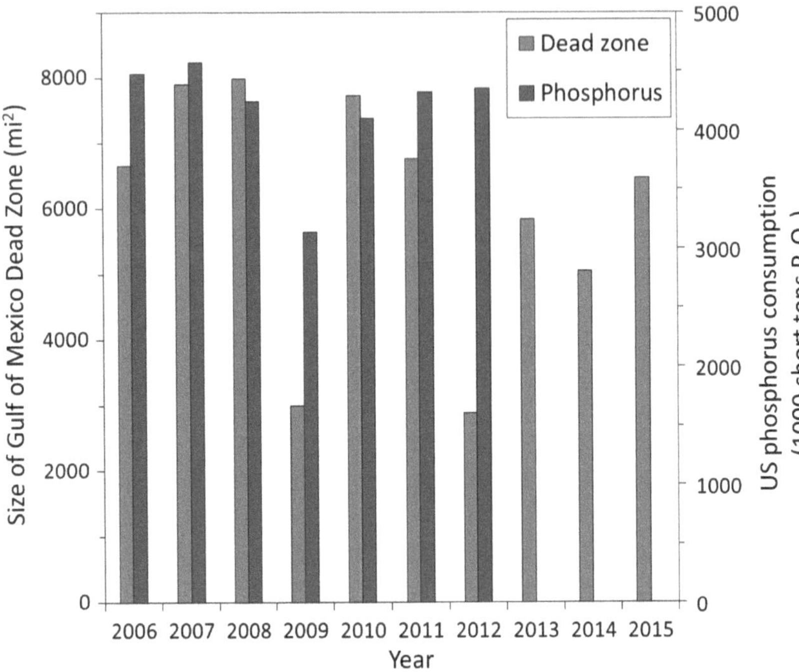

Figure 13.1 ***Size of the Gulf of Mexico dead zone for 2006–2015.*** The US consumption of phosphorus-based fertilizers is also plotted against time (up to 2012). *Data collected from LUMCON. Louisianna Universities Marine Consortium. http://www.gulfhypoxia.net. USDA. USDA economic research service food availability documentation, http://www.ers.usda.gov/data-products/food-availability-(per-capita)-data-system/food-availability-documentation.aspx, 2015.*

into estuarine systems from the ocean. Mid-Atlantic estuaries also exhibit high levels (>60%) of hypoxia, which is primarily driven by urbanization and intensive agricultural practices [44,45]. This condition is more severe in river-dominated estuaries. One example of a highly impacted, river-dominated estuary is the Chesapeake Bay, which is prone to eutrophication by the extensive shoreline in close proximity to intensive poultry production operations [46]. Nonpoint sources in the Chesapeake Bay watershed, such as agricultural runoff, contribute the majority of nitrogen (60%) and phosphorus (58%) inputs [46]. In the 1970s, aquatic life in the Chesapeake Bay was heavily impacted by nutrient enrichment, resulting in the loss of many submerged plant species [45]. In the last decade, regulations on nutrient loads from point (i.e., WWTPs) and nonpoint (e.g., application of nutrient-laden fertilizers and animal manures) sources have improved conditions, but hypoxic events still occur on an annual basis.

Inland water bodies, such as the Great Lakes, are also affected by eutrophication. Lake Erie is the smallest, shallowest, and most densely populated of the Great Lakes [47]. In addition, the basin is intensively farmed, with 63% of the watershed used for agricultural purposes [48]. Since the 1970s, Lake Erie has also been impacted by nutrient loading from wastewater. In the 1990s, eutrophication was primarily linked to DRP from agricultural runoff [23]. For these reasons, co-regulation of point and nonpoint sources of nutrients is needed to protect sensitive ecosystems. The continued growth of the Lake Erie hypoxic region, along with recent impacts on drinking water quality and availability [2], demonstrate the importance of this issue.

2.1.2 Biogeochemical Flux of Nutrients

Previous authors have examined nutrient flux in the environment. Falkowski et al. [49] documented +13%, +108%, +400%, and +113% change in carbon, nitrogen, phosphorus, and sulfur flux due to human activities. The increased flux of critical elements stems from fossil fuel combustion (C, N, S), land use changes (C), food crop cultivation (N), fertilizer production (N), mining (P), and biomass burning (S), among others. Clearly, phosphorus flux has outpaced other critical elements, including carbon and nitrogen, although the magnitude of phosphorus flux is much lower. Regardless, carbon and nitrogen have more options for remediation due to their favorable redox chemistry. For example, organic carbon can be transformed into biomass, oxidized to carbon dioxide (CO_2), or reduced to methane (CH_4), while nitrogen can be converted into nitrogen gas (N_2). For these reasons, the following discussion mostly focuses on phosphorus, which exhibits a more restricted chemistry due to the absence of thermodynamically favorable gaseous species and limited accumulation in microorganisms. Precipitation of phosphorus-containing minerals is favorable under certain conditions. As one example, struvite ($MgNH_4PO_4 \cdot 6H_2O$) readily precipitates in wastewater systems [50]. Intentional production and recovery of phosphorus-rich solids from animal manure is further explored in Sections 3 and 4.

Phosphorus is primarily introduced to the environment through natural weathering (i.e., biogeochemical processing) and anthropogenic activities (e.g., human production of phosphate rock). According to Bennett et al. [51], the mining flux of phosphorus (18.5 Tg/y) is similar to annual accumulation in continental ecosystems (10.5–15.5 Tg/y). This finding is unsurprising, since mined phosphorus is mostly used as a fertilizer. Overapplication of inorganic fertilizers and animal manure has resulted in high nutrient levels in agricultural soils around the world [52,53] and leaching of nutrients into adjacent water bodies [54–56]. In the coming decades, the regional and global flux of phosphorus is expected to be an increasingly important market due to exhaustion of natural deposits of phosphate rock.

2.1.3 Geographical Displacement of Nutrients in North America

In the US, phosphorus influx and distribution has been increased by anthropogenic activities, namely mining and food production. Phosphate rock mining operations in Florida and North Carolina account for 80% of US production, or 26 million tons (Mt) per year [57]. More than 95% of mined phosphorus is used to manufacture phosphoric acid, which is then used to produce phosphorus-based fertilizers, such as diammonium phosphate and monoammonium phosphate [57]. Annually, the US consumes approximately 5 Mt of phosphorus as fertilizer. The majority (i.e., 66%) of that demand stems from 12 midwestern states [58]. Intensive fertilizer application in this region has increased crop yields, but this practice has also resulted in excessive nutrient enrichment of soils and caused hypoxic events, such as those described in Section 2.1.1 for Lake Erie and the Gulf of Mexico. Furthermore, phosphorus is displaced through transportation of crops as animal feed. As discussed previously, phosphorus and nitrogen loads in animal manure are beneficial in conventional agriculture as a free fertilizer and soil amendment. However, the intensification of agricultural practices, namely establishment of concentrated animal feeding operations (CAFOs), has greatly increased nutrient loading from animal manure in relatively small land footprints.

Two particularly intensive animal production operations are the poultry and swine industries. Poultry operations are concentrated in the Southeastern and Mid-Atlantic regions. For example, the Delmarva Peninsula has the most concentrated poultry production in the US. Each year, these operations produce 568 million birds and over one million tons of poultry litter [59]. In fact, Delaware and Maryland are two of the top three states in terms of total nutrients generated per square kilometer of farmland (Fig. 13.2). Traditionally, poultry litter generated at CAFOs in these regions has been applied to adjacent farmland as a waste management strategy, but this practice has led to accumulation of excessive nutrients in soils [60]. Additionally, land application of poultry manure, which has a high P:N ratio, results in greater phosphorus loading to the environment compared to other livestock manure [61]. The major US swine operations

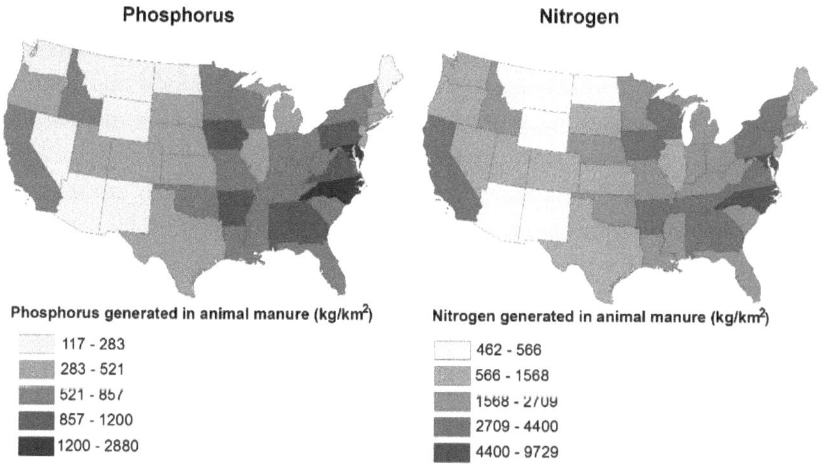

Figure 13.2 ***Maps showing the (left) phosphorus and (right) nitrogen generated in animal manure normalized to available farmland (per state).*** Note that Delaware and Maryland are ranked #1 and #3, respectively, for farmland-normalized nitrogen and phosphorus production, mainly due to the concentrated poultry operations in the Delmarva Peninsula. North Carolina, which ranks #2 for both nitrogen and phosphorus, has concentrated swine operations. *Data collected from: Animals per state (Nass U. Census of agriculture. Washington (DC): US Department of Agriculture, National Agricultural Statistics Service; 2007.); manure nutrient content (Ruddy BC, Lorenz DL, Mueller, DK. County-level estimates of nutrient inputs to the land surface of the conterminous United States, 1982–2001. US Department of the Interior, US Geological Survey; 2006.); farmland acreage (USDA. Farms and land in farms final estimates 2003–2007. National Agricultural Statistics Service; 2009.).*

are located in the Midwest and North Carolina [62]. Similar nutrient issues have arisen in these states due to land application of swine manure [63,64], along with related concerns about environmental justice [65].

2.2 Compromised Drinking Water Supplies

Although algae are important to ecosystem function as a food source for fish and other aquatic organisms, excessive algae can deplete dissolved oxygen. In addition, algal blooms often produce harmful toxins. Toxic algal blooms can affect human health in two ways: residual toxin levels in treated drinking water supplies and toxin accumulation in seafood, especially shellfish. In recent decades, increased occurrences of amnesic, paralytic, and neurotoxic shellfish poisoning have been observed [66]. Many coastal US water bodies, such as the Gulf of Maine [67], the southern coast of California [68], and the mid-Atlantic region [69,70], have experienced toxic algal blooms and have been designated as unsafe for recreational activities and seafood consumption. These designations affect not only water quality (and potentially human health), but also local economies, including the seafood and tourism industries. The economic impacts of algal blooms are further discussed in Section 2.4.

The reason for the "do-not-drink" ban in Toledo stemmed from detection of high levels of microcystins, which are toxins produced by *Microcystis* cyanobacteria, in the drinking water supply. The Toledo case study indicates that nutrient levels in agricultural runoff represent a clear risk to human health. This ongoing threat signifies an important concern for drinking water treatment plants and public health. In a 2009 World Health Organization (WHO) study, drinking water for approximately 3 million people in France did not meet WHO standards due to elevated nitrate levels [71], which can result in blue baby syndrome. These two examples demonstrate that elevated nutrient levels in water sources and treated supplies are an important issue for developed countries. However, the rapid population growth of other countries, along with more intensive agricultural practices, also suggests the importance of nutrient pollution issues around the world, especially for people without economic resources or access to other potable water supplies.

Such issues may pose a significant threat to rapidly developing areas, such as the Pearl River watershed (China, Vietnam), which serves over 455 million people, and the Ganges River basin (India, Bangladesh, Nepal), which comprises 400 million people. These regions represent some of the world's most agriculturally intensive watersheds. Furthermore, continued development will enable greater consumption and application of fertilizer, thereby increasing nutrient flux in these watersheds and their respective discharge points, namely the South China Sea and the Bay of Bengal, respectively.

Eutrophication of Chinese lakes has occurred over the last few decades. For instance, 91% of lakes in China were mesotrophic in the 1970s [72], indicating a moderate level of biological activity. By the 1980s, the percentage of eutrophic lakes increased from 5% to 55%. The three most eutrophic lakes in China are Dianchi Lake, Taihu Lake, and Chaohu Lake [72]. Dianchi Lake is the sixth largest lake in the country and a popular tourist destination. Increased pollution and eutrophication have resulted in *Microcystis* blooms, decreasing biodiversity and negatively impacting the fishing and tourism industries. Taihu Lake is the third largest lake in China, and the Taihu basin accounts for 12% of China's GDP [73]. Rapid urbanization and economic development have resulted in high nutrient loading. Algal blooms in Taihu Lake have intensified and now cover more than 50% of the lake [73]. The Chaohu watershed has been eutrophic for more than two decades. Watershed phosphorus flow analysis models have indicated that phosphorus loss from agricultural land increased from 5 Gg in 1978 to 38 Gg in 2012 [74], thereby increasing stress on the lake system. The marine ecosystem has also been severely affected by algal blooms in the East China Sea. The spread and duration of these blooms have increased over the last two decades; furthermore, recent work has shown that this growth is coincident with increased fertilizer usage [75,76].

Rapid industrialization in India has affected water quality in many freshwater and marine systems. Two major rivers, namely, the Ganges and the Yamuna, have been adversely impacted by minimally treated (or untreated) industrial and residential wastewater; unsurprisingly, high concentrations of phosphorus and nitrogen have been reported along river reaches near major cities [77]. Major harbor ports, such as Mumbai, have shown increased nitrate levels and lower dissolved oxygen levels in the last decade [78]. Inland freshwater lakes, including Dal Lake [79], Periyar Lake [80], and Chilka Lake [81], have gone eutrophic. These lakes serve as not only a freshwater supply for the adjacent communities, but also important tourist destinations; therefore, poor water quality poses a threat to both public health and regional economies. Nutrient loading and formation of dead zones in other developing countries is a topic of concern, as these scenarios may disproportionately affect subsistence farmers and other economically disadvantaged groups.

2.3 Depletion of Phosphorus Ores

The UN has deemed food security a global health challenge for the 21st century. Currently, the world's phosphorus supply is geographical skewed and mostly limited to three countries: Morocco (and Western Sahara), China, and the US [82]. Phosphorus-rich deposits are quickly being depleted. In fact, US production of phosphorus has peaked: mine production has decreased by approximately 730,000 tons per year since 2000 [57], and we are now importing 10% of our phosphorus from Morocco [83]. China mostly produces phosphorus for its domestic market [84]. A 2000 study by Buckley [85] predicted that global phosphate reserves would last approximately 120 years. However, population growth will cause phosphorus utilization to increase at an annual rate of 0.7–1.3%. Coincidentally, the average phosphorus content in raw ores dropped to 29.5% in 1996 from 32.7% in 1980, indicating that global reserves can only sustain demand for another 80 years [86]. Cordell [87] suggested a more dire scenario, namely that global phosphorus production would peak by 2033.

For these reasons, it is not surprising that the average price of phosphate rock has risen from $33/metric ton to $110/metric ton over the last 20 years, with a peak at $430/metric ton in August–September 2008 [88]. Technologies that advance the recovery of phosphorus from animal waste will provide incremental revenue streams to farmers, allowing them direct or tangential access to lucrative domestic, or even global, phosphorus trade in the coming decades.

2.4 Economic Impacts of Nutrient Pollution

The dead zone in the Gulf of Mexico eliminates a potential revenue stream from the fishing and shrimping industry, and reduces biodiversity. The Nature Conservancy [89] estimates that the Gulf of Mexico dead zone costs the US seafood and tourism industries $82 million per year. Multiple states have reported lakeside business closures and declines in restaurant and lodging sales due to algal blooms [90]. The fisheries industry has experienced increased operational and processing costs associated with reduced harvests. The 2005 red tide on the coast of Maine led to heavy economic losses: the quahog, soft shell clam, and mussel industries in Maine and Massachusetts lost approximately $2.5 million and $15.7 million, respectively [91]. Taxpayers bear much of the burden from treatment, remediation, and restoration of the aquatic ecosystem to safeguard drinking water supplies. The annual cost of eutrophication in US freshwaters has been estimated at $2.2 billion [92]. More work to understand the economic impacts of increasing eutrophication in a phosphorus-deficient global society will be an important driver for implementation of nutrient recovery operations; furthermore, economic analysis will facilitate increased adoption of new regulations to protect sensitive ecosystems and drinking water supplies.

2.5 Regulatory Action and Watershed Plans

In response to adverse impacts of pollutants on European water resources, the EU Water Framework Directive was implemented in 2000 [93]; furthermore, adoption of the Urban Wastewater Treatment Directive and Nitrate Directive provides additional control of nutrient pollution. Under this framework, nutrient pollution has been addressed through nitrate and phosphate regulations, and nutrient-sensitive water bodies have been declared protected areas [93]. In addition, a quarterly schedule has been implemented for monitoring nutrient levels in both inland and marine water bodies. Countries around heavily impacted regions, such as the Baltic Sea and the North Sea, have set joint targets to reduce nitrogen and phosphorus loads to 50% of the mid-1980s values [93].

Several regions in the US, including the Lake Erie, Mississippi River, and Chesapeake Bay watersheds, have taken regulatory action to reduce nutrient pollution. In 2011, Lake Erie experienced the largest algal bloom in its history. In response, the US and Canadian governments established the Lake Erie Ecosystem Priority program to control phosphorus influx to the lake [48]. This program promotes the following goals: to decrease hypoxic area in the central lake basin by 50%, to reduce the hypoxic period to 10 days each year, and to lower total phosphorus loads by 4300 Mt (and 550 Mt of DRP) [48]. Since agriculture is the primary source of DRP, outreach programs have been launched to educate farmers on improved fertilizer application methods. The committee has also recommended a ban on application of phosphorus-laden animal manure, biosolids, and commercial fertilizers onto frozen soils due to the lower permeability and increased runoff [48].

The Mississippi River watershed is the largest drainage basin in North America, covering 15% of the continent. Due to expansion of the Gulf of Mexico dead zone, the US EPA established the Mississippi River Gulf of Mexico Hypoxia Task Force. The 2008 action plan required that the 12 states located along the Mississippi River draft state-specific nutrient reduction strategies [94]. Iowa has targeted total nitrogen and phosphorus reductions of 45% from both point and nonpoint sources through better farming practices, fertilizer management, and creation of wetlands and buffer zones [94]. In Kentucky, farms larger than 10 acres are required to follow best management practices to control nitrogen and phosphorus [94]. Over 3130 CAFOs in the task force states have Nutrient Pollution Discharge Elimination System permits; however, another 3000+ CAFOs operate without permits [94]. Proper implementation and enforcement of these permits, along with adoption of best management practices, is required to achieve the total maximal daily load (TMDL) targets.

The Chesapeake Bay watershed has developed a TMDL plan that limits the mass of nitrogen, phosphorus, and sediment entering the Chesapeake Bay [95]. As implementation of this plan continues, individual states and counties must monitor nutrient loads and initiate best management practices to decrease discharges to the bay. While regulation and enforcement of point sources (e.g., WWTPs) of nitrogen and phosphorus are relatively straightforward, controlling nonpoint sources like agricultural runoff is more difficult. In June 2015, Maryland implemented the Phosphorus Management Tool [96] to directly tackle agricultural nutrient inputs. Under these regulations, manure containing phosphorus cannot be applied to an estimated 21% of farms on the Eastern Shore of the Chesapeake Bay [97]. In the first year of enforcement, 18% of Maryland fields, or approximately 230,000 acres, were flagged due to high phosphorus levels [98]. The cost of this regulation is estimated at $23—52 million [99]. We posit that innovative nutrient recovery technologies can fill the gap created by these new regulations and generate profits for farmers through production of high-purity, nutrient-laden fertilizers.

3. ANIMAL MANURE

The composition of manure from a variety of animals is available through state extension programs and the US Department of Agriculture. In Table 13.1, the inorganic composition of manure is shown for three major food-producing animals [100—102]. Consider that the N:P molar ratio for all three manure sources is approximately 3.6 mol N/mol P. However, the molar ratios for Ca:P and Mg:P vary as follows: 1.05 mol Ca/mol P and 0.34 mol Mg/mol P for poultry manure, 1.43 mol Ca/mol P and 0.50 mol Mg/mol P for swine manure, and 1.82 mol Ca/mol P and 1.05 mol Mg/mol P for dairy cow manure. These ratios are notable because they drive the precipitation of phosphorus-laden minerals. To further highlight the importance of the inorganic content of poultry, swine, and dairy cow manure, log C-pH diagrams were constructed to describe the chemistry of manure slurries. The following discussion focuses on the inorganic chemistry inherently present in manure slurries and, therefore, best relates to precipitation-based recovery systems.

In general, the log C-pH diagrams indicate that phosphorus is mostly soluble at low pH (i.e., less than 3); indeed, this behavior has been an integral aspect of phosphorus analysis since the 1940s [103—105]. At higher pH, a variety of phosphorus-containing solids can form based on the inorganic chemical composition of manure slurries. These potential precipitates (from the MINEQL+ version 4.62.3 database) include the following: active magnesium hydroxide, $Mg(OH)_2$ (active); active zinc oxide, ZnO (active); alpha-manganese sulfide, MnS (alpha); amorphous zinc hydroxide, $Zn(OH)_2$ (*am*); amorphous zinc sulfide, ZnS (*am*); beta-tribasic calcium phosphate, $Ca_3(PO_4)_2$ (beta); beta-manganese sulfide, MnS (beta); beta-zinc hydroxide, $Zn(OH)_2$ (beta); brucite, $Mg(OH)_2$; brushite, $CaHPO_4 \cdot 2H_2O$;

Table 13.1 Chemical Composition of Animal Manure

	Total N	NH_4-N	P_2O_5	K_2O	Ca	Mg	S	Na	Mn	Zn	Cu
Poultry											
lb/1000 gallons	62	42	59	37	35	6.8	8.2	5.3	0.42	0.43	0.08
mol/L	5.31E-01	3.59E-01	9.96E-02	9.41E-02	1.05E-01	3.35E-02	3.06E-02	2.76E-02	9.16E-04	7.88E-04	1.51E-04
Swine											
lb/1000 gallons	51.3	28.6	40.4	34.5	32.6	6.9	7.5	6.6	0.19	0.49	0.12
mol/L	4.39E-01	2.45E-01	6.82E-02	8.78E-02	9.75E-02	3.40E-02	2.80E-02	3.44E-02	4.14E-04	8.98E-04	2.26E-04
Dairy Cow											
lb/1000 gallons	12	5.5	7.8	7.7	8	2.8	1.5	2.4	0.1	0.12	0.09
mol/L	1.03E-01	4.71E-02	1.32E-02	1.96E-02	2.39E-02	1.38E-02	5.61E-03	1.25E-02	2.18E-04	2.20E-04	1.70E-04

These values were used as total concentrations of specific elements/species for chemical equilibrium modeling purposes [100–102].

copper(II) hydroxide, $Cu(OH)_2$; covellite, CuS; cupric orthophosphate, $Cu_3(PO_4)_2 \cdot 3H_2O$; dibasic calcium phosphate, $CaHPO_4$; epsilon-zinc hydroxide, $Zn(OH)_2$ (epsilon); gamma-zinc hydroxide, $Zn(OH)_2$ (gamma); hydroxyapatite, $Ca_5(PO_4)_3OH$; lime, CaO; magnesium phosphate tribasic, $Mg_3(PO_4)_2$; manganese(II) hydrogen phosphate, $MnHPO_4$; manganese(II) phosphate, $Mn(PO_4)_2$; newberyite, $MgHPO_4 \cdot 3H_2O$; octacalcium phosphate, $Ca_4H(PO_4)_3 \cdot 3H_2O$; periclase, MgO; portlandite, $Ca(OH)_2$; potassium struvite, $MgKPO_4 \cdot 6H_2O$; pyrochroite, $Mn(OH)_2$; sphalerite, (Zn,Fe)S; struvite, $MgNH_4PO_4 \cdot 6H_2O$; tenorite, CuO; wurtzite, (Zn,Fe)S; zinc hydroxide, $Zn(OH)_2$; zinc phosphate tribasic, $Zn_3(PO_4)_2$; and, zincite, ZnO. Regardless of the wide variety of potential solids, phosphate precipitation mostly occurs in the form of hydroxyapatite, monetite, manganese(II) hydrogen phosphate, and struvite.

3.1 Poultry Litter

Using the poultry manure composition provided in Table 13.1, log C-pH diagrams (Fig. 13.3) were constructed for pH 0–14. All log C-pH diagrams, including those in the following sections for swine and dairy cow manure, were constructed using MINEQL+ version 4.62.3. Convergence criteria were as follows: epsilon for mass balance = 1.0×10^{-6}, epsilon for ionic strength = 5.0×10^{-4}, threshold for reporting percentages = 0.01, and maximum number of iterations = 1000. Redox reactions were not included, as the electron activity varies between systems depending on storage and waste management practices. Activity corrections (using ionic strength–based relationships) were included, and a temperature of 25°C was employed.

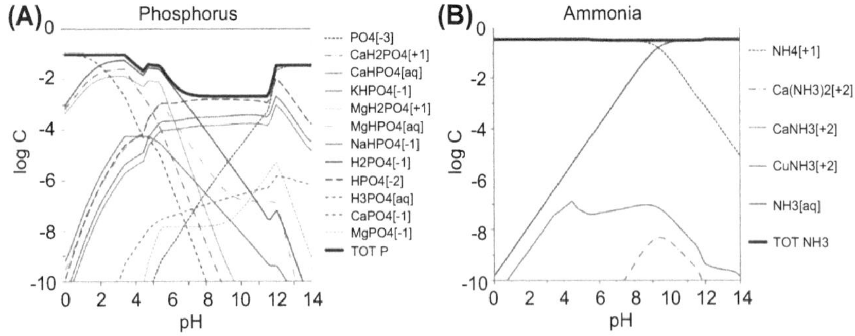

Figure 13.3 ***Log C-pH diagram for dissolved (A) phosphorus and (B) ammonia chemistry of poultry litter slurries, using the compositional information shown in Table 13.1.***

Here, 1401 separate problems were solved between pH 0 and 14, yielding a resolution of 0.01 pH units. Each problem involved 12 components, 37 aqueous complexes, and 36 dissolved solids; furthermore, the system was presumed to be closed to the atmosphere. Therefore, the carbonate system was not included in this analysis. Fig. 13.3A shows the log C-pH diagram for the dissolved phosphorus system in a poultry litter slurry; the dissolved ammonia system is presented in Fig. 13.3B.

The log C-pH diagram in Fig. 13.3A shows that phosphorus is completely dissolved at pH < 3.40. Furthermore, a local minimum for dissolved phosphorus occurs at pH 4.46 due to the precipitation of monetite. This local minimum is important for phosphorus release from poultry litter. To balance the acid demand with release efficiency, an operating pH of 4.74–5.04 is proposed for this particular manure composition. As indicated above, phosphorus precipitation occurs at pH > 3.40. Solid precipitates include hydroxyapatite (pH 4.47–14.00), monetite (pH 3.40–4.76), newberyite (pH 5.32–5.51), manganese monohydrogen phosphate (pH 0.49–12.22), and struvite (pH 5.51–12.01). Although manganese monohydrogen phosphate precipitates at most pH conditions, the production of this mineral is limited by the low manganese concentration (i.e., TOTMn ≈ 0.009 $TOTPO_4$). Similarly, while hydroxyapatite and monetite do precipitate, phosphorus recovery through these minerals is limited by the calcium concentration. Addition of external calcium forces precipitation of hydroxyapatite [106]. However, struvite precipitation can provide up to 97.8% recovery of phosphorus without addition of external cations. Optimal precipitation occurs at pH 8.22 to 11.49. The lower pH boundary is preferred, as it minimizes base demand and allows direct discharge or reuse of process effluent.

Ammonia chemistry is less convoluted, as indicated in Fig. 13.3B, because only three complexes [i.e., $Ca(NH_3)^{2+}$, $Ca(NH_3)_2^{2+}$, and $Cu(NH_3)^{2+}$] and one solid (i.e., $MgNH_4PO_4 \cdot 6H_2O$) can potentially form; note that this discussion only includes species included in the MINEQL+ thermodynamics database. Ammonia off-gassing during manure handling is a concern that requires attention [107–110], and some have proposed gas phase ammonia recovery systems [111]; however, this discussion focuses on the chemistry of aqueous nutrient recovery efforts. In this case, the optimal pH range identified above holds for ammonia recovery through struvite precipitation; however, only 9.2% of the ammonia in this system can be recovered as struvite due to magnesium and phosphorus limitations.

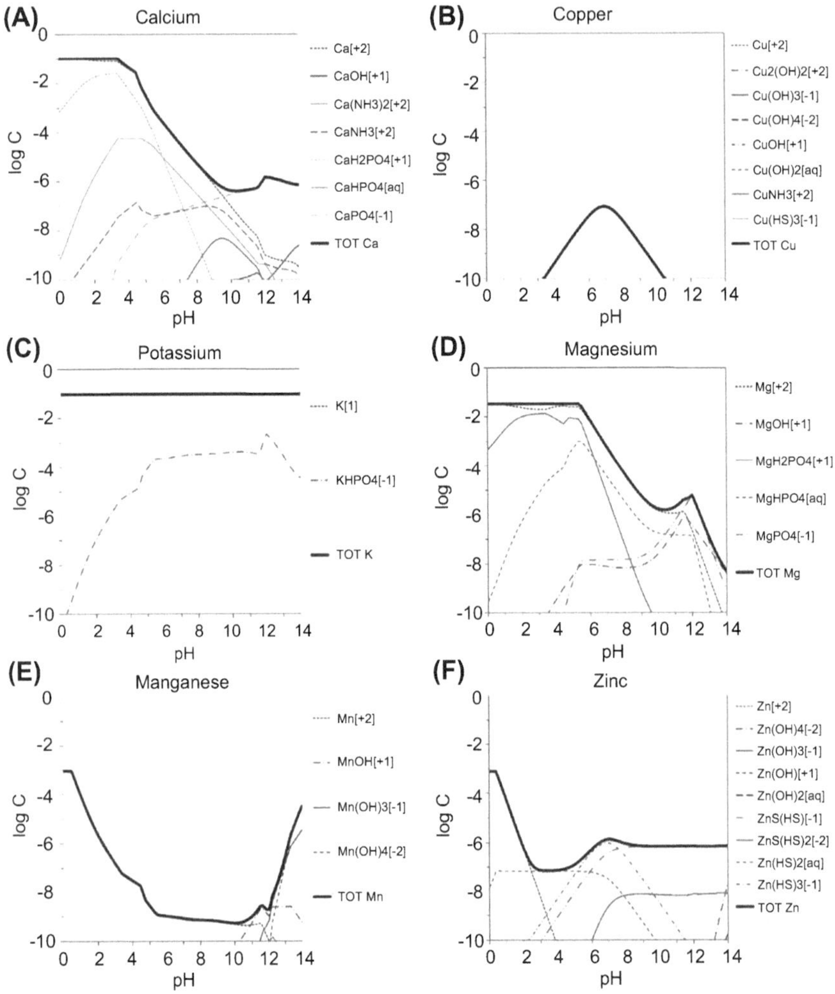

Figure 13.4 ***Log C-pH diagrams for the (A) calcium, (B) copper, (C) potassium, (D) magnesium, (E) manganese, and (F) zinc systems in poultry manure.***

The aqueous chemistry of the divalent cation systems, namely Ca^{2+}, Cu^{2+}, Mg^{2+}, Mn^{2+}, and Zn^{2+}, are shown in Fig. 13.4; in addition, potassium (K^{+}) chemistry is included. From these diagrams, struvite recovery at pH 8.22 coincides with production of hydroxyapatite (Ca^{2+} system), covellite (Cu^{2+} system), manganese hydrogen phosphate (Mn^{2+} system), and sphalerite (Zn^{2+} system). At pH 8.22, approximately >99.99%, 83%, 99.9%, >99.99%, and 99.9% of calcium, copper, magnesium, manganese, and zinc precipitate. As Fig. 13.4 demonstrates, high precipitation efficiencies are observed for all five divalent metals in the range of optimum phosphorus recovery. For that reason, trace contamination from copper, manganese, and zinc may be present in recovered nutrient-rich products. A significant fraction of calcium phosphate, as hydroxyapatite, will also be recovered. Depending on the intended use of recovered products, the copper, manganese, and zinc content should be fully vetted.

3.2 Swine Manure

Using the same protocols identified above for poultry manure, log C-pH diagrams were constructed for dissolved phosphorus and ammonia chemistry in swine manure (Fig. 13.5). Like the poultry manure scenario, phosphorus is completely dissolved at low pH (i.e., below 3.56), and a local maximum exists at pH 4.34; however, that maximum is narrow and does not offer potential as an operational set point for nutrient release from manure solids. For that reason, the optimal pH for nutrient release from swine manure is approximately 3.6. Above this pH, the aqueous phosphorus chemistry in swine manure demonstrates a distinctly different shape compared to that of poultry manure. This difference stems from the higher Mg:P molar ratio; recall that the Mg:P ratios for poultry and swine manures were 0.34 and 0.50 mol/mol, respectively. This scenario allows for more effective precipitation of struvite in swine manure. Regardless, a number of precipitates may form, including hydroxyapatite (pH 4.35−14.00), monetite (pH 3.56−4.49), newberyite (pH 5.32−5.51), manganese monohydrogen phosphate (pH 0.76−8.51, 11.47−11.96), and struvite (pH 5.76−11.76). As with poultry litter, manganese monohydrogen phosphate production is limited by the total manganese concentration (TOTMn ≈ 0.006 $TOTPO_4$); furthermore, recovery of hydroxyapatite and monetite is limited by the

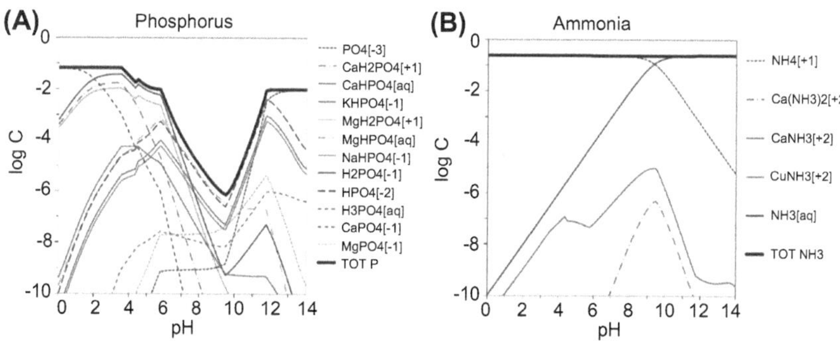

Figure 13.5 ***Swine manure log C-pH diagrams for dissolved (A) phosphorus and (B) ammonia chemistry.***

calcium concentration. Over 99.99% of phosphorus can be recovered as struvite at pH 8.21−10.26.

Because ammonia concentrations are much higher than dissolved phosphorus, nitrogen recovery from swine manure is limited. In this case, the optimal ammonia recovery occurs at the same pH range as phosphorus (pH 8.21−10.26), and 3.8% of the total ammonia can be recovered. However, higher recovery efficiencies may be observed in select swine manures, depending on compositional differences. For example, Romero-Guiza et al. [112] attained 25% ammonia recovery by struvite precipitation, although an external magnesium source was needed. This scenario suggests that for swine operations, upstream or downstream biological processes may be most applicable for nitrogen capture/treatment and waste minimization.

The aqueous chemistry of the Ca^{2+}, Cu^{2+}, K^{+}, Mg^{2+}, Mn^{2+}, and Zn^{2+} systems in swine manure slurries are shown in Fig. 13.6. From these diagrams, recovery at pH 8.21 results in coproduction of hydroxyapatite (Ca^{2+} system), covellite (Cu^{2+} system), struvite (Mg^{2+} system), manganese hydrogen phosphate (Mn^{2+} system), and sphalerite (Zn^{2+} system). At pH 8.21, approximately 99.8%, 83%, 28%, 99.9%, and 99.9% of calcium, copper, magnesium, manganese, and zinc precipitate. Potassium does not precipitate for the given conditions, as indicated in Fig. 13.6C. Like the poultry manure scenario, high precipitation efficiencies are observed for the divalent metals in the range of optimum phosphorus recovery, and therefore, trace levels of copper, manganese, and zinc will be present in recovered products. A significant amount of hydroxyapatite is also recovered. The agrochemical uses of hydroxyapatite, struvite, and other recovered solids are discussed further in Section 5.

3.3 Dairy Cow Manure

The log C-pH diagrams for dairy cow manure (Fig. 13.7) are much different than those shown above for poultry (Fig. 13.3) and swine (Fig. 13.5) manure. In this case, phosphorus is completely dissolved at pH < 4.49, and no local maxima/minima exist at acidic conditions. Above this pH, the aqueous phosphorus chemistry in dairy cow manure demonstrates a distinctly different shape compared to that of poultry and swine manure. This difference stems from the higher Ca:P molar ratio; recall that the Ca:P ratios for poultry, swine, and dairy cow manures were 1.05, 1.43, and 1.82 mol/mol, respectively. Because the available calcium exceeds the stoichiometric ratio (i.e., Ca:P $= 5/3 = 1.67$), hydroxyapatite formation is

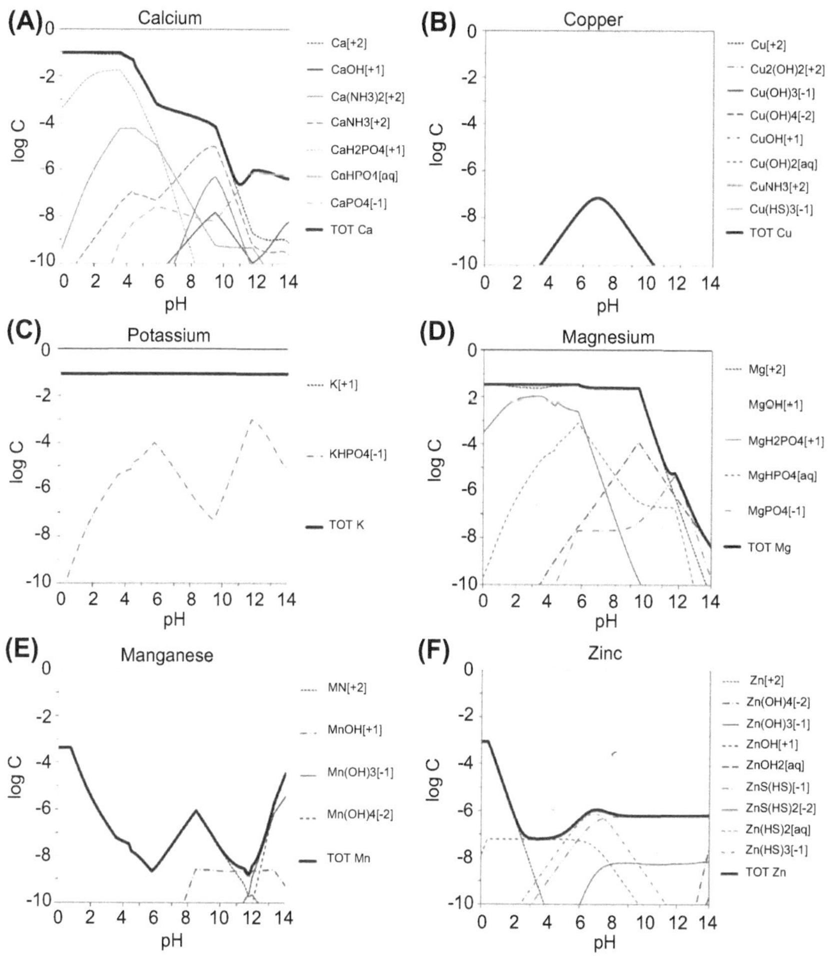

Figure 13.6 ***Swine manure log C-pH diagrams for the (A) calcium, (B) copper, (C) potassium, (D) magnesium, (E) manganese, and (F) zinc systems.***

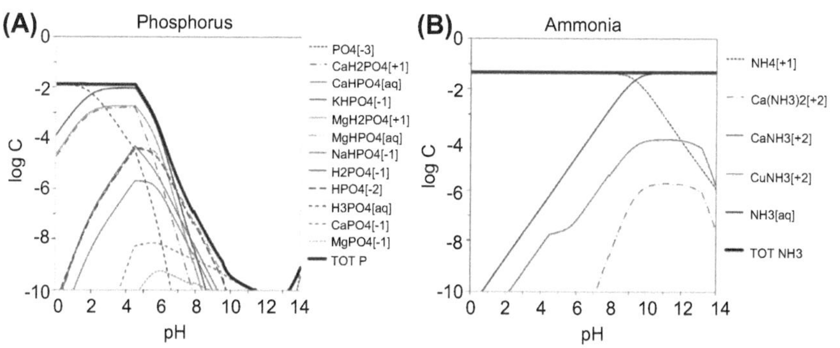

Figure 13.7 ***log C-pH diagrams for (A) phosphorus and (B) ammonia chemistry in dairy cow manure.***

limited by phosphorus. For that reason, significant phosphorus recovery efficiencies can be obtained. In particular, over 99.99% phosphorus recovery can be obtained at pH 7.12–14.00. Manganese monohydrogen phosphate also precipitates at pH 1.23–7.90; however, like the other scenarios, this product negligibly contributes (i.e., ~2.3%) to the total phosphorus recovery.

Because struvite does not precipitate in dairy cow manure, ammonia is not recovered. In actuality, the Mg:P ratio, which controls struvite precipitation, is highest in dairy cow manure (see Table 13.1); however, the high calcium levels force hydroxyapatite to dominate over struvite. Zhang et al. [113] addressed this issue by first releasing phosphorus from the calcium-based solid, complexing calcium with ethylenediaminetetraacetic acid (EDTA), and then dosing sufficient magnesium to drive struvite precipitation. Others have recovered 11–95% total ammonia in the form of struvite from anaerobically digested dairy cow manure [114,115].

The aqueous chemistry of the cation systems, namely Ca^{2+}, Cu^{2+}, K^+, Mg^{2+}, Mn^{2+}, and Zn^{2+}, in dairy manure are shown in Fig. 13.8. From these diagrams, phosphorus recovery at pH 7.12 results in production of hydroxyapatite (Ca^{2+} system), covellite (Cu^{2+} system), manganese hydrogen phosphate (Mn^{2+} system), and sphalerite (Zn^{2+} system). At pH 7.12, approximately 90%, 5%, 0%, 99.3%, and 99.95% of calcium, copper, magnesium, manganese, and zinc are recovered with phosphorus. As Fig. 13.8 demonstrates, high precipitation efficiencies are observed for Ca^{2+}, Cu^{2+}, and Zn^{2+} in the range of optimum phosphorus recovery, but incorporation of K^+, Mg^{2+}, and Mn^{2+} into recovery products is minimal.

More discussion of the primary recovered products, namely struvite for poultry and swine manure and hydroxyapatite for dairy cow manure, is provided as follows in Section 5.

3.4 Manure Composition of Less Consumed Animals

While poultry, swine, and cattle operations provide the greatest opportunity for nutrient recovery due to consumer preference for food products from these animals, potential application to other animal feeding (or rearing) operations exists. Table 13.2 shows the nutrient content as total N, P_2O_5, and K_2O for a variety of animal manures. From these aggregated data, phosphorus recovery from poultry and swine is most economical because of the higher phosphorus loads in manure from these animals. Based on the above observations, recovery of phosphorus as hydroxyapatite or struvite

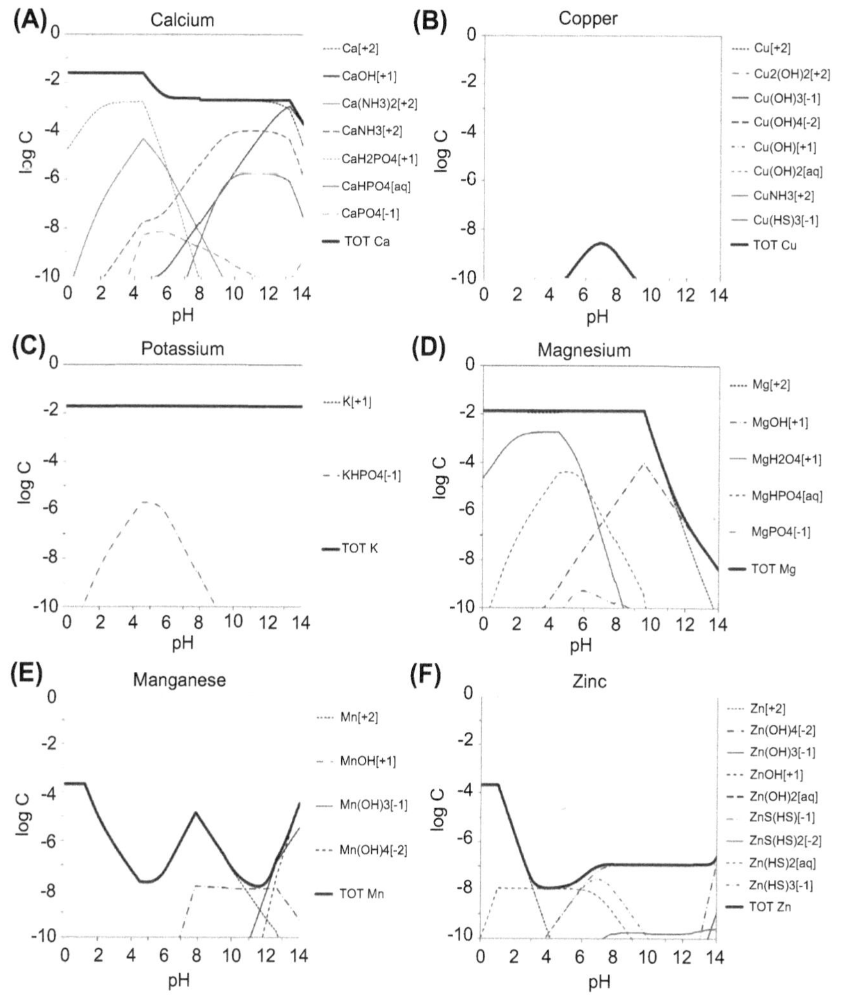

Figure 13.8 ***Log C-pH diagrams for the (A) calcium, (B) copper, (C) potassium, (D) magnesium, (E) manganese, and (F) zinc systems in dairy cow manure.***

from other types of animal manure should nevertheless be achievable through direct precipitation.

Pagliari and Laboski [116] reported phosphorus fractionation in beef cattle, dairy cattle, goat, sheep, horse, turkey, chicken, and swine manures and noted that phosphorus recovery operations are largely dependent on this speciation. While these lesser-consumed animals may not constitute a critical mass of nutrients at the national scale, specific farming/rearing operations may have impacts at microenvironments in sensitive watersheds. For that

Table 13.2 Nutrient Content in a Variety of Animal Manures [100–102,148,149]

Animal	Details	Total N	P_2O_5	K_2O	N:P:K (lb/lb/lb)
Cattle (beef)	Calf[a]	11	7	10	0.8:1:1.4
	Cow and calf[a]	11	7	10	0.8:1:1.4
	Finishing[a]	14	5	8	1.4:1:1.6
	No bedding[a]	21	14	23	0.8:1:1.6
	With bedding[a]	21	18	26	0.6:1:1.4
Cattle (dairy)	Calf[a]	10	3	4	1.7:1:1.3
	Dry cow[a]	9	3	7	1.5:1:2.3
	Fresh liquid manure[b]	12	8	8	0.8:1:1.0
	Fresh manure[a]	10	5	8	1.0:1:1.6
	Heifer[a]	10	3	7	1.7:1:2.3
	Lactating, liquid[b]	28	13	25	1.1:1:1.9
	Lactating, solid[a]	10	4	8	1.3:1:2.0
	No bedding[a]	9	4	10	1.1:1:2.5
	Scraped paved outside lot[a]	15	8	14	0.9:1:1.8
	Slurry[b]	23	14	21	0.8:1:1.5
	With bedding[a]	9	4	10	1.1:1:2.5
Cattle (veal)	[b]	19	13	25	0.7:1:1.9
Horse	[a]	12	5	9	1.2:1:1.8
	With bedding[a]	14	4	14	1.8:1:3.5
Poultry (chicken)	Breeder house litter[a]	7	54	31	0.1:1:0.6
	Broiler house litter[a]	11	78	46	0.1:1:0.6
	Fresh, broiler (no litter)[a]	10	17	11	0.3:1:0.6
	Fresh, layer (no litter)[a]	6	22	11	0.1:1:0.5
	Heavy broiler[a]	66	63	47	0.5:1:0.7
	High-rise stored[a]	18	56	30	0.2:1:0.5
	Layer[a]	37	55	31	0.3:1:0.6
	Layer, undercage scraped[a]	14	31	20	0.2:1:0.6
	Light broiler[a]	79	62	42	0.6:1:0.7
	No litter[a]	33	48	34	0.3:1:0.7
	Pullet[a]	43	46	26	0.5:1:0.6
	Roaster house litter[a]	12	75	45	0.1:1:0.6
	Stockpiled litter[a]	8	80	34	0.1:1:0.4
	With litter[a]	56	45	34	0.6:1:0.8
Poultry (duck)	Dry[a]	21	26	15	0.4:1:0.6
	Fresh (no litter)[a]	5	23	17	0.1:1:0.7
	House litter[a]	3	17	14	0.1:1:0.8
	Stockpiled litter[a]	5	42	22	0.1:1:0.5
	Wet[b]	33	23	16	0.7:1:0.7
Poultry (turkey)	Brooder house litter[a]	9	52	32	0.1:1:0.6
	Fresh (no litter)[a]	8	25	12	0.2:1:0.5
	Grower house litter[a]	16	72	40	0.1:1:0.6
	Hen[a]	73	88	46	0.4:1:0.5

Table 13.2 Nutrient Content in a Variety of Animal Manures [100–102,148,149]—cont'd

Animal	Details	Total N	P_2O_5	K_2O	N:P:K (lb/lb/lb)
	No litter[a]	27	20	17	0.7:1:0.9
	Stockpiled litter[a]	8	72	33	0.1:1:0.5
	Tom[a]	52	76	42	0.3:1:0.6
	With litter[a]	20	16	13	0.6:1:0.8
Sheep/goats	[a]	23	8	20	1.4:1:2.5
	Sheep, no bedding[a]	18	11	26	0.8:1:2.4
	Sheep, with bedding[a]	14	9	25	0.8:1:2.8
Swine	Farrow to wean[b]	18	18	11	0.5:1:0.6
	Fresh manure[b]	51	40	35	0.6:1:0.9
	Grow to finish[b]	31	24	22	0.6:1:0.9
	Manure from building[b]	17	13	14	0.7:1:1.1
	No bedding[a]	10	9	8	0.6:1:0.9
	Nursery[b]	19	8	14	1.2:1:1.8
	Wean to finish[b]	37	23	21	0.8:1:0.9
	With bedding[a]	6	7	7	0.4:1:1.0

[a]lb/ton.
[b]lb/1000 gallons.

reason, efforts to demonstrate the feasibility of nutrient recovery from a variety of animal manures are warranted. Given the ubiquitous presence of poultry, swine, and dairy farms throughout agriculturally active regions of the US, mixing manure from less consumed animals with primary (i.e., poultry, swine, and cow) manures is another option for sustainable waste management and optimization of nutrient recovery operations.

4. NUTRIENT RECOVERY PROCESSES

A variety of nutrient recovery processes have been developed, or are currently being explored, for municipal and agricultural waste. For the sake of brevity, this chapter does not describe the engineering design and operational parameters of individual processes. In general, these nutrient recovery systems can be broadly classified as chemical or thermochemical processes (see Table 13.3, adapted from information compiled by Nieminen [117]). The chemical processes (i.e., Airprex, Aquatron, BioCon, Crystalactor, Kurita, Ostara Pearl, PASH, PEARS, Phosnix, PRISA, P-RoC, Quick Wash, REM-NUT, Seaborne, and SEPHOS) are primarily based on recovery of phosphorus (and nitrogen in struvite-producing systems) through pH control. As detailed in Section 3, pH control is key to phosphorus release

Table 13.3 List of Reported Nutrient Rescovery Processes

Process	Principle	Scale	Location	Feed Material	Recovered Product
AirPrex	Precipitation	Industrial	Germany	Digested sludge, magnesium chloride, air, flocculent praestol 40560	Struvite
AQUA RECI	Oxidation precipitation	Pilot	Sweden	Sludge ash, lime	Calcium phosphate
Aquatron	Precipitation	Commercial	Sweden	Separated urine and household waste streams	Struvite
ASH DEC	Thermal	Pilot	Germany	Sewage sludge, calcium chloride, natural gas	Phosphorus-enriched ash
BioCon	Acid leaching ion exchange	Pilot	Denmark	Sewage sludge ash, sulfuric acid, sand, hydrochloric acid, ferric chloride, sodium hydroxide, potassium hydrogen sulfate, phosphoric acid	Phosphoric acid
Crystalactor	Precipitation	Industrial	Netherlands	Sand, phosphorus-rich wastewater, magnesium chloride, air	Calcium phosphate
Kurita	Precipitation	Bench	Japan	Secondary effluent, phosphate rock, calcium chloride, sodium hydroxide	Calcium phosphate
Mephrec	Thermal	Pilot	Germany	Dried sludge	Silico-phosphate
Ostara pearl	Precipitation	Industrial	Canada	Sludge liquor, magnesium chloride, sodium hydroxide	Struvite

PASH	Acid leaching precipitation	Pilot	Germany	Sewage sludge ash, hydrochloric acid, lime	Calcium phosphate
PEARS	Acid leaching precipitation	Bench	US	Poultry litter, carbon dioxide, hydrochloric acid, air, sodium hydroxide	Struvite Potassium struvite
Phosnix	Precipitation	Industrial	Japan	Sludge liquor, magnesium hydroxide, sodium hydroxide	Struvite
Phosphorus recovery system	Mechanical	Full	Denmark	Pig slurry, ferrous and aluminum salts	High P solids
PRISA	Precipitation	Pilot	Germany	Sludge liquor, sodium hydroxide, magnesium oxide	Struvite
P-RoC	Precipitation	Pilot	Germany	Phosphorus rich wastewater, calcium silicate hydrate	Calcium phosphate
Quick Wash	Acid leaching precipitation	Bench	US	Poultry litter, sulfuric acid, calcium hydroxide	Hydroxyapatite
REM-NUT	Ion exchange	Pilot	Italy	Nutrient rich effluent, zeolite (ionic exchange), magnesium chloride, sodium hydroxide	Struvite

(Continued)

Table 13.3 List of Reported Nutrient Rescovery Processes—cont'd

Process	Principle	Scale	Location	Feed Material	Recovered Product
Seaborne	Acid leaching precipitation	Industrial	Germany	Digested sludge, sulfuric acid, hydrogen peroxide, sodium sulfide, magnesium hydroxide, sodium hydroxide, flocculant praestol 40560	Struvite
SEPHOS	Acid leaching precipitation	Bench	Germany	Sewage sludge ash, sulfuric acid, sodium hydroxide, calcium	Calcium phosphate

More detail on each process, along with the process chemistry and current state of implementation, can be found through the process websites and corresponding publications. For the sake of brevity, these details are not included here.

and recovery in manure systems. In addition to pH control, two of the processes, namely REM-NUT and BioCon, use a train of cation and anion exchange columns to separate ammonium, ferric iron, phosphate, and sulfate from liquid waste streams for later elution and precipitation. The thermochemical processes (i.e., AQUA RECI, ASH DEC, and Mephrec) employ oxidation processes at high temperatures to produce phosphorus pentaoxide or silico-phosphates.

These nutrient recovery technologies are at different stages of development, but some processes have been incorporated at the industrial scale. For example, Ostara Nutrient Recovery Technologies, Inc. has installed a number of full-scale systems at WWTPs in Wisconsin (US), Saskatchewan (Canada), Slough (UK), Oregon (US), Pennsylvania (US), and Virginia (US). The incorporation of these systems into WWTPs is noteworthy; however, from a nutrient pollution vantage, nonpoint sources of nitrogen and phosphorus are equally important, as indicated previously for the Gulf of Mexico and Lake Erie. For that reason, continued advancement and development of new technologies that focus specifically on agricultural waste is needed to prevent eutrophication, hypoxia, and harmful algal blooms. To the extent possible, innovative technologies should aim to fully utilize the baseline chemistry of manure sources to decrease the operational costs and external chemical usage. In the next section, we further explore the variety of recovered solids and their potential use as agrochemicals.

5. PROPERTIES OF RECOVERED PRODUCTS

Various products, namely struvite, hydroxyapatite, monetite, newberyite, and potassium struvite, are derived from the nutrient recovery processes described above. The copper, manganese, and zinc products that precipitate (see Section 3) were not included here because the fraction of phosphorus bound in these minerals is minimal compared to the calcium- and magnesium-based minerals. The solubility reactions (Reactions 13.i−13.v) and associated log K_{sp} values (Eqs. 13.1−13.5), for these important minerals are shown below. Note that the K_{sp} values listed in Eqs. (13.2)−(13.4) stem from the MINEQL+ version 4.63.2 database. K_{sp} values for struvite (Eq. 13.1) and potassium struvite (Eq. 13.5) were collected from previous reports [118], and these reactions were added to the MINEQL+ database for the analysis presented in Section 3. Brief descriptions of the use and value of these minerals as agrochemicals are provided in the following subsections.

Struvite

$$MgNH_4PO_4 \cdot 6H_2O(s) \rightleftharpoons Mg^{2+} + NH_4{}^{+} + PO_4{}^{3-} + 6H_2O \tag{13.i}$$

$$K_{sp,MgNH_4PO_4 \cdot 6H_2O} = 10^{-13.2} = \left[Mg^{2+}\right]\left[NH_4{}^{+}\right]\left[PO_4{}^{3-}\right] \tag{13.1}$$

Hydroxyapatite

$$Ca_5(PO_4)_3(OH)(s) \rightleftharpoons 5Ca^{2+} + 3PO_4{}^{3-} + HO^{-} \tag{13.ii}$$

$$K_{sp,Ca_5(PO_4)_3(OH)} = 10^{-58.3} = \left[Ca^{2+}\right]^5\left[PO_4{}^{3-}\right]^3\left[HO^{-}\right] \tag{13.2}$$

Monetite

$$CaHPO_4(s) \rightleftharpoons Ca^{2+} + H^{+} + PO_4{}^{3-} \tag{13.iii}$$

$$K_{sp,CaHPO_4} = 10^{-19.3} = \left[Ca^{2+}\right]\left[H^{+}\right]\left[PO_4{}^{3-}\right] \tag{13.3}$$

Newberyite

$$MgHPO_4 \cdot 3H_2O\ (s) \rightleftharpoons Mg^{2+} + H^{+} + PO_4{}^{3-} \tag{13.iv}$$

$$K_{sp,MgHPO_4 \cdot 3H_2O} = 10^{-18.2} = \left[Mg^{2+}\right]\left[H^{+}\right]\left[PO_4{}^{3-}\right] \tag{13.4}$$

Potassium struvite

$$MgKPO_4 \cdot 6H_2O(s) \rightleftharpoons Mg^{2+} + K^{+} + PO_4{}^{3-} + 6H_2O \tag{13.v}$$

$$K_{sp,MgKPO_4 \cdot 6H_2O} = 10^{-10.6} = \left[Mg^{2+}\right]\left[K^{+}\right]\left[PO_4{}^{3-}\right] \tag{13.5}$$

5.1 Struvite

Struvite is produced in both natural (e.g., guano deposits in caves [119]) and engineered (i.e., WWTPs) systems. In engineered systems, struvite precipitation is a nuisance due to scale formation, which causes pressure drop and pipe blockages [120,121]. To eliminate formation in pipes, struvite recovery from municipal wastewater has been investigated and implemented. Struvite is one of the more widely recovered minerals from municipal and agricultural waste streams [122] through precipitation-based chemical processes. From Table 13.3, 8 of the 19 technologies provide struvite recovery; furthermore, the discussion in Section 3 demonstrated that struvite formation is expected for both poultry litter and swine manure. As struvite molecules are comprised of equal parts of magnesium, ammonium, and phosphate, this product effectively balances nitrogen and phosphorus and serves as an effective fertilizer for food crops.

Struvite crystals are most commonly formed at a pH of approximately 8.0–9.0 [123,124]. Operational conditions, including pH [125], solid retention time [126], saturation level, and the use of particle growth seeds [127], affect the morphology and size of recovered struvite. The purity of recovered struvite is also impacted by the presence of other dissolved species, such as calcium and ammonium [128]. One of the key advantages of struvite as a primary product of nutrient recovery operations stems from its classification as a slow-release fertilizer.

The agrochemical use of recovered struvite is paramount in Europe because of the continent's dependence on trade for phosphorus-based fertilizers. With rising concerns over the limited supply of phosphorus [85–87], the European Sustainable Phosphorus Platform has accepted the use of recovered struvite as a fertilizer in mainstream agriculture. Use of recovered struvite is dependent on mineral purity. In fact, specific requirements must be met with respect to nitrogen and phosphorus content; furthermore, maximum levels have been identified for concentrations of organic carbon and other contaminants [129]. Recovery of nutrients, such as phosphorus, through sustainable agricultural waste management strategies in Europe may also mitigate the stress on natural phosphate rock deposits.

5.2 Hydroxyapatite

Hydroxyapatite is a naturally occurring mineral of biological and agricultural importance. Human and animal bones are composed of hydroxyapatite. In nutrient recovery processes, hydroxyapatite is generated through treatment of waste streams containing calcium. From Table 13.3, 7 of the 19 technologies recover calcium phosphate, which is most likely to be in the form of hydroxyapatite based on the chemical equilibrium modeling in Section 3. Recall that hydroxyapatite precipitated in all manure types, but was dominant for calcium-rich dairy cow manure [101].

Characterization of hydroxyapatite crystals and chemical/operational factors that influence formation have been well studied at the lab scale. Dosing nutrient-rich manure extracts (see Figs. 13.3, 13.5, and 13.7) with seed crystals, such as calcite ($CaCO_3$), xonolite ($Ca_6Si_6O_{17}(OH)_2$), and tobermorite ($Ca_5Si_6O_{16}(OH)_2 \cdot 4H_2O$), can facilitate hydroxyapatite growth at the mineral surface, leading to larger particle sizes [130–132]. The use of tobermorite seeds also results in a higher purity of recovered hydroxyapatite;

furthermore, calcite formation has not been observed in previous studies, eliminating the need for decarbonation processes [133].

Hydroxyapatite is a valuable resource that serves as an important agrochemical fertilizer, although this use is limited due to the relatively limited dissolution (see Eq. 13.2). Innovative techniques for production of hydroxyapatite particles that provide slow release of phosphorus may increase the use of this mineral for fertilizer applications. Recently, hydroxyapatite nanoparticles have been fabricated in controlled environments and shown efficacy as a slow-release fertilizer [134]. Another example of novel strategies for increasing the potential use of hydroxyapatite fertilizers was reported by Kottegoda et al. [135], who generated urea-modified hydroxyapatite particles.

5.3 Monetite

In the natural environment, monetite formation has been reported through interaction of guano and limestone in caves [136]. Widespread use of guano as a soil amendment and fertilizer has, therefore, involved monetite recovery [137]. In nutrient recovery processes, the potential for monetite precipitation is fairly low due to unfavorable thermodynamics and conversion to hydroxyapatite above pH 7 [138]. However, as indicated in Section 3, monetite can form in poultry litter and swine manure over a relatively narrow pH domain. Various methods have been implemented to synthesize monetite particles using (1) monocalcium phosphate monohydrate and calcium hydroxide precursors [139], and (2) calcium nitrate tetrahydrate, sodium dihydrogen phosphate, and urea [140]. Production of more stable monetite particles using techniques similar to those described above for hydroxyapatite may prove useful, as this calcium phosphate mineral exhibits better dissolution chemistry, suggesting potential as a slow-release fertilizer.

5.4 Newberyite

Like monetite, newberyite naturally occurs in caves and has been found in bat guano deposits [141]. Newberyite synthesis in solution is a function of supersaturation chemistry, pH, and temperature [142]. Similar to monetite conversion to hydroxyapatite, newberyite transformation to struvite is driven by equilibrium pH and ammonium availability [143]. In general, newberyite is thermodynamically unstable and transforms to struvite at pH greater than 6.5 [143]. This behavior aligns with chemical equilibrium modeling findings from Section 3, which demonstrated that newberyite precipitation occurs within a narrow pH domain of 5.3—5.5. Thus, recovery of newberyite from animal manure is not effective, although continued use of natural newberyite deposits in bat guano is recommended due to the relatively slow release of phosphorus [137].

5.5 Potassium Struvite

In potassium struvite, the ammonium group in struvite is substituted with potassium. Thus, potassium struvite is composed of the three essential soil nutrients (i.e., N, P, and K), and serves as a slow-release fertilizer, allowing sustained delivery of nutrients to crops. Potassium struvite was discovered as a naturally occurring mineral in Binntal, Switzerland and Styria, Austria [144]. Unlike struvite, which has been exhaustively studied due to scale issues in WWTPs, relatively little work has focused on potassium struvite. Wilsenach et al. [145] reported potassium struvite formation from synthetic urine when ammonium concentrations were minimized. Shashvatt et al. [146] reported potassium struvite recovery from potassium-rich poultry litter slurries. In gel crystallization synthesis experiments, the morphology of potassium struvite crystals was influenced by pH [147]. Further exploration of this important mineral is warranted, given the balanced nutrient composition.

CONCLUSIONS

In this chapter, we have identified ecological, public health, and economic drivers for implementation of nutrient recovery operations in agriculture. The need for creating sustainable supplies of phosphorus is paramount in the coming decades due to population growth and diminishing supplies of phosphate rock. The potential for phosphorus recovery from poultry, swine, and dairy cow manure was described in detail using log C-pH diagrams to show the release of critical nutrients at acidic pH and the potential recovery at weakly basic pH. Phosphorus can be recovered at >97% from manure slurries without addition of external chemicals (i.e., ammonium, magnesium, calcium, etc.). The most common recovery products include struvite and hydroxyapatite, although a number of other minerals (e.g., monetite, newberyite, covellite, manganese hydrogen phosphate, and sphalerite) also precipitate under certain conditions. The recovered products have value as agrochemicals for both food crop and animal production operations. We posit that increased emphasis of nutrient recovery from animal waste will protect sensitive water bodies through the following outcomes: decreased eutrophication, hypoxia, and harmful algal blooms; improved fish/shellfish habitat; water security; food security; and political security.

ACKNOWLEDGMENTS

The authors gratefully acknowledge funding from the National Science Foundation (CBET 1511667) and the Maryland Industrial Partnerships Department of Natural Resources Manure-to-Energy program (#5428). We also thank colleagues at Triea Technologies, LLC and Chesapeake Bay farmers for helpful discussions on the complex web of issues involved with nutrient management and recovery.

REFERENCES

[1] EPA. Where nutrient pollution occurs. 2014. http://www2.epa.gov/nutrientpollution/where-nutrient-pollution-occurs.
[2] Troy. Mayor says water crisis is similar to 9/11, The Blade. August 19, 2014. http://www.toledoblade.com/Politics/2014/08/19/Mayor-says-water-crisis-is-similar-to-9-11.html#TxPggzUtPXb5pYu2.99.
[3] Federation WE. Moving towards water resource recovery facilities. Alexandria (VA): A special publication from the Water Environment Federation; 2014.
[4] Ostara. Gwinnett county and Ostara partner to protect precious water resources. October 23, 2015. http://ostara.com/project/gwinnett-county-and-ostara-partner-to-protect-precious-water-resources/.
[5] Ostara. Ostara and EPCOR partner to build nutrient recovery technology facility. September 30, 2014. http://ostara.com/project/ostara-and-epcor-partner-to-build-nutrient-recovery-technology-facility/.
[6] Ostara. Black & Veatch and Ostara to design-build new nutrient recovery system for World's largest water reclamation plant. September 12, 2013. http://ostara.com/project/black-veatch-and-ostara-to-design-build-new-nutrient-recovery-system-for-worlds-largest-water-reclamation-plant/.
[7] Van Epps A, Blaney L. Antibiotic residues in animal waste: occurrence and degradation in conventional agricultural waste management practices. Curr Pollut Rep 2016:1−21.
[8] Oenema O, Oudendag D, Velthof GL. Nutrient losses from manure management in the European Union. Livest Sci 2007;112(3):261−72.
[9] Larney FJ, Hao X. A review of composting as a management alternative for beef cattle feedlot manure in southern Alberta, Canada. Bioresour Technol 2007;98(17):3221−7.
[10] Tiquia SM, Tam NFY. Fate of nitrogen during composting of chicken litter. Environ Pollut 2000;110(3):535−41.
[11] Burton CH, Turner C. Manure management: treatment strategies for sustainable agriculture. Silsoe Research Institute; 2003.
[12] Rehm G, Schmitt M, Munter RC. Fertilizer recommendations for agronomic crops in Minnesota. 1994.
[13] EPA. People, prosperity and the planet—program. 2016. https://www.epa.gov/P3.
[14] Committee on Environmental and Natural Resources. Integrated assessment of hypoxia in the northern Gulf of Mexico. Washington (DC, USA): National Science and Technology Council, Committee on Environmental and Natural Resources; 2000.
[15] Unicef, World Health Organization. Progress on sanitation and drinking water: 2014 update. World Health Organization; 2014.
[16] Vaccari DA. Phosphorus famine: the threat to our food supply. Sci Am June 1, 2009.
[17] Nixon SW. Coastal marine eutrophication: a definition, social causes, and future concerns. Ophelia 1995;41(1):199−219.
[18] Art HW. Dictionary of ecology and environmental science. H. Holt; 1993.
[19] Carpenter SR. Eutrophication of aquatic ecosystems: bistability and soil phosphorus. Proc Natl Acad Sci USA 2005;102(29):10002−5.

[20] Carpenter SR. Phosphorus control is critical to mitigating eutrophication. Proc Natl Acad Sci USA 2008;105(32):11039–40.
[21] Schindler DW, Hecky RE, Findlay DL, Stainton MP, Parker BR, Paterson MJ, Beaty KG, Lyng M, Kasian SEM. Eutrophication of lakes cannot be controlled by reducing nitrogen input: results of a 37-year whole-ecosystem experiment. Proc Natl Acad Sci USA 2008;105(32):11254–8.
[22] Michalak AM, Anderson EJ, Beletsky D, Boland S, Bosch NS, Bridgeman TB, Chaffin JD, Cho K, Confesor R, Daloğlu I, DePinto JV, Evans MA, Fahnenstiel GL, He L, Ho JC, Jenkins L, Johengen TH, Kuo KC, LaPorte E, Liu X, McWilliams MR, Moore MR, Posselt DJ, Richards RP, Scavia D, Steiner AL, Verhamme E, Wright DM, Zagorski MA. Record-setting algal bloom in Lake Erie caused by agricultural and meteorological trends consistent with expected future conditions. Proc Natl Acad Sci USA 2013;110(16):6448–52.
[23] Scavia D, David Allan J, Arend KK, Bartell S, Beletsky D, Bosch NS, Brandt SB, Briland RD, Daloğlu I, DePinto JV, Dolan DM, Evans MA, Farmer TM, Goto D, Han H, Höök TO, Knight R, Ludsin SA, Mason D, Michalak AM, Peter Richards R, Roberts JJ, Rucinski DK, Rutherford E, Schwab DJ, Sesterhenn TM, Zhang H, Zhou Y. Assessing and addressing the re-eutrophication of Lake Erie: central basin hypoxia. J Gt Lakes Res 2014;40(2):226–46.
[24] Stumpf R, Wynne T. Experimental Lake Erie harmful algal bloom bulletin. National Centers for Coastal Ocean Science and Great Lakes Environmental Research Laboratory; November 10, 2015. Bulletin 27.
[25] Broecker WS. Chemical oceanography. Under the general editorship of Kenneth S. Deffeyes. 1974.
[26] Redfield AC. The biological control of chemical factors in the environment. Am Sci 1958;46(3):221A–30A.
[27] McCarthy JJ. Nitrogen. The physiological ecology of phytoplankton. 1980. p. 191–233.
[28] Howarth RW, Marino R, Lane J, Cole JJ. Nitrogen fixation in freshwater, estuarine, and marine ecosystems. 1. Rates and importance. Limnol Oceanogr 1988;33(4part2): 669–87.
[29] Paerl H. Physiological ecology and regulation of N_2 fixation in natural waters. In: Marshall KC, editor. Advances in microbial ecology. (US): Springer; 1990. p. 305–44.
[30] Malone T, Conley D, Fisher T, Glibert P, Harding L, Sellner K. Scales of nutrient-limited phytoplankton productivity in Chesapeake Bay. Estuaries 1996;19(2):371–85.
[31] Howarth RW, Jensen HS, Marino R, Postma H, Tiesen H. Transport to and processing of P in near-shore and oceanic waters. Scope 1995;54:323–45.
[32] Conley D. Biogeochemical nutrient cycles and nutrient management strategies. In: Garnier J, Mouchel J-M, editors. Man and river systems. (Netherlands): Springer; 1999. p. 87–96.
[33] Chave PA. The EU water framework directive: an introduction. IWA Publishing; 2001.
[34] Boesch D. Challenges and opportunities for science in reducing nutrient over-enrichment of coastal ecosystems. Estuaries 2002;25(4):886–900.
[35] Water Science and Technology Board, Ocean Studies Board, Committee on the Causes and Management of Eutrophication, National Research Council, Division on Earth and Life Studies, Commission on Geosciences, Environment and Resources. Clean coastal waters: understanding and reducing the effects of nutrient pollution. National Academies Press; 2000.
[36] Cloern JE, Schraga TS, Lopez CB, Knowles N, Grover Labiosa R, Dugdale R. Climate anomalies generate an exceptional dinoflagellate bloom in San Francisco Bay. Geophys Res Lett 2005;32(14):L14608.

[37] Moore MV, Hampton SE, Izmest'Eva LR, Silow EA, Peshkova EV, Pavlov BK. Climate change and the world's "sacred sea"—Lake Baikal, Siberia. BioScience 2009;59(5):405−17.

[38] Moore SK, Trainer VL, Mantua NJ, Parker MS, Laws EA, Backer LC, Fleming LE. Impacts of climate variability and future climate change on harmful algal blooms and human health. Environ Health 2008;7(2):S4.

[39] Diaz RJ, Rosenberg R. Spreading dead zones and consequences for marine ecosystems. Science 2008;321(5891):926−9.

[40] LUMCON. Louisianna Universities Marine Consortium. http://www.gulfhypoxia.net.

[41] Rabalais NN, Turner RE, Scavia D. Beyond science into policy: Gulf of Mexico hypoxia and the Mississippi river: nutrient policy development for the Mississippi river watershed reflects the accumulated scientific evidence that the increase in nitrogen loading is the primary factor in the worsening of hypoxia in the northern Gulf of Mexico. BioScience 2002;52(2):129−42.

[42] USGS. Streamflow and nutrient delivery to the Gulf of Mexico for October 2014 to May 2015 (preliminary). 2015. http://toxics.usgs.gov/hypoxia/mississippi/oct_jun/.

[43] USDA. USDA economic research service food availability documentation. 2015. http://www.ers.usda.gov/data-products/food-availability-(per-capita)-data-system/food-availability-documentation.aspx on Jan. 14, 2016.

[44] Bricker SB, Longstaff B, Dennison W, Jones A, Boicourt K, Wicks C, Woerner J. Effects of nutrient enrichment in the nation's estuaries: a decade of change. Harmful Algae 2008;8(1):21−32.

[45] Bricker SB, Clement CG, Pirhalla DE, Orlando SP, Farrow DRG. National estuarine eutrophication assessment: effects of nutrient enrichment in the nation's estuaries. 1999.

[46] Kemp WM, Boynton WR, Adolf JE, Boesch DF, Boicourt WC, Brush G, Cornwell JC, Fisher TR, Glibert PM, Hagy JD. Eutrophication of Chesapeake Bay: historical trends and ecological interactions. Mar Ecol Prog Ser 2005;303(21):1−29.

[47] Lake Erie, Lakewide Management Plan. Lake Erie binational nutrient management strategy: protecting Lake Erie by managing phosphorus. 2011. Prepared by the Lake Erie LaMP work group nutrient management task group.

[48] Loadings RP, Blooms HA. A balanced diet for Lake Erie. 2014.

[49] Falkowski P, Scholes RJ, Boyle E, Canadell J, Canfield D, Elser J, Gruber N, Hibbard K, Högberg P, Linder S, Mackenzie FT, Moore Iii B, Pedersen T, Rosenthal Y, Seitzinger S, Smetacek V, Steffen W. The global carbon cycle: a test of our knowledge of earth as a system. Science 2000;290(5490):291−6.

[50] Parsons SA, Doyle JD. Struvite scale formation and control. Water Sci Technol 2004;49(2):177−82.

[51] Bennett EM, Carpenter SR, Caraco NF. Human impact on erodable phosphorus and eutrophication: a global perspective: increasing accumulation of phosphorus in soil threatens rivers, lakes, and coastal oceans with eutrophication. BioScience 2001;51(3):227−34.

[52] Ma Y, Li J, Li X, Tang X, Liang Y, Huang S, Wang B, Liu H, Yang X. Phosphorus accumulation and depletion in soils in wheat−maize cropping systems: modeling and validation. Field Crops Res 2009;110(3):207−12.

[53] Motavalli P, Miles R. Soil phosphorus fractions after 111 years of animal manure and fertilizer applications. Biol Fertil Soils 2002;36(1):35−42.

[54] Yuan Y, Locke MA, Bingner RL, Rebich RA. Phosphorus losses from agricultural watersheds in the Mississippi Delta. J Environ Manag 2013;115:14−20.

[55] Watson CJ, Smith RV, Matthews DI. Increase in Phosphorus Losses from Grassland in Response to Olsen-P Accumulation. J Environ Qual 2007;36.

[56] Simard RR, Beauchemin S, Haygarth PM. Potential for preferential pathways of phosphorus transport. J Environ Qual 2000;29.
[57] USGS. Phosphate rock: statistics and information. Mineral commodity summaries. US Geological Survey; January 2016.
[58] Institute TF. Commercial fertilizers annual data, 2002−2011. The Association of American Plant Food Control Officials, Inc.; 2012.
[59] USDA. Broiler litter: waste or resource? USDA agricultural research service.
[60] Staver KW, Brinsfield RB. Agriculture and water quality on the Maryland eastern shore: where do we go from here? Long-term solutions to accelerated eutrophication must provide mechanisms for redistributing nutrients flowing into concentrated animal-producing regions. BioScience 2001;51(10):859−68.
[61] Sharpley A, Moyer B. Phosphorus forms in manure and compost and their release during simulated rainfall. J Environ Qual 2000;29.
[62] Hatfield JL, Brumm MC, Melvin SW. Swine manure management. In: Agricultural uses of municipal, animal, and industrial byproducts, 44; 1998. p. 78−90.
[63] Mallin MA, McIver MR, Robuck AR, Dickens AK. Industrial swine and poultry production causes chronic nutrient and fecal microbial stream pollution. Water Air Soil Pollut 2015;226(12):1−13.
[64] Heaney CD, Myers K, Wing S, Hall D, Baron D, Stewart JR. Source tracking swine fecal waste in surface water proximal to swine concentrated animal feeding operations. Sci Total Environ 2015;511:676−83.
[65] Nicole W. CAFOs and environmental justice: the case of North Carolina. Environ Health Perspect 2013;121(6):a182−9.
[66] Anderson DM, Burkholder JM, Cochlan WP, Glibert PM, Gobler CJ, Heil CA, Kudela RM, Parsons ML, Rensel JEJ, Townsend DW, Trainer VL, Vargo GA. Harmful algal blooms and eutrophication: examining linkages from selected coastal regions of the United States. Harmful Algae 2008;8(1):39−53.
[67] Anderson DM, Kulis DM, Doucette GJ, Gallagher JC, Balech E. Biogeography of toxic dinoflagellates in the genus *Alexandrium* from the northeastern United States and Canada. Mar Biol 1994;120(3):467−78.
[68] Trainer VL, Adams NG, Bill BD, Stehr CM, Wekell JC, Moeller P, Busman M, Woodruff D. Domoic acid production near California coastal upwelling zones, June 1998. Limnol Oceanogr 2000;45(8):1818−33.
[69] Lewitus AJ, Brock LM, Burke MK, DeMattio KA, Wilde SB. Lagoonal stormwater detention ponds as promoters of harmful algal blooms and eutrophication along the South Carolina coast. Harmful Algae 2008;8(1):60−5.
[70] Marshall HG, Burchardt L, Lacouture R. A review of phytoplankton composition within Chesapeake Bay and its tidal estuaries. J Plankton Res 2005;27(11):1083−102.
[71] The United Nations World Water. The United Nations world water development report 3−Water in a changing world. UNESCO Publishing/Earthscan; 2009.
[72] Liu W, Qiu R. Water eutrophication in China and the combating strategies. J Chem Technol Biotechnol 2007;82(9):781−6.
[73] Zhang L, Xia M, Zhang L, Wang C, Lu J. Eutrophication status and control strategy of Taihu Lake. Front Environ Sci Eng China 2008;2(3):280−90.
[74] Jiang S, Yuan Z. Phosphorus flow patterns in the Chaohu watershed from 1978 to 2012. Environ Sci Technol 2015;49(24):13973−82.
[75] Anderson D, Glibert P, Burkholder J. Harmful algal blooms and eutrophication: nutrient sources, composition, and consequences. Estuaries 2002;25(4):704−26.
[76] Zhang J. Atmospheric wet deposition of nutrient elements: correlation with harmful biological blooms in northwest Pacific coastal zones. Ambio 1994;23(8):464−8.
[77] Omair M. In: Ann Arbor MI, editor. Biological and chemical characteristics of the river Ganga (Ganges). MPublishing, University of Michigan Library; 2005.

[78] Sawant SS, Prabhudessai L, Venkat K. Eutrophication status of marine environment of Mumbai and Jawaharlal Nehru ports. Environ Monit Assess 2006;127(1):283–91.
[79] Vass KK. On the trophic status and conservation of Kashmir Lakes. Hydrobiologia 1980;68(1):9–15.
[80] Martin GD, Nisha PA, Balachandran KK, Madhu NV, Nair M, Shaiju P, Joseph T, Srinivas K, Gupta GVM. Eutrophication induced changes in benthic community structure of a flow-restricted tropical estuary (Cochin backwaters), India. Environ Monit Assess 2010;176(1):427–38.
[81] Nayak BK, Acharya BC, Panda UC, Nayak BB, Acharya SK. Variation of water quality in Chilika Lake, Orissa. Indian J Mar Sci 2004;33(2):164–9.
[82] Van Kauwenbergh SJ. World phosphate rock reserves and resources. IFDC; 2010.
[83] McCutcheon S. Phosphorus and Morocco – the future. Morocco news board. November 4, 2010. http://www.moroccoboard.com.
[84] Lifton J. Feeding the world's hunger for phosphorus. Resource investor. June 3, 2011. http://www.resourceinvestor.com.
[85] Buckley G. Fertilizer requirements in 2015 and 2030. FAO; 2000.
[86] Isherwood KF, International Fertilizer Industry Association and United Nations Environment Programme. Mineral Fertilizer Use and the Environment. IFA; 1998.
[87] Cordell D, White S. Peak phosphorus: clarifying the key issues of a vigorous debate about long-term phosphorus security. Sustainability 2011;3(10):2027–49.
[88] Bank W. GEM commodities – rock phosphate. 2014. http://www.indexmundi.com/commodities/?commodity=rock-phosphate&months=240.
[89] Conservancy N. The Gulf "dead zone" threatens seafood production, recreation and marine life. 2014. http://www.nature.org/ourinitiatives/regions/northamerica/areas/gulfofmexico/explore/gulf-of-mexico-dead-zone.xml.
[90] Larkin SL, Adams CM. Harmful algal blooms and coastal business: economic consequences in Florida. Soc Nat Resour 2007;20(9):849–59.
[91] Jin D, Hoagland P. The value of harmful algal bloom predictions to the nearshore commercial shellfish fishery in the Gulf of Maine. Harmful Algae 2008;7(6):772–81.
[92] Dodds WK, Bouska WW, Eitzmann JL, Pilger TJ, Pitts KL, Riley AJ, Schloesser JT, Thornbrugh DJ. Eutrophication of US freshwaters: analysis of potential economic damages. Environ Sci Technol 2009;43(1):12–9.
[93] Wfd, E.U. Directive 2000/60/EC of the European parliament and of the council establishing a framework for the community action in the field of water policy. Joint text approved by the conciliation committee provided for in article 251. 2000.
[94] EPA. Reassessment 2013: assessing progress made since 2008. Mississippi River, Gulf of Mexico watershed nutrient task force. 2015.
[95] EPA. Chesapeake Bay total maximum daily load for nitrogen, phosphorus and sediment. 2010.
[96] Phosphorus management tool. Annapolis (MD): Department of Agriculture; June 8, 2015.
[97] Cox J. Phosphorus management tool: what you need to know, in Delmarva now. 2015.
[98] Dance S. Excessive phosphorus levels measured on 18 percent of Maryland farmland. Baltim Sun March 14, 2016.
[99] A scenario analysis of the potential costs of implementing the phosphorus management tool on the eastern shore of Maryland. Salisbury University, Business, Economic and Community Network; 2014.
[100] Zublena JP, Barker JC, Carter TA. Poultry manure as a fertilizer source. North Carolina extension service. Publication AG-439-5. 1993.

[101] Chastain JP, Camberato JJ. Dairy manure production and nutrient content. Confined animal manure manager certification program manual dairy version. 2004. p. 1–16.

[102] Chastain JP, Camberato JJ, Albrecht JE, Adams J. Swine manure production and nutrient content. South Carolina confined animal manure managers certification program. Clemson University, SC; 1999. p. 1–17.

[103] Bertramson BR. Phosphorus analysis of plant material. Plant Physiosl 1942;17(3):447.

[104] Bartlett GR. Phosphorus assay in column chromatography. J Biol Chem 1959;234(3): 466–8.

[105] Pierzynski GM. Methods of phosphorus analysis for soils, sediments, residuals, and waters. North Carolina State University Raleigh; 2000.

[106] Szögi AA, Vanotti MB, Hunt PG. Phosphorus recovery from poultry litter. Trans ASABE 2008;51:1727–34.

[107] Bittman S, Jones K, Vingarzan R, Hunt DE, Sheppard SC, Tait J, So R, Zhao J. Weekly agricultural emissions and ambient concentrations of ammonia: validation of an emission inventory. Atmos Environ 2015;113:108–17.

[108] Li C, Salas W, Zhang R, Krauter C, Rotz A, Mitloehner F. Manure-DNDC: a biogeochemical process model for quantifying greenhouse gas and ammonia emissions from livestock manure systems. Nutr Cycl Agroecosyst 2012;93(2):163–200.

[109] Sánchez-García M, Alburquerque JA, Sánchez-Monedero MA, Roig A, Cayuela ML. Biochar accelerates organic matter degradation and enhances N mineralisation during composting of poultry manure without a relevant impact on gas emissions. Bioresour Technol 2015;192:272–9.

[110] Chowdhury MA, de Neergaard A, Jensen LS. Potential of aeration flow rate and biochar addition to reduce greenhouse gas and ammonia emissions during manure composting. Chemosphere 2014;97:16–25.

[111] Rothrock Jr MJ, Szögi AA, Vanotti MB. Recovery of ammonia from poultry litter using flat gas permeable membranes. Waste Manag 2013;33(6):1531–8.

[112] Romero-Güiza MS, Astals S, Chimenos JM, Martínez M, Mata-Alvarez J. Improving anaerobic digestion of pig manure by adding in the same reactor a stabilizing agent formulated with low-grade magnesium oxide. Biomass Bioenergy 2014;67(0):243–51.

[113] Zhang T, Bowers KE, Harrison JH, Chen S. Releasing phosphorus from calcium for struvite fertilizer production from anaerobically digested dairy effluent. Water Environ Res 2010;82(1):34–42.

[114] Uludag-Demirer S, Demirer GN, Frear C, Chen S. Anaerobic digestion of dairy manure with enhanced ammonia removal. J Environ Manag 2008;86(1):193–200.

[115] Uludag-Demirer S, Demirer GN, Chen S. Ammonia removal from anaerobically digested dairy manure by struvite precipitation. Process Biochem 2005;40(12): 3667–74.

[116] Pagliari PH, Laboski CAM. Investigation of the inorganic and organic phosphorus forms in animal manure. J Environ Qual 2012;41.

[117] Nieminen J. Phosphorus recovery and recycling from municipal wastewater sludge. Vesitalous 2010;3:15–7.

[118] Taylor AW, Frazier AW, Gurney EL. Solubility products of magnesium ammonium and magnesium potassium phosphates. Trans Faraday Soc 1963;59(0):1580–4.

[119] Hutchinson GE. The biogeochemistry of vertebrate excretion. Soil Sci 1950;70(4): 32s9.

[120] Borgerding J. Phosphate deposits in digestion systems. J Water Pollut Control Federation 1972;44(5):813–9.

[121] Bhattarai KK, Taiganides EP, Yap BC. Struvite deposits in pipes and aerators. Biol Wastes 1989;30(2):133–47.

[122] Rahman MM, Liu Y, Kwag J-H, Ra C. Recovery of struvite from animal wastewater and its nutrient leaching loss in soil. J Hazard Mater 2011;186(2):2026–30.

[123] Bouropoulos NC, Koutsoukos PG. Spontaneous precipitation of struvite from aqueous solutions. J Cryst Growth 2000;213(3):381–8.
[124] Ohlinger KN, Young TM, Schroeder ED. Predicting struvite formation in digestion. Water Res 1998;32(12):3607–14.
[125] Ronteltap M, Maurer M, Hausherr R, Gujer W. Struvite precipitation from urine—influencing factors on particle size. Water Res 2010;44(6):2038–46.
[126] Stratful I, Scrimshaw MD, Lester JN. Conditions influencing the precipitation of magnesium ammonium phosphate. Water Res 2001;35(17):4191–9.
[127] Britton A, Koch FA, Mavinic DS, Adnan A, Oldham WK, Udala B. Pilot-scale struvite recovery from anaerobic digester supernatant at an enhanced biological phosphorus removal wastewater treatment plant. J Environ Eng Sci 2005;4(4):265–77.
[128] Le Corre KS, Valsami-Jones E, Hobbs P, Parsons SA. Impact of calcium on struvite crystal size, shape and purity. J Cryst Growth 2005;283(3):514–22.
[129] Platform ESP. Proposed EU fertiliser regulation criteria for recovered struvite. 2015.
[130] Xuechu C, Hainan K, Deyi WU, Xinze W, Yongyong LIN. Phosphate removal and recovery through crystallization of hydroxyapatite using xonotlite as seed crystal. J Environ Sci 2009;21(5):575–80.
[131] Song Y-h, Donnert D, Ute B, Weidler PG, Nueesch R. Seed selections for crystallization of calcium phosphate for phosphorus recovery. J Environ Sci 2007; 19(5):591–5.
[132] Song Y, Weidler PG, Berg U, Nüesch R, Donnert D. Calcite-seeded crystallization of calcium phosphate for phosphorus recovery. Chemosphere 2006;63(2):236–43.
[133] Moriyama K, Kojima T, Minawa Y, Matsumoto S, Nakamachi K. Development of artificial seed crystal for crystallization of calcium phosphate. Environ Technol 2001; 22(11):1245–52.
[134] Liu R, Lal R. Synthetic apatite nanoparticles as a phosphorus fertilizer for soybean (glycine max). Sci Rep 2014;4.
[135] Kottegoda N, Munaweera I, Madusanka N, Karunaratne V. A green slow-release fertilizer composition based on urea-modified hydroxyapatite nanoparticles encapsulated wood. Curr Sci 2011;101(1):73–8.
[136] Frost RL, Xi Y, Millar G, Tan K, Palmer SJ. Vibrational spectroscopy of natural cave mineral monetite $CaHPO_4$ and the synthetic analog. Spectrosc Lett 2013;46(1): 54–9.
[137] Hadas A, Rosenberg R. Guano as a nitrogen source for fertigation in organic farming. Fertil Res 1992;31(2):209–14.
[138] Da Silva MHP, Lima JHC, Soares GA, Elias CN, De Andrade MC, Best SM, Gibson IR. Transformation of monetite to hydroxyapatite in bioactive coatings on titanium. Surf Coat Technol 2001;137(2):270–6.
[139] Jinawath S, Pongkao D, Suchanek W, Yoshimura M. Hydrothermal synthesis of monetite and hydroxyapatite from monocalcium phosphate monohydrate. Int J Inorg Mater 2001;3(7):997–1001.
[140] Jokić B, Mitrić M, Radmilović V, Drmanić S, Petrović R, Janaćković D. Synthesis and characterization of monetite and hydroxyapatite whiskers obtained by a hydrothermal method. Ceram Int 2011;37(1):167–73.
[141] Frost RL, Palmer SJ, Pogson RE. Thermal stability of newberyite Mg $(PO_3OH) \cdot 3H_2O$: a cave mineral from Skipton Lava Tubes, Victoria, Australia. J Therm Anal Calorim 2011;107(3):1143–6.
[142] Babić-Ivančić V, Kontrec J, Brečević L, Kralj D. Kinetics of struvite to newberyite transformation in the precipitation system $MgCl_2–NH_4\ H_2\ PO_4–NaOH–H_2O$. Water Res 2006;40(18):3447–55.
[143] Kontrec J, Babić-Ivančićand V, Brečević L. Formation and morphology of struvite and newberyite in aqueous solutions at 25 and 37°C. Coll Antropol 2005;29(1):289–94.

[144] Graeser S, Postl W, Bojar H-P, Berlepsch P, Armbruster T, Raber T, Ettinger K, Walter F. Struvite-(K), $KMgPO_4 \cdot 6H_2O$, the potassium equivalent of struvite—a new mineral. Eur J Mineral 2008;20(4):629—33.

[145] Wilsenach JA, Schuurbiers CAH, van Loosdrecht MCM. Phosphate and potassium recovery from source separated urine through struvite precipitation. Water Res 2007;41(2):458—66.

[146] Shashvatt U, Rogers N, Aris H, Blaney L. Recovering phosphorus from poultry litter: a step towards improving food security and protecting ecologically sensitive water bodies. In: 251st American chemical society annual meeting (San Diego, CA); March 13—17, 2016. p. 2016.

[147] Suryawanshi VB, Chaudhari RT. Synthesis and characterization of struvite-k crystals by agar gel. J Cryst Process Technol 2014;4(04):212.

[148] Penn State College of Agricultural Sciences. The agronomy guide. Penn State College of Agricultural Sciences; 2015.

[149] Rosen CJ, Bierman PM. Nutrient management for fruit and vegetable crop production. Univ. of Minnesota Ext. M; 2005. p. 1192.

[150] Nass U. Census of agriculture. Washington (DC): US Department of Agriculture, National Agricultural Statistics Service; 2007.

[151] Ruddy BC, Lorenz DL, Mueller DK. County-level estimates of nutrient inputs to the land surface of the conterminous United States, 1982—2001. US Department of the Interior, US Geological Survey; 2006.

[152] USDA. Farms and land in farms final estimates 2003—2007. National Agricultural Statistics Service; 2009.

CHAPTER FOURTEEN

Effect of Upflow Velocity on Nutrient Recovery from Swine Wastewater by Fluidized Bed Struvite Crystallization

R.R.M. Abarca[1], R.S. Pusta, Jr. [1], R.B. Labad[1], J.L.A. Andit[1], C.M. Rejas[1], M.D.G. de Luna[2], M.-C. Lu[3]
[1]Mindanao State University-Iligan Institute of Technology, Iligan City, Philippines
[2]University of the Philippines, Diliman, Quezon City, Philippines
[3]Chia-Nan University of Pharmacy and Science, Tainan, Taiwan

1. INTRODUCTION

Numerous large-scale swine breeding farms have been developed to satisfy the growing demand for meat. The increase in demand has resulted in increase in the amount of wastewater produced with substantial amount of phosphorus and nitrogen. These nutrients become a major cause of eutrophication to receiving bodies of water if disposed without prior treatment, making the swine breeding industry a major source of pollution around livestock farms [1]. Legislations on water quality have been passed concerning the reduction of nutrient level entering the surface waters [2]. Hence, removal of phosphate from swine wastewater is important in maintaining the water quality and lessens its impact to the environment. On the other hand, the main source of phosphate is mined as phosphate rock or phosphorite. The global supply is expected to become increasingly scarce and expensive due to its nonrenewability and the unavailability of high-purity phosphate rock in the future. In the next 50−100 years, the supply of this resource may become a serious problem as extrapolated through the current consumption of phosphate fertilizer. This situation urges researchers to investigate not only on phosphate removal, but more importantly on the recovery of phosphate from waste sources [3,4].

Technologies for nutrient removal and recovery mainly include biological, physical, and chemical precipitation and crystallization [5]. While relatively inexpensive, biological methods normally do not have removal efficiencies exceeding 30%. Physical methods, though efficient, are usually too expensive.

Chemistry and Water
ISBN 978-0-12-809330-6
http://dx.doi.org/10.1016/B978-0-12-809330-6.00014-3

Chemical precipitation and crystallization are generally highly efficient in treating wastewater with high concentrations of nutrients, but comparing the chemical precipitation and fluidized bed crystallization (FBC), FBC remain superior when it comes to efficiency and by-product recyclability [6,7]. Of the chemical processes that can be employed in a fluidized bed crystallizer, struvite crystallization is ideal as it can simultaneously remove P and N [8].

Struvite is a crystalline substance of magnesium, ammonium, and phosphate in equal molar concentrations ($MgNH_4PO_4 \cdot 6H_2O$). It is a slow-release fertilizer and has low heavy metal content compared to phosphate rock [9]. Struvite precipitation occurs according to Eq. (14.1) and under alkaline conditions.

$$Mg^{2+} + NH_4{}^{+} + HPO_4{}^{2-} + OH^{-} + 6H_2O \rightarrow MgNH_4PO_4 \cdot 6H_2O + H_2O \tag{14.1}$$

The recovered product from this process can be used as a substitute for phosphate mineral used in commercial application or as a slow-release agricultural fertilizer [10].

Many studies have been conducted and produced good results in removing nutrients via struvite crystallization in fluidized bed reactors [9,11–13]. Various parameters for struvite crystallization in fluidized bed reactors have been studied in order to have optimum removal of phosphorus [9]. Mixing energy or turbulence, is one of the parameters that affect not only the removal of phosphorus, but also its economics. Fluidized bed operation requires high flow rates or mixing energy to ensure that the bed is continuously fluidized [14].

However, there are only limited studies conducted on the effect of mixing energy in the crystallization of struvite in a fluidized bed. In this study, the effect of mixing energy, particularly by varying the upflow velocity in a fluidized bed for struvite crystallization from synthetic swine wastewater, was investigated.

2. EXPERIMENTAL

2.1 Seed Synthesis

Phosphate solution was prepared from KH_2PO_4 (Sharlau, 99.5%) and NH_4Cl (Sharlau, 99.5%), and the precipitant solution from $MgCl_2 \cdot 6H_2O$ (Sharlau, 99%) using deionized water. Struvite seed crystals were synthesized using semibatch crystallization process (Fig. 14.1) [15]. The synthesis was

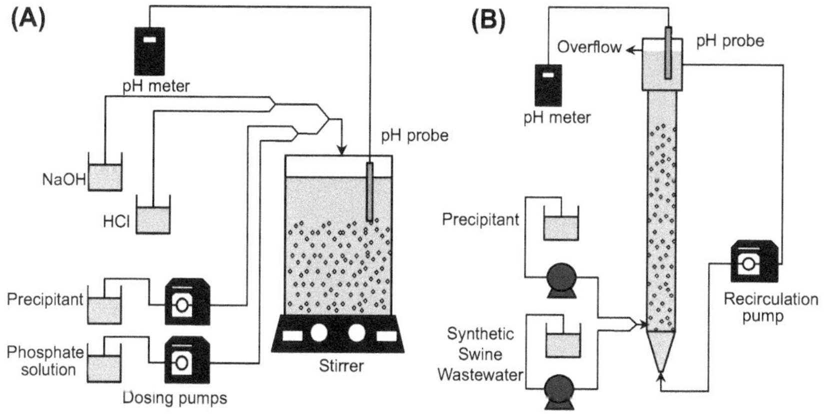

Figure 14.1 Experimental setup for (A) seed synthesis and (B) fluidized bed crystallization.

carried out using a 7-L glass reactor with a magnetic stirrer (PC-420D, Corning). The pH of the solution was monitored using a pH meter (Ultra-BASIC UB-5, Denver Instrument) introduced in the reactor and adjusted by addition of 1.0 M NaOH (RCI Labscan, 99.0%) or 1.0 M HCl (RCI Labscan, 37.0%). The seed synthesis was carried out by mixing 0.370 M of phosphate solution 0.370 M $MgCl_2$ and 0.555 M NH_4Cl. Both phosphate solution and precipitant were introduced to the reactor using peristaltic pumps (BT600-2J, Longer Pump).

2.2 Fluidized Bed Crystallization

A 0.48-L fluidized bed reactor with inlet, outlet, and recirculating sections was used as a struvite crystallizer. The fluidized bed reactor column was equipped with a pH meter and a peristaltic pump for recirculation.

The reactor was filled with distilled water prior to the addition of the struvite seed crystals. The pH of the precipitants and phosphate solutions were adjusted to the desired value before dosing at 6.0 mL/min. Simulated swine wastewater, together with the precipitant, was injected into the column. The system operated at ambient temperature of 25°C. The upflow velocity was varied from 20 to 50 cm/min. Samples were collected from the recirculation zone. Filtered and unfiltered samples were mixed with 30 μL of concentrated HNO_3 prior to phosphate analysis.

2.3 Analytical Methods

Spectrophotometric phosphate analysis was conducted using a UV-vis spectrophotometer (Genesys 20, Thermo Spectronic). Inorganic phosphate ion

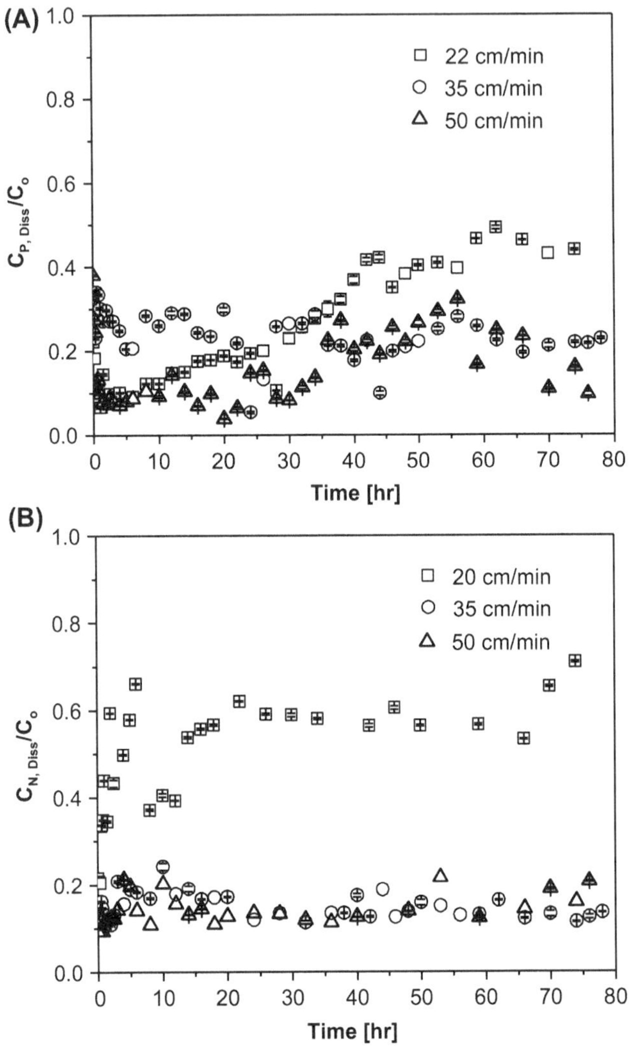

Figure 14.2 Phosphate concentration (A) and ammonium concentration (B) relative to influent concentration in the recirculation zone at different upflow velocity.

concentration was determined using molybdenum blue method scanned at 698 nm [16], while ammonium ion concentration was determined by forming an ammonium phenolate complex (indophenol blue method) at 640 [17].

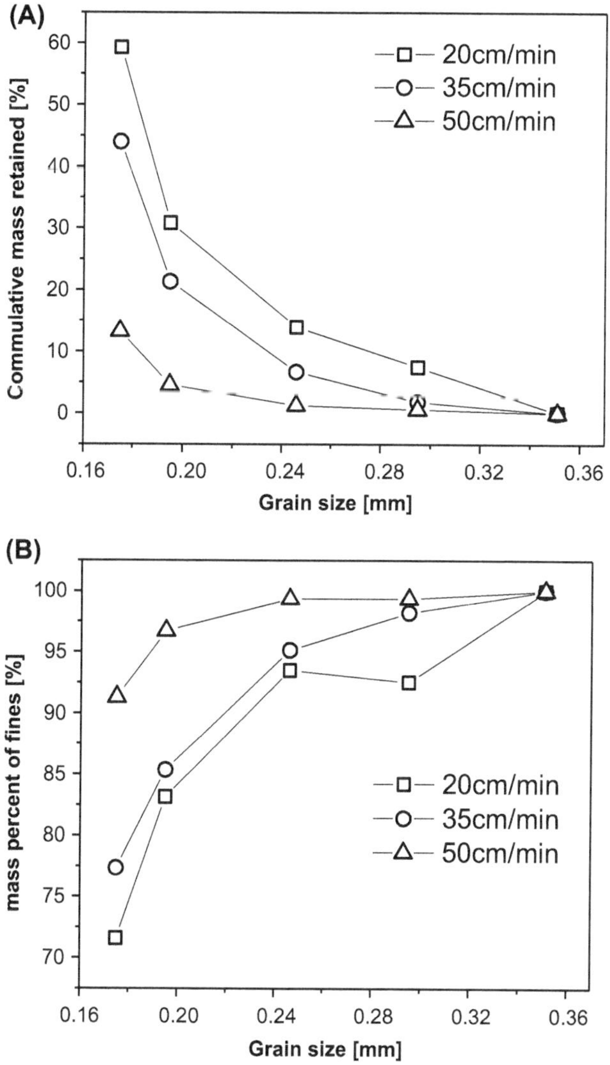

Figure 14.3 Sieve analysis of the collected solids at different upflow velocity: (A) cumulative mass retained versus particle size; (B) grain size analysis reported as percent passing versus particle size.

3. RESULTS AND DISCUSSION

3.1 Phosphate Removal and Conversion

Fig. 14.2 shows variation of dissolved phosphate and dissolved ammonium concentration relative to the initial phosphate and ammonium concentration for different upflow velocities. The phosphate and ammonium concentration were steady at the beginning of the process, and as the concentration increased with time, relative concentration dropped, implying that struvite crystallization has commenced.

The tested upflow velocities were observed to have a significant trend of phosphate recovery, up to 95% from the initial phosphate concentration occurring at upflow velocity of 35 cm/min. However, further increased in upflow velocity decreased the phosphate recovery to 94%. The same observation was noticed by Saidou et al. [18] in their study on struvite precipitation with varying mixing energy; phosphorus removal improved to a certain maximum degree of mixing but further increase would decrease removal.

Ammonium recovery increased with increasing upflow velocity. The trend may be explained by the fact that increase in upflow velocity will enhance transport of ammonium ions to struvite crystals, and a higher concentration of ammonium ions will shift the reaction to the formation of more struvite crystals [19].

3.2 Sieve Analysis

The sizes of struvite crystals were analyzed using Tyler Screen Scale, and Fig. 14.3 shows the sizes distribution of struvite crystals. In all levels of upflow velocities, considerable amounts of small crystals are obtained. However, the relative amount of larger crystals decreased with increase in the upflow velocity. Crystal size was affected by the increased in mixing energy or turbulence (by increasing the upflow velocities). Increase in turbulence will increase the liquid shear stress, which could disrupt crystal growth, promote discrete precipitation, affect the transport of struvite constituent ions, and increase crystal breakage [18,20].

CONCLUSIONS

FBC was used to recover phosphate and ammonium ions with struvite crystallization. The variation of the phosphate and ammonium concentration with time and the mass distribution of crystal with crystal sizes were investigated at different upflow velocities. For all tested upflow velocities, significant phosphate and ammonium recovery was achieved. It was observed that an increase in upflow velocity to 35 cm/min would increase phosphate recovery, but further increase would decrease the recovery. Ammonium recovery on the other hand, increased with increasing upflow velocity. Smaller crystals are of higher quantity than bigger crystals for all levels; however, the relative amount of small crystals is high for higher upflow velocities. Therefore, nucleation is more favored than crystal growth within the range of the tested upflow velocities.

REFERENCES

[1] Tong L, Li P, Wang Y, Zhu K. Analysis of veterinary antibiotic residues in swine wastewater and environmental water samples using optimized SPE-LC/MS/MS. Chemosphere 2009;74:1090–7. http://dx.doi.org/10.1016/j.chemosphere.2008.10.051.

[2] Liu Y, Kwag J-H, Kim J-H, Ra C. Recovery of nitrogen and phosphorus by struvite crystallization from swine wastewater. Desalination 2011;277:364–9. http://dx.doi.org/10.1016/j.desal.2011.04.056.

[3] Chen X, Kong H, Wu D, Wang X, Lin Y. Phosphate removal and recovery through crystallization of hydroxyapatite using xonotlite as seed crystal. J Environ Sci 2009;21: 575–80. http://dx.doi.org/10.1016/S1001-0742(08)62310-4.

[4] Cordell D, Rosemarin A, Schröder JJ, Smit AL. Towards global phosphorus security: a systems framework for phosphorus recovery and reuse options. Chemosphere 2011;84: 747–58. http://dx.doi.org/10.1016/j.chemosphere.2011.02.032.

[5] Zhang J, Shen Z, Mei Z, Li S, Wang W. Removal of phosphate by Fe-coordinated amino-functionalized 3D mesoporous silicates hybrid materials. J Environ Sci 2011; 23:199–205. http://dx.doi.org/10.1016/S1001-0742(10)60393-2.

[6] Morse G, Brett S, Guy J, Lester J. Review: phosphorus removal and recovery technologies. Sci Total Environ 1998;212:69–81. http://dx.doi.org/10.1016/S0048-9697(97)00332-X.

[7] Zhao Y, Xi B, Li Y, Wang M, Zhu Z, Xia X, et al. Removal of phosphate from wastewater by using open gradient superconducting magnetic separation as pretreatment for high gradient superconducting magnetic separation. Sep Purif Technol 2012;86: 255–61. http://dx.doi.org/10.1016/j.seppur.2011.11.014.

[8] Piekema P, Giesen A. Phospahte recovery by the crystallization process : experience and developments. In: Second International Conference on the Recover of Phosphorus from Sewage and Animal Wastes, Noordwijkerhout, The Netherlands; 2001.

[9] Forrest AL, Fattah KP, Mavinic DS, Koch FA. Optimizing struvite production for phosphate recovery in WWTP 2008;134:395–402.

[10] Wang J, Song Y, Yuan P, Peng J, Fan M. Modeling the crystallization of magnesium ammonium phosphate for phosphorus recovery. Chemosphere 2006;65:1182–7. http://dx.doi.org/10.1016/j.chemosphere.2006.03.062.

[11] de Luna MDG, Abarca RRM, Su C-C, Huang Y-H, Lu M-C. Multivariate optimization of phosphate removal and recovery from aqueous solution by struvite crystallization in a fluidized-bed reactor. Desalin Water Treat 2014:1–10. http://dx.doi.org/10.1080/19443994.2014.915584.

[12] Battistoni P, Angelis ADE, Pavan P, Prisciandaro M, Cecchi F. Phosphorus removal from a real anaerobic supernatant by struvite crystallization. Water Res 2001;35: 2167–78.

[13] Bhuiyan IMH, Mavinic D. Assessing struvite precipitation in a pilot scale fluidized bed crystallizer. Environ Sci Technol 2008;29:1157–67. http://dx.doi.org/10.1080/09593330802075452.

[14] Battistoni P, Paci B, Fatone F, Pavan P. Phosphorus removal from supernatants at low concentration using packed and fluidized-bed reactors. Ind Eng Chem Res 2005;44: 6701–7. http://dx.doi.org/10.1021/ie050186g.

[15] Su C-C, Abarca RRM, de Luna MDG, Lu M-C. Phosphate recovery from fluidized-bed wastewater by struvite crystallization technology. J Taiwan Inst Chem Eng 2014;45:2395–402. http://dx.doi.org/10.1016/j.jtice.2014.04.002.

[16] He Z, Honeycutt CW. A modified molybdenum blue method for orthophosphate determination suitable for investigating enzymatic hydrolysis of organic phosphates. Commun Soil Sci Plant Anal 2005;36:1373–83. http://dx.doi.org/10.1081/CSS-200056954.

[17] Afkhami A, Norooz-Asl R. Micelle-mediated extraction and spectrophotometric determination of ammonia in water samples utilizing indophenol dye formation 2008;19:1546–52.

[18] Saidou H, Korchef A, Ben Moussa S, Ben Amor M. Struvite precipitation by the dissolved CO_2 degasification technique: impact of the airflow rate and pH. Chemosphere 2009;74:338–43. http://dx.doi.org/10.1016/j.chemosphere.2008.09.081.

[19] de Luna MDG, Abarca RRM, Huang Y, Lu M. Phosphate recovery by crystallization process using magnesium ammonium phosphate crystals as seed material. In: International Proceedings of Chemical, Biological and Environmental Engineering70; 2014. p. 152–6. http://dx.doi.org/10.7763/IPCBEE.

[20] Stratful I, Scrimshaw M, Lester J. Conditions influencing the precipitation of magnesium ammonium phosphate. Water Res 2001;35:4191–9. http://dx.doi.org/10.1016/S0043-1354(01)00143-9.

CHAPTER FIFTEEN

Drought-Inspired Economic Use of Water in Wine Production

J.N. Seiber[1], K.A. Gibson[1], V. Tiangco[2], J. Real[2]
[1]University of California, Davis, CA, United States
[2]RST Cellars, Davis, CA, United States

1. INTRODUCTION

The drought in California, in its fourth year as of 2015, is having major impacts on users of water at all levels. Irrigated agriculture uses approximately 40% of applied water in the state, natural resources (stream flow, rivers, managed wetlands, and required delta outflow to the Pacific) use about 50%, and urban areas use about 10% [1]. Applied water refers to water delivered by an application to users including agriculture, urban settings, wetlands, and industry [2]. These allocations do not reflect actual usage. For example, the 41% to irrigated agriculture includes water that is used to produce food for consumption and also provide habitat for a number of wildlife species, including fish and waterfowl.

This chapter focuses largely on California wine-grape and wine-production issues. However, the drought is also affecting all areas associated with the Colorado River, from the Rockies to Southern Arizona and parts of northern Mexico and western Canada [3].

California is the state with the largest population in the United States (38,802,500 people estimated for 2014 according to the US Census Bureau) [4], the largest agricultural industry (farm gate value of $21.24 billion in 2013), and a large urban/suburban population. It is also the spawning ground for fish species, such as Pacific salmon, and provides a habitat for many bird species.

The California wine industry is large and growing. Wine grapes were the fourth valued agricultural and food commodity at 3.2 billion dollars, after milk, almonds, and all grapes in 2013 [5,6]. As with all commodities produced in California, the wine industry relies heavily on water from surface sources, private wells, rain, snowfall, and other irrigation and water sources. It takes on average 109 L of water to produce a 0.25-L serving

Chemistry and Water
ISBN 978-0-12-809330-6
http://dx.doi.org/10.1016/B978-0-12-809330-6.00015-5

Table 15.1 Liters of Water Required/250 mL Product Produced [8–10]

Common Beverages	Amount of Water (L)
Milk	255
Wine	109
Soft drinks	85–155
Beer	74
Tea	27
Bottled water	0.33

of wine [7]. Table 15.1 shows the volume of water used to produce common beverages.

In 2015, there was a water shortage of 2.7 million acre feet in California [11]. Some California state agencies and water districts were prompted to impose mandatory reductions in water usage. In 2015, Governor Jerry Brown issued the fourth in a series of executive orders mandating water reduction in urban jurisdictions across California to in order to reduce potable water usage by 25% statewide [12]. In 2008, the governor also ordered improvements in irrigation efficiency, leak detection, and other measures as appropriate [13]. Many growers in California have been forced to leave some acreage unplanted. In 2014, an estimated 428,000 acres were taken out of production [14]. Agricultural water suppliers are now required to file yearly drought management plans that describe the actions taken to manage and lessen water use. This reduction in irrigation based on water availability underscores the severity of the restrictions on irrigation water use [15].

In 2014 the California Sustainable Groundwater Management Act (SGMA) was implemented, which requires the adoption of a groundwater sustainability plan for all high or medium priority basins that provide more than half of the water Californians use in dry years [16]. In 2014, California voters passed Proposition 1, a $7.12 billion water bond that includes provision to increase the state's water storage capacity [17]. Without restrictions on water use, more drilling of wells will occur, potentially affecting land subsidence in areas like the San Joaquin Valley.

Vineyard water usage in 2010 was 1.6 million acre feet—roughly intermediate within California agricultural production acreage (Fig. 15.1).

Largely as a result of the drought, unprecedented efforts are underway to reduce water use in vineyards and wineries. This chapter will discuss some of

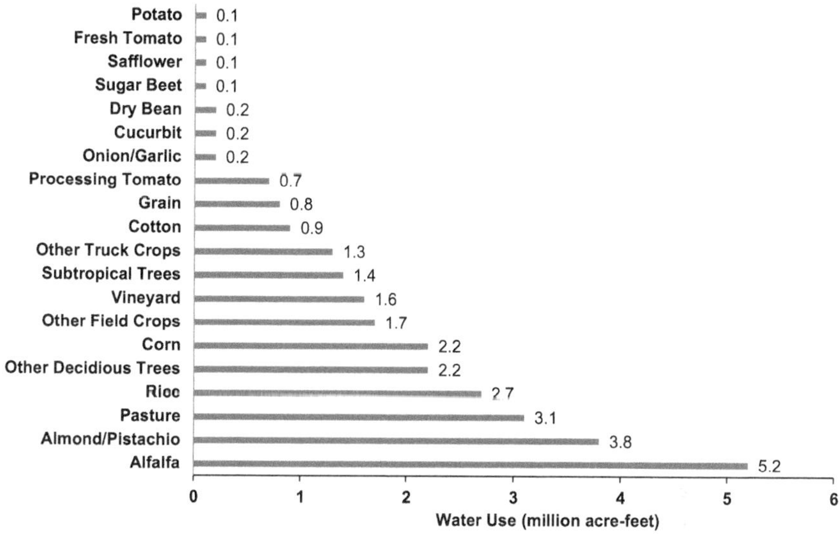

Figure 15.1 Applied water for California crops in 2010. *Reformatted from Cooley H. California agricultural water use: key background information. Pacific Institute; 2015. http://pacinst.org/wp-content/uploads/sites/21/2015/04/CA-Ag-Water-Use.pdf.*

those efforts and the impact of declining water on wine industry practices and its productivity. For additional sources of information on drought-inspired conservation measures in California, consult the Central Valley Project and State Water Project 2016 Drought Contingency Plan for Water Project Operations for February to November 2016 [19] and the State of California's current website [20]. Additional information on the California drought can be found in Ref. [21].

2. USE OF WATER IN VINEYARDS

Groundwater has provided about 30% of water for California farms and cities in most years, and around 60% in dry years, such as 2014 [22]. Water usage in agriculture and food/beverage production has generally turned from surface-water sources to groundwater sources, to the point where groundwater aquifers in the agriculturally vital Central Valley have been seriously depleted since the 1980s. Between 2003 and 2012 the draw-down in California's central valleys (San Joaquin and Sacramento River valleys, where much of the wine industry is located) was estimated to be equivalent to the entire water-storage volume of Lake Mead [23]. Estimates indicate that no amount of precipitation alone will be enough to restore the depleted groundwater in the near future [24]. The 2015–16 winter season brought relief and much needed rainfall to California and other southwest states. However, at the time of writing, many of California's largest storage reservoirs are well below their capacity [25–27].

Research also shows that with climate change, the average precipitation California receives will not be drastically different compared to prior years. However, there will be more severe extremes. There will be more storms instead of constant rain, and dry areas will get drier while wet areas get wetter. But the net precipitation California receives as a state on average is not expected to change drastically [28]. It should be noted that, as with all long-term projections, there is uncertainty about future precipitation and a transition to long-term dry conditions remains a possibility.

What can vintners do to reduce their water consumption? The following list and discussion are adapted from the San Luis Obispo Tribune and other sources.

1. Install modern drip irrigation systems, including variable volume emitters, varied in response to actual plant needs. Where feasible, subsurface drip irrigation systems [29,30] can improve efficiency of water use and reduce weed growth and evaporative loss (see Fig. 15.2).
2. Eliminate excessive weed or cover crop growth to reduce competition for water within the vineyard. This should be balanced with some plant growth between the vine rows, which is beneficial as it reduces the rate of evaporation from top soil and/or improves soil tilth, water penetration, and the habitat and decor of vineyards. There is also interest in intercropping—using the areas between rows of vines to produce a second crop for forage or feed, reducing pest pressure and soil erosion, diversifying the vegetative structure, and thus increasing the yield of each crop acreage relative to primary crop yields [31,32].
3. Plant grape vines with drought-tolerant rootstocks. There is renewed interest in breeding more drought-tolerant plants by conventional and biotechnology-based breeding methods as a way to adapt to climate change. For binary rootstock crops such as grapes, this approach can be reflected in both the rootstock and on the vines, which are grafted (scions). Vine cuttings, slips, or scions are usually grafted onto rootstock that has been specially cultivated to combine growth vigor with resistance to disease, water use, and other traits. They are then put into sand for one season, called bench grafting, and once the graft forms part of the vine the scion is often referred to as the *fruiting wood* or *bud wood*, as

Figure 15.2 Dr. Suat Irmak's subsurface drip irrigation and surface energy flux research project in Nebraska. *Copyright © Suat Irmak. Printed with Permission.*

differentiated from the *rootstock*. [33] Both rootstock and grafted vine stock can be targets for improved drought resistance [34].

4. Use deficit irrigation, i.e., apply less water than the vines actually need. Some water stress is considered advantageous for wine quality by adapting the grapes to use less water, but there are important questions regarding sustainability of deficit irrigation and the resulting stresses it imposes on vines over multiple years [29,35].

   Knights Bridge Winery and other vineyards use what is known as the "big drink" method of irrigation, which has resulted in roughly a 50% decrease in water use. Instead of frequent applications of smaller volumes of water, they apply less frequent but deeper applications of water. This forces the vines to grow deeper and stronger. By applying water more frequently, roots can come to the surface and be exposed to heat and pathogens. This method was aided through the use of drought-resistant rootstock, compost to increase the soil's water-holding capacity, and planting plow-down crops between rows to improve water penetration [36].

5. Reduce excessive use of sprinkler-supported frost protection, substituting wind machines or other devices, for a water-based frost-management system. Although there are different frost-management systems, the main principles of any such system are avoiding grapevine injury by reducing the risk of frost accumulation and protecting the vines during cold temperature periods [37].
6. Expand use of soil-moisture sensors to avoid overwatering and in general deviate from a fixed watering schedule to one that responds to climatic conditions and growth stage, and Brix development in maturing grapes. With modern equipment, including sensors, data loggers, and targeted and/or variable irrigation systems, growers can potentially tailor their irrigation to meet the geophysical and microclimate characteristics of all parts of their vineyards. More information on taking geography into account can be found in Ref. [30].

   These and other practices in use in California areas that are heavily dependent on scarce groundwater also include the practices executed by the wineries below [29]:

- J. Lohr Wineries uses pressure bombs to measure sap flow and water needs for vines. A pressure bomb is a pressure chamber where the water potential in a plant can be determined by measuring the pressure at which the xylem exudes fluid [38].
- Turley Wine Cellars practices dry farming, i.e., relying on rainfall for deep-rooted vines.
- Pear Valley Vineyards monitors soil moisture in all sections of the vineyard, and uses that data to optimize irrigation.
- Pear Valley Vineyards uses recycled wastewater from the winery plus rainwater collection for irrigation.
- Pipestone Vineyards uses cover crops to enhance soil organic matter and microbiological function, which allow water and nutrients to penetrate without the excessive use of water alone due to less competition with weeds [39]. Winter cover crops also increase the infiltration rate of rain, reduce runoff, increase soil moisture, decrease erosion, and provide nutrients when mowed in the spring [36]. One study indicated that grass, especially California native perennial grass, is a good cover crop [40]. Another found that alfalfa (in Canada), followed by tall fescue (in the United States), are also excellent cover crops [41].

A common practice in most California vineyards now is to use drip irrigation and/or microsprinkler systems, which dramatically reduce water use overall and are amenable to varying the amounts of water delivered with weather conditions, rows, stress requirements, spatial distribution, etc. [30]. Only a few vineyards are set up with subsurface drip irrigation, which must be installed at the time of planning the vineyard layout before vines are planted and cannot easily be retrofitted for existing vineyards. These systems deliver water directly to the root zone, reducing evaporation losses and discouraging the growth of weeds [42]. As with other forms of irrigation, drip irrigation and microsprinklers can be used to deliver nutrients and pesticides, reducing the losses that occur in a general broadcast application of fertilizers or pesticides. These agrochemicals are often applied in a separate water solution or suspension, versus irrigation water as the delivery medium.

A systems approach to irrigation water use should be based on recording water use by date, block, or row, to pinpoint areas of excessive watering. J. Lohr estimates that the amount of water fine-tuned to produce wine can be as little as 1 gallon of applied water per gallon of wine. With the optimized irrigation system in place, J. Lohr has reduced its water use per gallon of wine from 3.5 gallons (2003) to about 1.2 gallons of applied water (2012)—approximately a 70% decrease over that period of time [43].

Use of deep-rooted rootstock, or drought-tolerant rootstock, has increased substantially to the point where their use represents the majority of new vines now planted in California. Examples of such rootstock include Ramsey, 140Rug, 1103P, and 110R [44,45]. The targets in rootstock selection include drought and salt tolerance, disease and pest resistance, and vigor and other growth characteristics [46]. Soil preparation, choice of irrigation method, and soil nutrient addition and method of control can all influence plant growth, to benefit efficient use of water by the vines.

3. PESTICIDES

A sound program of pest control using herbicides to eliminate unwanted vegetation and use of fungicides, insecticides, and fumigants to improve plant health will affect water use. In the herbicide category, mineral oil is the leading herbicide in quantity (for contact control of weeds in vineyards), and glyphosate (Roundup) is second—a contact systemic material that can be used during the dormant season often in combination with a postemergent herbicide. Examples of the latter in current use are sulfentrazone, bentazone, and atrazine [47].

A list of top pesticides, excluding herbicides, used on wine grapes in California is given in Table 15.2, and the top herbicides are given in Table 15.3. Fumigants are frequently applied to the soil before planting the

Table 15.2 Top Ten Pesticides, Excluding Herbicides, Used on Wine Grapes in California (2012) [48]

Name	Class	Use	Amount in 2012 (lbs)
Sulfur	Inorganic	Fungicide, insecticide	20,351,800
Mineral oil (total)	Petrol derivative	Insecticide, plant growth regulator (PGR), fungicide, adjuvant, solvent	1,711,761
1,3-Dichloropropene	Halogenated	Plant growth regulator (PGR), fungicide, nematicide	666,605
Lime sulfur	Inorganic	Fungicide, insecticide	464,613
Copper hydroxide	Inorganic	Fungicide, nematicide	153,972
Potassium bicarbonate	Inorganic	Fungicide	151,331
Nonylphenoxy poly(ethyleneoxy) ethanol	Synthetic organic	Fungicide, surfactant	149,415
p-nonylphenyl polyoxyethylene (4–14 moles) dihydrogen phosphate ester	Polyalkyloxy compound	Adjuvant, soap/surfactant	65,707
Methyl soyate	Fatty acid ester	Adjuvant	63,056
Ammonium sulfate	Inorganic	pH adjustment	59,048

Table 15.3 Top Ten Herbicides Used on Wine Grapes in California (2012) [48]

Name	Class	Use	Amount in 2012 (lbs)
Mineral oil (total)	Petrol derivative	Herbicide	1,711,761
Glyphosate salts	Synthetic organic	Phosphonoglycine herbicide	646,000
Pendimethalin	Synthetic organic	Dinitroaniline herbicide	143,225
Paraquat dichloride	Synthetic organic	Bipyridylium herbicide	99,172
Orzalin	Synthetic organic	Dinitroaniline herbicide	91,878
Simazine	Synthetic organic	Triazine herbicide	74,313
Oxyfluorfen	Synthetic organic	Diphenyl ether herbicide	71,232
Glufosinate ammonium	Phosphonoglycine	Herbicide	70,683
Ammonium sulfate	Inorganic	Herbicide	59,048
Flumioxazine	Unclassified	Herbicide	19,818

vineyard to control nematodes, insects, weed seeds, and other undesirable organisms in the soil. The leading fungicide in quantity used is sulfur, and the leading fumigant is 1,3-dichloropropene. Biopesticides, such as spinosyn and avermectin, are increasing in use, in part as a result of policy in California to encourage use of biopesticides and reduce dependence on conventional synthetic chemicals.

Care must be taken in the use of chemical treatments so as not to harm the vines or impact the quality of the wine.

Sulfur has a variety of uses and benefits in wine grape production [49]. It is generally not used within 5 to 6 weeks of harvest to avoid sulfur residues including (sulfites and mercaptans) in the wine, which can impact flavor and odor [50]. A recent study showed that ceasing the application of sulfur at least 35 days prior to harvest keeps the sulfur-residue concentrations low enough to not impact wine quality. While some uses of sulfur led to a measurable increase in H_2S production, this did not noticeably decrease the wine quality. However, there are many variables that still have not been accounted for so it is safer to adhere to a flexible timetable and experience rather than a set number of days [51].

Glyphosate is not used on vines with foliage, which can absorb the chemical and transport it systemically, thus damaging the vines. Contact postemergent herbicides, like pendimethalin, must be chosen and used to not harm the vines or leave residues in the grapes at harvest. Restrictions on the use of specific pesticides on wine grapevines are spelled out in detail on the chemical's label [52,53].

Cultural methods such as tilling and hand removal of weeds and removing leaves and/or trellising vines to increase air circulation in the vine canopy may reduce chemical use. Reducing tractor and truck use in and around vineyards reduces dust that settles on the foliage, which discourages mites and other pests. This can be effective in reducing chemical use, thus reducing the use of water as a diluent for pesticide delivery.

Use of pheromone traps is an important adjunct to pest control, particularly with invasive or exotic pests, such as the grape leaf skeletonizer, the European grapevine moth (EGVM), and light-brown apple moth (LBAM) [54]. In many cases the use of pheromone traps is required in susceptible vineyards to aid in timing pest control to meet periods of infestation. The skeletonizer and glassy winged sharp shooter are problems in just some areas, and these must be monitored frequently. A blue-green sharpshooter transfers bacteria from infected plants to healthy plants, and has more than a 90% chance of transmitting bacteria. Due to limited effectiveness of available insecticides, research designed to select barrier crops, as well as other means of control such as resistant vines, is in process. The glassy-winged sharpshooter carries Pierce's disease (PD) and is a concern for multiple agricultural crops. In 2003 a joint state-federal plan dedicated $36 million to preventing the spread of this pest [55].

Pierce's disease symptoms do not get severe unless there is an environmental factor involved. These factors can include drought, root pruning with cultivation equipment, other diseases, overproduction of fruit, and normal plant senescence. The disease symptoms in grapevines include chlorosis, abscission of leaves, and acropetal development of symptoms, which are all characteristics of senescence in grapevines. However, because the bacteria responsible for PD, *X. fastidiosa*, prefers older tissue over younger tissue, practices that help grapevines grow actively and healthy can reduce the severity of disease symptoms, especially if the grapes already have a genetic tolerance to PD [56]. Merry Edwards, who makes Pinot Noir in the Russian River Valley and Sonoma Coast, expressed how the yield of wine grapes was impacted by the number of vines that had PD. Also, a warm winter and lack of rain encouraged a population boom in insect/pest populations, including the blue-green sharpshooter, which is responsible for PD [57].

4. WINERIES—CLEANERS

Cleaning equipment such as crushers, destemmers, tanks, bins, and pumps account for the majority of water use in wineries, so modifications and conservation in the normal practice of water rinsing of equipment including pads, tanks, and floors can lead to more efficient use of water while maintaining sanitation. In general, water is not added intentionally to wine.

In-depth information on washing and rinsing practices for wine production, can be found in *Wine Science: Principles, Practice, and Perception* [58]. A summary of common water conservation methods in wineries include:

1. High-pressure washing can use less water and more effectively clean and rinse equipment than conventional rinsing. A modification is to use high-pressure nozzles with automatic shutoff devices. These nozzles are costly ($10–25), but have been shown to save up to 40% of water used for cleaning [59,60].
2. Use of cleaning and sanitizing chemicals to remove deposits is an important adjunct to cleaning equipment, since they can reduce heavy water washing and rinsing.

3. Collecting rinse waters and recycling them for irrigation or other uses in and around the winery, often with filtration or other modest upgrades of water quality, prior to recycling for use. Use of nanofiltration and reverse osmosis to upgrade winery wastewater is becoming more accepted and economical as the cost of water increases [61,62].
4. Use of water meters throughout the winery to pinpoint excessive use of water, and timers that automatically shut off water as well as related aids, further reduces excessive use of water for cleaning equipment.
5. Educating and/or providing incentives for workers to conserve water works wonders, including the use of awards, bonuses, and parking perks.

Water management has been a priority for E. & J. Gallo Winery for decades. Water-conservation efforts reduced water consumption by 25% in 2012, compared to the previous 3-year average. During each cleaning and sanitization process, bottle-washing efficiencies have eliminated 90% of water used, and changes to the sanitization process for the wine press have resulted in savings of up to 12,000 gallons of water [63].

Chemical cleansers are used to remove dirt and grime from surfaces of containers, including fermentation bins, transfer containers, aging containers, and any equipment that comes into contact with wine, from crushing and destemming through fermentation, barreling, and bottling. A list of popular cleaning chemicals is given in Table 15.4, and more thorough information is available in *Wine Microbiology: Practical Applications and Procedures* [64].

5. WINERIES—SANITIZERS

Sanitizers are used to prevent the growth of spoilage organisms including molds, wild yeasts, and bacteria. A list of popular cleaning agents and sanitizers is given in Table 15.4. An example is sodium percarbonate, which is used as a disinfectant and bleaching agent as a one-step, no-abuse cleaner/sanitizer, on glass, stainless steel, food-grade plastic surfaces, polyethylene terephthalate carboys, silicone equipment (tubing, etc.), and oak barrels. Percarbonates make use of oxygen and a mild alkali to help lift grime. The active ingredients are sodium carbonate and hydrogen peroxide (sanitizer). This cleaner and sanitizer, environmentally and septic system friendly, is formed from sodium percarbonate in water by the following reaction:

$$2Na_2CO_3 \cdot 3H_2O_2(aq) \xrightarrow{H_2O(1)} 2Na_2CO_3(aq) + 3H_2O_2(aq) \qquad (15.1)$$

Potassium metabisulfite, $K_2S_2O_5$, can be used for sulfating prior to the addition of yeast (to kill wild yeast), as well as after fermentation to stop

Table 15.4 Popular Cleaning and Sanitizing Chemicals for Wine Equipment [64–66]

Chemical	Description
Percarbonates	Combination of sodium carbonate and hydrogen peroxide. Removes dirt and deposits from beer and wine-making equipment. Work with active oxygen and a mild alkali. Hydrogen peroxide also acts as a sanitizer. Compatible with septic systems.
Powdered Brewery Wash (PBW)	High-strength polycarbonate. Soak overnight for stubborn stains.
Straight A	Similar to PBW, but not as strong.
Prozyme	Enzyme-enhanced detergent for removing protein deposits.
Chlorine bleach	For glass containers and equipment. Requires rinsing to remove residues and prevent odor.
Potassium metabisulfite	Excellent for sulfating must prior to adding yeast. Also effective in sanitizing equipment that is ready to use after drip-dry.
Iodophor	Iodine detergent germicide and sanitizer. No rinse.
Sparkle Brite or Diversol	Contains trisodium phosphate and potassium bromide. Short contact time required.
Chlorine bleach	Use only for glass. 5 min soak. Must rinse heavily to remove residual chlorine and odor.
Potassium bisulfite	Used to sanitize must, bottles, and corks.
Star San	Blend of phosphoric acid and dodecylbenzenesulfonic acid.

unwanted additional fermentation, and for some routine sanitation of equipment and surfaces. Equipment and containers (such as bottles) are ready to use after drip-dry, without any rinsing. This is crucial to save water in wine-making and during bottling.

$$K_2S_2O_5(s) \xrightarrow{H_2O(1)} K_2SO_3(s) + SO_2(g) \tag{15.2}$$

Ethyl alcohol is an effective sanitizer, often used in winemaking as a 70% solution with water. It leaves no residue after use but precautions include cost, lack of ready availability, and flammability.

There is also a shift occurring in wine sanitation where corks are being replaced by synthetic stoppers or plastic corks that resemble natural corks. Using a bottle with a natural cork can often lead to a musty smell when the wine bottle is uncorked. This is due to "cork taint," generally caused by trichloroanisole (TCA, Fig. 15.3), which occurs when there are interactions between mold and bacteria in the cork with traces of bleach that were used to sanitize the cork. Many countries are moving toward synthetic

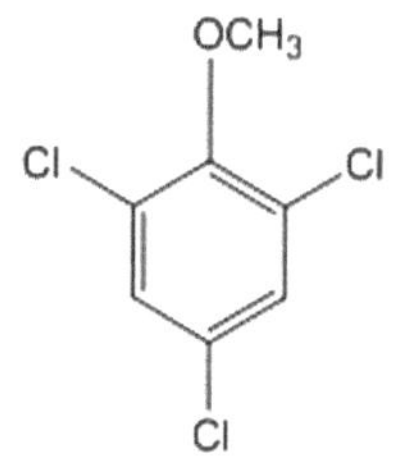

Figure 15.3 Structure of trichloroanisole.

stoppers, allowing a decrease in off-flavors and off-odors from anisole derivatives. Though corks play a vital role in the economies of Spain and Portugal, there is also an interest to preserve the cork tree, *Quercus suber*, as the materials for corks come from one of the biggest wildlife habitats in Europe, *Dehesa* in Spanish, and *Montado* in Portuguese, stretching approximately 13,000 square miles across western Spain and eastern Portugal [67–69].

6. SUSTAINABILITY

The wine industry consumes large amounts of energy as well as water. This reality, coupled with the drought and economic costs, have stimulated interest in conserving both water and energy [70].

The word "sustainable" became common among wineries in 2008. Though it could mean anything from adding a "green dimension" to conserving the environment, the general definition for wineries usually is an increased emphasis on energy, water, chemical, and/or packaging management. In 2009, wineries added an economic component to the idea of sustainability [70].

The University of California, Davis, a world leader in wine-making research and education, demonstrates water conservation and reuse at its new Jackson Leadership in Energy and Environmental Design (LEED) winery pilot plant. The pilot plant makes use of rainwater, tank cleaning, green chemistry in the choice of cleaning and sanitizing agents, and also recycles and recovers both water and carbon dioxide [71,72]. The latter can be used to make chalkboard for the construction industry.

Several wineries are developing protocols for sustainability in areas such as conservation of water, reduction and reuse of carbon dioxide, energy efficiency, temperature control, recycling, and other parameters of winemaking, aging, and storage. For example, Far Niente, one of the wineries in the Napa Valley, follows an integrated program of sustainable measures

Figure 15.4 Solar panels at Far Niente Winery [73].

affecting vineyard, winery, and day-to-day practices, starting with a physical footprint that minimizes land use and excess traffic flow.

Far Niente's sustainable wineries:

1. Are solar-powered, net-zero users of electricity. A substantial part of the solar array is floating on the irrigation pond, as shown in Fig. 15.4, reducing evaporative water loss.
2. Use organically farmed vineyards.
3. Reuse process water from winery operations for vineyard irrigation and frost protection.
4. Farm with a fleet of Tier 4 tractors powered primarily with propane, replacing diesel fuel.
5. Replace conventional company cars with electric-motor or hybrid-electric-engine vehicles.
6. Conduct regular energy- and water-use audits to determine conservation opportunities.
7. Regularly train staff on resource-management practices.

Other ways to reduce water and energy consumption, adapted from the Washington State University Extension service, include [74]:

1. Install skylights in winery structures to minimize electricity.

2. Install lights with occupancy sensors and staged lighting levels to allow for lower intensity for light foot traffic and higher intensity when working with heavier equipment such as forklifts.
3. Insulate wine-storage tanks to reduce heat gain from the surrounding air.
4. Implement programs to detect leaks in compressed air and steam lines.
5. Ensure that refrigeration units are well-maintained and insulate the refrigeration supply piping to maintain a more consistent temperature, and install variable speed refrigerant compressors.

Guides to sustainable water and energy measurements include:

- The Efficiency Savings Tool made by Lawrence Berkeley National Lab. This software application was designed to evaluate the energy and water efficiency at wineries, and to help assess the environmental and financial impacts of potential improvement strategies [75].
- A guide to sustainable management of winery water and associated energy has been prepared on behalf of the American Vineyard Foundation (AVF) and the California Wine Institute, with support and guidance from the National Grape and Wine initiative (NGWI). It provides a set of tools for wineries of all sizes to use in realigning existing facilities or designing new facilities to achieve goals for sustainable management of winery source water and wastewater with the ancillary benefits of increasing energy efficiency and reducing greenhouse gas generation [76].

7. DROUGHTS ACROSS THE WORLD

There is an increase in drought frequency and severity in many parts of the world [77]. This section discusses the drought situations in some countries with significant wine production as well as their policies for reducing the effects of drought.

7.1 Australia

Australia is one of the most drought-prone countries on the planet, with the worst droughts in its history including the Federation Drought (1895–1902), World War II Drought (1937–45), and "Big Dry" (1997–2009) [78].

Australia has been a major wine exporter in the world, being the major supplier of imported wine for England, and only slightly behind France and Italy in supplying the United States [78]. However, as water became more expensive for wine irrigation, in 2012 wine industry groups estimated that about 1000 of the country's 7000 wine growers would stop growing wine grapes [78]. In 2013, the drought resulted in record-low commodity values for agricultural crops, including wine grapes. Due to this fact, wine growers have diversified into passionfruit production, fish farming, dates, and pomegranate production [79]. The Commonwealth Scientific and Industrial Research Organization (CSIRO) researchers have also hypothesized that 44% of viticultural areas will be unsuitable for growing grapes by the year 2050 [78].

As a large wine producer in a chronically semiarid continent, Australia has made some significant changes in grape-growing and/or fermentation-related water uses that include many of the same features mentioned in California such as reducing water demand by increasing water efficiency and altering water use practices (e.g., switching to drip irrigation or micro-sprinklers), stormwater capture, and water recycling. Currently, the Australian national government also has more power in reaching water-policy decisions, and the development of water-trading markets has accelerated. Urban water suppliers are also undertaking projects to increase water supplies including desalination [78].

7.2 Israel

Israel has consistently been a leader in innovative ways to extend meager freshwater resources, and in particular for the use of desalination and recovery and/or reuse of wastewater.

The Israeli desert makes up 60% of the country's total area [80]. In the history of Israel and its development in sustainable agriculture, the most important key factor has been developing water resources. This includes irrigation systems that save water (e.g., drip irrigation), enclosing some agricultural production (e.g., greenhouses and plastic tunnels), and substitute soil with sand and artificial growth media. Controlled drip irrigation and drainage has solved salinity problems while improved pest-management practices have also helped in plant protection [80]. Controlled drip irrigation that is integrated with fertilizer is used to practice agriculture on Israel's sand dunes, which can be flattened and cultivated [80].

The north part of Israel also plays a large part in the country's sustainability practices. Water from the Sea of Galilee is transported to the south, and recycled sewage water from Israel's largest cities is also used in agriculture after undergoing treatment. This sewage treatment consists of biological treatment before being filtered through sand dunes down to a depth of 80 m where it can be stored for a long period of time. When the treated sewage water reaches a quality that is comparable to drinking water, the water is pumped through a distribution system to the south where the agricultural sector uses the water, or it is stored in large reservoirs [80].

Israel also uses saline water for crop irrigation—a motivation behind Israel's development of the drip-irrigation system back in the 1970s. By using drip irrigation with saline water, the long and frequent irrigation applications at a low water discharge rate allow the permanent leaching of salts from the root-zone system of the plants. Research has shown that many crops are tolerant and resistant to saline water including grapes. In fact, vineyards that use saline water in their irrigation have shown that grapes attain a higher level of dry matter, which can result in high-quality wine [80].

The blending of desalinated water with water transported from the Sea of Galilee can reduce the levels of total dissolved solids (TDS), chloride, sodium, nitrate, and hardness levels, which provide healthy water for consumption [81]. In Israel desalination was limited to the reverse osmosis of brackish water for many years, and these desalination plants continue to provide water. However, in some parts of Israel, the conservation and transportation of water from the northern Sea through aquifers and wells, combined with increased agricultural water efficiency and the use of recycled and treated wastewater, had succeeded in offsetting the need for seawater desalination. However, in the mid-20th century, the Israeli Water Commission reported that the lower cost alternatives were exhausted and decided to introduce and integrate large-scale seawater desalination plants for the country's water supply [81].

Further information on desalination of seawater and brackish water, and Israel's water practices, can be found in Hessy Taft's work on optimizing water resources in Israel [82].

For more information regarding the droughts and water shortages around the world, consult *Let There be Water* by Seth Siegel.

CONCLUSIONS

As with other beverage producers, the wine industry is taking sustainability seriously. Efficient use of water is vital, particularly in regions experiencing episodes of sustained drought such as the western United States and, previously, Israel and Australia. Other sustainability measures include energy efficiency, reduction and/or reuse of emissions such as CO_2, prudent choices of fertilizers/nutrients, pesticides, and cleaning/sanitizing agents. As illustrated by several of the leading wine producers, everyone can benefit from sustainability. A popular bumper sticker related to water efficiency reads: Save Water, Drink Wine.

ACKNOWLEDGMENTS

The authors would like to thank Roger Boulton, Loreen Kleinschmidt, John Finley, Jenny Vo, and Robert M. Hansen for their assistance and input.

REFERENCES

[1] Hanack E, Lund J, Burt C, Howitt R, Medellín-Azuara J, Parker D, Summer D, Zoldoske D. California's water: water for farms. California: Public Policy Institute of California; 2015.

[2] Water Education Foundation. Applied water. California: Water Education Foundation; 2016. http://www.watereducation.org/aquapedia/applied-water.

[3] Wines M. Colorado river drought forces a painful reckoning for states. New York: New York Times; January 5, 2014. http://www.nytimes.com/2014/01/06/us/colorado-river-drought-forces-a-painful-reckoning-for-states.html?_r=0.

[4] United States, census, bureau, state & country QuickFacts. United States Census Bereau; 2015. http://quickfacts.census.gov/qfd/states/06000.html.

[5] California Agricultural Statistics. In: Crop year. Sacramento (California): United States Department of Agriculture, National Agricultural Statistics Service, C. Pacific Regional Field Office; 2013. 2015, https://www.cdfa.ca.gov/statistics/PDFs/CropYearStats2013_NASS.pdf.

[6] Wine Institute. California wine profile 2013. San Francisco: Wine Institute; 2013. http://www.wineinstitute.org/files/CA%20Wines%20Stat%20Profile%202013.pdf.

[7] Institute of Mechanical Engineers. Processing, global food, waste not, want not. Institution of Mechanical Engineers; 2013. p. 12.

[8] Sedghi A. How much water is needed to produce food and how much do we waste? The Guardian January 10, 2013. http://www.theguardian.com/news/datablog/2013/jan/10/how-much-water-food-production-waste.

[9] Product water footprints: soft drinks. Water Footprint Network; 2015. http://temp.waterfootprint.org/?page=files/Softdrinks.

[10] Environmental footprint. Virginia: International Bottled Water Association; 2015. http://www.bottledwater.org/education/environment-impact/environmental-footprint.

[11] Howitt R, MacEwan D, Medellín-Azuara J, Lund J, Sumner D. Economic analysis of the 2015 drought for California agriculture. Davis: UC Davis Center for Watershed Sciences; ERA Economics; UC Agricultural Issues Center, University of California; 2015. https://watershed.ucdavis.edu/files/biblio/Final_Drought%20Report_08182015_Full_Report_WithAppendices.pdf.

[12] California Water Boards. In: Board SWRC, Agency CEP, editors. Emergency conservation regulation implementing 25% conservation statewide. California: California Water Board; 2015. p. 1–2.

[13] Office of Governor Edmund F. Brown Jr . Executive order S-06-08 state of California, California. 2010. https://www.gov.ca.gov/news.php?id=9797.

[14] Howittt R, Medellín-Azuara J, MacEwan D, Lund J, Sumner D. Economic analysis of the 2014 drought for California agriculture. Davis (California): Center for Watershed Sciences; University of California, Davis; UC Agricultural Issues Center; ERA Economics; 2014. https://watershed.ucdavis.edu/files/biblio/Economic_Impact_of_the_2014_California_Water_Drought_1.pdf.

[15] California drought restrictions FAQ: What the governor's executive water order means for you. California: Southern California Public Radio; April 2, 2015. http://www.scpr.org/news/2015/04/02/50747/california-drought-restrictions-faq-what-the-gover/.
[16] SGM sustainable groundwater management. California Department of Water Resources; 2015. US.
[17] California Proposition 1, Water Bond. The encyclopedia of American politics, 2015. 2014.
[18] Cooley H. California agricultural water use: key background information. Pacific Institute; 2015. http://pacinst.org/wp-content/uploads/sites/21/2015/04/CA-Ag-Water-Use.pdf.
[19] CVP and SWP Operations. Central Valley project and state water project 2016 drought contingency plan for water project operations February–November 2016. 2016. http://www.water.ca.gov/waterconditions/docs/2016-DroughtContingencyPlan-CVP-SWPOOperations-Feb-Nov_1.19.16-FINAL.pdf.
[20] State of California. California drought. 2014. http://ca.gov/drought/#content1.
[21] Siegel S. Let there be water. New York: Thomas Dunne Books; 2015.
[22] Rogers P. California drought: after years of overpumping growndwater, state may be ready for reforms, San Jose. Mercury News March 30, 2014. http://www.mercurynews.com/science/ci_25453902/california-drought-after-years-overpumping-groundwater-state-may?source=infinite.
[23] McNutt M. The drought you can't see. 2013. http://www.grac.org/Science-2014-McNutt.pdf.
[24] Manrique J. To fight the drought in California, we need more than just rain. New America Media; March 19, 2015.
[25] California Department of Water Resources. California data exchange center, Folsom lake – storage conditions as of February 6, 2016, state of California. 2012. http://cdec.water.ca.gov/cdecapp/resapp/resDetailOrig.action?resid=FOL.
[26] California Department of Water Resources. California data exchange center, conditions for selected reservoirs, state of California. 2013. http://cdec.water.ca.gov/cdecapp/resapp/getResGraphsMain.action.
[27] California Department of Water Resources. California data exchange center, Lake Shasta – storage conditions as of February 6, 2016, state of California. 2012. http://cdec.water.ca.gov/cdecapp/resapp/resDetailOrig.action?resid=SHA.
[28] Wang H, Schubert S. Causes of the extreme dry conditions over California during early 2013. Bull Am Meteor Soc 2014;95:S7–11.
[29] Lynem J. Wineries working to conserve water. The Tribune; June 18, 2013. http://www.sanluisobispo.com/2013/06/18/2551008_wineries-working-to-conserve-water.html?rh=1.
[30] Schwankl LJ, Hanson BR. Surface drip irrigation. In: Lamm FR, Ayars JE, Nakayama FS, editors. Microirrigation for crop production: design, operation, and management. Italy: Elsevier; 2007. p. 431–5.
[31] Muenchrath D. Lesson 9: Cropping intersystems 2: intercropping. Iowa State University; 2006. http://masters.agron.iastate.edu/Content/Extras/lessons/classes/sample/lesson01/sampleL01.pdf.
[32] Intercropping in Practice. In: Pimentel D, editor. Encyclopedia of pest management. New York: Marcel Dekker, Inc.; 2002.
[33] LaMar J. Viniculture. Professional Friends of Wine; 2011. http://www.winepros.org/wine101/viniculture.htm.
[34] Christensen PL. Rootstock selection, wine grape varieties in California. Extension 2003:12–5.

[35] Fereres E, Soriano MA. Deficit irrigation for reducing agricultural water use. J Exp Bot 2006;58:147–59.
[36] Lambert K. Dry farming wine grapes: a best management practice guide for California growers, community alliance with family farmers. 2015.
[37] Hellman E. Frost injury, frost avoidance, and frost protection in the vineyard. Texas AgriLife Extension; 2015. http://articles.extension.org/pages/31768/frost-injury-frost-avoidance-and-frost-protection-in-the-vineyard.
[38] The Oxford companion to wine. 4th ed. United Kingdom: Jancis Robinson and Oxford University Press; 2015.
[39] Steenwerth K, Belina KM. Cover crops enhance soil organic matter, carbon dynamics and microbiological function in a vineyard agroecosystem. Appl Soil Ecol 2008;40:359–69.
[40] Ingels CA, Scow KT, Whisson DA, Drenovsky RE. Effects of cover crops on grapevines, yield, juice composition, soil microbial ecology, and gopher activity. Am J Enol Vitic 2005;56:19–29.
[41] Xi Z, Tao Y, Zhang L, Li H. Impact of cover crops in vineyard on the aroma compounds of *Vitis vinifera* L. *cv* Cabernet Sauvignon wine. Food Chem 2011;127:516–22.
[42] Payero JO, Yonts CD, Irmak S. Advantages and disadvantages of subsurface drip irrigation. University of Nebraska; 2005. http://www.ianrpubs.unl.edu/live/ec776/build/ec776.pdf.
[43] Lohr Wineries J. Water conservation and sustainability: an Interview with Jeff Zucker safety and environmental coordinator. J. Lohr Vineyards & Wineries; 2015. http://www.jlohr.com/sustainability/water-conservation/.
[44] Nikolaou N, Angelopoulos K, Karagiannidis N. Effects of drought stress on mycorrhizal and non-mycorrhizal Cabernet Sauvignon grapevine, grafted onto various rootstocks. Exp Agric 2003;39:241–52.
[45] White RE. Understanding vineyard soils. 2nd ed. Oxford University Press; 2015.
[46] Ray PK. Breeding tropical and subtropical fruits. India: Narosa Publishing House; 2002.
[47] Burgess C. Managing weeds in warm-season lawns. Clemson University; 2004. http://www.clemson.edu/extension/hgic/pests/weeds/hgic2310.html.
[48] Kegley SE, Hill BR, Omme S, Choi AH. Pesticide use on wine grapes in 2012. PAN pesticides Database – California Pesticide Use; 2014. http://www.pesticideinfo.org/DS.jsp?sk=29143.
[49] Griffith CM, Woodrow JE, Seiber JN. Environmental behavior and analysis of agricultural sulfur. J Pest Manag Sci June 2015;71(11):1486–96.
[50] Schilder A. Late-season fungicide sprays in grapes and potential effects on fermentation. Michigan State University Extension; 2011. http://msue.anr.msu.edu/news/late_season_fungicide_sprays_in_grapes_and_potential_effects_on_fermentatio.
[51] Kwasniewski MT, Sacks GL, Wilcox WF. Persistence of elemental sulfur spray residue on grapes during ripening and vinification. Am J Enol Vitic 2014;65:453–62.
[52] Bisson LF, Waterhouse AL, Ebeler SE, Walker AM, Lapsley JT. The present and future of the international wine industry. Nature 2002;418:696–9.
[53] Goodell PB, Zalom FG, Strand JF, Wilen CA, Windbiel-Rojas K. Maintaining Long-Term Management: over 35 years, integrated pest management has reduced pest risks and pesticide use. Calif Agric 2014;68:153–7.
[54] Vasquez SJ. Invasive pests severely impact California agriculture. Western Farm Press; May 4, 2011. http://westernfarmpress.com/grapes/invasive-pests-severely-impact-california-agriculture?page=1.

[55] Cash SB, Zilberman D. Environmental issues in California agriculture. In: Seibert J, editor. California agriculture dimensions and issues. University of California Giannini Foundation of Agricultural Economics; 2003. p. 215–40.
[56] Hopkins DL, Purcell AH. *Xylella fastidiosa*: cause of Pierce's disease of grapevine and other emergent diseases. Plant Dis 2002;86:1056–66.
[57] Daniel L. A conversation with Merry Edwards. Wines & Vines; 2016. p. 68–9.
[58] Jackson RS. Wine science: principles, practice, and perception. 2nd ed. United States of America: Academic Press; 2000.
[59] Beddow V. Best practice water management in wineries. IWA Water Wiki: Information Resource and Hub for the Global Water Community; 2011. http://www.iwawaterwiki.org/xwiki/bin/view/Articles/WaterManagementinWineries.
[60] Harris R. Process water, watersmart guidebook: a water-use efficiency plan review guide for new businesses, east bay municipal utility district. 2015.
[61] Massot A, Mietton-Peuchot M, Peuchot C, Milisic V. Nanofiltration and reverse osmosis in winemaking. Desalination 2007;231:283–9.
[62] Loannou LA, Micheal C, Vakondios N, Drosou K, Xekoukoulotakis NP, Diamadopoulos E, Fatta-Kassinos D. Winery wastewater purification by reverse osmosis and oxidation of the concentrate by solar photo-Fenton. Sep Purif Technol 2013;118:659–69.
[63] Highlights. California Sustainable Winegrowing Alliance News; Summer 2012. http://www.sustainablewinegrowing.org/docs/Water_Management_May2012.pdf.
[64] Fugelsang KC, Edwards CG, editors. Wine microbiology: practical applications and procedures. 2nd ed. New York: Springer; 2007.
[65] Bader S. Simplifying cleaning and sanitizing for home brewers. Washington: Bader Beer and Wine Supply. https://www.baderbrewing.com/content/simplifying-cleaning-and-sanitizing-home-brewers.
[66] Vandergrift T. Simplifying Cleaning and Sanitizing for Home winemakers. Bader Beer and Wine Supply. https://www.baderbrewing.com/content/simplifying-cleaning-and-sanitizing-home-brewers.
[67] Smith M. The increasing popularity of synthetic wine stoppers and screw caps is threatening wildlife-rich cork forests on the Iberian Peninsula, where for centuries people and nature have lived in harmony. National Wildlife Federation. https://www.nwf.org/News-and-Magazines/National-Wildlife/News-and-Views/Archives/2006/Put-a-Cork-in-It.aspx.
[68] Vlachos P, Kampioti K, Kornaros M, Lyberatos G. Development and evaluation of alternative processes for sterilization and deodorization of cork barks and natural cork stoppers. Eur Food Res Technol 2007;225:653–63.
[69] Prat C, Trias R, Cullure L, Escudero A, Antico E, Baneris L. Off-odor compounds produced in cork by isolated bacteria and fungi: a gas chromatography–mass spectrometry and gas chromatography–olfactometry study. J Agric Food Chem 2009; 57:7473–9.
[70] Winery water and energy use, and new technologies. New World Winemaker; 2013. http://www.newworldwinemaker.com/article/winery-water-and-energy-use/.
[71] Roger B. Boulton research homepage. UC Davis Department of Viticulture and Enology; 2005. http://boulton.ucdavis.edu/index.htm.
[72] LEED platinum facilities. UC Davis College of Agricultural and Environmental Sciences, Robert Mondavi Institute; 2012. http://robertmondaviinstitute.ucdavis.edu/about/facilities.
[73] Sustainable practices, far Niente. The Napa Valley Wine Estate; 2015. http://farniente.com/napa-wine-estate/sustainability/.
[74] Energy efficiency in wineries. Washington State University Extension Energy Program; 2011. http://www.energy.wsu.edu/Documents/Industrial_Wineries_Feb2011.pdf.

[75] Galitsky C, Worrell E, Radspieler A, Healy P, Zechiel S. Best winery guidebook: benchmarking and energy and water savings tool for the wine industry. Lawrence Berkeley National Laboratory; 2005.
[76] Comprehensive guide to sustainable management of winery water and associated energy. Wine Institute; 2008.
[77] Xiao-jun W, Jian-yun Z, Shahid S, ElMahdi A, Rui-min H, Zhen-xin B, Ali M. Water resources management strategy for adaptation to droughts in China. Mitig Adapt Strat Glob Change 2012;17:923−37.
[78] Heberger M. Australia's millennium drought: impacts and responses. In: Gleick PH, Allen L, Christian-Smith J, Cohen MJ, Cooley H, Heberger M, Morrison J, Palaniappan M, Schulte P, editors. The World's water. Washington, D.C.: Pacific Institute for Studies in Development, Environment, and Security; 2012. p. 97−125.
[79] Kiem AS, Austin EK. Drought and the future of rural communities: opportunities and challenges for climate change adaptation in regional Victoria, Australia. Glob Environ Change 2013;23:1307−16.
[80] Katzir R. Advance Agriculture in the desert: the Israeli case story. Sci Cold Arid Reg 2014;7:1−6.
[81] Dreizin Y, Tenne A, Hoffman D. Integrating large scale seawater desalination plants within Israel's water supply system. Desalination 2008;220:132−49.
[82] Taft HL. Sustaining water resources: a global imperative. In: Ahuja S, et al., editors. Water challenges and solutions on a global scale. American chemical society symposium; 2015. p. 413−33.

CHAPTER SIXTEEN

Developing a Global Hydrohub: Singapore's Leadership in Water Innovation

A.M. Dhalla
Nanyang Technological University, Singapore

1. SEEMINGLY INTRACTABLE CHALLENGES

As with many densely urban centers, with limited ability to store natural water resources, water management in Singapore has always been a huge challenge, and correspondingly, one of the key priorities for its government.

It may seem difficult to imagine today, but Singapore in the 1960s was a country with a low-end agriculture, commerce, and service-based economy, including pig and poultry farming, boat building and repair, chandlery (supply of victuals to docked ships), and other shipping- and trade-related services. Its main water bodies were heavily contaminated due to these activities, as well as a lack of sanitation in many areas [1–3].

Though Singapore receives an average rainfall of close to 2330 mm per annum, there is limited land to store the water. Hence, another key challenge was that Singapore depended, very heavily, on the neighboring state of Johor, Malaysia, for the supply of its potable water.

2. APPROACH TO SUSTAINABLE SOLUTIONS

In order to address these challenges, a multifaceted approach toward sustainable solutions was urgently needed. This was a complex situation with all options having significant long-term implications. Over a period of time, Singapore planned, and successfully overcame this challenge, through a holistic approach—not just tackling the water challenge, but simultaneously enacting long-term policies that would be viable for sustenance and future growth.

Chemistry and Water
ISBN 978-0-12-809330-6
http://dx.doi.org/10.1016/B978-0-12-809330-6.00016-7

A decision was made to move away from the existing occupations, over the years, and to create jobs in value-adding industries like refining, petrochemicals, electronics, and biomedical manufacture, to name a few, while also growing its services and financial sectors.

One of the other key decisions made was to rezone and relocate residential, industrial, and commercial activity. In order to change from the then existing urban sprawl, many of the residents were rehoused in high-rise public housing. This also ensured that there were more areas for "green spaces," making Singapore a true "City in a Garden." Some industrial and commercial activities were gradually relocated to new zones as needed.

Both residential and commercial buildings were designed with water- and sewage-management systems. This made it possible to keep Singapore's catchments clean and allow improved harvesting of rainwater. It also opened up possibilities for collection and treatment of used water for reuse.

3. DIVERSIFIED WATER SOURCES

Singapore's water supply today is based on four "National Taps."

3.1 Local Catchment Water

Singapore, being a tropical country, gets significant rainfall each year (around 2330 mm). However, being an island nation, land area, and correspondingly, the ability to collect and store rainwater, is limited. Over the years, Singapore has steadily increased its catchment area to two thirds of its land mass, and continues to look at ways to expand this in coming years (http://www.pub.gov.sg/water/Pages/singaporewaterstory.aspx).

One of the keys to doing this was to not only develop reservoirs in protected catchments, but in land-scarce Singapore, also develop urbanized water catchments [2]. This was extremely challenging, since rainwater flow could pick up litter and nutrients on its way to collection. Details of how this was achieved are explained very well in various publications [1–3].

Blue Map of Singapore

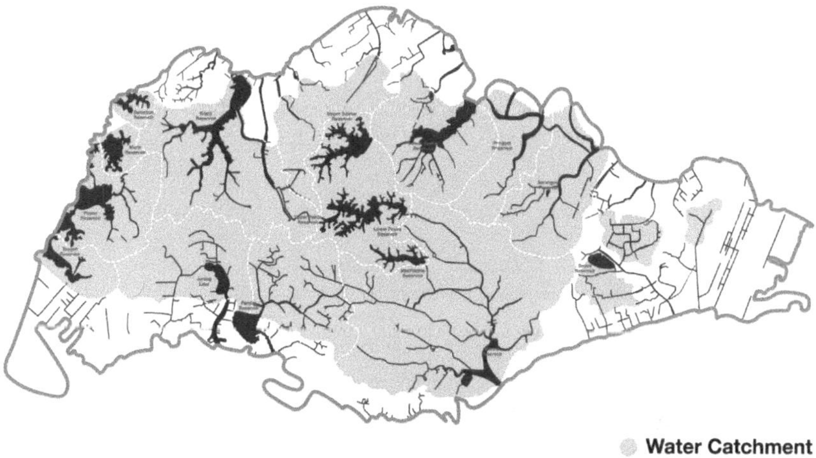

Besides the measures taken to transform the key economic drivers to nonpolluting industries, the herculean task of not only cleaning up Singapore's rivers, but equally importantly, maintaining their long-term viability as pristine water bodies, was initiated. With comprehensive planning, and the drive for execution, this once challenging goal was achieved well within the 10-year timeframe.

Early industries that were sited along the banks of the Singapore River, ship building and repairs, street hawkers and squatters, had contributed to the water pollution. These rivers were essentially open sewers and became extremely polluted by the 1960s. The Singapore River cleanup took place from 1977 to 1987. This enormous undertaking involved the development of infrastructure such as housing, industrial workshops, and sewage systems; massive resettlement of squatters, backyard trades, industries, and farmers; re-siting of street hawkers to food centers; and phasing out of pollutive activities along or close to the river banks. With the joint efforts of several agencies, results were seen and enjoyed in 1987. The river was flowing freely and its banks were transformed into riverside walkways and landscaped parks.

After River Cleanup Effort.

Besides the existing inland reservoirs, Singapore has also created additional freshwater storage reservoirs by damming up the mouths of its rivers. Of course, this would not have been possible without pollution-control measures within urban water catchments. As an example, consider the mouth of the Singapore River, an area known as Marina Bay. With the cleanup of the Singapore and Kallang rivers, the late Mr. Lee Kuan Yew envisioned putting a dam across the Marina channel to form a freshwater lake. This vision has been realized today with the completion of Marina Barrage and the creation of Marina Reservoir.

Opened on October 31, 2008, the Marina Barrage serves three functions—water supply, flood control, and lifestyle attraction. Built across the mouth of the Marina Channel, Marina Barrage creates Singapore's 15th reservoir, and the first in the heart of the city. Desalting began in 2009 through natural replacement by rainwater and the reservoir was commissioned as a freshwater reservoir in 2010. With a catchment area of 10,000 hectares, the Marina catchment is the island's largest and most urbanized catchment.

The Marina Barrage at the mouth of the Singapore River.

The Barrage also helps alleviate flooding in these low-lying areas by keeping out the tidal influence and allowing excess stormwater to be discharged into the sea. As the water in the Marina Basin is unaffected by the tides, its water level can be kept constant all year round, making it ideal for all kinds of recreational activities such as canoeing, kayaking and dragon boating, etc.

3.2 Imported Water

Singapore was a British colony from 1824 until 1959, when it became self-governed. Two years after this, in 1961, the City Council of the then State of Singapore signed an agreement (which has since expired) with the State of Johor in Malaysia to give Singapore full and exclusive access to water in certain catchments and rivers in Johor, until 2011. A second agreement, which is currently active, was signed in 1962, for the supply of a specified amount of water from the Johor River, until 2061. These became known as the 1961 and 1962 Water Agreements [1].

In 1963, Singapore merged with Malaya, Sabah, and Sarawak to form Malaysia, but after 2 years, Singapore separated from Malaysia in 1965. Before separation, the two Water Agreements were enshrined in the Separation Agreement with Malaysia [1].

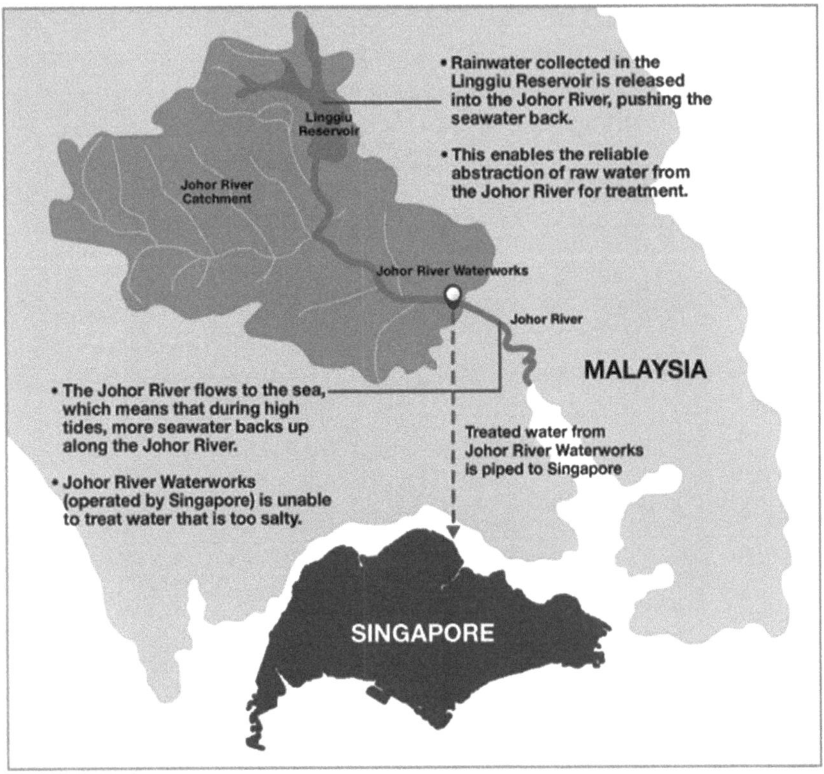

One of these agreements has expired, and Singapore continues to import water under the 1962 agreement. Concurrently, Singapore is planning ahead for water security, including drought resilience from climate change and other challenges.

3.3 The New Water Taps—Advances in Technology

As part of the plan to meet this challenge, Singapore has developed two new "National Taps," both made possible by significant advances in the field of membrane-based water purification, especially reverse osmosis.

Osmosis is a natural phenomenon used to transport salts and nutrients in plant and animal systems. The basic principle is that if two solutions with different concentrations are separated by a semipermeable membrane, the dilute solution will flow through the membrane toward the concentrated solution due to osmotic pressure.

When purifying water containing salts (saline or brackish water), the flow is in exactly the opposite direction, by taking an impure (concentrated) solution and passing it through a membrane to a more dilute, or "pure" form. Since this is the reverse of natural osmosis (hence called "reverse osmosis"), there is a significant energy requirement to push the water through the membrane in the opposite direction to what nature intends and thus to overcome osmotic pressure. Advances in this field have reduced the energy requirement significantly; however, the base entitlement cost of overcoming osmotic pressure is still significant.

Reverse Osmosis System.

Because of this cost, it is imperative to have an efficient pretreatment system, both to protect the reverse osmosis membranes and also to reduce the operational costs. There have been significant advances in this field (e.g., microfiltration and ultrafiltration), especially using hollow fiber membranes.

Ultrafiltration Membranes in use at a desalination plant.

3.4 Ultraclean, High-Grade Reclaimed Water (Known as NEWater)

Singapore has two separate systems for collecting rainwater and used water. Stormwater drains collect the runoff from rainwater, while all wastewater is collected separately and sent for treatment at various water reclamation plants for treatment to standards fit for sea discharge. The treated water then goes through a rigorous three-step purification process, namely microfiltration or ultrafiltration, followed by reverse osmosis, and disinfection by ultraviolet light (the last is an added safety barrier). More details on this process are available at the PUB website (http://www.pub.gov.sg/water/newater/newatertech/Pages/default.aspx).

The quality of water produced by this rigorous method is ultraclean, and safe to drink, surpassing WHO guidelines and other international standards (e.g., USEPA), and is regularly audited by an international panel of experts.

With its high quality, NEWater is mainly used for industries requiring high-grade water and for cooling. During dry weather, a small percentage of the NEWater production is added to the raw water in Singapore's reservoirs, which then undergoes further treatment at the waterworks before becoming part of the domestic water supply (http://www.pub.gov.sg/water/newater/Pages/default.aspx).

Bedok NEWater Visitor Center, Singapore.

NEWater, which currently supplies ~30% of Singapore's water needs, is one of the biggest success stories in water technology. The plan is to increase NEWater capacity to meet 55% of Singapore's future water needs by 2060. It will continue to play a significant part in Singapore's water planning for the future.

3.5 Desalinated Water

This technique of removing salt from seawater is generally associated with arid and water-scarce regions, because of its high energy costs. While Singapore is a tropical country with significant rainfall, desalination is part of its holistic approach to ensure a diversified and sustainable water supply for its population.

The reason for the relatively high costs relates to the concentration of salt in seawater, and correspondingly, the higher osmotic pressure to be overcome for its removal. As discussed earlier, with significant advances in membrane-based separation technologies, the overall costs of these plants have decreased significantly in recent decades.

Singapore currently has two desalination plants, both on its western side, providing ~25% of its water needs. Two additional 30 mgd desalination plants will be completed in the next 4 years, one at Tuas and one at Marina East. PUB is also building a fifth desalination plant on Jurong Island. It is envisioned that desalinated water will meet 30% of Singapore's future water needs by 2060.

Top view of one of Singapore's Desalination Plants.

3.6 Leadership in Water Purification R&D

Singapore has long established itself as one of the world's leaders in the development of innovative separation technologies, especially in the area of water, where it is recognized as a global hydrohub. Its environment and water R&D ecosystem is rated among the best in the world. The two major universities, Nanyang Technological University (NTU) and National University of Singapore (NUS), are recognized as world leaders in the field.

In addition to these two universities, there are also various government and industrial research laboratories engaged in cutting-edge research in the field, leading to several novel technologies being developed by these institutions. Singapore's next aim is to accelerate the translation of these technologies into unique commercially viable products.

One of the biggest challenges for commercialization of novel technologies, even if the intellectual property is protected, is the gap between laboratory processes, results and testing, and the full-scale final product. Hence, translation of lab-scale technologies directly to the final application would involve, in most cases, significant risk. This gap is based on several key challenges, including scale-up of component materials and equipment, systems-level thinking, and testing at pilot scale in an actual application setting.

The Separation Technologies Applied Research and Translation (START) Center is being set up as a national-level facility to bridge this gap and provide significant risk mitigation for the commercialization of materials, equipment, and processes related to separations. The benefits would be multiple, providing academic and research institutions a much needed platform to take their innovative technologies forward to actual products, in partnership with key companies, both global and local.

3.7 Looking Into the Future

Singapore's path to water sustainability for the future includes various innovative strategies including:

- Increasing catchment areas
- Increasing NEWater capacity and output
- Increasing desalination output
- Water demand management

In addition to these strategies, which are related to increasing the efficacies of the current sources of water, an increasingly important area of focus is industrial wastewater reuse.

Unlike domestic wastewater, which is fairly uniform, and hence can be consolidated and treated, industrial wastewater, coming from different industries, varies significantly depending on its point of origin. For example, process wastewater from refining would be very different from pharmaceutical process wastewater. Hence, consolidation and treatment would be limited to industrial clusters with similar products, and even this would lead to a fairly complex mixture of ingredient species. Hence, a key focus is to encourage individual industrial companies to reduce their water demand and increase their water reuse and recycling rate within their facilities.

Of course, economics plays a big part in any industry's decision on whether to recycle and reuse its process water, or to treat it to the extent mandated by law and subsequently discharge it. While there is a certain amount of industrial wastewater that is currently treated and reused, Singapore is actively exploring opportunities to reduce industrial sector consumption, correspondingly saving on the supply of fresh water to industry, as part of water-demand management. There are already a number of innovative water-efficiency grants available, and significant engagement with both the technology providers, and potential users, to encourage recycle and reuse.

CONCLUSIONS

With admirable foresight and planning, and equally importantly, its pristine execution, Singapore has not only overcome the significant challenges associated with its water security, but has emerged as a world leader in the field. It has been at the forefront of adopting cutting-edge technologies, and is constantly developing innovations to take its global hydrohub to the next level.

ACKNOWLEDGMENTS

All photographs and illustrations used in this chapter are courtesy of PUB, Singapore's national water agency.

REFERENCES

[1] Centre for Liveable Cities and PUB Singapore. Water – from scarce resource to national asset; Singapore's urban systems studies booklet series. Cengage Learning Asia Pte Ltd.; 2012.
[2] Soon TY, Jean LT, Tan K. Clean, green and blue – Singapore's journey towards environmental and water sustainability. Institute of South East Asian Studies; 2009.
[3] Tortajada C, Joshi Y, Biswas AK. The Singapore water story. Routledge; 2013.

CHAPTER SEVENTEEN

Water Quality and Public Health: Role of Wastewater

S. Tongesayi[1], T. Tongesayi[2]
[1]Walden University Public Health, Minneapolis, MN, United States
[2]Monmouth University, West Long Branch, NJ, United States

1. INTRODUCTION—THE PRECIOUS LIQUID

Did You Know?

- Of the approximately 1400 million cubic km world water reserve [1–3]:
  - only about 0.003% (45,000 cubic km) is freshwater
  - only about 9000–14,000 cubic km of the freshwater is economically available for use in domestic, agricultural, and industrial applications
  - about 70% of all freshwater withdrawals are used in agriculture
- About 3000 liters of water are required to produce a person's daily food requirements [3].
- Around 2–4 liters are required per person per day for drinking [3].
- Almost 80% of the current world population is under a significant threat of water security [4].
- In some poor countries people are barely getting a liter of water per person per day for drinking, cooking, and washing, compared to the normal daily requirement of 20 L per person [3].
- Global water demand is projected to increase by 55% by 2050 compared to demand levels in 2000, and will subject at least 40% of the world's population to severe water stress [5].
- About 1.2 billion people lack access to safe drinking water around the world [6].

The role of water in providing and sustaining health and life is uncontested. Humans need water for almost all activities including drinking, agriculture, food preparation, recreation, sanitation, hygiene, medicine, and industry. However, besides giving life, water can also be a source of human suffering through injuries and the spread of environmental contaminants that can cause human disease and death. In addition to being one of the major distributors of environmental contaminants, water is also one of the principal routes of exposure of contaminants to humans. Scientific research

Chemistry and Water
ISBN 978-0-12-809330-6
http://dx.doi.org/10.1016/B978-0-12-809330-6.00017-9

links 85% of all diseases, particularly chronic ones such as obesity, diabetes, heart disease, and cancer, to environmental contaminants [7]. It is also known that the majority of infectious diseases in developing nations are the direct result of the use of unsafe and unclean water for drinking, hygiene, and sanitation [8]. According to Brian Richter, a director of the Global Freshwater Program at the Nature Conservancy, a US nongovernmental organization, at least a billion people have no access to clean drinking water and at least half of hospital beds in the world are occupied by people with preventable waterborne diseases [8]. The World Health Organization (WHO) considers emerging and reemerging water-related diseases as the leading cause of morbidity and mortality worldwide [9]. Such water-related diseases include microbiological, chemical, hygienic, contact, vector habitat, excretal, disposal, and aerosol diseases [10]. Diseases such as cholera, typhoid fever, dysentery, and hepatitis A, which were eliminated by filtration and chlorination at the beginning of the 20th century, are reemerging primarily due to poor and/or aging infrastructure, struggling economies, and poor planning in the face of increased population-driven demand for water [11]. Apart from the effects of direct contact with contaminated water through drinking, sanitation, and hygiene, water contamination can also impact human health and life by contaminating food and other consumer goods through the use of contaminated water in agriculture and the manufacturing industry.

Water can be contaminated from both natural and anthropogenic sources, with the latter being the principal contributor in most cases. As shown in Fig. 17.1, the major anthropogenic sources of pollution are, ironically, the principal drivers of modern civilization. These sources include industrialization, urbanization, agriculture, commercial establishments, and waste-management sites. Contaminants from these sources enter aquatic systems primarily through (1) the pumping and/or overflows of raw sewage and raw effluents from agricultural, industrial, and commercial establishments into urban drainage systems and surrounding water bodies; (2) urban stormwater-drainage systems; and (3) leachates and runoff from solid waste sites, acid-mine drainage, and contaminated watersheds. The practice of combining raw sewage, industrial and commercial effluents, and storm water into a large effluent flow of wastewater and then pumping it into traditional sources of water for domestic, agriculture, and industrial purposes or using it directly for irrigation in agriculture has become the norm in a large number of countries [11]. Such a practice is potentially making wastewater one of the principal anthropogenic sources of water contamination. By the nature of its origin, wastewater and wastewater-contaminated water can contain a slew

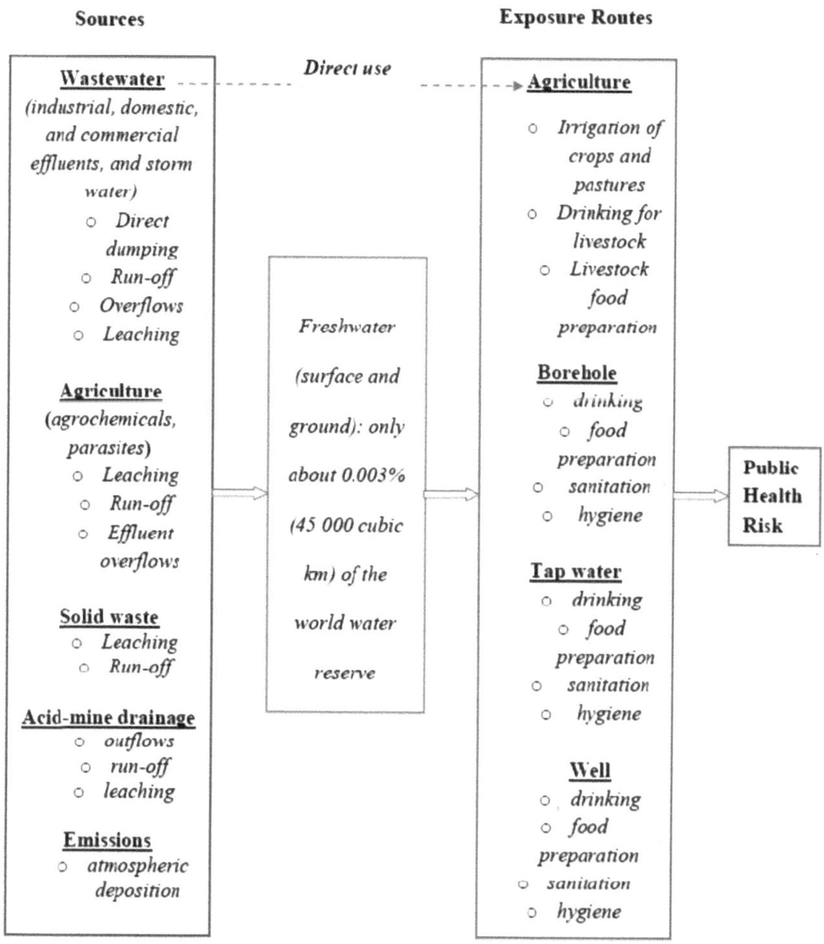

Figure 17.1 Major sources of water pollution and the principal routes of exposure of waterborne contaminants to humans.

of chemical pollutants and pathogens in levels that may be detrimental to human health and life.

Fig. 17.1 also shows that water for drinking, sanitation, and hygiene can be delivered to consumers via the tap or can be obtained from a borehole or a well, among other sources. Tap water is usually the preferred source but the water can be just as good as diluted wastewater or water from an unprotected well if it is not adequately treated prior to distribution. In most parts of the developing world, local authorities are failing to cope with the demand for clean and safe water for domestic as well as commercial, agricultural, and industrial purposes [12]. As a result, it is now common for city dwellers to supplement tap water with borehole and well water. It is not uncommon for borehole and well water to be used untreated, and wells can also be unprotected. In some cases, tap water is also unsafe because of insufficient treatment. As a result, the situation depicted in Fig. 17.2, which shows

Women Carrying Water[13]

"Illness from drinking dirty water and the time lost fetching it robs entire communities of their futures."[15]

Poverty and Clean Water[16]

"Throughout parts of the developing world, some people work hard just to get dirty water for drinking, cooking, and for other personal needs...." Donald Hopkins, the Carter Center, Atlanta, Georgia[1]

These images are not a graphic misrepresentation of facts but reveal a true but sad reality that is certainly negating the gains of civilization. To millions of people this is the new "normal".

"There are a number of reasons why poverty has become an epidemic in Africa. Poverty can be the result of political instability, ethnic conflicts, climate change and other man-made causes. But one of the greatest causes of poverty in Africa is also the most overlooked.....the lack of access to clean drinking water."[14]

"She dreams of spending time school rather than walking miles to find only dirty water that will likely make her sick. She dreams tomorrow could be different than today"[16]

Figure 17.2 Typical water situation in some parts of the developing world. *Photos African Women Carrying Water (n. d.) jpegthewaterproject. org. http://www.bing.com/images/search?q=African+Women+Carrying+Water+On+Their+Head+thewater project&go=Search&qs=bs&form=QBIR; The Water Project. (n. d.). Poverty in Africa begins with a lack of clean water: poverty and water are related. https://thewaterproject.org/poverty; The Water Project. (n. d.). Why Water? Dirty water causes needless suffering. https://thewaterproject.org/why-water/; The Water Project. (n. d.). Safe and reliable water changes lives. https://thewaterproject.org/?gclid=CM2ojPTdvssCFcQjgQodYlQPng; Courtesy of TheWaterProject.org.*

the typical water situation in the developing world [13–16], is no longer confined to rural areas where communities traditionally rely on protected or unprotected wells, streams, and boreholes for their daily water needs. The rural communities, however, fare much better in terms of getting "safe" water compared to their city counterparts because there are fewer anthropogenic pollution sources in rural areas than in cities. Regardless, the consumption of *untreated* water is always a health hazard and explains the frequent outbreaks of waterborne diseases in developing countries [17].

The situation shown in Fig. 17.2 should not be happening in the 21st century. The situation is, unfortunately, getting worse and certainly threatening to negate the gains in public health in recent decades. According to the Natural Resources Defense Council (NRDC), "Clean and plentiful water provides the foundation for prosperous communities. We rely on clean water to survive, yet right now we are heading toward a water crisis. Changing climate patterns are threatening lakes and rivers, and key sources that we tap for drinking water are being overdrawn or tainted with pollution" [18]. This assertion by the NRDC is mostly true except that we are not heading toward a water crisis—we already have a water crisis! For instance, a study on the quality of drinking water in the United States reported the presence of more than 300 chemicals in the tap water supplied to the public between 2004 and 2009 in 45 states [18]. Only 114 of the detected chemicals were part of the US Environmental Protection Agency (US EPA) regulations on water quality [19]. In fact, the US EPA concurred with the findings and acknowledged that at least 3 million Americans were exposed to drinking water with illegal concentrations of carcinogenic environmental pollutants such as arsenic and radioactive elements [19]. The source of the pollutants was identified as mainly untreated wastewater, a combination of sewage, effluents from commercial, agricultural, and industrial establishments, and storm water. It is therefore not surprising that illness in about 20 million Americans every year as well as an increase in the incidence of breast and prostate cancers over the past 30 years are believed to be associated with environmental contaminants in drinking water [18,19].

It is common knowledge that many sewer systems across the United States frequently overflow, spilling raw sewage into surrounding freshwater sources in the process. In fact, more than 9400 out of the nation's 25,000 sewage systems were reported to have violated the law by dumping untreated or partly treated human waste, chemicals, and other hazardous materials into drinking water sources between 2006 and 2009 [20].

Figure 17.3 Sewage overflows into Newtown Creek in Brooklyn. *Credit: Fred R. Conrad/ The New York Times/Redux., Duhigg C. As sewers fill, waste poisons waterways. 2009; http://www.nytimes.com/2009/11/23/us/23sewer.html.*

New York's sewage system reportedly overflows almost every time it rains; see Fig. 17.3. Such sewage overflows in the United States, just like in other countries, are the direct result of large urban populations, storms, and aged infrastructure [20]. The US EPA once acknowledged that such overflows pose significant public health risks and that it has been the agency's priority to enforce the elimination or reduction of the overflows since the mid-90s [20].

Besides the contamination of freshwater sources by wastewater, numerous complaints and confirmed cases of groundwater contamination have been reported at and around oil or gas drilling sites in Pennsylvania, Ohio, West Virginia, and Texas [21]. This is probably a result of the hundreds of thousands of gallons of water, sand, and chemicals that are pumped into the ground during fuel extraction from the shale formation inadvertently mixing with groundwater. Some of the groundwater may find its way to the surface together with the acquired high levels of salts, drilling chemicals, heavy metal(loid)s, and naturally occurring radiation [19]. The increasing demand for energy, together with the increasing wastewater production and the climate change phenomena, will, unfortunately, remain impediments to the provision of clean and safe water to citizens across the world. According to the NRDC, "dirty water is the world's biggest health risk, and continues to threaten both quality of life and public health in the United States. When water from rain and melting snow runs off roofs and roads into our rivers, it picks up toxic chemicals, dirt, trash and disease-carrying organisms along the way. Many of our water resources also lack basic protections, making them vulnerable to pollution from factory farms, industrial plants, and activities like fracking" [22].

History has taught mankind that water has to be sufficiently treated before use, particularly for drinking and other domestic use. As a result, incompetence and poor planning on the part of those responsible for treating the water can have serious and lasting health impacts on the general public. A typical example is what happened in the Michigan city of Flint in the United States. The city's drinking water became contaminated with lead (Pb) in April 2014 after the city opted to draw and treat its drinking water from the known-to-be-contaminated Flint River instead of the usual practice of distributing to its residents water from Lake Huron that was treated by the Detroit Water and Sewerage Department. The Flint River water was not sufficiently treated and, as a result, leached Pb from distribution pipes and fixtures. The incident exposed all those who drank the city's water, including over 8600 children, to Pb during the entire period they were supplied with water from the Flint River [23]. In addition to Pb, the water also violated the Safe Drinking Water Act by having elevated levels of carcinogenic chlorine disinfection by-products such as trihalomethane (TTHM or THM) [24]. Residents reportedly complained about the color, smell, and taste of the water and the city responded by issuing several boil advisories to kill coliform bacteria [25].

The foregoing discussion clearly shows that wastewater, as defined earlier, is one of the principal sources of water contamination. Among its slew of contaminants are microplastics, whose presence in the water is emerging as one of the major water-quality challenges. Microplastics are comprised of (1) synthetic polymer products manufactured as additives in various consumer products such as hand, facial, and body cleansers; (2) small pieces from degrading industrial and domestic polymer products; (3) polymeric fibers released by washing of synthetic clothing and plastic abrasion during dishwashing; and (4) preproduction pellets that are used in plastic production [26,27]. They occur in various shapes that include spheres, fibers, and fragments, and enter the aquatic environment primarily via improper waste disposal, insufficient waste management, and urban runoffs [26,27]. In addition to being pollutants, microplastics adsorb other chemical pollutants and harbor pathogenic organisms, which they then distribute thanks to their water buoyancy, chemical inertness, and durability [26–32]. As a result, microplastics can introduce pathogenic organisms and chemical toxicants from raw wastewater to freshwater and marine habitats. By adsorbing chemical pollutants and microorganisms on their surfaces, microplastics can influence the biogeochemistry of metal(loid)s and hence alter the speciation, mobility, bioavailability, and toxicity of chemical pollutants in water. In this chapter, the role of microplastics in the biogeochemistry of metal(loid)s will be illustrated using the chemistry of arsenic and mercury.

2. CHEMICAL CONTAMINANTS

Based on its origin, wastewater can contain a wide range of environmental chemical contaminants that include toxic heavy metal(loid)s, chlorinated dibenzo-*p*-dioxins (CDDs), polychlorinated biphenyls (PCBs), polycyclic aromatic hydrocarbons (PAHs), phthalates, chlorinated organic solvents, steroids, prescription drugs, nonprescription drugs, fragrances and flavors, antibiotics, pesticides, fire retardants, detergents, disinfectants, cosmetics, dyes, resins, and fuels [33–35]. Some of these contaminants appear on the US EPA list of priority pollutants but a significant proportion remains unregulated [36]. In addition to these common pollutants, engineered nanomaterials (ENMs) and microplastics are also entering the aquatic system at unprecedented rates thanks to recent advances in technology research, innovation, and development [37–39]. The rate and scope of water contamination by these environmental contaminants is undoubtedly hampering efforts by a large number of nations to provide clean and safe drinking water to their citizens. The following discussion will focus on the health impacts of selected heavy metal(loids) and persistent organic pollutants (POPs) as well as the role of microplastics on the distribution and geochemistry in selected chemical contaminants.

2.1 Metal(loid)s

Did You Know?

- Natural elements in the earth's crust.
- Used in a wide range of consumer products.
- Nondegradable but convert from one form to the other.
- Redistributed in the environment by natural and anthropogenic activities.

- Are emerging as an important class of human carcinogens.
- Arsenic, cadmium, chromium (VI), beryllium, and nickel are accepted human carcinogens.
- Arsenic, cadmium, lead, and mercury have no nutritional value and are potentially the most toxic; they are associated with obesity, diabetes, and cancers as well as neurological and heart diseases.
  - Toxicity depends on chemical form and dose—thresholds do not exist for carcinogens.
  - Typically do not require bioactivation.
  - Many follow the metabolic pathways of similar essential metals.
- Cobalt, chromium (III), copper, iron, manganese, molybdenum, selenium, and zinc are essential micronutrients
- Aluminum, antimony, barium, beryllium, silver, strontium, and thallium—no known effects, widely used in industry—higher chances of exposure to humans

Heavy metal(loid)s such as arsenic (As), cadmium (Cd), lead (Pb), mercury (Hg), and chromium (Cr) and their compounds are some of the most toxic yet common chemicals found in wastewater. Their presence in wastewater is a direct result of them being integral components of a large number of consumer products. In addition to entering the natural aquatic system via wastewater, heavy metal(loid)s can leach into both surface and groundwater from solid-waste sites that include electronic waste (E-waste). E-waste includes disused computers, computer monitors, television sets, and cell phones and contains toxic and carcinogenic heavy metals such as Pb, Hg, Cd, hexavalent Cr, Ni, and Sb, as well as organic pollutants such as polybrominated di-Phenyl ethers and PCBs [40].

2.1.1 Arsenic

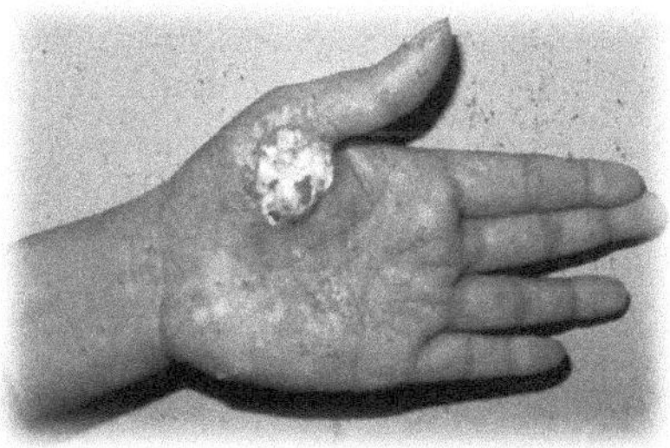

Lesions on hand, cancer [41].

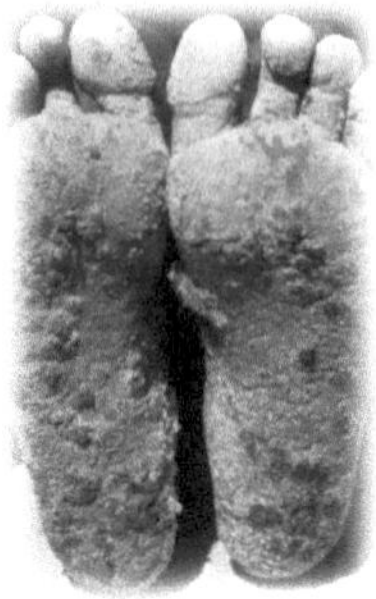

Lesions on feet, extreme [41].

Did you know?

- A neurotoxin, carcinogen, and cocarcinogen that is linked to lung, bladder, kidney, liver, and nonmelanoma skin cancers, as well as cardiovascular diseases, diabetes, and reproductive toxicity [42–44].
- Exposure in the womb and early childhood is associated with development of chronic diseases such as bladder, lung, and skin cancers, and cardiovascular and respiratory diseases in early life or later in life [45–49].
- Oral ingestion of high doses (about 300 mg) can kill an adult.
- Single or repeated oral high doses (0.04 mg/kg/day) for can cause diarrhea, stomach cramping, anemia, leukopenia, peripheral neuropathy, and cardiovascular diseases.
- Chronic exposure (inhalation or oral) to small doses (0.01 mg/kg/day or higher) can cause diffuse or spotted hyperpigmentation of the skin, benign skin lesions (hyperkeratosis), skin cancer, lung cancer, and liver disease.
- Chronic exposure in *drinking water* to levels as low as 2 μg/L can cause bladder, lung, liver, and kidney cancers.
- Symptoms of arsenic exposure include nausea, vomiting, diarrhea, abnormal heart rhythm, blood vessel damage, and a pins and needles sensation in hands and feet [50].

Arsenic is toxic to both plants and animals and accumulates in living tissues because of its high affinity for proteins, lipids, and other cellular components [51]. As noted, arsenic is a neurotoxin, carcinogen, and cocarcinogen that is linked to lung, bladder, kidney, liver, and nonmelanoma skin cancers [56–58]. The metalloid is also associated with cardiovascular, pulmonary, immunological, neurological, and endocrine diseases and diabetes

[42,43,52]. According to current knowledge, the metalloid induces tumorigenicity via (1) cellular proliferation, apoptosis, and differentiation; (2) reduction in the expression of the phosphatase and tensin homolog tumor suppressor gene; (3) oxidative stress; (4) genotoxic damage; (5) chromosomal abnormalities; and (6) epigenetics [43]. Epigenetics mechanisms involve alteration of the methylation levels of global DNA and gene promoters; acetylation, methylation, and phosphorylation of histone; miRNA expression; and acceleration and exacerbation of plaque formation in arteries which increases the risk of cardiovascular diseases [42,43]. At sufficiently high levels, arsenic can cause dyspigmentation, keratosis, peripheral vascular diseases, reproductive toxicity, and neurological effects [43,44]. Prenatal as well as early childhood exposure to the metalloid has been associated with the development of chronic diseases that include bladder, lung, and skin cancers, as well as cardiovascular and respiratory diseases in early life or later in life [45–49]. Prenatal exposure to levels considered low or moderate has also been linked to neurological and cognitive dysfunction, as well as memory and learning impairments [44,53].

Symptoms of arsenic exposure include nausea, vomiting, diarrhea, abnormal heart rhythm, blood vessel damage, and a pins and needles sensation in hands and feet [50]. Current literature suggests that the two inorganic forms of arsenic, As(III) and As(V), are more toxic than the organic forms, with As(III) being the most toxic of all forms of the metalloid [54]. In the form As(III), $\left(AsO_2^{-} \text{ or } AsO_3^{3-}\right)$, the metalloid inhibits various dehydrogenases like pyruvate, α-ketoglutarate, and dihydrolipoate and binds to sulfhydryl groups of proteins and dithiols such as glutaredoxin [55]. On the other hand, in the form As(V), $\left(AsO_4^{3-}\right)$, it inhibits oxidative phosphorylation due to its structural similarity to phosphate $\left(PO_4^{3-}\right)$ [55]. As(V) also manifests its toxicity via reduction to As(III) first [56].

2.1.2 Cadmium

Did you know?

- Is a carcinogen associated with liver, bladder, stomach, renal, pulmonary, and hematopoietic cancers, as well as kidney and cardiovascular diseases [42,57,58].
- Increases cancer susceptibility in type-2 diabetes patients [59].
- Causes decreased performance in attention- and perception-related tasks in adults and learning disabilities in children [57,60].
- High exposure to cadmium fumes causes acute bronchitis, emphysema, or pulmonary fibrosis and lung cancer [61].

- Chronic inhalation of cadmium can cause lung cancer [61].
- Chronic exposure to low doses of cadmium (cigarette smoking or daily ingestion of cadmium-contaminated rice) can cause kidney tubular dysfunction and osteoporosis [62].

Just like arsenic, cadmium is a classified human carcinogen that is linked to cancers of the liver, bladder, and stomach, as well as cancers of the renal, pulmonary, and hematopoietic systems [42,57]. It is also associated with kidney and cardiovascular diseases, as well as bone fractures following long-term exposure to low levels of the metal [58]. The metal is also associated with increased cancer susceptibility in type-2 diabetes patients, with prolonged exposure reportedly leading to prediabetes, diabetes, and overall increased sex-dependent cancer mortality in some types of cancer [59]. Cadmium also causes decreased performance in attention- and perception-related tasks in adults exposed to levels considered low over long periods while in children and teens exposure to levels considered safe increases chances of developing learning disabilities [57,60]. The toxicity of cadmium manifests via epigenetic mechanisms that include hypo- and hypermethylation of DNA [42].

2.1.3 Lead

Did you know?

- Is a neurotoxin that affects the nervous, renal, skeletal, and hematopoietic systems [42].
- Pre, peri-, and postnatal exposures cause irreversible mental retardation, learning disabilities, and in some instances coma and death in children [42].
- Early-life exposure increases susceptibility to hypertension, cardiovascular disease, diabetes, schizophrenia, and neurological diseases later in life [42,63].
- Causes hypertension, cardiovascular diseases, reproductive defects, and calcium deficiency in adults [42,63].
- Recently linked to lung and stomach cancers [64].

Lead is a known human neurotoxin that affects the central nervous system as well as the renal, skeletal, and hematopoietic systems. The major target organs for Pb in the human body include blood plasma, nervous system, and soft tissues, but the metal subsequently accumulates in bone. Pre, peril-, and postnatal exposures to the metal are known to cause irreversible mental retardation, learning disabilities, and in some instances coma and death in children [42]. Lead induces toxicity by causing oxidative stress, disruption of cell signaling and neurotransmission pathways, and influencing the methylation of DNA [42]. Exposure to Pb in early life is also reported to induce epigenetic mechanisms, which increase susceptibility to diseases such as hypertension, cardiovascular disease, diabetes, schizophrenia, and neurological diseases later in life [42,63]. In adults, Pb exposure is linked to increased chances of hypertension, cardiovascular diseases, reproductive defects, and calcium deficiency. Recent studies have linked the metal to lung and stomach cancers [64].

2.1.4 Mercury

Did you know?

- Is a neurotoxin but there is no current data that link Hg to human cancers [65,66].

2.1.4.1 Elemental mercury (Hg^0)

- Exposure occurs via mercury vapor inhalation.
- Target organs are the central nervous system and the kidneys.
- The toxicity is believed to be due to mercuric mercury.
- Readily crosses the blood–brain barrier and is oxidized to mercuric mercury, which then binds to macromolecules in the brain.
- Causes tremor, psychiatric disturbances, altered behavior, glomerulonephritis, and renal failure.

2.1.4.2 Methyl Mercury (CH_3Hg^+)

- Exposure occurs via consumption of fish and shellfish
- The maximum allowable concentration of CH_3Hg^+ in seafood (edible portions) is 1 mg/kg.
- The bodyweight-based limit on CH_3Hg^+ consumption of 0.1 mg/kg bodyweight per day.
- The target organ—the brain
- Crosses the placenta resulting in exposure and toxicity to the developing fetal brain.
- Prenatal exposure impairs cognitive thinking, memory, attention, language, as well as fine motor and visual spatial skills.
- Low levels of exposure cause impaired development of motor and language skills during neonatal life and early childhood.
- Higher exposures can cause severe cognitive effects, including paresthesia, blindness, and deafness.

- More severe exposures can cause fetal death and abortion.
- Symptoms of exposure include impairment of peripheral vision, disturbances in sensations, lack of coordination of movements, impairment of speech, hearing, walking, and muscle weakness.

2.1.4.3 Inorganic Mercury (Hg^{2+})

- Damages the gastrointestinal tract, nervous system, and kidneys.
- Symptoms of exposure include skin rashes and dermatitis, mood swings, memory loss, mental disturbances, and muscle weakness.

Like lead, mercury is a human neurotoxin [65,66]. In water, it exists in various forms that include methyl mercury (CH_3Hg^+) and inorganic mercury (Hg(II)), with the former being the most toxic of all the forms of the metal. Fetuses and children are more susceptible to the neurological effects of CH_3Hg^+ than adults. Symptoms of exposure to CH_3Hg^+ include impairment of peripheral vision, disturbances in sensations, lack of coordination of movements, impairment of speech, hearing, walking, and muscle weakness. Prenatal exposure to CH_3Hg^+ leads to impairments in cognitive thinking, memory, attention, language, as well as fine motor and visual spatial skills. Also, prenatal exposures to CH_3Hg^+ at levels that do not induce symptoms to the mother have been reported to be harmful to the infant. Hg(II) is reported to cause damage to the gastrointestinal tract, the nervous system, and the kidneys. Symptoms of exposure to this form of the metal include skin rashes and dermatitis, mood swings, memory loss, mental disturbances, and muscle weakness. Current knowledge does not associate mercury with any of the known human cancers even though mercuric chloride and methyl mercury are reported to have caused tumor development in laboratory mice [65,66].

According to the US Food and Drug Administration (FDA), the maximum allowable concentration of CH_3Hg^+ in seafood (edible portions) is 1 mg/kg [67]. Any product reaching or exceeding this limit is withdrawn from the market [67]. The US EPA also established a guideline that recommends a bodyweight-based limit on CH_3Hg^+ consumption of 0.1 mg/kg bodyweight per day [68]. The CODEX Alimentarius general standard for contaminants and toxins in foods puts the maximum concentration of CH_3Hg^+ in processed or fresh noncarnivorous fish and crustaceans involved in international trade at 0.5 mg/kg wet weight. while in carnivorous fish such as shark, swordfish, tuna, and pike it stands at 1 mg/kg wet weight [69]. In the United States, the majority of the states have one or more advisories on fish as shown in Fig. 17.4. The recognition of the global health

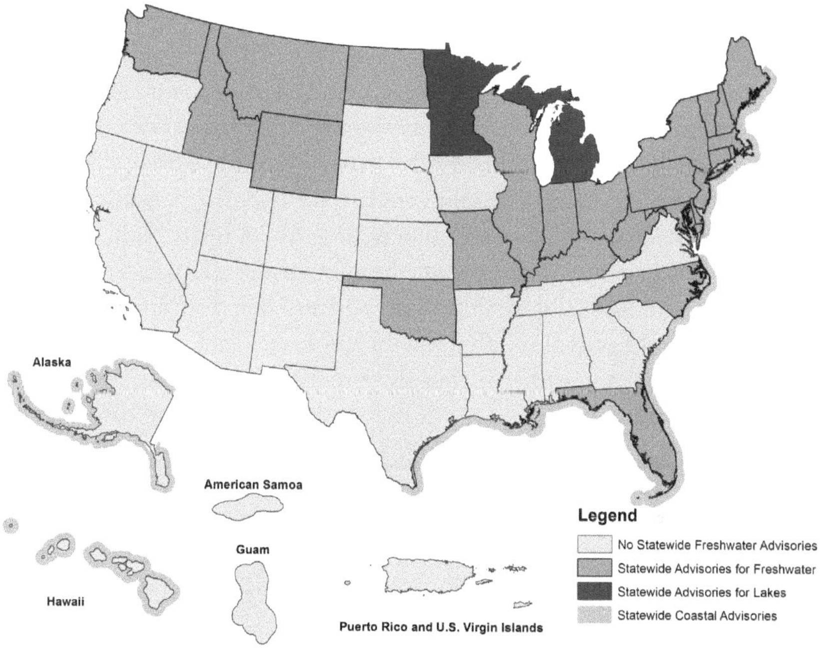

Figure 17.4 States with statewide advisories in effect in 2011 [71].

threats posed by mercury gave birth to the recent Minamata Convention on Mercury, a global treaty meant to protect human health and the environment from the effects of mercury [70]. The major highlights of the treaty include the banning of new Hg mines and phasing out of existing mines as well as the implementation of mechanisms for the control of Hg emissions from new power plants within 15 years of the treaty coming into effect. The treaty also calls for international regulations of the informal sector for artisanal and small-scale gold mining and the banning of a wide range of consumer products that contain Hg by 2020 [70].

2.1.5 Chromium

Did You Know?

2.1.5.1 Chromium (VI)

- Is a carcinogen associated with liver, bladder, stomach, renal, pulmonary, and hematopoietic cancers, as well as kidney and cardiovascular diseases [42,57,58].
- Increases cancer susceptibility in type-2 diabetes patients [59].
- Causes decreased performance in attention- and perception-related tasks in adults and learning disabilities in children [57,60].

2.2 Heavy Metal Biogeochemistry and Toxicity: The Role of Microplastics

It is a known fact that for noncarcinogenic metal(loid)s, toxicity is dependent on chemical form, bioaccessibility, bioavailability, and dose. For carcinogenic metal(loid)s, no thresholds exist and any dose can trigger tumorigenesis [72]. As noted, other metals are essential poisons; they are important for certain biochemical functions in the human body at certain thresholds but become toxic beyond those thresholds [73]. Essential metal requirements are normally met through food and therefore any inadvertent exposure to the metals through contamination can lead to essential metal overdose and toxicity. Other metals are essential in one form and toxic in the other. A typical example is Cr; it is essential as Cr(III) and carcinogenic in the form of Cr(VI). The mobility, bioaccessibility, bioavailability, and toxicity of metal(loid)s are dependent on the biogeochemistry of the metals in specific environmental compartments as dictated by prevailing biological, chemical, physical, and geological processes. As pointed out earlier, the biogeochemistry of metal(loid)s in the aquatic system in this chapter will be illustrated using arsenic and mercury.

2.2.1 Arsenic Biogeochemistry

As shown in Figs. 17.5 and 17.6, contamination of groundwater by arsenic is global and many countries in South Asia including Bangladesh, India, and Nepal have thousands of drinking water wells sunk in areas with concentrations of arsenic that exceed 10 ppb, the maximum contaminant level of the metalloid in drinking water [74,75]. However, it should be noted that arsenic is still toxic at levels as low as 2 ppb. According to the USGS, drinking water for a significant proportion of the US population has arsenic levels that exceed the maximum contaminant level of the metalloid in drinking water as shown in Fig. 17.7 [76]. The USGS data is reportedly based on samples from 31,350 wells, and as shown on the map, high concentrations appear to be more prevalent in the West, the Midwest, parts of Texas, and the Northeast [76].

Natural activities such as biological processes, weathering, and volcanic eruption as well as anthropogenic activities such as mining and agriculture are responsible for the input and distribution of arsenic in the environment. For example, the use of arsenic-containing pesticides such as Roxarsone (3-nitro-4-hydroxyarsonic acid, ROX) at poultry farms has significantly contributed to the loading of the metalloid in the environment. According to current scientific literature, most of the ROX ingested by chicken is

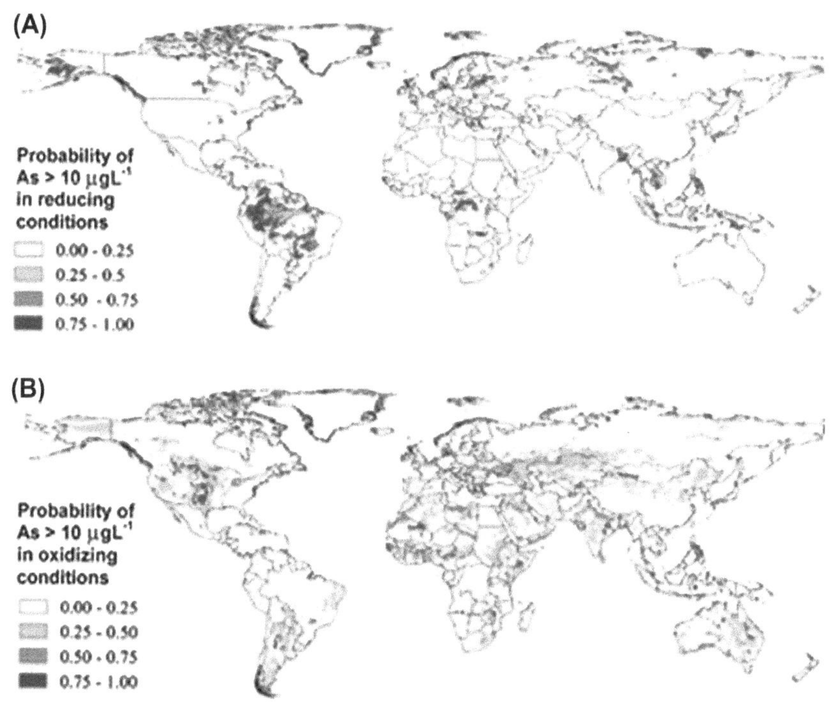

Figure 17.5 Modeled global probability of geogenic arsenic contamination in groundwater for (A) reducing groundwater conditions and (B) high-pH/oxidizing conditions where arsenic is soluble in its oxidized state [74].

excreted unchanged in the manure bedding, and poultry litter (PL) collected from a typical 42-day feeding period can contain 30–50 mg/kg^{-1} of total arsenic [77]. As a result, the practice of amending soils with arsenic-containing PL and the use of uncovered windrows for litter storage could be contributing to the contamination of drinking water sources.

The biogeochemistry of arsenic in the natural environment is better understood in terms of a dynamic balance of biological, chemical, physical, and geological processes on the various forms of the metalloid [78,79]. The inorganic forms (i.e., arsenite, As(III) and arsenate, As(V)) are the most prevalent in the environment, with the predominant form of the two being dependent on the prevailing chemical and/or microbiological conditions. As(III) is more prevalent under reducing conditions in natural water systems as arsenous acid species, H_3AsO_3, $H_2AsO_3^-$, or $HAsO_3^{-2}$, while As(V) is more stable in oxygenated waters as H_3AsO_4, $H_2AsO_4^-$, $HAsO_4^{-2}$,or AsO_4^{-3}, depending on pH conditions [51,56]. As(III) is more mobile than As(V) due to the higher affinity of the latter for insoluble species

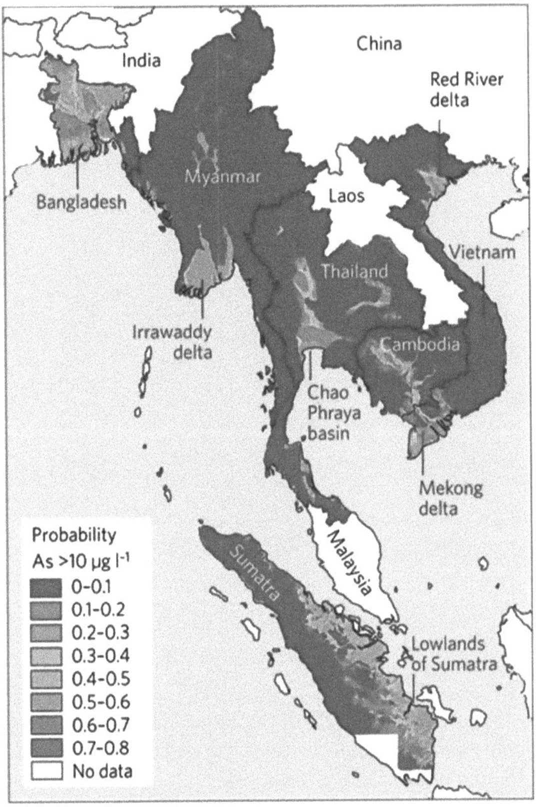

Figure 17.6 Geological map of Southeast Asia showing the probability of arsenic contamination in the groundwater. *Reproduced with permission from Nature Publishing Group, Winkel L, et al. Probablity of Aarsenic contamination in groundwater. Nature Geosci. 2008; http://dx.doi.org/10.1038/ngeo254. http://www.nature.com/news/2008/080714/full/454263a/box/1.html.*

such as hydrous ferric or manganese oxides and the uncharged molecular state of the former in common soil/sediment pH range 5–8 [80]. Organic forms of arsenic such as methanearsonic acid and dimethylarsonic acid that have been extensively used as herbicides in agriculture can readily undergo aerobic oxidation and demethylation to As(V) [51,56,81]. The As(V) produced can then be readily reduced to As(III) by dissolved organic matter, iron or manganese oxyhydroxides, or microorganisms [51,56,81]. According to scientific literature, organic matter containing semiquinone free radicals, carboxylate, and phenolate is the dominant reductant in redox reactions involving metals in soil/sediment and water environments [77]. Dissolved organic matter (DOM) found in natural waters (e.g., fulvic acid,

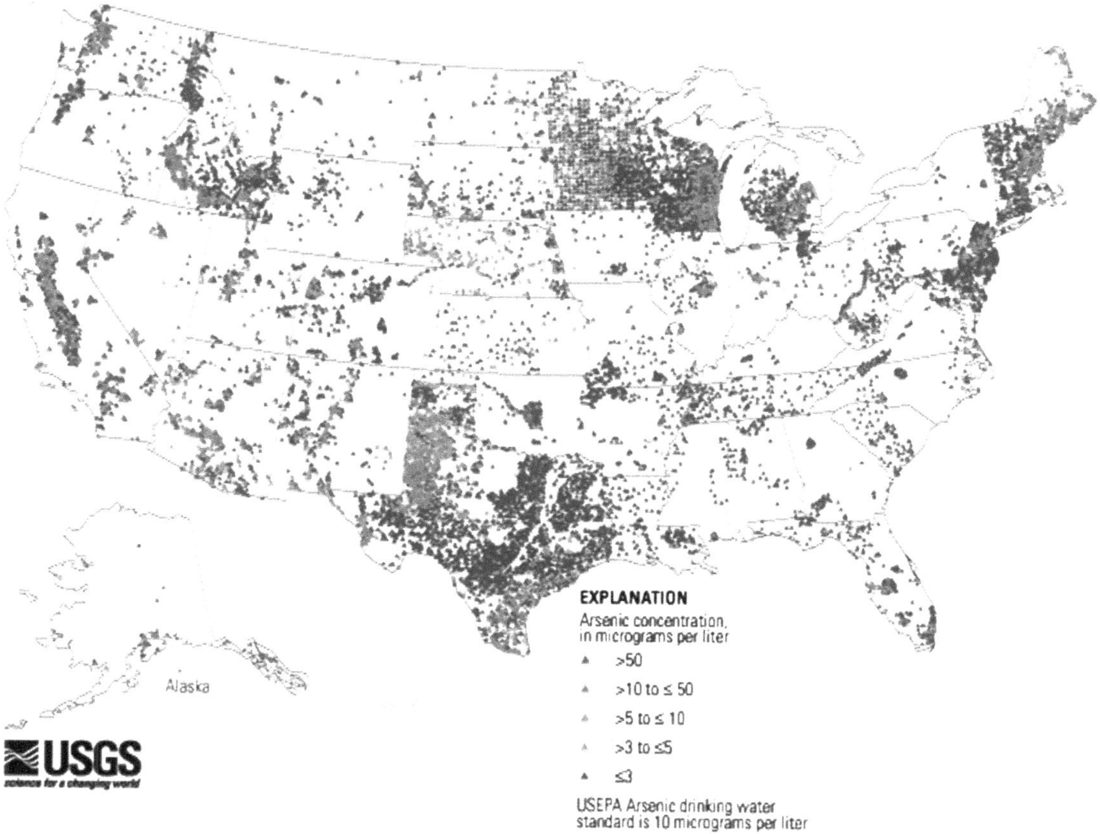

Figure 17.7 Arsenic in groundwater of the United States. *Credit: US Geological Survey, Department of the Interior/USGS, U.S. Geological Survey (2011). Arsenic in groundwater of the United States: featured publications. http://water.usgs.gov/nawqa/trace/arsenic/.*

FA and humic acid, HA) contains both electron-rich and electron-deficient sites that are responsible for its redox properties [77].

According to work by Tongesayi and Smart, fulvic acid (FA) can reduce As(V) to As(III) as well as oxidize As(III) to As(V) depending on pH, with lower pH favoring As(III) oxidation and high pH favoring As(V) reduction [77]. Their work also showed that the FA-mediated redox cycling of the arsenic species can occur in both light and dark, with reduction being accelerated by Fe(III). The reduction of As(V) by FA was hypothesized to occur via complexation, which is reportedly enhanced by the presence of iron through intermetallic bridging. It was postulated that Fe(III) facilitates the binding of the negatively charged As(V) ($H_2AsO_4^-$, $HAsO_4^{-2}$, or AsO_4^{-3}) by FA resulting in the reduction of As(V) to As(IV) and Fe(III) to Fe(II). The Fe(II) produced then reduces As(IV) to As(III). In the absence of Fe(III), binding of the negative As(V) by the negative FA was thought to occur through intermetallic bridging by cationic metals inherent in the FA. Competition from H^+ ions for the binding sites on FA was thought to be the reason for the observed diminished reduction of As(V) at pH 2 [77].

Microorganisms are also known to influence the geochemistry of arsenic directly as well as indirectly through dissimilatory iron reduction. In the latter, bacteria gain energy by coupling reduction of Fe(III) to Fe(II) with oxidation of an electron donor such as organic carbon or hydrogen under anaerobic conditions [52,82]. For instance, Zobrist et al. showed that *Sulfurospirillum barnesii* can grow anaerobically using Fe(III) or As(V) as electron acceptors [83]. Some bacteria have also been shown to rapidly oxidize As(III) via arsenic detoxification rather than energy generation [55,82]. There are also known anaerobic bacteria that utilize As(V) as an alternative electron acceptor during detoxification or energy generation under anaerobic conditions [55]. Arsenic species can also be biomethylated by certain fungi, yeast, and bacteria species. Microorganisms also play a significant role in remobilizing arsenic species from sediments into the water column [78]. Microbial activity can also influence the mobility of arsenic in subsurface soils in a way that can significantly impact groundwater contamination. Arsenic is immobilized in soils by sorption and/or coprecipitation onto solids such as iron, manganese, and aluminum oxides. However, the stabilities of the solid arsenic sequesters are dependent on prevailing redox conditions. For instance, iron(III) oxides become unstable under reducing conditions while sulfide minerals become unstable under oxidizing conditions [78]. Reducing conditions, as discussed earlier, tend to favor the more mobile and more toxic As(III) while oxidizing conditions tend to favor the less mobile and less toxic As(V). A review by Korte and Fernando suggests that As(III) is more prevalent in groundwater than As(V). This is probably because groundwater conditions are mainly reducing due to the lack of oxygen. The above discussion on the biogeochemistry of arsenic at the sediment–water interface clearly favors dissolved arsenic compared to sediment-bound arsenic [88].

2.2.1.1 Potential Role of Wastewater and Microplastics

Wastewater contains all the biological, chemical, and physical ingredients involved in the biogeochemistry of arsenic across the sediment/water interface. Fig. 17.8 shows a simplified schematic of the biogeochemistry of arsenic in water that is contaminated with wastewater laden with microplastics. Upon entering wastewater, microplastics will adsorb a slew of microorganisms and chemical species that include As, HA, FA, Fe, Mn, Al, Ca, Ba, and S. As a result, the reactions responsible for the speciation of arsenic will occur on the surface of the microplastics in the wastewater itself and in any water that gets contaminated with the wastewater. Due to the concentration effect on the surface of the microplastics, reaction rates on their surfaces will likely be higher than in the water. Reduction conditions will favor the more mobile and more toxic As(III) while oxidizing conditions will favor As(V). The high biochemical oxygen demand (BOD) in wastewater will promote reducing conditions, hence As(III). The produced As(III) and/or As(V) can dislodge from the microplastic surface into the water, adsorb onto the microplastic surface, adsorb onto Fe, Mn, and Al oxides [78] on the surface of the microplastics, and/or precipitate as sulfides and arsenates of calcium and barium [78] on the surface of the microplastics. Some of the arsenic species adsorbed on Fe, Mn, and Al oxides and that precipitated as sulfides and arsenates of calcium and barium may dislodge from the microplastics and then remain suspended in the water column or sink to the bottom sediments. As(III) will tend to go into the water more than As(V) because of its higher mobility as explained earlier. The higher concentration of complexing agents in wastewater will promote the remobilization of the sediment-bound arsenic species into the water column [78] where it will be converted to mostly As(III) due to wastewater-induced reducing conditions. Overall, wastewater and microplastics seem to favor the most toxic form of arsenic in the water column, i.e., As(III). Thus wastewater and microplastics can significantly increase human health risk from arsenic toxicity will increase through drinking water, the seafood chain, and agricultural food as depicted in Fig. 17.8.

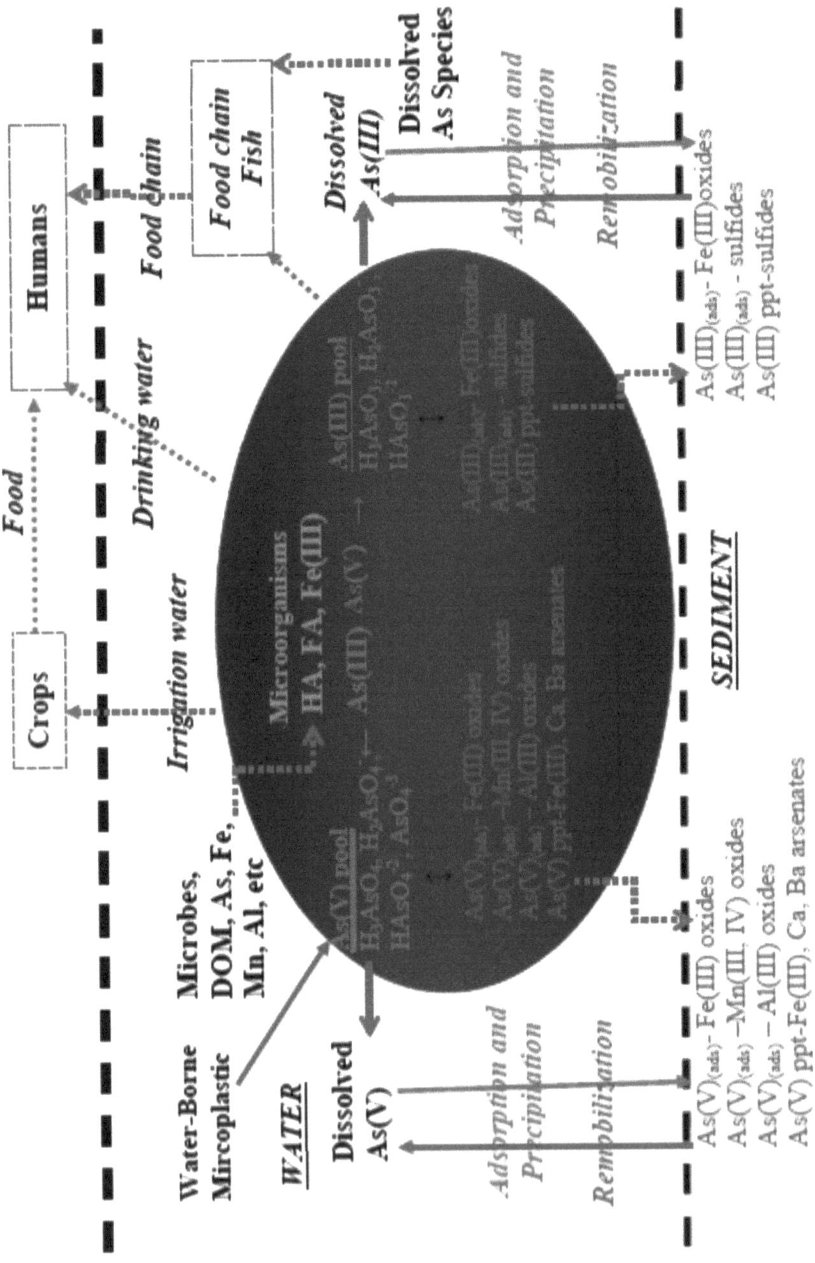

Figure 17.8 Schematic of the biogeochemistry of arsenic on the surface of a microplastic in an aqueous environment.

2.2.2 Mercury Biogeochemistry

Mercury exists naturally in the environment but a large proportion of it is from anthropogenic sources. For instance, coal-fired power plants, waste incineration, metal processing, and cement production account for 75% of the total mercury released into the atmosphere annually and the United States' thermal power plants are responsible for about half of the global emissions [85]. The airborne Hg eventually deposits into rivers, lakes and oceans where it is converted to methyl mercury (CH_3Hg^+) by specific microorganisms under certain natural water conditions. Through biomagnification, the CH_3Hg^+ builds up to at least 90% of the total mercury in fish such as shellfish, swordfish, tuna, king mackerel and shark [86]. In the United States, 30% of lakes, estuaries, and wetlands and about 473,000 miles of streams, rivers, and coasts are reportedly polluted with Hg [86].

Fig. 17.9 is a simplified schematic of the Hg cycling between the sediment/water/air interfaces and some of the major sources of the metal. As shown in the figure, Hg is loaded into the aquatic system via atmospheric deposition, direct inflow of wastewater, leaching from contaminated solid waste sites, and leaching and runoff from contaminated watersheds. Mercury is released into the atmosphere in the form of reactive gaseous mercury (RGM), $HgCl_2$, total particulate mercury (TPM), Hg^{2+}-PM, HgO-PM, and/or gaseous elementary mercury (GEM), Hg^0. These forms of Hg are produced by the following reactions [87]:

$$\text{RGM: Hg(II)} + \text{Cl, Cl}_2\text{, HCl} \rightarrow \text{HgCl}_2 \tag{17.i}$$

$$\text{TPM: Hg}^0 + \text{O}_2 + \text{PM} \rightarrow \text{HgO-PM} \tag{17.ii}$$

$$\text{GEM: Hg(II)} + 2e \rightarrow \text{Hg}^0 \tag{17.iii}$$

The atmospheric deposition of RGM, TPM, and GEM occurs by gravity and in raindrops and dust. Due to its reactivity and water solubility, RGM has an atmospheric lifetime of just a few days to a week and is deposited very close to the source in rain and heavy particulate matter [87]. On the other hand, TPM spends weeks in the atmosphere before being deposited together with particulate matter not far from the source but relatively further away than RGM. Gaseous elementary mercury has the longest atmospheric lifetime of months to a year and is the form that is subject to long-range transport [87]. It is deposited after oxidation to the water-soluble mercuric ion, Hg^{2+}, very far away from the source [87]. Oxidation of GEM is achieved by reaction with oxidants such as the hydroxyl radical and ozone in the atmosphere according to the following reaction:

$$\text{Hg}^0 + {}^{\bullet}\text{OH or O}_3 \rightarrow \text{Hg}^{2+} \text{ (water soluble)} \tag{17.iv}$$

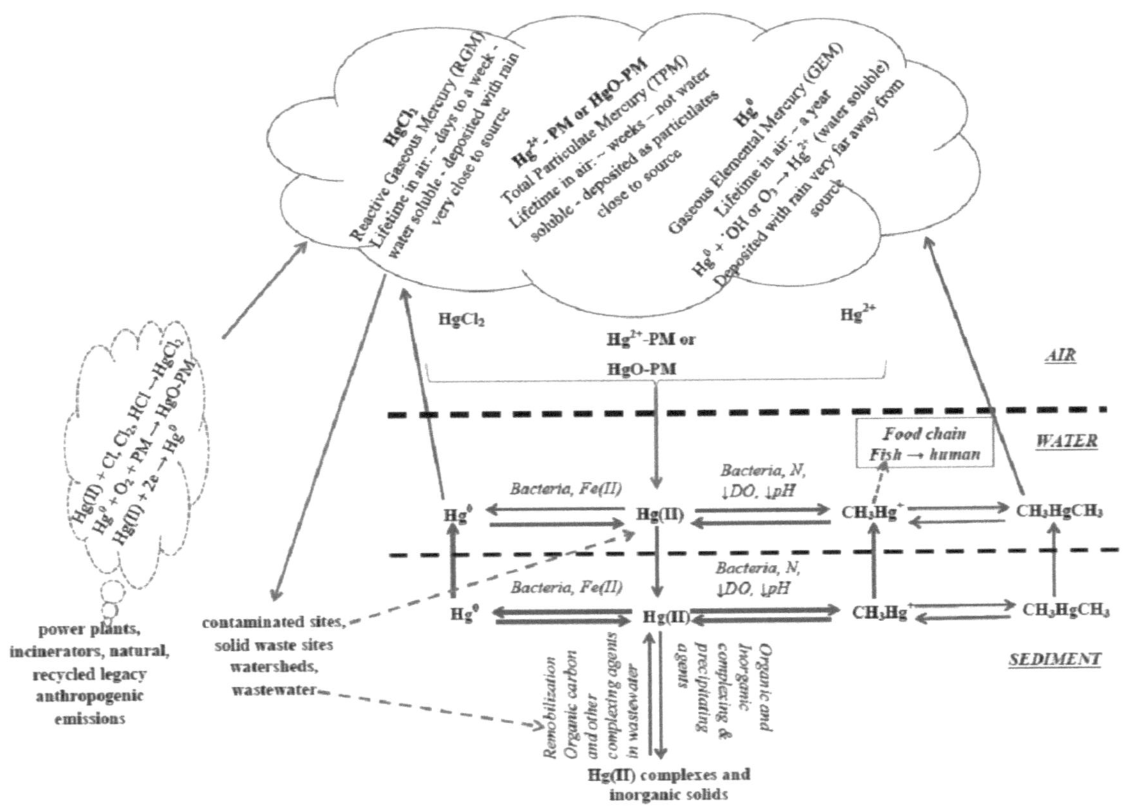

Figure 17.9 Schematic of the biogeochemistry of mercury across the sediment, water, and air interfaces as influenced by sources and environmental factors.

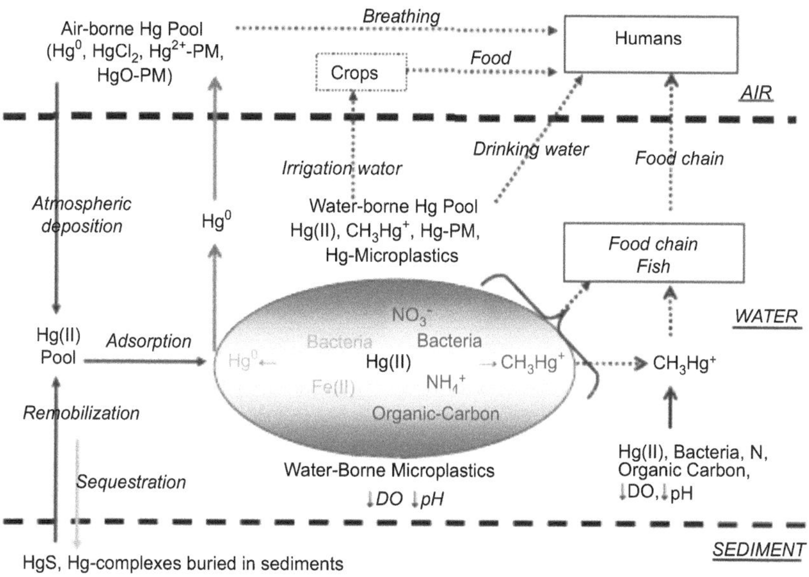

Figure 17.10 A simplified schematic showing the potential role of waterborne microplastics on the biogeochemistry, mobility and bioavailability of mercury across the sediment/water/air interfaces.

As shown in Fig. 17.10, once deposited as Hg^{2+} in the aquatic system, mercury undergoes biogeochemical transformations that depend on the prevailing biological, chemical, and physical conditions. It can be sequestered into bottom sediments through complexation and precipitation, processes that remove the metal from the water column, thus making it less bioavailable. Hg^{2+} can also be reduced to elemental mercury, Hg^0, or methylated to monomethyl mercury, CH_3Hg^+, two forms of the metal that are the most mobile, bioavailable, and toxic. Reduction to Hg^0 is achieved by Fe(II) and/or dissimilatory Fe(II) reducing bacteria under anaerobic conditions [88]. Once formed, Hg^0 will find its way back to the atmosphere and the circle is repeated. It can also get oxidized to Hg^{2+} under suitable conditions. Methylation of Hg^{2+} is carried out by bacteria under acidic, eutrophic, and anoxic conditions where nitrate and ammonium are abundant [88]. The produced monomethyl mercury can persist in the water column where it can get bioaccumulated and biomagnified in the food chain. It can also be further methylated into dimethyl mercury, which can traverse the water—air interface or demethylated to Hg^{2+} in the presence of denitrifying bacteria [88]. Sediment-bound Hg can be remobilized into the water column as Hg^{2+} by excess organic carbon and water-soluble complexing agents [88] where it can be reduced to Hg^0 or methylated to CH_3Hg^+ depending on the water conditions. The remobilization of natural and/or anthropogenic Hg from the sediments and its subsequent reduction or methylation as well as the reduction and/or methylation of atmospherically deposited Hg^{2+} promote the most toxic forms of the heavy metal.

2.2.2.1 Role of Wastewater and Microplastics

By its nature, wastewater contains all the ingredients necessary for the existence of Hg in its most mobile, bioavailable, and toxic forms, i.e., Hg^0 and CH_3Hg^+. In addition to potentially containing substantial amounts of Hg, wastewater also contains (1) large amounts of organic carbon complexing agents that can remobilize sediment-bound mercury into the water column; (2) nitrogen, organic carbon, high biochemical oxygen demand (BOD), and bacteria required for the methylation of Hg^{2+} to CH_3Hg^+; and (3) Fe(II) and other reductants required for the reduction of Hg^{2+} to Hg^0. In addition, wastewater contains microplastics, which, as pointed out earlier, adsorb other chemical species and microbes in water. As a result, all the ingredients required for the biogeochemistry and speciation of Hg can be present on the surface of the microplastics at levels much higher than in the water. Due to the preconcentration effect on microplastic surfaces, reactions that produce Hg^0 or CH_3Hg^+ can occur much faster on the surface of the microplastics than in the water. Since the reactions will also be taking place in the water column, more Hg^0 or CH_3Hg^+ will be produced in the presence of microplastics than in their absence.

A simplified schematic of the biogeochemistry and speciation of Hg on microplastics, as well as the distribution of the Hg species and exposure routes to humans, is shown in Fig. 17.10. The reactions are expected to occur in any type of water as long as the required conditions exist.

As shown in the figure, the increased rates production of Hg^0 and CH_3Hg^+ brought about by wastewater and microplastics means increased risk to humans via the seafood chain, drinking water, agricultural food, and absorption via the respiratory and dermal systems. Levels of CH_3Hg^+ in fish and other forms of seafood are bound to increase from the increased levels of CH_3Hg^+ in the water column as well as from the ingestion of microplastics laden with Hg species on their surfaces. Drinking water is normally water treated before consumption, but, unless the water is nanofiltered, microplastics, with their adsorbed mercury species, can remain in the treated drinking water. As a consequence, water containing both microplastics and Hg can be a major source of Hg exposure to humans. An increase in the amount of CH_3Hg^+ in water that can come into contact with humans during normal work or recreational activities can also contribute to human exposure through dermal exposure and inadvertent ingestion. An increase in the Hg^0 in the atmosphere also increases exposure to humans through adsorption in the respiratory system during breathing. The widespread use of wastewater and wastewater-contaminated freshwater for irrigation in agriculture also exposes humans to Hg toxicity through the consumption of contaminated agricultural food.

2.3 Persistent Organic Pollutants

Did you know? [89–91]

- Include pesticides and industrial chemicals some of which were manufactured in large quantities during the 20th century.
- Include hexachlorocyclohexanes (HCHs), dichlorodiphenyltrichloroethane (DDT), and PCBs, polybrominated diphenyl ethers (PBDEs), perfluorooctane sulfonate (PFOS), and polycyclic aromatic hydrocarbons (PAHs).
- Persist in the environment for decades.
- The world's freshwater ecosystems are generally suffering from POPs pollution.
- Bioaccumulate and biomagnify in food chains that include humans.
- Dispersed globally by air, water, and in living organisms.
- Accumulate in fatty tissues of living organisms due to their high fat solubilities.
- Associated with serious health effects in humans.
  - Cancers
  - Damage to the nervous system—neurodevelopmental problems
  - Reproductive disorders—decreased sperm quality and count
  - Disruption of the immune system
  - Learning and behavior disabilities
  - Children have suffered adverse effects from high-level exposure
- Exposure comes from food such as fish, meat, dairy products, inhalation, and dermal absorption

Discussion of POPs will be limited to CDDs, PCBs, PAHs, phthalates, and chlorinated organic solvents. These POPs are a significant component of wastewater and hence some of the major contaminants in both ground and surface freshwater. Chlorinated dibenzo-*p*-dioxins consist of 75 chemically related compounds commonly known as chlorinated dioxins [91]. They are carcinogenic to humans and affect the skin and organ development, as well as the immune and reproductive systems. While CDDs are not manufactured for commercial purposes, they are formed as by-products during some industrial processes and also during chlorination in wastewater and drinking water treatment. For instance, TCDD (shown below) is one of the most toxic CDDs, and is formed during chlorine bleaching at pulp and paper mills [92].

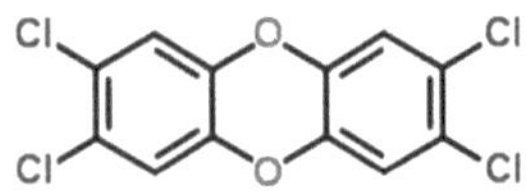

2,3,7,8-Tetrachlorodibenzo-*p*-dioxin (TCDD). *Credit: RSC ChemSpider, 2,3,7,8-Tetrachlorodibenzo-p-dioxin (TCDD) structure. http://www.chemspider.com/Chemical-Structure.14865.html*

Chlorinated dibenzo-*p*-dioxins can also occur as contaminants in the manufacture of certain organic chemicals [92] and can, through wastewater, contaminate drinking water sources via wastewater.

PCBs consist of up to 209 chlorinated compounds with no known natural sources that were manufactured and extensively used as coolants and lubricants in transformers, capacitors, and other electrical gadgets [94]. Even though they are no longer being produced, PCBs still get into freshwater sources from hazardous waste sites and leaks from old electrical transformers. However, because of their low water solubilities, only small amounts of PCBs remain dissolved in water, with the bulk getting deposited onto bottom sediments. The lipophilic nature of PCBs results in them bioaccumulating and biomagnifying in fish and marine animals in levels several times higher than in water [94]. The human health effects that are linked to PCBs include acne-like skin conditions in adults and neurobehavioral and immunological disorders in children, as well as liver damage. PCBs are currently not linked to human cancers but are classified as animal carcinogens [94].

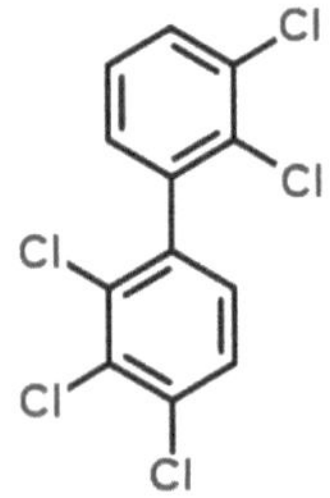

2,2′,3,3′,4-PCB. *Credit: RSC ChemSpider, Polychlorinated biphenyl: 2,2′,3,3′,4-PCB. http://www.chemspider.com/Chemical-Structure.36975.html?rid=16e63350-822c-4d6a-bdec-69e220a6479c*

Polycyclic aromatic hydrocarbons are produced during the burning of tobacco and charbroiled meat and during the incomplete combustion of fossil fuels such as coal, oil, gas, wood, and garbage [96]. Polycyclic aromatic hydrocarbons are also manufactured for research and for use in the manufacture of dyes, plastics, and pesticides, as well as for use in medicine. Polycyclic aromatic hydrocarbons are also found in asphalt that is used in road construction, crude oil, coal, coal tar pitch, creosote, and roofing tar, and are ubiquitous throughout the environment in the air, water, and soil. Polycyclic aromatic hydrocarbons enter surface water sources through discharges from industrial plants and wastewater treatment plants, and are released in soils at hazardous waste sites. Polycyclic aromatic hydrocarbons in soils can also leach into underground water. These compounds can be bioaccumulated by both plants and animals from contaminated soil or water [96]. Polycyclic aromatic hydrocarbons that are currently classified as probably or possibly human carcinogens by both the US EPA and the International Agency for Research on Cancer (IARC) include benzo[a]anthracene and benzo[a]pyrene, benzo[b]fluoranthene, benzo[j]fluoranthene, benzo[k]fluoranthene, and indeno[1,2,3-c,d]pyrene [96]. The IARC has also added chrysene and dibenzo[a,h]anthracene to the list. The US Department of Health and Human Services (DHHS) also classifies all the above-mentioned PAHs as animal carcinogens [96].

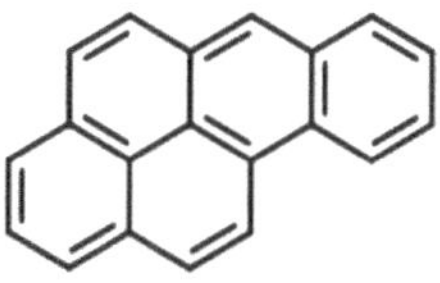

benzo[*a*]pyrene. *Credit: RSC ChemSpider, Benzo[a]pyrene. http://www.chemspider.com/Chemical-Structure.2246.html?rid=64a6f40a-6aea-4fe2-aeab-81b2b919453e.*

Phthalates are aromatic compounds that are used as plasticizers in the manufacture of plastics to render the plastics flexible and resilient. They are present in a wide-range of consumer products that include toothbrushes, automobile parts, tools, toys, packaging, cosmetics, insecticides, aspirin, wall coverings, floor tiles, furniture upholstery, shower curtains, garden hoses, swimming pool liners, rainwear, baby pants, shoes, sheathing for wire and cable, medical tubing, blood storage bags, carpet backings, paints, glue, and rocket fuel, among other products. They do not form part of the polymer (plastic) chain and they can therefore easily leach from the products. Examples of phthalates include di(2-ethylhexyl) phthalate (DEHP), di-*n*-butyl phthalate (DNBP), and di-*n*-octylphthalate (DNOP). No harmful effects have yet been reported in humans but some phthalates are known to impair reproduction as well as cause developmental effects and deaths in laboratory animals. They enter the aquatic system as part of industrial, agricultural, and domestic effluents and leachates from hazardous waste sites [98,99].

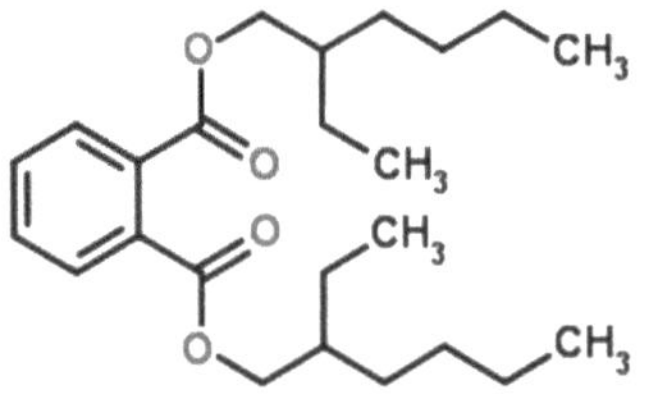

Bis(2-ethylhexyl) phthalate. *Credit: RSC ChemSpider, Bis(2-ethylhexyl) phthalate. http://www.chemspider.com/Chemical-Structure.21106505.html?rid=c265a4bf-6285-4da7-8878-2f65b7edf304.*

Other POPs commonly found in wastewater that were not discussed above include human carcinogens such as trichloroethene (TCE) and perchloroethylene (PCE) [101].

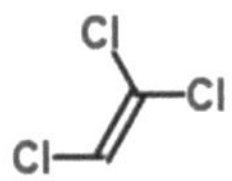

TCE. *Credit: RSC ChemSpider, Trichloroethene (TCE). http://www.chemspider.com/Chemical-Structure.13837280.html?rid=d8d86fff-e36f-41d5-8385-e70095b2147b.*

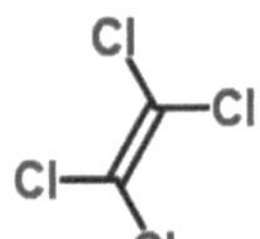

PCE. *Credit: RSC ChemSpider, Perchloroethene (PCE). http://www.chemspider.com/Chemical-Structure.13837281.html?rid=b524917e-7b34-4c2a-a0a2-400c1f99bea1.*

The presence of microplastics in the aquatic system enhances the distribution of POPs by adsorbing and carrying them far and wide within the aquatic system. Filter feeders and fish can ingest microplastics laden with POPs, which are then transferred and biomagnified in the food chain. Thus the presence of microplastics increases the presence of POPs in the water column, which, in turn, increases the increase exposure of POPs to humans through the food chain and drinking water.

3. PATHOGENIC ORGANISMS IN THE AQUATIC SYSTEM

Waterborne pathogens can naturally exist in all types of water but wastewater remains the major source of these pathogens in the natural aquatic environment. In addition to an increase in the variety of pathogens in water, there has also been a notable increase in the types of diseases that are associated with waterborne pathogens. Instead of just diarrhea and intestinal diseases, there are waterborne pathogens that are now associated with a variety of chronic diseases such as hepatitis, myocarditis, insulin-dependent diabetes, aseptic meningitis, and ocular and respiratory infections [104]. The increase in the types of diseases could be the direct result of the increase in the variety of pathogens and/or improved knowledge of the diseases. The increase in the variety of pathogens in water is thought to be a result of the emergence of new environments, the current boom in technology, and changes in human behavior and vulnerability [9]. For example, climate change and intensive livestock rearing are reportedly creating new environments that are promoting the emergence and reemergence of pathogenic organisms [9]. Intensive livestock rearing in particular has been linked to the discharge of pathogens that include *Cryptosporidium*, *E. coli* O157, and *Campylobacter* into water sources via concentrated animal wastes and waste from feeding operations [9]. The presence of pharmaceuticals such as antibiotics in water may also be contributing to emergence and reemergence of water-related pathogens as a result of the development of antimicrobial resistance [103]. In fact, waterborne illnesses are reported to have significantly increased over the past two decades primarily due to the emergence and spread of drug-resistant pathogens [9].

Did You Know?

- Water-related diseases are a major cause of morbidity and mortality worldwide [9].
- Chlorine is not effective for treating water-related pathogens such as *giardia* and *cryptosporidium* at concentrations required for water disinfection [6].
- Developing countries disproportionately suffer from the severe consequences of water-related diseases due to poor infrastructure [106].

- The frequency and the range of water-related microbial diseases is increasing around the world [9].
- Water-related diseases are still reported in both developing and developed countries.
- Pathogens naturally exist in natural water but untreated wastewater remains the major source of a large number of them.
- Waterborne pathogens are associated with diarrhea and intestinal illnesses as well as a variety of chronic diseases that include hepatitis, myocarditis, insulin-dependent diabetes, aseptic meningitis, and ocular and respiratory infections [104].

3.1 Pathogenic Organisms Commonly Found in Wastewater

3.1.1 Burkholderia pseudomallei *(*Pseudomonas pseudomallei*)*

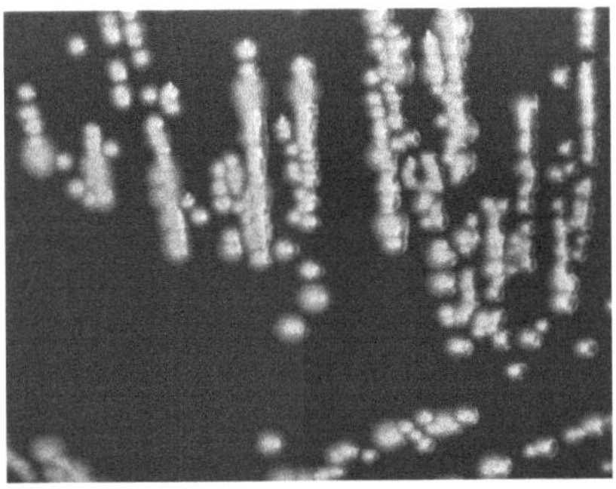

Gram-negative *Achromobacter xylosoxidans* bacteria (*Pseudomonas pseudomallei*). *Credit: Dr. Todd Parker. Ph.D.; Assoc. Director for Lab. Science/DPEI (Acting) and LRN Training Coordinator., Parker T. Center for Disease Control and Prevention (2014). Gram-negative* Achromobacter xylosoxidans *bacteria (*Pseudomonas pseudomallei*). http://phil.cdc.gov/phil/details.asp.*

Did you know?

- Is a gram-negative bipolar bacterium that infects humans, animals, and plants [108,109].
- Causes melioidosis, also called Whitmore's disease, in both humans and animals [108,109].
- It is found in contaminated water and soil and spreads to humans and animals from water and soil via oral (drinking), respiratory (breathing), and dermal (open wounds) systems [108,109].
  - Predominately a disease of tropical climates, especially in southeast Asia and northern Australia [108].

- Symptoms include localized pain or swelling, fever, ulceration, abscess, cough, chest pain, high fever, headache, anorexia, respiratory distress, abdominal discomfort, joint pain, disorientation, weight loss, muscle pain, and seizures [108].
- Symptoms generally appear 2 to 4 weeks following exposure [108].
- Major risk factors include diabetes, liver disease, renal disease, thalassemia, cancer, or another immune-suppressing condition other than HIV, and chronic lung disease [108].

3.1.2 Cryptosporidium Parvum

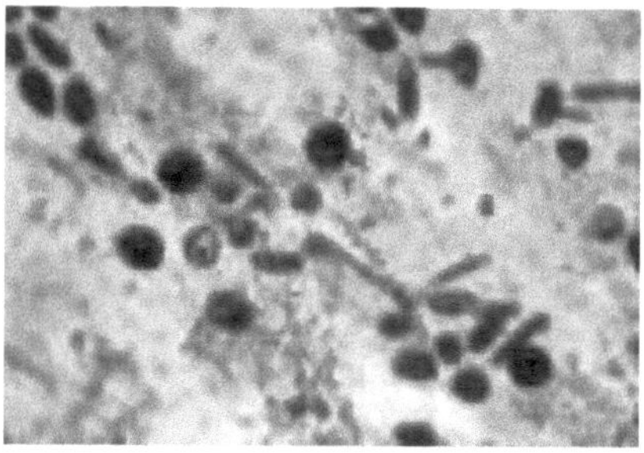

Cryptosporidium. *Credit: CDC, Center for Disease Control and Prevention. Three-step stool examination for cryptosporidiosis in 10 homosexual men with protracted watery Diarrhea. J Infect Dis. 1983;147(5),824–8. http://phil.cdc.gov/phil/details.asp.*

Did you know? [111]

- Is a parasitic protozoan that causes gut cryptosporidiosis, as well as respiratory cryptosporidiosis, hepatitis, and cholecystitis when it gets to the lungs, liver, and gallbladder, respectively.
- Both the parasite and the disease cryptosporidiosis are commonly known as "crypto."
- The parasite has a lifecycle stage called an oocyst whose outer shell protection enables it to survive outside the body for extended periods of time and to survive chlorination.
- It is found in every region of the United States and throughout the world.
- It is commonly spread via contaminated water (drinking and recreational) and is a leading cause of waterborne disease among humans in the United States.

- Symptom include watery diarrhea, stomach cramps or pain, dehydration, nausea, vomiting, fever, and weight loss.
- People with AIDS, inherited diseases that affect the immune systems, cancer patients, and transplant patients taking immunosuppressive drugs may develop serious, chronic, and sometimes fatal illness.

3.1.3 Giardia lamblia (Giardia intestinalis, *or* Giardia duodenalis)

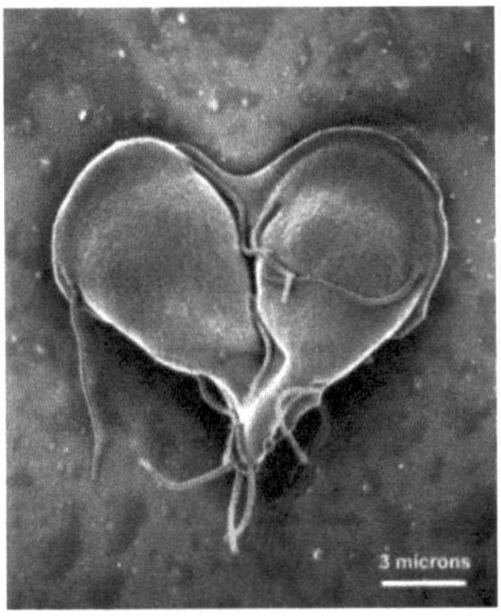

Giardia lamblia protozoan about to become two separate organisms. *Credit: CDC/Dr. Stan Erlandsen Erlandsen S, Center for Disease Control and Prevention. Giardia lamblia protozoan. 1999. http://phil.cdc.gov/phil/details.asp.*

Did you know? [113]

- It is a parasitic flagellated protozoan that causes a diarrheal illness called giardiasis.
- It is found in water, food, soil, and surfaces that has been contaminated with feces from infected humans or animals.
- It has a lifecycle stage called an oocyst whose outer shell protection enables it to survive chlorination as well as survive outside the body for extended periods of time.
- Drinking water and recreational water remain the two most common modes of transmission.
- It can be found within every region around the world.
- Symptoms include diarrhea, gas or flatulence, greasy stool that can float, stomach or abdominal cramps, upset stomach or nausea, dehydration, and weight loss.

3.1.4 Salmonella

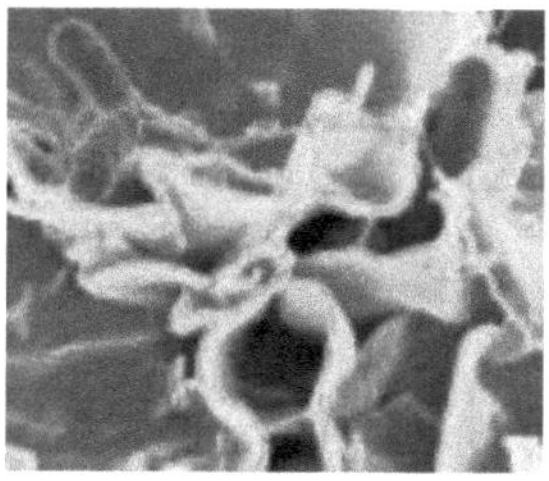

Salmonella. *Credit CDC/National Institute of Allergy and Infectious Diseases (NIAID), Center for Disease Control and Prevention/National Institute of Allergy and Infectious Diseases (NIAID)* Salmonella *2011. http://phil.cdc.gov/phil/details.asp.*

Did You Know? [115]

- It is a gram-negative bacteria that causes illnesses such as typhoid fever, paratyphoid fever, and food poisoning.
- Symptoms include diarrhea, fever, and abdominal cramps 12–72 h following infection and lasting 4–7 days.
- In severe cases, infection may spread to the bloodstream and other body sites that may lead to death if treated promptly.
- The elderly, infants, and individuals with compromised immune systems are more vulnerable to severe illness.
- According to the CDC, *Salmonella* causes about 1.2 million illnesses and 450 deaths in the United States annually.
- Contaminated drinking water remains one of the major sources of human exposure to *Salmonella*.

3.1.5 Norovirus (Norwalk Virus)

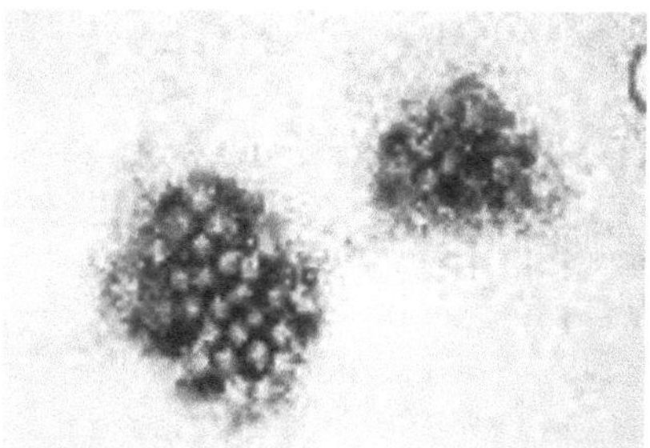

Norovirus, with 27–32 nm-sized viral particles. *Credit: CDC/E.L. Palmer E.L., Center for Disease Control and Prevention. Norovirus, with 27–32 nm-sized viral particles. 1981; http://phil.cdc.gov/phil/details.asp.*

Did You Know? [117]

- It is a single-stranded RNA virus which is the most common cause of viral gastroenteritis in humans, and affect people of all ages.
- The virus reportedly affects around 267 million people, causing over 200,000 deaths annually.
- It is very contagious and can be contracted from an infected person, contaminated water or food or by touching contaminated surfaces.
- It inflames the stomach which leads to stomach pain, nausea, and diarrhea, vomiting, headache, and body aches
- Symptoms usually develop 12–48 h following exposure and can be serious in the young and the elderly.
- It is the most common cause of acute gastroenteritis in the United States and causes 19–21 million illnesses and 570–800 deaths annually.

3.1.6 Helminths

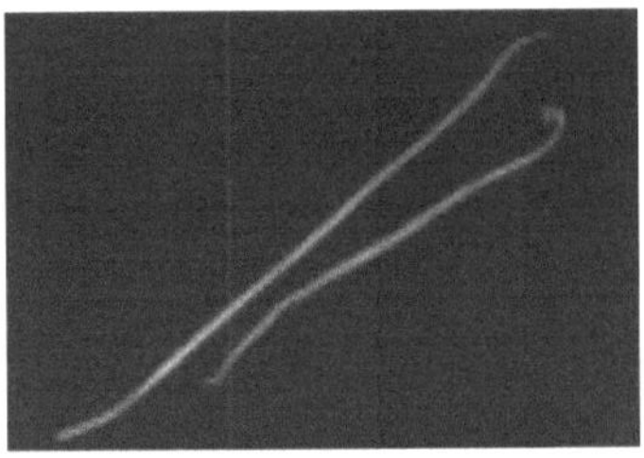

Ascaris lumbricoides nematodes (roundworms). *Credit: CDC Center for Disease Control and Prevention—Div of Parasitic Diseases (1960). Ascariasis. http://www.cdc.gov/dpdx/ascariasis/index.html.*

Did You Know? [119]

- Are eukaryotic parasitic worms that live and feed on living hosts while causing malnutrition, weakness, and disease to the host.
- They include flatworms (platyhelminths), thorny-headed worms (acanthocephalans), and roundworms (nematodes).

According to WHO, the emergence and reemergence of water-related diseases are partly the direct result of global complacence that followed the eradication of the majority of some of the most devastating pathogenic water-related diseases [9]. For example, *Cryptosporidium* emerged in Milwaukee, Wisconsin, US, in 1993 as a result of flaws in the water-treatment process and reportedly killed 100 out of the more than 400,000 people who were affected [6,121]. In 2006, 18 cryptosporidiosis outbreaks were reported in the United States, 11 more outbreaks than those reported in the same country in 2003 [121]. The major challenge with cryptosporidium is that it is resistant to chlorination and can survive several days in chlorinated water containing recommended residual chlorine levels [105]. According to Water Quality and Health Council (2003), cryptosporidium remains a major threat to the US water supply [6]. Other waterborne diseases are reemerging despite the effectiveness of chlorination because of the formation of chlorine-resistant pathogenic biofilms in aging and cracked distribution pipes [9].

Despite the availability of advanced diagnostic technologies for waterborne diseases in developed nations, the etiology of about 40% of drinking water outbreaks that occurred between 1991 and 2000 in the United States was unknown [9]. This shows that public health professionals should always be vigilant in looking for emerging and reemerging water-related disease outbreaks. Developing countries will likely remain under more threat from waterborne diseases than developed countries because of poor infrastructure, lack of financial resources, lack of regulations on water pollution, and corruption. International nongovernmental organizations that have helped and continue to help developing countries with initial funding for water infrastructure but the lack of resources for operational costs and refurbishments will remain impediments to the provision of clean and safe water in most regions of the developing world.

CONCLUSIONS

The foregoing discussion shows that water contamination and the resulting public health threats are real. The issue of water contamination is also global—it's no longer a developing country problem. Developed countries are also facing water-quality challenges. Humans are being exposed to increasing numbers of toxic chemicals and pathogens, emerging and reemerging diseases, at unprecedented rates and levels through drinking water, hygiene water, recreational water, agricultural water, seafood, and agricultural food. Ironically, water contamination is mostly the direct result of anthropogenic activities that are the hallmarks of modern civilization—energy, industry, mining, technology, medicine, agriculture, and urbanization, just to name a few. Effects of natural phenomena such as climate change are exacerbating the water situation. To preserve and sustain public health and life, pollution of water has to be avoided or minimized; water infrastructure has to be constantly innovated in line with demand and challenges; periodic risk analyses have to be conducted to update existing laws or enact new laws on pollution to deal with emerging threats; and reporting on outbreaks has to be improved to ensure public awareness.

The Safe Drinking Water Act in the United States is an example of an environmental law that needs alignment with current knowledge and trends in pollution. The law was passed in 1974 after tests discovered carcinogens, Pb, and dangerous bacteria in drinking water in New Orleans, Pittsburgh, and Boston, among other places in the United States [2]. At the time of its enactment, the law required local systems to monitor only 20 chemical substances in waters reportedly due to the little knowledge at the time about chemicals in water. Research-informed amendments to the law were made over the following decade, expanding the list of regulated chemicals to 91. However, the law has not been revised since the year 2000 despite the fact that over 60,000 chemicals are currently in use in the United. States [2]. Data from the US EPA and the US Geological Survey that show that the majority the 830 contaminants most commonly found in water supplies across the United States are also associated with cancer and/or otherx diseases even at trace levels [2]. This means that water can be harmful yet legal in one of the most developed and prosperous nations on earth. Local authorities in the country may not feel obligated to upgrade water infrastructure and water-treatment technologies in order to provide safe water outside current regulations even when resources are available. In addition to being outdated, the law does not cover private wells, despite the amount of scientific data that shows unprecedented levels of groundwater contamination [2].

Risk analysis involves the quantification of risk in terms of probability and severity for the purpose of evaluating the economic viability of available options for alleviating the risk relative to the benefits of risk reduction [72,73]. Also crucial to risk analysis is risk communication, which deals with the dissemination of information about risk assessment and management to ensure that the general public understands the risks, policies, and options available [72,73]. Communication about outbreaks has not been particularly effective, and, in most cases, it is the most severe and wide-scale outbreaks that are recorded or reported [10]. Municipal authorities and national governments need updated risk analysis data to allocate adequate resources to the provision of clean and safe water to their people.

REFERENCES

[1] Fox JL. Current topics: at 48th ICAAC, several novel agents target gram-negative pathogens. Microbe 2009;4(1):4–10.

[2] Duhigg C. That tap water is legal but may be unhealthy. 2009. http://www.nytimes.com/2009/12/17/us/17water.html.

[3] FAO. Water—the most basic resource but also the most essential: how much water do we use for…?. 2014. http://www.fao.org/zhc/detail-events/en/c/231215/.

[4] Voeroesmarty CJ, McIntyre PB, Gessner MO, Dudgeon D, Prusevich A, Green P, et al. Global threats to human water security and river biodiversity. Nature (Lond UK) 2010;467(7315):555–61.

[5] Organisation for Economic Cooperation and Development (OECD). (n. d.). Water chapter of the OECD environmental outlook to 2050: the consequences of inaction http://www.oecd.org/environment/indicators-modelling-outlooks/waterchapteroftheoecdenvironmentaloutlookto2050theconsequencesofinaction.htm.

[6] Water Quality and Health Council. Drinking water chlorination: a review of disinfection practices and issues. 2003. http://www.waterandhealth.org/drinkingwater/wp.html.

[7] National Institute of Environmental Health Sciences: Child development and environmental toxins. https://www.niehs.nih.gov/health/assets/docs_a_e/child_development_and_environmental_toxins_508.pdf.

[8] Richter B. Water policies suffer sinking feeling. 2009. http://news.bbc.co.uk/2/hi/science/nature/8205132.stm.

[9] World Health Organization. Emerging issues in water and infectious disease. 2003. http://www.who.int/water_sanitation_health/emerging/emerging.pdf.

[10] Stanwell-Smith R, (n. d.). Classification of water-related disease. Water Health, 1:1–21.

[11] Raschid-Sally L, Jayakody P. Drivers and characteristics of wastewater agriculture in developing countries: results from a global assessment. 2008. IWMI Research Report 127, http://www.iwmi.cgiar.org/Publications/IWMI_Research_Reports/PDF/PUB127/RR127.pdf.

[12] Chartres C. Water everywhere and not a drop to grow. 2008. http://news.bbc.co.uk/2/hi/science/nature/7564140.stm.

[13] African Women Carrying Water (n. d) jpegthewaterproject. org. http://www.bing.com/images/search?q=African+Women+Carrying+Water+On+Their+Head+thewaterproject&go=Search&qs=bs&form=QBIR.

[14] The Water Project. (n. d.). Poverty in Africa begins with a lack of clean water: poverty and water are related. https://thewaterproject.org/poverty.

[15] The Water Project. (n. d.). Why Water? Dirty water causes needless suffering. https://thewaterproject.org/why-water/.

[16] The Water Project. (n. d.). Safe and reliable water changes lives. https://thewaterproject.org/?gclid=CM2ojPTdvssCFcQjgQodYlQPng.

[17] Center for Disease Control and Prevention (CDC). Global wash-related diseases and contaminants. 2012. http://www.cdc.gov/healthywater/wash_diseases.html.

[18] Luntz T. U.S. drinking water widely contaminated: EPA finds 202 unregulated chemicals in 45 states. Scientific American; 2009. http://www.scientificamerican.com/article/tap-drinking-water-contaminants-pollutants/.

[19] Duhigg C. Millions in U.S. drink dirty water, records show; 2009. http://www.nytimes.com/2009/12/08/business/energy-environment/08water.html.

[20] Duhigg C. As sewers fill, waste poisons waterways; 2009. http://www.nytimes.com/2009/11/23/us/23sewer.html.

[21] Begos K. Four states confirm water pollution from drilling; 2014. http://www.usatoday.com/story/money/business/2014/01/05/some-states-confirm-water-pollution-from-drilling/4328859/.

[22] Natural Resources Defense Council (NRDC) (n. d.). Water. http://www.nrdc.org/water/.

[23] Durando J. How water crisis in Flint, Michigan, became federal state of emergency. 2016. http://www.usatoday.com/story/news/nation-now/2016/01/19/michigan-flint-water-contamination/78996052/.

[24] Srivastava R. A "House of horrors": thousands of Michigan residents subject to contaminated drinking water; 2015. http://thinkprogress.org/climate/2015/06/05/3666238/flint-michigan-drinking-water-contamination/.

[25] Brewer J. Drinking water crisis in Flint, Michigan. 2015. http://www.wsws.org/en/articles/2015/01/12/flin-j12.html.

[26] Tagg AS, Sapp M, Harrison JP, Ojeda JJ. Identification and quantification of microplastics in wastewater using focal plane array-based reflectance Micro-FT-IR Imaging. Anal Chem 2015;2015(87):6032–40.

[27] Klein S, Worch E, Knepper TP. Occurrence and spatial distribution of microplastics in river shore sediments of the Rhine-Main Area in Germany. Environ Sci Technol 2015;2015(49):6070–6.

[28] Yonkos LT, Friedel EA, Perez-Reyes AC, Ghosal S, Arthur CD. Microplastics in four estuarine rivers in the Chesapeake Bay, U.S.A. Environ Sci Technol 2014; 2014(48):14195–202.

[29] Bouwmeester H, Hollman PCH, Peters RJB. Potential health impact of environmentally released micro- and nanoplastics in the human food production chain: Experiences from nanotoxicology. Environ Sci Technol 2015;2015(49):8932–47.

[30] Zhang W, Ma X, Zhang Z, Wang Y, Wang J, Wang J, et al. Persistent organic pollutants carried on plastic resin pellets from two beaches in China. Mar Pollut Bull 2015. xxx (2015) xxx–xxx.

[31] Napper IE, Bakir A, Rowland SJ, Thompson RC. Characterisation, quantity and sorptive properties of microplastics extracted from cosmetics. Mar Pollut Bull 2015; xxx–xxx.

[32] Andrady AL. Microplastics in the marine environment. Mar Pollut Bull 2011; 62(2011):1596–605.

[33] Bansal OP. Heavy metals in soil, plants as influenced by irrigation with sewage effluents: a 20 year study. J Indian Assoc Environ Manag 2008;35(3):143–8.

[34] Mueller B, Scheytt T, Asbrand M, de Casas AM. Pharmaceuticals as indictors of sewage-influenced groundwater. Hydrogeol J 2012;20(6):1117–29.

[35] Focazio MJ, Kolpin DW, Barnes KK, Furlong ET, Meyer MT, Zaugg SD, et al. A national reconnaissance for pharmaceuticals and other organic wastewater contaminants in the United States—(II) Untreated drinking water sources. Sci Total Environ 2008;402(2–3):201–16.

[36] 21 EPA. Priority pollutants. 2013. http://water.epa.gov/scitech/methods/cwa/pollutants.cfm.

[37] Judy JD, Unrine JM, Bertsch PM. Evidence for biomagnification of gold nanoparticles within a terrestrial food chain. Environ Sci Technol 2011;45(2): 776–81.

[38] BBC. Who, what, why: Why do people want to ban "microbeads"?. 2014. http://www.bbc.com/news/blogs-magazine-monitor-27440188.

[39] Steinmetz K. States are cracking down on face wash. 2014. http://time.com/#74956/states-arecracking-down-on-face-wash/.

[40] Robinson BH. E-waste: an assessment of global production and environmental impacts. Sci Total Environ 2009;408(2):183–91.

[41] Wilson R. Chronic arsenic poisoning: picture of sufferers. 2008. http://users.physics.harvard.edu/~wilson/arsenic/pictures/arsenic_project_pictures2.html.
[42] Fragou D, Fragou A, Kouidou S, Njau S, Kovatsi L. Epigenetic mechanisms in metal toxicity. Toxicol Mech Methods 2011;21:343–52.
[43] Ren X, McHale CM, Skibola CF, Smith AH, Smith MT, Zhang L. An emerging role for epigenetic dysregulation in arsenic toxicity and carcinogenesis. Environ Health Perspect 2011;119(1):11–9.
[44] Cronican AA, Fitz NF, Carter A, Saleem M, Shiva S, et al. Genome-wide alteration of Histone H3K9 acetylation pattern in mouse offspring prenatally exposed to arsenic. PLoS One 2013;8(2):e53478.
[45] Farzan SF, Karagas MR, Chen Y. In utero and early life arsenic exposure in relation to long-term health and disease. Toxicol Appl Pharmacol 2013;272(2):384–90.
[46] Fei DL, Koestler DC, Li Z, Giambelli C, Sanchez-Mejias A, Gosse JA, et al. Association between in utero arsenic exposure, placental gene expression, and infant birth weight: a US birth cohort study. Environ Health 2013;12:58.
[47] Moon K, Guallar E, Navas-Acien A. Arsenic exposure and cardiovascular disease: an updated systematic review. Curr Atheroscler Rep 2012;14(6):542–55.
[48] Liaw J, Marshall G, Yuan Y, Ferreccio C, Steinmaus C, Smith AH. Increased childhood liver cancer mortality and arsenic in drinking water in Northern Chile. Cancer Epidemiol Biomarkers Prev 2008;17(8):1982–7.
[49] Yorifuji T, Tsuda T, Doi H, Grandjean P. Cancer excess after arsenic exposure from contaminated milk powder. Environ Health Prev Med 2011;16(3):164–70.
[50] National Institute of Environmental Health Sciences: Health Homes = Health Kids. http://www.niehs.nih.gov/health/assets/docs_f_o/health-kids-water.pdf.
[51] Harrington JM, Fendorf SE, Rosenzweig RF. Biotic generation of arsenic(iii) in metal(loid)-contaminated freshwater lake sediments. Environ Sci Technol 1998;32:2425–30.
[52] Tadanier CJ, Schreiber ME, Roller JW. Arsenic Mobilization through microbially mediated Deflocculation of Ferrihydrite. Environ Sci Technol 2005;39:3061–8.
[53] Pilsner JR, Hall MN, Liu X, Ilievski V, Slavkovich V, et al. Influence of prenatal arsenic exposure and newborn sex on global methylation of cord blood DNA. PLoS One 2012;7:e37147.
[54] Piscator M. Life Sciences research Report 33(Importance of Chemical Speciation Environmental Processes). 1996. p. 59–70.
[55] Anderson CR, Cook GM. Isolation and characterization of arsenate-reducing bacteria from arsenic-contaminated sites in New Zealand Isolation and characterization of arsenate-reducing bacteria from arsenic-contaminated sites in New Zealand. Curr Microbiol 2004;48:341–7.
[56] Ferguson JF, Gavis J. A review of the arsenic cycle in natural waters. Water Res 1972;6(11):1259–74.
[57] Ciesielski T, Weuve J, Bellinger DC, Schwartz J, Lanphear B, Wright RO. Cadmium exposure and neurodevelopmental outcomes in U.S. children. Environ Health Perspect 2012;120:758–63.
[58] Riederer AM, Belova A, George BJ, Anastas PT. Urinary cadmium in the 1999–2008 U.S. National health and nutrition Examination Survey (NHANES). Environ Sci Technol 2013;47(2):1137–47.
[59] Satarug S, Moore MR. Emerging roles of cadmium and heme oxygenase in type-2 diabetes and cancer susceptibility. Tohoku J Exp Med 2012;228(4):267–88.
[60] Ciesielski T, Bellinger DC, Schwartz J, Hauser R, Wright RO. Associations between cadmium exposure and neurocognitive test scores in a cross-sectional study of US adults. Environ Health (Lond, UK) 2013;12:13.

[61] CDC. National biomonitoring program: Cadmium. 2013; http://www.cdc.gov/biomonitoring/Cadmium_BiomonitoringSummary.html.

[62] Jarup L, Berglund M, Elinder CG, Nordberg G, Vahter M. Health effects of cadmium exposure—a review of the literature and a risk estimate. Scand J Work Environ Health 1998;24(Suppl. 1):1—51.

[63] Cory-Slechta DA, Virgolini MB, Rossi-George A, Thiruchelvam M, Lisek R, Weston D. Lifetime consequences of combined maternal lead and stress. Basic Clin Pharmacol Toxicol 2008;102(2):218—27.

[64] Technology Planning and Management Corporation. Report on Carcinogens background document for lead and lead compounds; Contract Number N01-ES-85421. Durham, NC: U.S. Department of Health and Human Services; May 8, 2003. http://ntp.niehs.nih.gov/ntp/newhomeroc/roc11/Lead-Public.pdf#search=lead.

[65] EPA. Mercury health effects. 2014. http://www.epa.gov/mercury/effects.htm.

[66] EPA. Mercury. 2014. http://www2.epa.gov/region8/mercury.

[67] U.S. Food and Drug Administration, Industry Activities Staff Booklet. Action levels for poisonous or Deleterious substances in human food and animal feed. August 2000. http://www.fda.gov/Food/GuidanceRegulation/GuidanceDocumentsRegulatoryInformation/ChemicalContaminantsMetalsNaturalToxinsPesticides/ucm077969.htm.

[68] Public Affairs Television, Science and Health section, "The Mercury Story." http://www.pbs.org/now/science/mercuryinfish.html.

[69] CODEX General standards for contaminants and toxins in foods and feed (CODEX STAN 193—1995). http://www.fao.org/fileadmin/user_upload/livestockgov/documents/1_CXS_193e.pdf.

[70] United Nations Environment Programme (UNEP). Countries: Signatories. http://mercuryconvention.org/Countries/tabid/3428/Default.aspx.

[71] EPA.. 2011 National listing of fish advisories. 2013. http://www.epa.gov/sites/production/files/2015-06/documents/technical-factsheet-2011.pdf.

[72] Hammitt JK. Risk assessment and economic evaluation. In: Rom WN, editor. Environmental and occupational medicine. 4th ed. Philadelphia: Lippincott—Raven Publishers; 2006.

[73] Wilson R, Crouch EAC. Risk-benefit analysis. Harvard University Press; 2001.

[74] Amini M, Abbaspour KC, Berg M, Winkel L, Hug EH, Yang H, Johnson CA. Statistical modeling of global geogenic arsenic contamination in groundwater. Environ Sci Technol 2008;42(10):3669—75.

[75] Winkel L, et al. Probablity of Arsenic contamination in groundwater. Nat Geosci 2008. http://dx.doi.org/10.1038/ngeo254. http://www.nature.com/news/2008/080714/full/454263a/box/1.html.

[76] U.S. Geological Survey. Arsenic in groundwater of the United States: featured publications. 2011. http://water.usgs.gov/nawqa/trace/arsenic/.

[77] Tongesayi T, Smart RB. Arsenic speciation: reduction of Arsenic(V) to Arsenic(III) by Fulvic acid. Environ Chem 2006;3:137—41.

[78] Resinger HJ, Burris DR, Hering JG. Remediating subsurface arsenic contamination with monitored natural attenuation. Environ Sci Technol 2005;39:458A—64A (xxxxxx same as 94).

[79] Cullen WR, Reimer KJ. Arsenic speciation in the environment. Chem Rev 1989;89:713—64.

[80] Arai Y, Sparks DL, Davis JA. Arsenate adsorption mechanisms at the Allophane—Water interface. Environ Sci Technol 2005;39:2537—44.

[81] Woolson EA, Kearney PC. Persistence and reactions of [14C]-cacodylic acid in soils. Environ Sci Technol 1973;7:47—50.

[82] Macur RE, Jackson CR, Botero LM, McDermott TR, Inskeep WP. Bacterial populations associated with the oxidation and reduction of arsenic in an unsaturated soil. Environ Sci Technol 2004;38:104–11.
[83] Zobrist J, Dowdle PR, Davis JA, Oremland RS. Mobilization of arsenite by dissimilatory reduction of adsorbed arsenate. Environ Sci Technol 2000;34: 4747–53.
[84] Korte NE, Fernando Q. A review of arsenic (III) in ground water. Crit Rev Environ Control 1991;21(1):1–39.
[85] U.S.EPA. Basic Information about Mercury. http://www2.epa.gov/mercury/basic-information-about-mercury.
[86] Dickson H. Measuring mercury levels in fish. 2009. http://www.labmanager.com/laboratory-technology/2009/12/measuring-mercury-levels-in-fish#.VubtNlP2Zjo.
[87] Cohen M. Atmospheric mercury: emissions, transport/fate, source-receptor relationships. 2006. http://www.arl.noaa.gov/documents/reports/frostburg_overview.pdf.
[88] Lamborg CH, Kent DB, Swarr GJ, Munson KM, Kading T, O'Connor AE, Fairchild GM, LeBlanc DR, Wiatrowski HA. Mercury speciation and mobilization in a wastewater-contaminated groundwater plume. Environ Sci Technol 2013;47: 13239–49.
[89] World Health Organization. Persistent organic pollutants: impact on child health. 2010. http://apps.who.int/iris/bitstream/10665/44525/1/9789241501101_eng.pdf.
[90] Xu F-L, Jorgensen SE, Yoshihisa SY, Silow E. Persistent organic pollutants in fresh water ecosystems. 2013. http://www.hindawi.com/journals/tswj/2013/303815/.
[91] Monosson E. Health effects of chlorinated Dibenzo-p-dioxins (CDDs). Agency for Toxic Substances and Disease Registry (ATSDR); 2011. http://www.eoearth.org/view/article/153350/.
[92] Agency for Toxic Substances and Disease Registry (ATSDR). Toxicological profile for chlorinated Dibenzo-p-dioxins (CDDs). Atlanta, GA: U.S. Department of Health and Human Services, Public Health Service; 1998. http://www.atsdr.cdc.gov/toxfaqs/tf.asp?id=363&tid=63.
[93] 2,3,7,8-Tetrachlorodibenzo-p-dioxin (TCDD) structure. http://www.chemspider.com/Chemical-Structure.14865.html.
[94] Agency for Toxic Substances and Disease Registry (ATSDR). Toxicological profile for polychlorinated biphenyls (PCBs). Atlanta, GA: U.S. Department of Health and Human Services, Public Health Service; 2000. http://www.atsdr.cdc.gov/toxfaqs/tf.asp?id=140&tid=26.
[95] Polychlorinated biphenyl: 2,2′,3,3′,4-PCB. http://www.chemspider.com/Chemical-Structure.36975.html?rid=16e63350-822c-4d6a-bdec-69e220a6479c.
[96] Agency for Toxic Substances and Disease Registry (ATSDR). Toxicological profile for polycyclic aromatic hydrocarbons (PAHs). Atlanta, GA: U.S. Department of Health and Human Services, Public Health Service; 1995. http://www.atsdr.cdc.gov/PHS/PHS.asp?id=120&tid=25.
[97] Benzo[*a*]pyrene. http://www.chemspider.com/Chemical-Structure.2246.html?rid=64a6f40a-6aea-4fe2-aeab-81b2b919453e.
[98] Agency for Toxic Substances and Disease Registry (ATSDR). Toxicological profile for Di-n-butyl phthalate. Update. Atlanta, GA: U.S. Department of Health and Human Services, Public Health Service; 2001. http://www.atsdr.cdc.gov/toxfaqs/tf.asp?id=858&tid=167.
[99] Agency for Toxic Substances and Disease Registry (ATSDR). Toxicological profile for Di(2-ethylhexyl)phthalate (DEHP). Atlanta, GA: U.S. Department of Health and Human Services, Public Health Service; 2002. http://www.atsdr.cdc.gov/phs/phs.asp?id=376&tid=65.

[100] Bis(2-ethylhexyl) phthalate. http://www.chemspider.com/Chemical-Structure.21106505.html?rid=c265a4bf-6285-4da7-8878-2f65b7edf304.

[101] Agency for Toxic Substances and Disease Registry, Division of Community Health Investigations (n. d.). Possible health risks from exposure to TCE & PCE. http://www.atsdr.cdc.gov/HAC/pha/NavyYardMillsSite/FINALNavyYardMillsfactsheet508.pdf.

[102] Trichloroethene (TCE). http://www.chemspider.com/Chemical-Structure.13837280.html?rid=d8d86fff-e36f-41d5-8385-e70095b2147b.

[103] Perchloroethene (PCE). http://www.chemspider.com/Chemical-Structure.13837281.html?rid=b524917e-7b34-4c2a-a0a2-400c1f99bea1.

[104] Crockett CS. Cross-cutting drinking water wastewater issues. 2010. http://www.slideshare.net/crockettc/emerging-pathogens-in-wastewater.

[105] Centers for Disease Control and Prevention. Addressing emerging infectious disease threats: a prevention strategy for the United States. 1994 (Executive Summary). MMWR, 43, RR-5).

[106] National Academies Press. Global issues in water, sanitation, and health: workshop summary. 2009. http://www.ncbi.nlm.nih.gov/books/NBK28463/.

[107] Parker T, Center for Disease Control and Prevention. Gram-negative *Achromobacter xylosoxidans* bacteria (*Pseudomonas pseudomallei*). 2014. http://phil.cdc.gov/phil/details.asp.

[108] Center for Disease Control and Prevention. Melioidosis. 2012. http://www.cdc.gov/melioidosis/.

[109] Lee YH, Chen Y, Ouyang X, Gan YH. Identification of tomato plant as a novel host model for *Burkholderia pseudomallei*. BMC Microbiol 2010;10(28). http://dx.doi.org/10.1186/1471-2180-10-28.

[110] Center for Disease Control and Prevention. Three-step stool examination for cryptosporidiosis in 10 homosexual men with protracted watery Diarrhea. J Infect Dis 1983;147(5):824–8. http://phil.cdc.gov/phil/details.asp.

[111] Center for Disease Control and Prevention. Parasites-Cryptosporidium (also known as Crypto): Illness & symptoms. 2015. http://www.cdc.gov/parasites/crypto/illness.html.

[112] Erlandsen S, Center for Disease Control and Prevention. *Giardia lamblia* protozoan. 1999. http://phil.cdc.gov/phil/details.asp.

[113] Center for Disease Control and Prevention. What is Giardiasis?. 2015. http://www.cdc.gov/parasites/giardia/general-info.html.

[114] Center for Disease Control and Prevention/National Institute of Allergy and Infectious Diseases (NIAID). Salmonella. 2011. http://phil.cdc.gov/phil/details.asp.

[115] Center for Disease Control and Prevention. Salmonella. 2015. http://www.cdc.gov/salmonella/general/technical.html.

[116] Palmer EL, Center for Disease Control and Prevention. Norovirus, with 27–32nm-sized viral particles. 1981. http://phil.cdc.gov/phil/details.asp.

[117] Norovirus 2016. https://en.wikipedia.org/wiki/Norovirus.

[118] Center for Disease Control and Prevention—Div of Parasitic Diseases. Ascariasis. 1960. http://www.cdc.gov/dpdx/ascariasis/index.html.

[119] UNICEF. Pneumonia and Diarrhoea: tackling the deadliest diseases for the world's poorest children. 2016. http://rehydrate.org/diarrhoea/#Health%20hazards.

[120] Schneider M-J. Introduction to public health. 2nd ed. Sudbury, MA: Jones and Bartlett; 2006.

[121] Centers for Disease Control and Prevention. Cryptosporidiosis outbreaks associated with recreational water use—five states, 2006. Morb Mortal Wkly Rep (MMWR) 2007b;56(29):729–32.

CHAPTER EIGHTEEN

Challenges to Achieving the Sustainable Development Goals: Water Treatment

J. Luh, J. Bartram
University of North Carolina at Chapel Hill, Chapel Hill, NC, United States

1. INTRODUCTION

The adoption of the Sustainable Development Goals (SDGs) in 2015 included a target "to achieve universal and equitable access to safe and affordable drinking water for all" [1]. The emphasis on *safe* drinking water is not unique to the SDGs and has appeared in international development targets for the Second and Third United Nations Development Decades [2–4]; International Drinking Water Supply and Sanitation Decade [5,6]; World Declaration on the Survival, Protection and Development of Children [7]; and Millennium Development Goals (MDGs) [8], and is recognized in the Human Right to Water and Sanitation [9,10], (see Fig. 18.1).

During the MDG period, the indicator used to monitor progress toward increasing access to safe drinking water was the "proportion of the population using water from an *improved* [emphasis added] source" [8]. This indicator implicitly assumes that water from an improved source is protected from outside contamination [12] and thus is higher in microbial quality than water from an unimproved source. However, while some studies support this assumption, there is also evidence to show that not all water from improved sources are safe [13,14]. In addition, regardless of whether the drinking water source is safe, water collected from community-level improved water sources (including from piped systems) and stored in households can deteriorate in microbial and/or chemical quality between collection and consumption (e.g., introduction of lead through corrosion of pipes) [15]. To address the fact that use of an improved source does not ensure access to safe water, the indicator to monitor progress toward the SDG water target is the "percentage of the population using safely managed drinking water services" where *safely managed drinking water services* is comprised of

Chemistry and Water
ISBN 978-0-12-809330-6
http://dx.doi.org/10.1016/B978-0-12-809330-6.00018-0

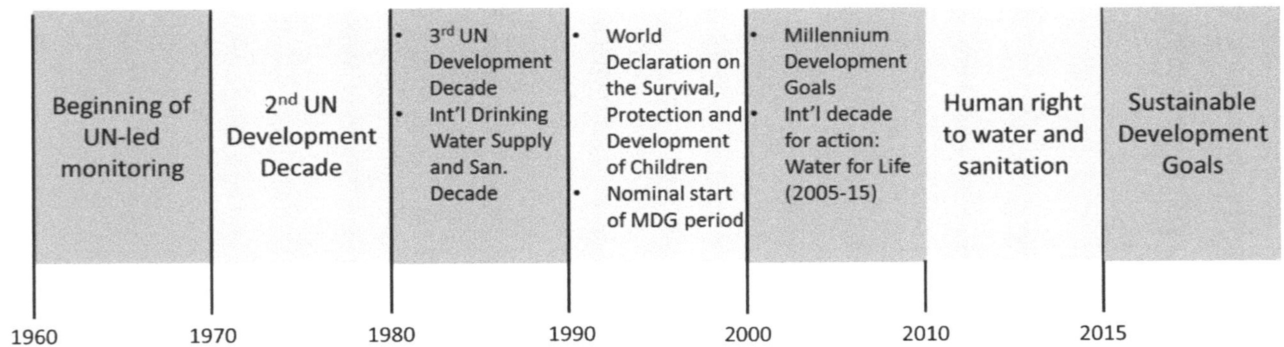

Figure 18.1 Timeline of international drinking water targets and actions. *Modified from Bartram J, Brocklehurst C, Fisher MB, Luyendijk R, Hossain R, Wardlaw T, Gordon B. Global monitoring of water supply and sanitation: history, methods and future challenges. Int J Environ Res Public Health 2014;11:8137–65.*

an improved source that is accessed on premises, available when needed, and, importantly, compliant with fecal and priority chemical standards [16].

Throughout history, two general approaches have been taken to securing safe water: (1) use of preferred sources, typically groundwater, and (2) treatment of water to improve its chemical, physical, microbiological, and radiological characteristics. The preference for groundwater is based on the lower levels of microbial contamination/presence in groundwater from deep aquifers [17,18]. Groundwater from shallow aquifers or karst aquifers is more readily contaminated with pathogenic microorganisms [17,18], and, as such, requires treatment to prevent waterborne diseases. Similarly, surface waters, which are open to contamination, require treatment before consumption. Forms of treatment vary from the household level to centralized treatment and distribution with the aim of reducing the health effects of contaminants. For example, in the US, the introduction of chlorination alone in 1908 led to significant decreases in waterborne diseases such as cholera, typhoid, and dysentery [19]. Household water treatment has also been shown to reduce the incidence of diarrheal disease with proper safe storage and handling to prevent recontamination [20]. For chemical contaminants, treatment can also vary. For example, membranes, coagulation/sand filtration, anion exchange, and precipitation can be used to remove arsenic, a heavy metal found particularly in groundwater that can cause severe health effects (e.g., increased risk of cancers) at low concentrations [21]. Household water treatment can also be used to remove arsenic. A study using household sand filters for 2 years to treat raw groundwater for arsenic in Vietnam showed, using hair analysis, that arsenic exposure was lower in people drinking treated water as compared to untreated water [22].

With the increasing global population, and with it a growing population without access to safe water, as well as other factors, such as decreasing availability of freshwater resources and increasing pollution of existing water sources from industrial and agricultural activities, the global scale of water in need of treatment is large. In this chapter, we describe the challenges faced in providing safe drinking water, and we characterize the scale of treatment needed to achieve the SDG drinking water target.

2. QUALITY VERSUS SAFETY

Quality and safety are concepts that are often used interchangeably in relation to drinking water, but they have very different meanings. Quality describes the composition of water and refers to its physical, chemical, biological, and radiological characteristics. Safety compares the quality of water to a specific set of standards for human consumption in order to determine its suitability. For substances that negatively affect the organoleptic properties of water (e.g., taste, smell), these two concepts may overlap. Table 18.1 provides water quality standards for an illustrative range of countries.

Due to the fact that changes in water quality can be rapid, testing and comparing water quality results to water quality standards is not sufficient to ensure safety against microbial contaminants. This is in contrast to most chemical contaminants, where it is chronic cumulative exposure that is of health concern for exposure through drinking water. For microbial contaminants, monitoring of water quality must be supplemented with management measures that seek to prevent, and not only react, to hazards that may adversely affect water quality. One systematic risk management approach that combines both management measures and water quality monitoring is Water Safety Plans (WSPs), which is an approach that has

Table 18.1 Examples of Drinking Water Quality Standards From Different Countries

Contaminant	Australia [23]	European Union [24]	Uganda [25]	United States [26]	World Health Organization [27]
Benzene (μg/L)	1	1	10	5	10
Fluoride (mg/L)	1.5	1.5	1.5	4	1.5
Mercury (μg/L)	1	1	1	2	6
Iron (μg/L)	[a]	200	300	300[b]	[c]
Manganese (μg/L)	500	50	100	50[b]	[c]
Escherichia coli (count)	0 in 100 mL	0 in 100 mL	0 in 100 mL	0	0 in 100 mL
Enterococci (count)	0	0 in 100 mL	Not mentioned	Not mentioned	Not mentioned

[a]Insufficient data to set a guideline value based on health considerations. A value of 300 μg/L was set based on aesthetic considerations.
[b]Secondary drinking water standards are not mandatory and are not enforced. These are guidelines for aesthetic considerations (taste, color, odor).
[c]Not of health concern at levels causing acceptability problems in drinking water. For example, a health-based value of 0.4 mg/L can be derived for manganese; however, this is much higher than concentrations of manganese normally found in drinking water.

been used in both developing countries (e.g., Bangladesh, South Africa, Uganda [28–30]) and developed countries (e.g., Australia, Canada, Germany, Slovenia, the UK [31–35]). The implementation of WSPs has been shown to improve drinking water quality and regulatory compliance [36] in addition to other benefits, such as improved communication and collaboration between stakeholders and asset management [37,38].

The *Guidelines for Drinking-water Quality* defines safe drinking water as water that "...does not represent any significant risk to health over a lifetime of consumption..." [27]. The definition of quality, on the other hand, varies and may be determined by societal demands that include nonhealth-related factors. Table 18.1 presents examples of water quality standards. Thus, in this chapter, we focus on safety and only the water quality parameters that are relevant to safety. The word "potable" is not considered in this chapter, as it is used in the literature to mean both "safe" and/or "intended for human consumption" and is therefore a less useful term for our purposes.

3. CONTAMINANTS OF CONCERN

Three general types of contaminants can be found in drinking water: chemical, biological, and radioactive. We briefly describe each type below, with the exposure routes of these contaminants summarized in Section 4.

3.1 Chemical Contaminants

The World Health Organization (WHO) has established guideline values for chemical contaminants in the *Guidelines for Drinking-water Quality* [27]. Classification of chemical contaminants according to the source-categories described in the following is useful because the source-categories determine optimal monitoring approaches and options for prevention and remediation.

Naturally occurring inorganic chemical contaminants are found in rocks and soil through which water flows, and include arsenic, chromium, and selenium [27]. Sources of naturally occurring organic chemical contaminants are vegetation and algae in the water, for example, microcystin-LR, which is produced by cyanobacteria (i.e., blue–green algae) [27].

Anthropogenic chemical contaminants can be classified into four main sources: industry, agriculture, water treatment, and households. Industrial activities, such as mining and chemical and fuel storage, are a source of chemical contaminants when materials and products containing the chemicals are stored, discharged, or disposed of in a manner that leads to drinking water contamination. Examples of chemical contaminants resulting from industrial sources are trichloroethylene (TCE), and tetrachloroethylene (PCE), both of which are used in dry cleaning and as industrial solvents, and benzene, which is a gasoline additive [27].

Contamination from agricultural activities occurs from activities such as the application of fertilizer and pesticides to crops, which enter drinking water through rainfall runoff, or point-source pollution due to inappropriate cleaning of equipment and disposal of products. Pesticides include herbicides, insecticides, and fungicides, and examples of pesticides are atrazine and carbofuran [27]. While water treatment and distribution are designed to ensure that the quality of a water meets standards for human consumption, the processes can inadvertently create chemical contaminants through the formation of disinfection by-products (DBPs) (e.g., *N*-nitrosodimethylamine), the corrosion of lead and copper pipes, and the leaching of chemicals from materials in contact with the water (e.g., vinyl chloride monomer) [27]. Finally, households are considered a source of chemical contamination, as household waste can contain heavy metals (e.g., zinc from household activities) or pharmaceutical products (e.g., carbamazepine) [39–41].

3.2 Biological Contaminants

Biological contaminants of health concern in drinking water primarily result from exposure to human and animal excreta, which contain bacteria, viruses, protozoan cysts, and helminths that are pathogenic to humans (e.g., *Campylobacter jejuni*, rotavirus, *Cryptosporidium parvum*, *Schistosoma* spp.), and from agents that multiply in water under suitable conditions. Multiplication of these pathogens (e.g., *Legionella* spp., *Mycobacterium avium* complex) throughout the drinking water system is becoming of increasing concern with the expansion of complex piped water systems within buildings [42,43]. While many microorganisms can infect a wide variety of hosts, some may have host specificity. Given that there are pathogens with strict human host specificity, exposure to excreta from humans may pose a greater risk to humans than exposure to excreta from other host species, such as cattle or pigs [44]. However, there have been cases where contamination of waters with animal excreta has resulted in waterborne disease, for example, the contamination of water by cattle manure resulted in an outbreak of *Escherichia coli* O157 in Swaziland [45].

3.3 Radioactive Contaminants

Radioactive contaminants in drinking water arise from anthropogenic and naturally occurring sources. Anthropogenic sources of radioactive substances, such as medical diagnostic activities, are usually controlled so that waste is disposed of in an appropriate manner to avoid contamination of drinking water [27]. Naturally occurring sources of radioactive contaminants (e.g., radium, thorium) occur through the dissolution of radioactive substances from rocks and soil to water [27]. Ingestion of radioactive contaminants through drinking water typically accounts for a minor proportion of total radiation received from all sources [27]. In general, the concentration of radioactive contaminants in drinking water is low and not a major global concern, with the exceptions of uranium and radium, and therefore, radioactive contaminants are not considered beyond this section of the chapter.

It should be noted that while uranium has been found in drinking water in various countries ranging from <0.1 μg/L up to 700 μg/L, the primary health effect of concern is its chemical toxicity, specifically nephritis, and not radiological toxicity [46]. Due to the fact that the chemical toxicity of uranium is much higher than its radiological concern, the provisional guidelines value of 30 μg/L provided by the WHO for uranium in drinking water is determined by its chemical toxicity [27,46].

Some drinking water sources may contain radon, a radioactive gas that is soluble in water. Concentrations of radon are generally higher in groundwater from wells and boreholes and lower in surface waters [27]. Turning on the shower or a tap may release radon into the air, although other sources of radon in indoor air exist, such as the soil and building materials. Exposure to radon has been shown to occur primarily through inhalation, rather than ingestion, and thus, guidance levels for radon in drinking water are usually not necessary [27].

4. EXPOSURE THROUGH DRINKING WATER

Understanding exposure is fundamental to determining the desired quality of water and the corresponding degree of treatment required. For chemical contaminants, it is generally chronic exposure, as opposed to acute exposure, that determines health risk, and therefore, periodic measurements of water quality are appropriate. In contrast, it is acute exposure to biological contaminants that typically causes disease. As short-term microbial exposure can lead to adverse health effects, there is no water sampling and testing regime that can guarantee that the drinking water is safe from microbial contaminants. Thus, an effective management approach would be to combine monitoring of hazards with continuous monitoring of the control measures that are critical to preventing contamination.

There are a limited number of chemical contaminants for which there is evidence that suggests drinking water is a significant source of exposure at concentrations of health concern. The WHO provides guideline values for these chemicals, depending on the type of chemical contaminant. For "non-threshold chemicals," which are largely genotoxic carcinogens, guideline values are calculated as the concentrations in drinking water that would result in an estimated upper-bound excess lifetime cancer risk of 10^{-5} [27]. For "threshold chemicals," which are the majority of chemicals, the guideline value is determined based on its tolerable daily intake, which is derived using a no observed adverse effect level (NOAEL). The NOAEL is determined using a variety of approaches depending on data availability, although the preference is to use long-term studies of ingestion of drinking water [27].

While no globally representative data are available, the principal chemical contaminants of concern, in decreasing order based on population exposure, are likely to be fluoride and arsenic. Additional chemicals of concern with established WHO guideline values include selenium and chromium, among others. Both fluoride and arsenic occur naturally in groundwater and are common in various parts of the world. Arsenic, known to cause cancer after exposure through drinking water and also known to have other adverse health effects [21,47], is prevalent in many regions in the world, particularly in Bangladesh and also West Bengal, Vietnam, and China [22]. Daily consumption of fluoride above 14 mg per day leads to skeletal fluorosis and an increased risk of bone fractures [48]. Fluoride is problematic in the East African Rift Valley from the Jordan Valley through Sudan, Ethiopia, Uganda, Kenya, and the United Republic of Tanzania as a result of volcanic rocks, and also from Iraq and Iran through Syria and Turkey to the Mediterranean region due to sediments of marine origin [49]. In affected areas, water tends to be the principal source of exposure to both fluoride and arsenic.

Lead is introduced into drinking water as a result of corrosion of lead pipes [50,51] and can cause neurological, renal, and hematological problems [51,52]. However, lead exposure via ingestion of drinking water is typically low when compared to lead exposure via ingestion of lead-contaminated dust (e.g., from paint containing lead) and via inhalation of lead particles (e.g., generated by burning materials containing lead, present in fumes from gasoline containing lead additives) [53,54]. Other important routes of exposure to lead are ingestion of contaminated food and soil [53,54].

A large body of evidence exists that demonstrates that consumption of water from contaminated community sources, which include improved sources such as hand pumps and protected springs, and contaminated distribution systems may be a route of exposure for biological contaminants associated with a substantive disease burden. Both the point of water collection/receipt and household stored water need to be considered, as water quality deteriorates due to contamination during its collection, transportation, and storage [13–15]. Studies that assess the presence of fecal contamination in drinking water generally focus on *E. coli* or thermotolerant coliforms (TTC), which serve as indicators of fecal contamination. While these two indicators are useful in identifying populations that have been exposed to pathogens, the absence of either do not necessarily indicate an absence of fecal contamination. For example, *E. coli* is much more susceptible to free chlorine as compared to protozoan cysts, and therefore, chlorinated waters in which *E. coli* is absent may still contain infective protozoan cysts. As such, the use of only *E. coli* or TTC does not provide a reliable estimate of risk. Instead, alternative methods, such as quantitative microbial risk assessment (QMRA) [55,56], should be used to determine treatment requirements.

5. TREATMENT LOCATION

Drinking water treatment can occur at one or more of three locations: centralized prior to reticulated distribution, centralized at the point of collection (i.e., community sources), and in households [i.e., point of entry (PoE), point of use (PoU)/household water treatment and safe storage (HWTS)]. The treatment location is important, depending on the hazard of concern and its management. For example, for chemical contaminants (e.g., lead from corrosion of lead pipes, vinyl chloride monomer from unplasticized polyvinyl chloride used in drinking water pipes) and biological contaminants (e.g., *Legionella* spp. that colonize building plumbing) that are introduced during transportation, distribution, and storage, prevention, and/or household water treatment would be needed. Depending on the location of treatment, the design, operation, and management of the treatment technique/process varies.

5.1 Centralized Treatment Prior to Reticulated Distribution

For centralized treatment prior to reticulated distribution, experience has shown that large, centralized utility water supplies can successfully apply treatment to render water safe. The success of these large utilities is in part due to greater resources in terms of well-engineered, constructed facilities, financial resources, as well as access to technical expertise for continuous professional management. However, availability of resources alone does not necessarily ensure the production of safe water, with data showing that drinking water quality improves with active management. As an example, analysis of health and water quality data from municipal water utilities in Iceland showed that when WSPs, a systematic preventive management and risk assessment approach, were implemented, a significant decrease in incidence of clinical cases of diarrhea, an improvement in water quality as measured by heterotrophic plate counts, and a significant decrease in regulatory noncompliance were observed [36]. While no studies are available in the literature on the effect of WSP implementation on chemical contaminants, WSPs also have the potential to assist water providers in better controlling for chemical contaminants.

Community-managed systems also have centralized treatment prior to reticulated distribution; however, they differ from large utilities in the operating and management challenges they face. Specifically, the WHO noted that "the precise definition of a 'community water supply' will vary. While a definition based on population size or the type of supply may be appropriate under many conditions, it is often administration and management that set community supplies apart, and this is especially true in developing countries. The increased involvement of ordinary, often untrained and sometimes unpaid, community members in the administration and operation of water-supply systems is characteristic of small communities; this provides a ready distinction between community water supplies and the supply systems of major towns and cities." [57]. As such, community-managed systems may have problems managing complex and failure-sensitive treatment systems, and waterborne outbreaks can result (e.g., Walkerton, Canada).

The community-managed systems require professional management and are often associated with certification schemes, as safety is a key consideration for these types of business models. One example of a business model for community-managed systems is the circuit rider model. The circuit rider model allows community-managed systems to contract with an association that provides technical expertise, assistance, and training in the supply, treatment, and distribution of drinking water for community-managed systems. One technical advisor visits several communities on a frequent basis, and thus, communities have access to technical expertise without incurring the expense of having a full-time expert on staff.

5.2 Centralized Treatment at the Point of Collection

Centralized treatment at the point of collection refers to community sources, such as water kiosks where water can come from a spring to piped network or from a borehole and hand pump. These are generally found in peri-urban areas for both developing and developed countries, although with the SDG requirement for on-premises water sources and ongoing development favoring higher levels of service, the prevalence of community sources is likely to decline. Due to the relatively low volumes of water produced at community sources, advanced treatment processes such as reverse osmosis can be applied. An additional example would be the use of ultrafiltration followed by UV disinfection. However, it must be noted that recontamination of the water can occur after collection during the transportation and storage process. Community sources often require professional management and certification schemes.

5.3 Packaged Water

One specialized case of centralized treatment is that of packaged, including bottled and sachet, water. Consumption of packaged water has grown in the last decade, particularly in low- and middle-income countries, and is used in urban, rural, and peri-urban areas. As an example, 18% of households in Ghana used packaged water as their primary source of water in 2011 [58], with one study showing that in a peri-urban area of Accra, this proportion increased to 47% [59]. In theory, professional management and governance should ensure that packaged water meets specific certification; however, this is not always achieved. With the rapid increase in use of packaged water, effective regulation does not always occur in many low- and middle-income countries, resulting in outbreaks of cholera [60] and other waterborne diseases. While studies have shown that packaged water in West African countries may contain microbial, chemical, and radioactive contamination [60–67], a recent study in Freetown, Sierra Leone [68] showed that these risks must be weighed against alternative water sources and that packaged water may be the least unsafe option compared to stored household water for consumption. Microfranchising has been used as a business model for the packaging and selling of water where startup costs are significantly lower than those associated with a traditional franchise model, thereby making it affordable for individuals of low income, but the benefits of brand recognition and marketing are available.

5.4 Household Water Treatment

Management of household water treatment (i.e., PoE, PoU/HWTS) is at the individual or household level. PoE devices can be plumbed directly into the sink faucet or tap, where the device is installed in the main water pipe and all water used in the household undergoes treatment (i.e., device installed at PoE) [69]. Examples of PoE devices include water softeners and reverse osmosis systems.

PoU devices, also referred to as HWTS, generally refer to free-standing devices or treatment techniques that are not plumbed in and that are widely used in both developed and developing countries. Many HWTS devices are countertop, although some devices are portable, allowing them to be used in various settings, such as disaster relief when drinking water infrastructure has been compromised, for outdoor camping and hiking, and in areas where households must collect and carry water back to their homes. Examples of HWTS practices would be boiling, chlorination, and filtration, while examples of HWTS devices would be ceramic pot filters and activated carbon filters, including filters that are installed on the tap(s) of interest [69–71]. The success of HWTS devices in preventing disease is dependent on various factors, such as the efficacy of the treatment method to remove chemicals or inactivate pathogens, user behavior in terms of consistent and correct use, and also effective governance in ensuring product certification. Proper household storage, including use of a closed or narrow-necked container to prevent recontamination, is essential to the success of HWTS [69–71]. Similar to PoE treatment, HWTS does not increase access to safe water and focuses only on improving water quality. It should be noted that not all HWTS devices are designed to remove/inactivate microorganisms and provide microbially safe water. Instead, these devices are intended to either provide incremental improvement or address nonmicrobial concerns (e.g., taste).

6. TREATMENT OPTIONS

Numerous sources describe the different options available for treating drinking water [18,72]. Here we provide a brief summary of the treatment processes used for addressing biological and chemical contaminants, regardless of whether they are applied to centralized or household systems.

For biological contaminants, treatment efficacy in reducing microbes of concern is, in broad terms, achieved by the physical removal of microbes from water and/or their inactivation, where the latter refers to rendering the microbes unable to reproduce. It is common to use a multiple barrier approach in the form of a sequence of processes to provide higher security against failure [18,72–74]. Physical removal of microbes from water is performed through processes such as sedimentation and filtration. As microbial size is an important parameter for successful removal, the order of decreasing difficulty for removal is typically viruses > bacteria > protozoa > helminthes [73]. Helminthes are easily removed through gravity settling and/or filtration. During the sedimentation process, particles settle out due to gravity, with larger particles settling out more efficiently than smaller particles. Prior to sedimentation, water can also undergo coagulation and flocculation to promote the association of microbes into larger particles so that they can be physically removed. Filtration relies on a physical barrier to remove microbes and particles. Types of filtration vary widely and include granular media, ceramic, and membrane filters [73]. One filtration technique that relies on both physical and biological processes for microbial removal is slow sand filtration, where removal occurs in the top layer, the schmutzdecke, which consists of a biofilm of microorganisms including bacteria, protozoa, and fungi, and also throughout the filter bed [18,74]. For microbes that are small enough to pass through sedimentation and filtration, chemical or UV treatment is used to inactivate these microorganisms. Common chemical disinfectants include chlorine, ozone, and chlorine dioxide, all of which react with and oxidize the microorganism in order to prevent their reproduction. UV disinfection uses UV light to destroy or disrupt the DNA of microorganisms, thus preventing them from replicating. In developing regions, UV disinfection can be in the form of solar disinfection where bottles of water are exposed to sunlight by placing them on rooftops.

Though not a part of the so-called "primary disinfection," which refers to the use of disinfectant to inactivate microorganisms present in the source water, "residual disinfection" is also used to ensure the microbiological quality of drinking water [18]. Residual disinfection refers to the addition of disinfectant in the already treated water so that microorganism regrowth does not occur and so that any microorganisms that enter the water after it has already been treated are also inactivated before consumption. One of the reasons why free chlorine is a preferred disinfectant for primary disinfection is because it carries a residual, while other chemical disinfectants, such as ozone and chlorine dioxide, and UV disinfection, do not. In the cases where a residual disinfectant is desired, free chlorine or monochloramine is usually added at the end of the water treatment process.

Removal of chemical contaminants is achieved through various processes, depending on the contaminant, its concentration in water and the target concentration to be achieved, and water quality characteristics, among other factors [74]. Table 18.2 lists examples of treatment processes and corresponding chemical contaminants addressed. Processes can range from chemical processes that transform the contaminant such as oxidation to physical processes that remove the contaminant such as membrane treatment. Chemical processes target either contaminant modification to make it more easily removed in a subsequent step (e.g., Fe^{2+} to Fe^{3+}, where the latter is more easily precipitated) or mineralization. In some cases, however, by-products may be created that are of health concern; for example, NO_2^- can be formed when attempting to mineralize NO_3^- to N_2 and water. To control for DBPs, which are produced during chemical disinfection, in addition to changing process conditions (e.g., changing the pH which may affect the DBP concentrations formed) and removing DBPs after their formation (e.g., via air stripping), an additional strategy is to remove DBP precursors prior to disinfection via, for example, enhanced coagulation. For chemical contaminants, such as lead and copper that are introduced into drinking water as a result of corroding pipes, corrosion control is needed. Commonly used methods to control corrosion include pH adjustment, increasing the alkalinity or hardness, or adding corrosion inhibitors (e.g., polyphosphates, silicates, and orthophosphates) [27].

7. SCALE OF CHALLENGE TO REACH SUSTAINABLE DEVELOPMENT GOAL WATER TARGET

In the SDG era, the indicator used to assess progress toward the SDG water target now requires compliance with fecal and priority chemical (e.g., arsenic, fluoride) standards [16]. Globally, the main contaminants of concern are fecal contamination and in certain countries, arsenic and fluoride. *E. coli* is considered the most reliable indicator of fecal contamination, with WHO guidelines of no detectable *E. coli* in a 100 mL sample [27]. As water quality data was not available, water quality was not included in the indicators of previous international water targets, nor was it monitored. As such, there

Table 18.2 Examples of Treatment Processes and Contaminants Addressed

						Membrane processes			Ion exchange		Adsorption		
	Aeration and stripping	Coagulation, sedimentation or DAF, filtration	Precoat filtration	Lime softening	Chemical oxidation and disinfection	Nanofiltration	Reverse osmosis	Electrodialysis/ ED reversal	Anion	Cation	Granular activated carbon	Powdered activated carbon	Activated alumina
Inorganics													
Antimony							X	X					
Arsenic (+3)		XO		XO			X	X	X				X
Arsenic (+5)		X		X			X	X	X				X
Barium				X			X	X		X			
Beryllium		X		X			X	X					
Cadmium		X		X			X	X		X			
Chromium (+3)		X		X			X	X		X			
Chromium (+6)							X	X	X				
Cyanide					X								
Fluoride				X			X	X					X
Lead													
Mercury (inorganic)				X			X	X					
Nickel				X			X	X		X			
Nitrate							X	X	X				
Nitrite							X	X	X				
Selenium (+4)		X					X	X	X				X
Selenium (+6)							X	X	X				X
Thallium							X	X					
Organic contaminants													
Volatile organics	X										X		
Synthetic organics							X				X	X	
Pesticides/Herbicides						X	X				X	X	
Dissolved organic carbon		X				X	X				X	X	

DAF, dissolved air flotation; *X*, appropriate process for this contaminant; *XO*, appropriate when oxidation used in conjunction with this process; Lead is generally a product of corrosion and is controlled by corrosion control treatment rather than removed by water treatment processes.

Taken from Table 3.1 of American Water Works Association. Water quality & treatment: a handbook of community water supplies. 5th ed. McGraw-Hill, Inc.: New York (USA); 1999.

is little data available to assess the current status of the global population with access to water that meets fecal and priority chemical standards. Using water quality data collected by the WHO and UNICEF from five countries (Ethiopia, Jordan, Nicaragua, Nigeria, and Tajikistan), Onda et al. [76] estimated that in 2010, 1.8 billion people (28% of the global population) used unsafe water, where "unsafe" water was that which tested positive for TTC or had greater than two sanitary risks on the sanitary risk inspection. Of these 1.8 billion people, 1.2 billion people (18% of the global population) used an improved water source that had presence of TTC or significant sanitary risks.

Assuming that the same percentage (28%) of the global population will not have access to safe water free from fecal contamination and sanitary risk, and using the 2030 global population estimate at constant fertility of 8.7 billion people [77], approximately 2.4 billion people would not have access to safe water in 2030. Given that the Onda et al. study looked at national estimates without further breakdowns such as differentiating between rural and urban regions, groundwater versus surface water, or utility-managed versus community-managed versus household-managed sources, future estimates for 2030 at a subnational level are not provided in this chapter. However, it is likely that the majority of this 2.4 billion people will be located in rural areas. In 2015, access to improved water sources was 85% for the global rural population, which was lower than 96% for the global urban population [78]. These percentages were estimated using the MDG indicator for access to improved water sources, which only required access to the improved water source to be at the community level. With the SDG indicator now requiring that improved drinking water must be delivered on premises, this reflects the changed approach that coverage must be at the household rather than community level. Accordingly, both urban and rural populations with access to improved sources will decrease, but more so in rural areas where community sources are more prevalent [79]. Countries will therefore need to allocate resources to not only ensure that water quality meets fecal and priority chemical standards, but also to focus on increasing access from community to household level.

While it is not possible to estimate subnational breakdowns (e.g., rural versus urban) for the proportion of the population with access to safe water, we can examine country clusters to identify groupings that will have the greatest difficulty in meeting the SDG target. Regional and income groupings are often used when making comparisons; however, this is not always the logical choice, as some countries are vastly different. For example, while Japan and the Democratic People's Republic of Korea are both located in

Eastern Asia, they are vastly different in most other respects. Therefore, instead of using regional or income groupings, we use the five country clusters developed by Onda et al. [80], where countries are grouped based on similarities across nine water- and sanitation-related indicators. These five clusters shown in Table 18.3 include 156 countries, or 6.75 billion people.

Table 18.3 Country Clusters developed by Onda et al. [80] Based on Similarities Across Water- and Sanitation-Related Indicators and the Proportion of Their Population Without Access to Improved Water

Cluster	Countries	% of Population Without Access to Improved Sources
1 (n = 33)	Australia, Austria, Belgium, Canada, Cyprus, Czech Republic, Denmark, Estonia, Finland, France, Germany, Greece, Iceland, Ireland, Israel, Japan, Latvia, Lithuania, Luxembourg, Malta, Netherlands, New Zealand, Norway, Portugal, Singapore, Slovak Republic, Slovenia, South Korea, Spain, Sweden, Switzerland, United Kingdom, United States	0.3
2 (n = 15)	Argentina, Belarus, Brazil, Bulgaria, Chile, Colombia, Cuba, Iran, Kazakhstan, Mexico, Oman, Russia, Ukraine, Uruguay, Venezuela	3.4
3 (n = 28)	Albania, Algeria, Armenia, Azerbaijan, Bosnia and Herzegovina, Costa Rica, Croatia, Dominican Republic, Egypt, El Salvador, FYR Macedonia, Georgia, Iraq, Jordan, Kyrgyzstan, Lebanon, Maldives, Mauritius, Moldova, Mongolia, Sri Lanka, Syria, Tajikistan, Tunisia, Turkey, Turkmenistan, Uzbekistan, Vietnam	5.8
4 (n = 24)	Belize, Bhutan, Bolivia, Botswana, China, Ecuador, Gabon, Guatemala, Guyana, Honduras, India, Indonesia, Jamaica, Namibia, Nicaragua, Panama, Paraguay, Peru, Philippines, Sao Tome and Principe, South Africa, Suriname, Thailand, Trinidad and Tobago	6.0

(*Continued*)

Table 18.3 Country Clusters developed by Onda et al. [80] Based on Similarities Across Water- and Sanitation-Related Indicators and the Proportion of Their Population Without Access to Improved Water—cont'd

Cluster	Countries	% of Population Without Access to Improved Sources
5 (n = 51)	Afghanistan, Angola, Bangladesh, Benin, Burkina Faso, Burundi, Cambodia, Cape Verde, Central African Republic, Chad, Comoros, Congo, Cote d'Ivoire, Democratic Republic of the Congo, Djibouti, Equatorial Guinea, Ethiopia, the Gambia, Ghana, Guinea, Guinea-Bissau, Haiti, Kenya, Laos, Lesotho, Liberia, Madagascar, Malawi, Mali, Mauritania, Morocco, Mozambique, Myanmar, Nepal, Niger, Nigeria, Pakistan, Papua New Guinea, Rwanda, Senegal, Sierra Leone, Sudan, Swaziland, Tanzania, Timor-Leste, Togo, Uganda, Yemen, Zambia, Zimbabwe	27.7

For each of the five clusters, the proportion of the population without access to improved sources is listed in Table 18.3. Assuming that these percentages stay constant during the SDG era (although as mentioned previously, these percentages will likely decrease with the additional SDG requirements for safe water), the countries in Cluster 5 will face the greatest challenge to achieving safe water in 2030.

Cluster 1 consists of developed and recently industrialized countries (e.g., Canada, Germany, Japan) that are characterized by high per capita GDP and low levels of inequality as given by the Gini index. These countries have near-universal access to improved water (at 99.7% of the cluster population) and of the four components of the SDG indicator: (1) improved drinking water facility, (2) available when needed, (3) located on premises, and (4) compliant with fecal and priority chemical standards, the challenge for these countries will likely be ensuring water quality compliance, as the other three factors have mostly been met. In these countries, outbreaks of waterborne disease may still occur occasionally, and one approach to address this problem is to use a preventive risk management approach, such as WSPs and sanitation inspections. WSPs can also be used to plan for the possibility of chemical spills in drinking water sources by identifying potential hazards, conducting risk assessments, and employing control measures to ensure that the effect on the operation of drinking water systems is minimal. In addition, there is also a need to ensure proper management of small community-managed systems, as past examples have shown that waterborne outbreaks can occur when improperly managed (e.g., Walkerton, Canada *E. coli* O157 and *Campylobacter* outbreak in 2000; North Battleford, Canada cryptosporidium outbreak in 2001).

Cluster 2 countries are characterized by high urbanization and high, though not universal, access to water (96.6% of the cluster population). Example countries in this cluster include Russia, Argentina, and Iran. Similar to the countries in cluster 1, the main focus of cluster 2 countries will be on water treatment to ensure compliance with fecal and priority chemical standards. One challenge in these countries is the existence of small residual populations that continue to use community sources (centralized treatment at the point of collection); for these populations, there will be a need to move toward household-level access.

Cluster 3 countries are mainly from Eastern Europe and the Middle East, with some Latin American and Asian countries (e.g., Albania, Lebanon, Costa Rica). These countries are middle income, with ~50% urbanization, and generally have the lowest renewable freshwater resources per capita. Cluster 4 consists of countries from Latin America and South Africa, with some Asian countries (e.g., Honduras, South Africa, Philippines). The GDP per capita, urbanization, and expected years of education are similar to those countries from cluster 3; however, the cluster 4 countries have a higher income inequality and receive less official development assistance per capita. Both cluster 3 and 4 countries will have the challenge of continuing to extend services to the unserved while ensuring water is treated to meet water quality standards. With ~50% urbanization, there will need to be a balance between focusing on upgrading services in urban areas to meet SDG requirements (and which may seem the easier task to achieve) versus installing new water services in rural areas where there may be a lack of infrastructure, management, funds, and technical expertise in place to sustain services.

Cluster 5 is mainly composed of countries from Sub-Saharan Africa and South/Southeast Asian countries, plus Haiti (e.g., Mozambique, Pakistan, Cambodia). These countries have the lowest per capita GDP, lowest years of expected education, and lowest urbanization. The challenge for these countries will be in all four components of the SDG indicator, as they will need to work on increasing on premises household level access as well as ensuring water quality and availability. For areas where populations do not have access to even an improved source, the focus would be on providing an improved source before addressing issues of availability (i.e., available when needed).

CONCLUSIONS: PRESENT, EMERGING, AND FUTURE CHALLENGES

Monitoring water quality remains a challenge, as the long-term goal is for countries to have independent surveillance and regulation. Currently, water quality data is not available for all countries. Depending on the country, for example, countries that are high income as classified by the World Bank, water quality data may be collected by regulatory agencies. However, even for the most developed nations, data on all supply types that provide a full population overview are typically not available, particularly for small/private systems, which are not monitored. The challenge will be for all low- and middle-income countries to establish and further develop drinking water regulatory agencies tasked with collecting data on water quality, availability, and accessibility, among other parameters.

For utility- and community-managed systems, current water treatment operations focus on steady state performance. However, modeling suggests that a short-term deterioration in water quality can negate the benefits of long periods of safety. This suggests that instead of focusing on steady state, assessments should focus on failure mode analysis. One such analysis is QMRA, which has been used for various purposes, such as the determination of health-based targets, the evaluation of alternative treatment strategies, and to determine situations where the risk of exposure may be increased [55].

One difficulty in assessing progress toward the SDG target is reporting on vulnerable groups, such as nomadic people and people living with HIV and AIDS (PLHA). These groups are often difficult to capture in standard national surveys, and they possess special needs. In the case of nomadic groups and PLHA, PoU treatment may be the best option. In the case of nomadic groups, they may roam and temporarily settle in areas without access to improved water, and PoU treatment provides a solution to treating their drinking water. While PLHA may live in areas with centralized water treatment prior to reticulated distribution, they may have compromised immune systems that may require that an additional step of PoU treatment be applied prior to water consumption to ensure that contaminants introduced to the water in the distribution system are removed.

During the SDG era, an additional challenge will be the provision of safe drinking water in response to emergencies and disasters. Climate-related events, such as floods and droughts, are projected to increase in certain areas of the world, and building climate-resilient water systems is a necessity. In addition to increasing resilience, there is also a need to increase disaster response capacity, as the result of emergencies and disasters is often degradation in water quality, a disruption in treatment, and/or a complete loss of access to water. For example, during flooding in June and July 2007, the Mythe drinking water treatment plant in the UK was inundated with floodwater and shut down for 17 days. Approximately 140,000 households were affected, and 50 million liters of bottled water were distributed [81].

Finally, the SDG target is not only confined to households, but also extends to institutional settings, such as schools and healthcare facilities. Both of these settings have large buildings with their own challenges. Healthcare facilities face unique challenges in that their populations are already vulnerable, and there are multiple uses for water in these settings. For example, nosocomial infections from microorganisms that grow in pipes are a concern. While limited data is available on drinking water access in schools and healthcare facilities, additional data and monitoring is required [82–84].

REFERENCES

[1] United Nations General Assembly. Draft outcome document of the United Nations summit for the adoption of the post-2015 development agenda. A/69/1.85. August 12, 2015.

[2] World Health Organization. EB55/44. United Nations Second Development Decade: mid-term review and appraisal. January 17, 1975.

[3] World Health Organization. A29/12. community water supply and Wastewater disposal (mid-decade progress report). March 29, 1976.

[4] United Nations General Assembly. A/RES/35/56. International development strategy for the Third United Nations development decade. December 5, 1980.

[5] O'Rourke E. The international drinking water supply and sanitation decade: dogmatic means to a debatable end. Water Sci Technol 1992;26:1929–39.

[6] World Health Organization and United Nations Children's Fund. Global water supply and sanitation assessment 2000 report. 2000.

[7] Adopted at the World Summit for Children Plan of action for implementing the world declaration on the survival, protection and development of children in the 1990s. September 30, 1990.

[8] United Nations Development Group. Indicators for monitoring the millennium development goals. New York: United Nations; 2003.

[9] United Nations General Assembly A/RES/64/292. The human right to water and sanitation. August 3, 2010.

[10] United Nations General Assembly A/HRC/RES/15/9. Human rights and access to safe drinking water and sanitation. October 6, 2010.

[11] Bartram J, Brocklehurst C, Fisher MB, Luyendijk R, Hossain R, Wardlaw T, Gordon B. Global monitoring of water supply and sanitation: history, methods and future challenges. Int J Environ Res Public Health 2014;11:8137–65.

[12] WHO/UNICEF Joint Monitoring Programme (JMP) for Water Supply and Sanitation. Improved and unimproved water sources and sanitation facilities. 2016. Available: www.wssinfo.org/definitions-methods/watsan-categories/.

[13] Bain R, Cronk R, Wright J, Yang H, Slaymaker T, Bartram J. Fecal contamination of drinking-water in low- and middle-income countries: a systematic review and meta-analysis. PLoS Med 2014;11(5):e1001644.
[14] Bain R, Cronk R, Hossain R, Bonjour S, Onda K, Wright J, Yang H, Slaymaker T, Hunter P, Pruss-Ustun A, Bartram J. Global assessment of exposure to faecal contamination through drinking water based on a systematic review. Trop Med Int Health 2014;19:917–27.
[15] Shields KF, Bain RES, Cronk R, Wright JA, Bartram J. Association of supply type with fecal contamination of source water and household stored drinking water in developing countries: a bivariate meta-analysis. Environ Health Perspect 2015;123(12): 1222–31.
[16] WHO/UNICEF Joint Monitoring Programme (JMP) for Water Supply and Sanitation. JMP Green Paper-Zero Draft: global monitoring of water, sanitation and hygiene post-2015. 2015. Available: www.wssinfo.org/fileadmin/user_upload/resources/JMP-Green-Paper-15-Oct-2015.pdf.
[17] Baird C, Cann M. Chapter 14: The pollution and purification of water. In: Environmental chemistry. (USA): W.H. Freeman and Company; 2012. p. 601–60.
[18] Crittenden JC, Trussell RR, Hand DW, Howe KJ, Tchobanoglous G, editors. MWH water treatment: principles and design. 2nd ed. Hoboken (NY, USA): John Wiley & Sons; 2012.
[19] Centers for Disease Control and Prevention (CDC). A centurty of U.S. water chlorination and treatment: one of the ten greatest public health achievements of the 20th century. Available at: http://www.cdc.gov/healthywater/drinking/history.html.
[20] Quick RE, Kimura A, Thevos A, Tembo M, Shamputa I, Hutwagner L, Mintz E. Diarrhea prevention through household-level water disinfection and safe storage in Zambia. Am J Trop Med Hyg 2002;66(5):584–9.
[21] Choong TSY, Chuah TG, Robiah FL, Koay G, Azni I. Arsenic toxicity, health hazards and removal techniques from water: an overview. Desalination 2007;217:139–66.
[22] Berg M, Luzi S, Trang PTK, Viet PH, Giger W, Stuben D. Arsenic removal from groundwater by household sand filters: comparative field study, model calculations, and health benefits. Environ Sci Technol 2006;40:5567–73.
[23] HMRC, NRMMC. Australian drinking water guidelines Paper 6 National Water Quality Management Strategy. Canberra: National Health and Medical Research Council, National Resource Management Ministerial Council, Commonwealth of Australia; 2011.
[24] Council Directive 98/83/EC of 3 November 1998 on the quality of water intended for human consumption. Official J Eur Communities. 5.12.98, L 330/32–330/54; 1998.
[25] Uganda standard potable water – specification. 2014. US EAS 12, 2014–10-15.
[26] United States Environmental Protection Agency. Drinking water contaminants – standards and regulations. Available at: https://www.epa.gov/dwstandardsregulations.
[27] World Health Organization (WHO). Guidelines for drinking-water quality. 4th ed. Geneva (Switzerland): WHO Press; 2011.
[28] Howard G, Godfrey S, Tibatemwas A, Niwagaba C. Water safety plans for piped urban supplies in developing countries: a case study from Kampala, Uganda. Urban Water J 2005;2(3):161–70.
[29] Mahmud SG, Shamsuddin AJ, Ahmed MF. Development and implementation of water safety plans for small water supplies in Bangladesh: benefits and lessons learned. J Water Health 2007;05(4):585–97.
[30] Viljoen FC. The World Health Organization's water safety plan is much more than just an integrated drinking water quality management plan. Water Sci Technol 2010;61(1): 173–9.

[31] Malzer HJ, Staben N, Hein A, Merkel W. Identification, assessment, and control of hazards in water supply: experiences from Water Safety Plan implementation in Germany. Water Sci Technol 2010;61(5):1307–15.
[32] Reid DC, Abramowski K, Beier A, Janzen A, Lok D, Mach H, Radhakrishnan H, Rahman MM, Schroth R, Vatcher R. Implementation of Alberta's drinking water safety plans. Water Qual Res J Can 2014;49(1).
[33] Martel K, Kirmeyer G, Hanson A, Stevens M, Mullenger J, Deere D. Application of HACCP for distribution system protection. Denver (CO, USA): AWWA Research Foundation; 2006.
[34] Davidovits Z. The effects of changes in legal regulations on planning drinking water safety, vol. 9. Hadmernok; 2014. p. 234–40.
[35] Brauer F, Sturm S. European strategic workshop on water safety planning. Berlin, Germany. 2014.
[36] Gunnarsdottir MJ, Gardarsson SM, Elliot M, Sigmundsdottir G, Bartram J. Benefits of water safety plans: microbiology, compliance, and public health. Environ Sci Technol 2012;46:7782–9.
[37] Gunnarsdottir MJ, Gardarsson SM, Bartram J. Icelandic experience with water safety plans. Water Sci Technol 2012;65(2):277–88.
[38] Gelting RJ, Delea K, Medlin E. A conceptual framework to evaluate the outcomes and impacts of water safety plans. J Water Sanit Hyg Dev 2012;02(2):103–11.
[39] Comber SDW, Gunn AM. Heavy metals entering sewage-treatment works from domestic sources. Water Environ J 1996;10(2):137–42.
[40] Vogler S, Leopold C, Zuidberg C, Habl C. Medicines discarded in household garbage: analysis of a pharmaceutical waste sample in Vienna. J Pharm Policy Pract 2014;7:6.
[41] National Association of Clean Water Agencies. Pharmaceuticals and personal care products in the environment: a white paper on options for the wastewater treatment community. 2005. Available at: http://www.dep.state.fl.us/waste/quick_topics/publications/sh.
[42] Bartram J, Chartier Y, Lee JV, Pond K, Surman-Lee S, editors. *Legionella* and the prevention of Legionellosis. Geneva: World Health Organization; 2007.
[43] Pedley S, Bartram J, Rees G, Dufour A, Cotruvo J, editors. Pathogenic mycobacteria in water: a guide to public health consequences, monitoring and management. Geneva: IWA Publishing, London and World Health Organization; 2004.
[44] Dufour A, Bartram J, Bos R, Gannon V, editors. Animal waste, water quality and human health. Geneva: IWA Publishing, London and World Health Organization; 2012.
[45] Effler E, Isaäcson M, Arntzen L, Heenan R, Canter P, Barrett T, Lee L, Mambo C, Levine W, Zaidi A, Griffin PM. Factors contributing to the emergence of *Escherichia coli* O157 in Africa. Emerg Infect Dis 2001;7(5):812–9.
[46] World Health Organization. Uranium in drinking-water: background document for development of WHO guidelines for drinking-water quality. 2004. WHO/SDE/WSH/03.04/118.
[47] Nicomel NR, Leus K, Folens K, Van Der Voort P, Du Laing G. Technologies for arsenic removal from water: current status and future perspectives. Int J Environ Res Public Health 2016;13:62.
[48] Fawell J, Nieuwenhuijsen M. Contaminants in drinking water. Br Med Bull 2003;68: 199–208.
[49] Fawell J, Bailey K, Chilton J, Dahi E, Fewtrell L, Magara Y. Fluoride in drinking-water. Geneva: IWA Publishing, London and World Health Organization; 2006.
[50] National drinking water regulations. EPA 816-F-09–004. United States Environmental Protection Agency (US EPA); 2009.

[51] Centers for Disease Control and Prevention. Lead in drinking water and human blood lead levels in the United States. Morb Mortal Wkly Rep 2012;61(Suppl.):1−9.
[52] Toxicological profile for lead. Atlanta (Georgia): Agency for Toxic Substances and Disease Registry; 2007.
[53] Sannolo N, Carelli G, De Lorenzo G, Castellino N. Chapter 14: Environmental exposure. In: Castellino N, Castellino P, Sannolo N, editors. Inorganic lead exposure. (USA): CRC Press; 1995.
[54] Lead poisoning and health, Fact sheet N°379, Reviewed. World Health Organization; August 2015.
[55] Tfaily R, Papineau I, Andrews RC, Barbeau B. Application of quantitative microbial risk assessment at 17 Canadian water treatment facilities. J Am Water Works Assoc 2015;107:E497−508.
[56] Petterson SR, Ashbolt NJ. QMRA and water safety management: review of application in drinking water systems, J Water Health 2016;14(4):571−89.
[57] World Health Organization. Guidelines for drinking-water quality. In: Surveillance and control of community supplies. 2nd ed., vol. 3. Geneva (Switzerland): WHO Press; 1997.
[58] Ghana statistical service. Ghana multiple indicator cluster survey with an enhanced malaria module and biomarker, 2011, Final Report. Accra, Ghana. 2011.
[59] Stoler J, Weeks JR, Fink G. Sachet drinking water in Ghana's Accra-Tema metropolitan area: past, present, and future. J Water Sanit Hyg Dev 2012;2(4).
[60] Oluwafemi F, Oluwole ME. Microbiological examination of sachet water due to a cholera outbreak in Ibadan, Nigeria. Open J Med Microbiol 2012;2:115.
[61] Dada A. Sachet water phenomenon in Nigeria: assessment of the potential health impacts. Afr J Microbiol Res 2009;3(1):015−21.
[62] Kwakye-Nuako G, Borketey P, Mensah-Attipoe I, Asmah R, Ayeh-Kumi P. Sachet drinking water in Accra: the potential threats of transmission of enteric pathogenic protozoan organisms. Ghana Med J 2007;41(2).
[63] Obiri-Danso K, Okore-Hanson A, Jones K. The microbiological quality of drinking water sold on the streets in Kumasi, Ghana. Lett Appl Microbiol 2003;37(4):334−9.
[64] Ackah M, Anim A, Gyamfi E, Acquah J, Nyarko E, Kpattah L, et al. Assessment of the quality of sachet water consumed in urban townships of Ghana using physico-chemical indicators: a preliminary study, Adv Appl Sci Res 2012;3(4):2120−7.
[65] Ajayi A, Sridhar M, Adekunle L, Oluwande P. Quality of packaged waters sold in Ibadan, Nigeria. Afr J Biomed Res 2008;11(3).
[66] Orisakwe OE, Igwilo IO, Afonne OJ, Maduabuchi J-MU, Obi E, Nduka JC. Heavy metal hazards of sachet water in Nigeria. Arch Environ Occup Health 2006;61(5): 209−13.
[67] Ajayi O, Adesida G. Radioactivity in some sachet drinking water samples produced in Nigeria. Iran J Radiat Res 2009;7(3):151−8.
[68] Fisher MB, Williams AR, Jalloh MF, Saquee G, Bain RES, Bartram JK. Microbiological and chemical quality of packaged sachet water and household stored drinking water in Freetown, Sierra Leone. PLoS One 2015;10(7):e0131772.
[69] Point-of-use or point-of-entry treatment options for small drinking water systems. United States Environmental Protection Agency; 2006. EPA 815-R-06−010.
[70] Sobsey M, Brown J. Evaluating household water treatment options: health-based targets and microbiological performance specifications. World Health Organization; 2011.
[71] Ojomo E, Elliott M, Goodyear L, Forson M, Bartram J. Sustainability and scale-up of household water treatment and safe storage practices: enablers and barriers to effective implementation. Int J Hyg Environ Health 2015;218(8):704−13.
[72] Edzwald JK, editor. Water quality & treatment: a handbook on drinking water. 6th ed. (USA): American Water Works Association; 2011.

[73] Sobsey M. Managing water in the home: accelerated health gains from improved water supply. World Health Organization; 2002. WHO/SDE/WSH/02.07.
[74] Hager LS, Smith SM, editors. Water treatment plant design. 5th ed. (USA): American Water Works Association and American Society of Civil Engineers; 2012.
[75] American Water Works Association. Water quality & treatment: a handbook of community water supplies. 5th ed. New York (USA): McGraw-Hill, Inc.; 1999.
[76] Onda K, LoBuglio J, Bartram J. Global access to safe water: accounting for water quality and the resulting impact on MDG progress. Int J Environ Res Public Health 2012; 9:880–94.
[77] United Nations Population Division, Department of Economic and Social Affairs. World Population Prospects: the 2015 revision. Available at: https://esa.un.org/unpd/wpp/.
[78] WHO/UNICEF Joint Monitoring Programme (JMP) for Water Supply and Sanitation. Data and estimates. Available at: http://www.wssinfo.org/data-estimates/.
[79] Cumming O, Elliott M, Overbo A, Bartram J. Does global progress on sanitation really lag behind water? An analysis of global progress on community- and household-level access to safe water and sanitation. PLoS One 2014;9(12):e114699.
[80] Onda K, Crocker J, Kayser GL, Bartram J. Country clustering applied to the water and sanitation sector: a new tool with potential applications in research and policy. Int J Hyg Environ Health 2014;217:379–85.
[81] Charles K, Pond K, Pedley S, Hossain R, Jacot-Guillarmod F. Vision 2030-The resilience of water supply and sanitation in the face of climate change: technology projection study. Available at: http://www.who.int/water_sanitation_health/publication.
[82] Jordanova T, Cronk R, Obando W, Medina OZ, Kinoshita R, Bartram J. Water, sanitation, and hygiene in schools in low socio-economic regions in Nicaragua: a cross-sectional survey. Int J Environ Res Public Health 2015;12(6):6197–217.
[83] Cronk R, Bartram J. Water, sanitation and hygiene in health care facilities: status in low and middle income countries and way forward. Geneva: World Health Organization; 2015. Available at: http://www.who.int/water_sanitation_health/publications/wash.
[84] Cronk R, Slaymaker T, Bartram J. Monitoring drinking water, sanitation, and hygiene in non-household settings: priorities for policy and practice. Int J Hyg Environ Health 2015;218(8):694–703.

INDEX

'*Note*: Page numbers followed by "f" indicate figures, "t" indicate tables and "b" indicates boxes.'

A

D

E

H

I

J

K

L

M

N

Q

R

S

T

U

X

Y

Z

FSC
www.fsc.org
MIX
Papier | Fördert gute Waldnutzung
FSC® C083411

Kolibri 360 GmbH
Ferdinand-Jühlke-Straße 7
99095 Erfurt, Deutschland
produktsicherheit@kolibri360.de